Frontiers in Earth Sciences

Serge Lallemand • Francesca Funiciello
Editors

Subduction Zone Geodynamics

Springer

Editors
Serge Lallemand
Laboratoire Géosciences Montpellier
Université Montpellier 2, CNRS/INSU
CC. 60, place E. Bataillon
34095 Montpellier cedex 5
France
lallem@gm.univ-montp2.fr

Francesca Funiciello
Dipartimento Scienze Geologiche
Università degli Studi «Roma TRE»
Largo S. Leonardo Murialdo 1
00146 Roma
Italy
ffunicie@uniroma3.it

ISBN 978-3-540-87971-8 e-ISBN 978-3-540-87974-9

DOI: 10.1007/978-3-540-87974-9

Library of Congress Control Number: 2008940148

Cover design: deblik, Berlin

Printed on acid-free paper

9 8 7 6 5 4 3 2 1

springer.com

Preface

Subduction is a major process that plays a first-order role in the dynamics of the Earth. The sinking of cold lithosphere into the mantle is thought by many authors to be the most important source of energy for plates driving forces. It also deeply modifies the thermal and chemical structure of the mantle, producing arc volcanism, and is responsible for the release of most of the seismic energy on Earth. There have been considerable achievements done during the past decades regarding the complex interactions between the various processes acting in subduction zones. This volume contains a collection of contributions that were presented in June 2007 in Montpellier (France) during a conference that gave a *state of the art* panorama and discussed the perspectives about "Subduction Zone Geodynamics". The conference was held under the patronage of the "Société Géologique de France" in the frame of the International Year of Planet Earth sponsored by UNESCO and the International Union of Geological Sciences. During the conference, a volume of 172 abstracts has been published in the "Mémoires Géosciences Montpellier" vol. 41 (2007) and is available at Géosciences Montpellier Laboratory (http://www.gm.univ-montp2.fr). Twelve keynotes launched the discussion on the following topics: Geodynamics and Physical Models (S. Sobolev and T. Becker), Geodesy, Seismogenic Zone and Seismic Hazard (W. Thatcher and S. Das), Seismology (D. Zhao and P. Silver), Structure and Tectonics (R. von Huene and R. Wortel), Geochemistry and Metamorphism (Y. Tatsumi and S. Guillot), and Petrology and Mineral Physics (M. Schmidt and H. Green). We favoured the multidisciplinary approach to debate on remaining key questions.

The chapters included in this special volume provide a sampling of the presentations given in Montpellier and offer a unique multidisciplinary picture of the recent research on subduction zones geodynamics. They were organized into five main topics: Subduction zone geodynamics, Seismic tomography and anisotropy, Great subduction zone earthquakes, Seismogenic zone characterization and Continental and ridge subduction processes. Each of the 13 chapters collected in the present volume is primarily concerned with one of these topics. However, it is important to highlight that chapters always treat more than one topic so that all are related lighting on different aspects of the complex and fascinating subduction zones geodynamics.

Subduction Zone Geodynamics

Seismological data as Wadati-Benioff zones and the distribution of tomographic anomalies illustrate short-term snapshot of the subduction process; the cold lithosphere sinks into the fluid-like mantle with different dips and shapes. However, the long-term evolution of subduction is still uncertain due to its transient character. Hence, to better understand the dynamics of subduction, seismic data need to be integrated with petrological, geochemical and structural constraints as well as numerical and laboratory models. Only modelling, in fact, can provide a dynamic view of the slab behaviour giving the opportunity to insert all the direct and indirect observables into a comprehensive picture. Recognizing the importance of modelling, *Becker and Faccenna (this volume)* offer a detailed and updated review on the subduction (numerical and laboratory) modelling and, in turn on current understanding of subduction dynamics, analyzing the process from a regional to a global view.

It is generally accepted that slab negative buoyancy provides the primary driving force for subduction. In this view, the subduction of areas of over-thickened oceanic crust (seamount chains, oceanic plateaus, island arcs), isostatically more buoyant than normal oceanic lithosphere, potentially affects the subduction behaviour influencing its kinematics, slab shape, seismic activity, arc volcanism and coastal geomorphic features. *Royden and Husson (this volume)* use three-dimensional semi-analytic "unforced" subduction models, in which trench kinematics is controlled only by slab buoyancy, with the aim of systematically analyzing the relationships between slab density, slab geometry and subduction velocity rates.

Subduction dynamics depends on the existence and the distribution of slab windows. *Wortel et al. (this volume)* address this question by focussing on STEP (*subduction-transform-edge-propagator*) faults that control the final stages of evolution of subduction zones. A STEP fault is a tear in the slab which decouples oceanic subducting lithosphere from continental buoyant lithosphere. This tectonic element allows the lateral lithospheric segmentation and, in turn, the continuation of an active subduction process once continental lithosphere is forced at depth. The central Mediterranean Sea is used as a case study because of its peculiar recent history.

240 scientists have attended the three-days meeting among which 180 Europeans and 60 from USA, Asia and other countries. (First row from the left: Francesca Funiciello with her daughter (12th place) and Serge Lallemand (5th place))

Seismic Tomography and Anisotropy

The seismology is the most effective method to explore the structure of subduction zones at great depths. A major characteristic that makes the subduction zones different from other regions comes from the release of water from the sinking slab. This process leads to seismogenesis, melting and viscosity reduction and thus enhances the mantle flow. *Mainprice and Ildefonse (this volume)* explore the contribution of hydrous phases in seismic anisotropy by reporting the elastic properties and characteristics of the wave propagation in anisotropic media for a number of hydrated minerals.

Calò et al. (this volume) present seismic tomographic inversions of P- and S-wave velocity in the southern Tyrrhenian region based on the combined use of the Double-Difference technique for inversions, a recently developed algorithm for local earthquake tomography, and a statistical Weighted Average Model method. This innovative tomographic approach is capable to strongly improve the resolution of the final model giving the opportunity to highlight also heterogeneities within the Calabrian subducting lithosphere never visualized before.

Great Subduction Zone Earthquakes

Among the biggest challenges of this century, the earthquake prediction is highly ranked, especially those which focus on the great event occurring along some subduction zones. One can address the problem at different scales and using various approaches. In this section, the emphasis is first given at a macroscopic scale either from rupture characteristics or kinematics and structural considerations.

Based on several well-studied rupture processes during great earthquakes, *Das and Watts (this volume)* observe that subducting seamounts should play a significant role. The approach chose by *Gutscher and Westbrook (this volume)* is substantially different. They observe that many of the margins with broad accretionary wedges have produced strong earthquakes (M9) in the past, as well as giant tsunamis. They propose that these wide accretionary wedges promote larger co-seismic slips and rupture duration for shallow earthquakes because of their lower rigidity.

Seismogenic Zone Characterization

Before having the opportunity to drill, sample and monitor the seismogenic zone, we have no other choice than improving our imagery of the margin's structure and mechanical behaviour using either detailed bathymetry, reflection and refraction seismics, seismology or geodesy. It is clear that marine and land observations have notably contributed in our understanding during the past decades.

Von Huene et al. (this volume) present a review of the current models based on high quality geophysical images of active margins. He refers to the Kelin Wang's concept of dynamic Coulomb wedge model to explain the differences in strain during inter-seismic locking and co-seismic slip in a converging plate environment. The imaging of inter-seismic locking is a challenge for geodesists and only densely monitored regions can help us to elucidate, for example the occurrence of slow slip episodes and non-volcanic tremors along the brittle-plastic transition zones of the subduction interface. Such study is exemplified at the Nankai subduction zone and the results are presented in this volume by *Aoki and Scholtz*. They have used the vertical component of the GPS to discriminate the rigid plate motion from the deformation due to interplate locking.

Continental and Ridge Subduction Processes

Subduction processes partly differ from "classical" oceanic subduction when a continent or a spreading ridge subducts. The buoyancy, the compositional structure and the magmatic activity in the case of a spreading ridge obviously interfere with the "regular" subduction mechanisms.

At first, *Guillot et al. (this volume)* present a review of the exhumation processes in both oceanic and continental contexts based on petrological and geochronological data on HP-UHP/LT metamorphic rocks presently sliced in meta-sediments or embedded in "mélange" formations. Comprehensive two-dimensional thermo-mechanical models are proposed to account for the different settings of occurrence of these rocks. Then, a history of subduction episodes that produced the Taiwan orogen is developed by *Chang et al. (this volume)* based on constraints provided by the two "mélanges" recognized on the island.

Scalabrino et al. (this volume) review in their chapter the main events that occurred on land at the Chile Triple Junction where the Chile spreading ridge subducts beneath the Patagonian Cordillera. Tectonic, sedimentary and magmatic features are described chronologically and feed an evolutionary model of the area.

Finally, *Mantilla-Pimiento et al. (this volume)* present a detailed study based on two-dimensional seismic reflection data and potential field interpretations about the complex history of the active Colombian margin since late Cretaceous. They are able to image a huge oceanic basement complex sandwiched between an ancient continental tectonic wedge and the present active accretionary wedge.

Serge Lallemand
Francesca Funiciello

Acknowledgements

The editors gratefully acknowledge the institutions that sponsored the "Subduction Zone Geodynamics Conference" among which were the Ministère de l'Enseignement Supérieur et de la Recherche, CNRS/INSU, National Science Foundation (Margins Program), European Science Foundation (EURYI Program – Roma TRE University), Montpellier 2 University, Conseil Régional Languedoc-Roussillon, Agglomération de Montpellier, Conseil Général de l'Hérault, Société Géologique de France and the International Year of Planet Earth.

We warmly thank all the reviewers for their time and efforts in increasing the scientific value of the volume by thoroughly reviewing the manuscripts. We are also indebted to Goetz Bokelmann (France), Claudio Faccenna (Italy), Clint Conrad (USA), Raoul Madariaga (France), David Mainprice (France), René Maury (France), Onno Oncken (Germany) and Cesar Ranero (Spain), all members of the Scientific Committee, for their contribution in elaborating the scientific program and handling the sessions during the conference.

Contents

Contributors

Philippe Agard University Pierre et Marie Curie-Paris VI-CNRS, 4 Place Jussieu 75252 Paris cedex 05, France

Carlos Arturo Alfonso-Pava Ecopetrol S.A (now at Saudi Aramco), P.O Box 6347, Dhahran 31311, Saudi Arabia, alfonsca@aramco.com

Jacques Angelier Géosciences Azur, Observatoire Océanologique de Villefranche, Université Pierre-et-Marie-Curie, Villefranche-sur-Mer, France, angelier@geoazur,obs-vlfr.fr

and

Institut Universitaire de France, France

Yosuke Aoki Earthquake Research Institute, University of Tokyo, 1-1 Yayoi 1, Bunkyo-ku, Tokyo 113-0032, Japan, yaoki@eri.u-tokyo.ac.jp

Thorsten W. Becker Department of Earth Sciences, University of Southern California, 3651 Trousdale Parkway, Los Angeles, CA 90089-0740, USA, twb@usc.edu

Marco Caló CFTA Dipartimento di Chimica e Fisica della Terra, Università di Palermo, Via Archirafi 36, 90100, Palermo, Italy, marco.calo@unipa.it or co.cal@tiscali.it

Chung-Pai Chang Center for Space and Remote Sensing Research, National Central University, Chungli, Taiwan

and

Institute of Geophysics, National Central University, Chungli, Taiwan, cpchang@csrsr.ncu.edu.tw

Giuseppe D'Anna INGV, Istituto Nazionale di Geofisica e Vulcanologia, Via di Vigna Murata 605, 00143, Rome, Italy

Shamita Das Department of Earth Sciences, University of Oxford, Parks Road, Oxford OX1 3PR, UK, das@earth.ox.ac.uk

Aude de la Rupelle CNRS-UMR 5243 Géosciences Montpellier, Université Montpellier 2, CC 60, Place Eugène Bataillon, 34095 Montpellier Cedex 5, 34095, France

Catherine Dorbath IRD-IPGS, Institut de Physique du Globe de Strasbourg, 5 Rue René Descartes, Strasbourg 67084, Cedex, France, catherine.dorbath@eost.u-strasbg.fr

Felipe Espinoza Departamento de Geología, Universidad de Chile, Casilla 13518, Correo 21, Santiago, Chile

Claudio Faccenna Dip. Scienze Geologiche University "Roma TRE", L.San Leonardo Murialdo, 1 00146 Rome, Italy, faccenna@uniroma3.it

Rob Govers Department of Earth Sciences, Faculty of Geosciences, Utrecht University, Budapestlaan 4, 3584 CD Utrecht, The Netherlands, govers@geo.uu.nl

Stéphane Guillot University of Grenoble, OSUG – CNRS, 1381 rue de la Piscine, 38041 Grenoble Cedex 9, France, sguillot@ujf-grenoble.fr

Marc-André Gutscher IUEM, University of Brest, UMR 6538, Plouzané, France, gutscher@univ-brest.fr

kéiko Hattori Department of Earth Sciences, University of Ottawa, 140 Louis Pasteur, Ottawa, K1N 6N5, Canada

Chi-Yue Huang Department of Earth Sciences, National Cheng Kung University, Tainan, Taiwan

Laurent Husson Department of Earth, Atmospheric and Planetary Sciences, M.I.T., Cambridge, MA 01890, USA

and

Géosciences Rennes, CNRS UMR6118, Université Rennes-1, 35042 Rennes, France, lhusson@univ-rennes1.fr

Benoit Ildefonse Université Montpellier 2, Géosciences Montpellier, CNRS/INSU, UMR 5243, CC 60, Place Eugène Bataillon, 34095 Montpellier Cedex 5, France, ildefonse@gm.univ-montp2.fr

Gerhard Jentzsch Friedrich-Schiller Universität Jena, Burgweg 11, 07749 Jena, Germany, Gerhard.Jentzsch@uni-jena.de

Jonas Kley Friedrich-Schiller Universität Jena, Burgweg 11, 07749 Jena, Germany, Jona.Kley@uni-jena.de

Yves Lagabrielle CNRS-UMR 5243 Géosciences Montpellier, Université Montpellier 2, CC 60, Place Eugène Bataillon, Montpellier Cedex 5, 34095, France, lagabrielle@gm.univ-montp2.fr

Dario Luzio CFTA Dipartimento di Chimica e Fisica della Terra, Università di Palermo, Via Archirafi 36, 90100, Palermo, Italy

David Mainprice Université Montpellier 2, Géosciences Montpellier, CNRS/INSU, UMR 5243, CC 60, Place Eugène Bataillon, 34095 Montpellier, Cedex 5, France, David.Mainprice@gm.univ-montp2.fr

Jacques Malavieille CNRS-UMR 5243 Géosciences Montpellier, Université Montpellier 2, CC 60, Place Eugène Bataillon, 34095 Montpellier Cedex 5, France, malavieille@gm.univ-montp2.fr

Adriana Maria Mantilla-Pimiento Fugro Gravity & Magnetics Servuces (Fugro Middle East) P.O. Box 2863 Dubai, United Arab Emirates, AMantilla@fugro.com, adriana_maria_mantilla@yahoo.com

Diego Morata Departamento de Geología, Universidad de Chile, Casilla 13518, Correo 21, Santiago, Chile, dmorata@cec.uchile.cl

Mireille Polvé OMP, Université Paul Sabatier, Toulouse, France

Cesar R. Ranero ICREA at Instituto de Ciencias del Mar, CSIC, Pg Maritím de la Barceloneta 37-49, 08003, Barcelona Spain, cranero@icm.csic.es

Silvio G. Rotolo CFTA Dipartimento di Chimica e Fisica della Terra, Università di Palermo, Via Archirafi 36, 90100, Palermo, Italy

Leigh H. Royden Department of Earth, Atmospheric and Planetary Sciences, M.I.T., Cambridge, MA 02139, USA, lhroyden@mit.edu

Bruno Scalabrino CNRS-UMR 5243 Géosciences Montpellier, Université Montpellier 2, CC 60, Place Eugène Bataillon, Montpellier Cedex 5 34095, France, scalabrino@gm.univ-montp2.fr

Christopher H. Scholz Lamont-Doherty Earth Observatory and Department of Earth and Environmental Sciences, Columbia University, Route 9W, Palisades, NY 10964, USA, scholz@ldeo.columbia.edu

Dave W. Scholl US Geological Survey and Stanford University, 345 Middlefield Rd, Menlo Park, CA 95025, USA, dscholl@usgs.gov

Stéphane Schwartz University of Grenoble, OSUG – CNRS, 1381 rue de la Piscine, 38041 Grenoble Cedex 9, France

Wim Spakman Department of Earth Sciences, Faculty of Geosciences, Utrecht University, Budapestlaan 4, 3584 CD Utrecht, The Netherlands

Manuel Suarez Servicio Nacional de Geología y Minería, Avenida Santa María 0104, Santiago, Chile, msuarez@sernageomin.cl

Olivier Vidal University of Grenoble, OSUG – CNRS, 1381 rue de la Piscine 38041 Grenoble Cedex 9, France, Olivier.Vidal@obs.ujf-grenoble.fr

Roland von Huene US Geological Survey and University of California, Davis, 4300 Carlson Way., Diamond Springs, CA 95619, USA, rhuene@mindspring.com

A.B. Watts Department of Earth Sciences, University of Oxford, Parks Road, Oxford OX1 3PR, UK

Graham K. Westbrook University of Birmingham, Edgbaston, Birmingham, B15 2TT, UK, G.K.Westbrook@bham.ac.uk

Rinus Wortel Department of Earth Sciences, Faculty of Geosciences, Utrecht University, Budapestlaan 4, 3584 CD Utrecht, The Netherlands, wortel@geo.uu.nl

The Editors

Dr. Serge Lallemand is research scientist at the CNRS-INSU (Centre National de la Recherche Scientifique – Institut National des Sciences de l'Univers) based in Montpellier 2 University (France). He obtained his Ph.D. in marine geology and geophysics in 1987 and is presently at the head of Geosciences Montpellier laboratory.

His main research interests concern the geodynamics of subduction zones including Subduction dynamics at the scale of the plates, Mantle wedge dynamics and interaction with plates, Genesis of great earthquakes in relation with subduction of oceanic highs, Strain partitioning in oblique convergence and coupling between plates, Mechanics of arc/back-arc strain, Material transfer in subduction zones : tectonic erosion versus accretion and Stability and deformation of active margins.

Dr. Francesca Funiciello is a geodynamic modeller specialized in the study of the subduction process. She gained a Ph.D. at the ETH-Institut für Geophysik in Zürich in 2002, proceeding to a postdoc position at the "Roma TRE" University. Winner of the EURYI (European Young Investigator) Award 2006, she is currently researcher at the Department of Geology of the University "Roma TRE". Alongside her current work, she is involved in outreach programmes, including her role of Italian representative in the Committee of Education of the EGU (European Geosciences Union).

Part I

Subduction Zone Geodynamics

A Review of the Role of Subduction Dynamics for Regional and Global Plate Motions

Thorsten W. Becker and Claudio Faccenna

Abstract Subduction of oceanic lithosphere and deep slabs control several aspects of plate tectonics. We review models of subduction dynamics that have been studied over the last decade by means of numerical and analog experiments. Regional models indicate that trench rollback, trench curvature, and back-arc deformation may be explained by fluid slabs that are ~250–500 times stiffer than the upper mantle. Slab width and, more importantly, rheology determine the role of viscous bending, poloidal-sinking flow and toroidal-rollback stirring, and interactions of the slab with the higher viscosity lower mantle. Several of these contributions can be represented by a local sinking velocity. Back-arc deformation may then result from an imbalance if larger-scale plate forcing leads to deviations of the convergence rate from the local equilibrium. Lateral viscosity variations (LVVs) are also key for understanding plate driving forces. The realism of global circulation computations has advanced and such models with weak zones and other LVVs have lead to an improved match to observed plate tectonic scores. Those include the correlation with plate motions, the magnitude of intraplate deformation, and oceanic to continental plate velocity ratios. Net rotation of the lithosphere with respect to the lower mantle may be caused jointly by regional slab forcing and the stirring effect of cratonic keels. However, slab models have so far only produced net rotations that are small compared to recent hotspot reference-frame models. Progress in the next years will likely come from a better understanding of slab strength, which is still uncertain since large-scale subduction zone observables and laboratory results do not put strong constraints on slab rheology. Importantly, circulation models with an improved representation of convergent margins will help to close the gap between regional and global approaches to subduction, and to better understand the potential role of the overriding plate.

Keywords Numerical geodynamics • Plate velocities • Subduction • Trench rollback

T.W. Becker
Department of Earth Sciences,
University of Southern California,
3651 Trousdale Parkway, Los Angeles CA90089-0740, USA,
twb@usc.edu

S. Lallemand and F. Funiciello (eds.), *Subduction Zone Geodynamics*,
DOI 10.1007/978-3-540-87974-9, © Springer-Verlag Berlin Heidelberg 2009

1 Introduction

Models of Earth's mantle dynamics can be tested with the surface observables that are provided by plate tectonic motions. Such analysis can address questions including those about the nature of the plate driving forces, the magnitude of plate–mantle coupling, and the strength of the lithosphere and asthenospheric mantle. Several lines of evidence point toward the importance of the subduction process for all of these issues, implying that models of slab dynamics may provide important clues for understanding the workings of plate tectonics. Sinking of dense material in a Stokes fluid captures a range of subduction zone observations, but trench kinematics are complicated by significant dissipation in the bending oceanic lithosphere, moving trenches, and lateral viscosity variations (LVVs). While a comprehensive, theoretical description of such effects is, notably, still missing, this paper strives to summarize the state of knowledge and may so provide a basis for future analysis.

Models of subduction have so far concentrated on two, complementary aspects: The first approach is global and focused on reproducing first-order observations, including force-balanced plate motions, the geoid, and tomographically-mapped, deep slab anomalies. Long-range forces due to global mantle circulation can be accounted for, but numerical modeling has not quite advanced enough to fully include regional plate-boundary dynamics at the required level of detail. It is therefore useful to pursue the second, "regional," approach to study how an isolated slab enters into the mantle and, by doing so, directly or indirectly exerts forces on the attached plates. Given the ease of analyzing the detailed processes inside a subduction zone, the regional approach is still most suitable for analysis of a specific geological setting.

Here, we wish to provide an overview of what has been learned about subduction dynamics from both approaches. We start with a review of observations of surface velocities and the implications of their vectorial character and spatial gradients (Sect. 2). If relative motions and regional plate boundary and slab geometries are considered in isolation, Earth may be considered as providing "natural laboratories." The degree to which individual subduction zones represent different realizations of the same general subduction experiment for different parameter values such as plate age is, however, debatable. It is not always clear if global or regional dynamics lead to local deviations from general trends.

Section 3 reviews results from global models for the role of slabs in driving plate motions. Important issues are the role of LVVs, such as due to continental roots, for affecting relative plate speeds and absolute reference frames; both are relevant if regional plate boundary dynamics are to be inferred from kinematic parameters. Subsequently, we discuss regional modeling (Sect. 4), including some of the current technical limitations, and then tie those results back into the global plate tectonic picture.

Table 1 Symbols used

Parameter	Symbol
slab dip angle	δ
radius of Stokes' sphere	a
bending radius	R
slab/plate width	W
along-slab length in mantle	L
plate extent at surface	l
convective scale (mantle thickness)	H
area factor	$A = \frac{2\sqrt{HW}+\sqrt{Wl}}{H}$
velocity vector	$\vec{v}$
trench normal velocities defined in Fig. 3	V_T, V_P, V_{OP}, V_C, V_B, and V_S
Stokes velocity	V_{Stokes}
modified Stokes (bending) velocity (7)	V'
slope of the V_T/V_P anticorrelation	α
Stokeslet slowing factor	s
gradient operator	$\vec{\nabla}$
pressure	p
deviatoric stress tensor	τ
gravitational acceleration	$g = \lvert\vec{g}\rvert$
density	ρ
density anomaly	$\Delta\rho$
viscosity	η
mantle viscosity	η_m
slab/Stokes sphere viscosity	η_s
slab/mantle viscosity ratio	$\eta' = \frac{\eta_s}{\eta_m}$
lower/upper mantle viscosity ratio	$\eta_l = \frac{\eta_{lm}}{\eta_{um}}$
effective visco-plastic viscosity	η^{eff}
spherical harmonic degree	ℓ

Given the subject matter, we have to be subjective in the choice of material but refer the reader to subduction reviews on different aspects of the problem in the text. "Lithosphere" will be used synonymously with "plate" for simplicity, and contrasted to the weaker "mantle" including the asthenosphere; abbreviations and symbols used are given in Tables 1 and 2. We wish to distinguish between relatively high and low viscosity (and density) fluids, glossing over complications such as thermo–chemical effects. Those are often neglected in the experiments, either by necessity in the lab, or for simplicity in the numerics. The hope is that more realistic systems will obey the same rules as long as their averaged properties match the continuum, effective behavior. We refer to both numerical computations and laboratory analog work as "experiments," as they ideally follow the same, basic laws of physics (i.e. conservation of mass, momentum, and energy), and both can be used to discover new ("emergent") phenomena, or higher-level rules on system behavior.

Table 2 Abbreviations used

Abbreviation	Meaning
APM	absolute plate motions
FE	finite elements
GSRM	global strain rate model of Kreemer et al. (2003), NNR surface velocities used here
HS-3	hotspot reference frame velocities in the rigid-plate model of Gripp and Gordon (2002)
LVVs	lateral viscosity variations
NR	net rotation component of surface velocities
NNR	no net rotation (reference-frame)
TPR	toroidal to poloidal velocity field component RMS ratio for $\ell \geq 2$

2 Kinematic Constraints from Plate Tectonics

From a dynamical point of view, the motions of the lithospheric plates are part of the mantle convection system, namely that of the upper boundary layer. However, this boundary layer is not only cold, but also stiff, and weakened at the plate boundaries. Plate tec-

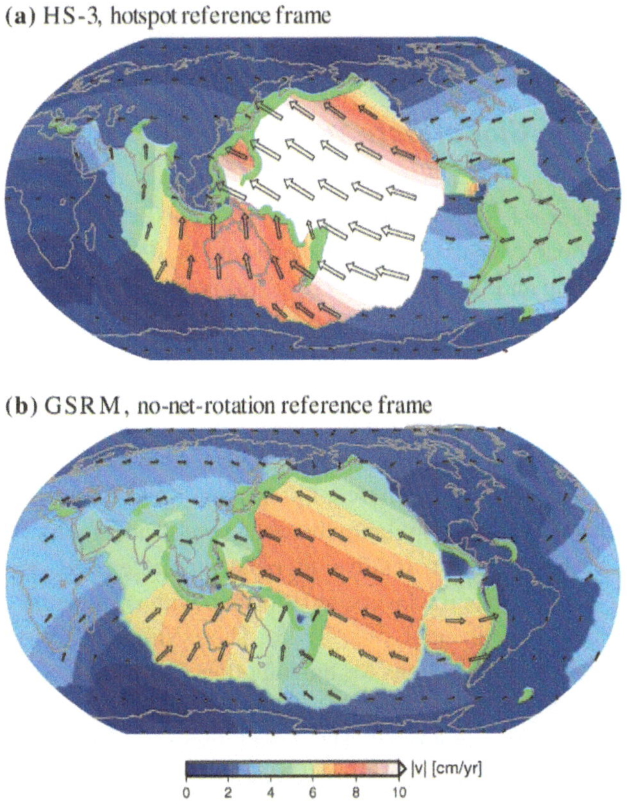

Fig. 1 Crustal velocity amplitudes (background shading) and directions (vectors). (**a**) From the rigid-plate, hotspot reference frame model HS-3 by Gripp and Gordon (2002) (expanded on a 0.25° × 0.25° grid). (**b**) From the deforming, no-net-rotation reference frame model GSRM by Kreemer et al. (2003) (on a 1° × 1° grid). Green contour lines indicate slab seismicity from Gudmundsson and Sambridge (1998)

tonics is therefore unlike isochemical and isoviscous convection (e.g. Bercovici et al., 2000; Tackley, 2000a). A simplified view of plates as a thermal boundary layer is probably closest to the truth within the oceanic plates. Plates with large continental areas behave differently, partly because of their thicker, positively buoyant crustal packages and partly because of the underlying tectosphere (Jordan, 1978). There, volatile depletion due to melting in the past may be even more important in increasing plate strength than within oceanic plates (e.g. Lee et al., 2005).

The long-term cooling of the Earth is mainly through the oceanic plate system, which is in turn dominated by the subduction process. However, oceanic plate convection is likely codetermined by continental formation and motion cycles (e.g. Lenardic et al., 2005; Zhong et al., 2007). An example of such interactions highlighted below are the lateral viscosity variations between stiff keels and weak suboceanic asthenosphere. Those affect relative plate speeds, observations of which have been used to infer slab force transmission. Moreover, LVVs due to keels also induce net rotations of the lithosphere with respect to the lower mantle. Such global, absolute reference frame motions are, in turn, important when regional subduction kinematics are to be extracted.

2.1 Global Plate Motions

Figure 1 shows two endmember representations of global crustal velocities at present. Model HS-3 by Gripp and Gordon (2002) (Fig. 1a) is an example of a rigid, plate-tectonics model in the sense of McKenzie and Parker (1967) and Morgan (1968). The surface is subdivided into 15 major plates with relatively well constrained Euler vectors. When velocities are plotted as in Fig. 1, it is apparent that the plates that have slabs attached to them move faster than those without. Moreover, the correlation between speed and length of attached trench, relative to plate circumference, is stronger than the inverse relationship with continental area (Forsyth and Uyeda, 1975). This is still one of the strongest indications for the importance of subduction for driving plates.

Relative motions in HS-3 are representative of the last 5.8 Ma and are from NUVEL-1A (DeMets et al., 1994). When continental regions with active deformation are excluded, satellite geodetic measurements, available for the last ~25 years, typically yield similar Euler vectors (e.g. Sella et al., 2002). However, any such geologic and geodetic data can only constrain relative motions, and further assumptions are required to define an absolute reference frame. For HS-3, it is of the hotspot kind, and was inferred from ten Pacific ocean island age progressions (Gripp and Gordon, 2002). The idea behind hotspot reference-frames is that oceanic islands may be caused by thermal plumes which rise from the deep mantle (Morgan, 1971; Wilson, 1973). In the most common approach, strictly stationary hotspots then define a reference frame with respect to the lower mantle (e.g. Minster and Jordan, 1978). Relative motions between hotspots are in fact ~one order of magnitude smaller than plate motions (Molnar and Stock, 1987; Tarduno et al., 2003). Such motions are quantitatively consistent with plumes if the lower mantle convects more sluggishly than the upper mantle because of its higher viscosity (Steinberger et al., 2004; Boschi et al., 2007).

Spatio–temporal deviations from the assumption of constant motion of rigid plates, i.e. intraplate deformation (e.g. Gordon, 2000), can be most easily constrained with satellite geodetic measurements. If one chooses to represent the velocities by subdivision of the large plates into microplates, this often yields a regionally sufficient description (e.g. McClusky et al., 2000), but many microplate Euler vectors are still not well constrained (Bird, 2003). Fig. 1b displays an example of an alternative approach, the GSRM model by Kreemer et al. (2003). Intraplate deformation is allowed in certain, data-rich regions (compare, e.g., SE Asia in Figs. 1a and b), and inferred from geodetic measurements and geologic fault offset rates. Treated as a medium with viscous, continuous deformation, the crustal strain-rates can then be used to infer lithospheric and mantle rheology (e.g. England and Molnar, 1997).

2.2 Poloidal and Toroidal Velocities and Reference Frames

It is instructive to decompose velocity fields into poloidal, $\vec{v}_p$, and toroidal, $\vec{v}_t$, components. Poloidal flow corresponds to pure sources and sinks in the horizontal plane ($\vec{\nabla} \times \vec{v}_p = \vec{0}$), and is associated with vertical mass transport. Toroidal motion corresponds to vortex-like flow and rigid body rotations ($\vec{\nabla}.\vec{v}_t = 0$). The types of plate boundaries where poloidal and toroidal velocity amplitudes are largest are spreading centers and subduction zones, and transform faults, respectively (*cf.* O'Connell et al., 1991; Dumoulin et al., 1998; Tackley, 2000b). In a minimum viscous-dissipation configuration, Cartesian isoviscous convection will only contain poloidal flow. However, both spherical geometry and lateral viscosity variations will make a substantial toroidal flow component energetically favorable (O'Connell et al., 1991; Olson and Bercovici, 1991; Bercovici et al., 2000).

If plate velocities are expressed with spherical harmonics in terms of poloidal and toroidal fields, the degree one toroidal harmonic represents an average, rigid-body motion of the entire lithospheric shell (net rotation, NR, component). In Fig. 1b, velocities from GSRM are in the no-net-rotation (NNR) reference-frame, defined by the requirement that the NR is zero. Hotspot reference-frames often show nonzero NR, and the relatively strong net-rotation component of HS-3 is apparent for Figs. 1a and b. The NR corresponds to a mean velocity of 3.8 cm per year, with maximum values reaching 4.9 cm per year, enhances NW-ward velocities in the Pacific, and reduces the motion of Africa.

NR components for several hotspot reference-frames are shown in Fig. 2. The sense of motion is similar for all models, but amplitudes depend on parameters such as the Euler poles for slow-moving plates (e.g., Africa) and, importantly, the geographic selection of hotspots (e.g. Ricard et al., 1991; O'Neill et al., 2005). HS-3 shows the strongest NR of the models in Fig. 2, and we consider it as an endmember case that, by focusing on the Pacific hotspots, might overestimate NR motion. Model SB04 (Steinberger et al., 2004) allows for motions of hotspots with respect to each other; this reduces the amount of net rotation compared to HS-3. The NR is then slightly larger for SB04 than for GJ86 (Gordon and Jurdy, 1986) and of the same order as R01h (Ricard et al., 1991). The excitation mechanism(s) for net rotations are still not universally agreed upon (e.g. Doglioni et al., 2007), but LVVs are a necessary condition (Ricard et al., 1991; O'Connell et al., 1991). The most likely explanation is that a combination of weak zone geometry (Bercovici, 2003) and strength contrasts between subcontinental and suboceanic asthenosphere (Ricard et al., 1991; Ribe, 1992) contributes to NR. Global convection computations by Zhong (2001) and Becker (2006) show that the right sense of motions is predicted (blue symbols in Fig. 2) if LVVs and stiff cra-

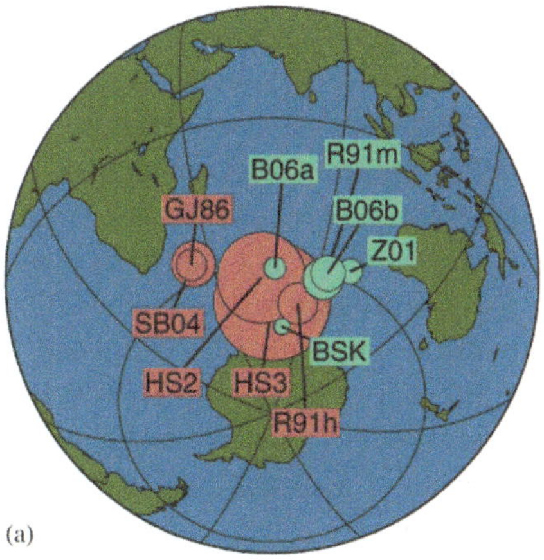

(a)

model	lon [°E]	lat [°N]	rate [°/Ma]
HS3	70	−56	0.44
HS2	65	−49	0.33
SB04	38	−40	0.17
GJ86	37	−40	0.12
R91h	84	−56	0.15
R91m	93	−47	0.15
Z01	103	−42	0.09
B06a	71	−46	0.08
B06b	94	−45	0.13
BSK	76	−63	0.06

(b)

Fig. 2 NR Euler poles, circle radius scales with magnitude. *Red* symbols are from plate reconstructions with the stationary hotspot assumption (GJ: Gordon and Jurdy (1986), R91h: Ricard et al. (1991) using NUVEL, HS-2: Gripp and Gordon (1990), and HS-3: Gripp and Gordon (2002)), and allowing for flow-advected plume conduits (SB04: Steinberger et al. (2004) and pers. comm., B. Steinberger). The *blue* symbols are from the simplified continent/ocean shell model of Ricard et al. (1991) (R91 m), and from full circulation models; Z01: global slab model with deep cratonic keels by Zhong (2001), B06a and B06b: using shallower keels, tomographic anomalies, and a temperature-dependent, power-law (**a**) and lab-derived, effective diffusion/dislocation creep rheology (**b**, η_{eff}), respectively (from Becker, 2006), and BSK: a computation that has no keels and whole mantle slabs (*cf.* Fig. 6). For the B models, NR components were scaled such that the NNR velocities match NUVEL-1 RMS values (see Becker, 2006, for details and unscaled representation) tonic keels are accounted for. We will return to net rotation excitation in Sect. 3.5. From a kinematic point of view, it is apparent that several plate motion characteristics are affected by the reference frame. The overall toroidal–poloidal partitioning is a trivial case with the $\ell = 1$ NR component (O'Connell et al., 1991). For $\ell \geq 2$, the ratio between global toroidal and poloidal RMS power (TPR; e.g. Tackley, 2000b) is ≈0.53 for all NUVEL-type models such as HS-3 and ≈0.57 for GSRM as in Fig. 1. TPR has been at comparable levels (between 0.49 and 0.64) since 120 Ma (Lithgow-Bertelloni et al., 1993). Further, the ratio of the mean velocities within oceanic or continental regions depends somewhat on the net rotation component. For HS-2 (Gripp and Gordon, 1990), HS-3 and GSRM-NNR, for example, the values are ~2.55, 2.05, and 1.55, respectively. This ratio, or that of the plates with predominantly oceanic or continental area, is of interest, as it has been used to infer the strength of slab pull and plate driving forces (Forsyth and Uyeda, 1975; Conrad and Lithgow-Bertelloni, 2002; Becker, 2006).

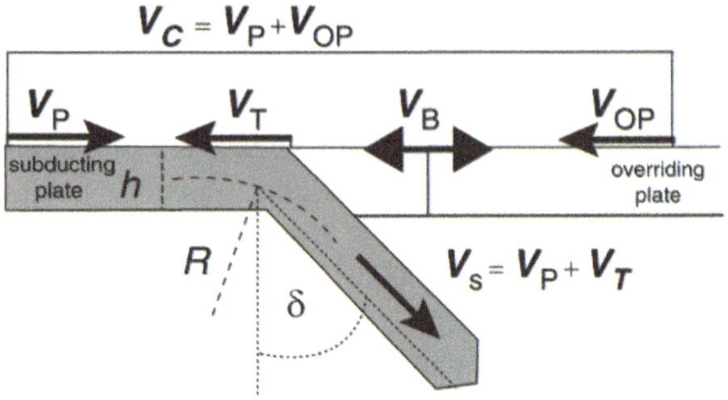

Fig. 3 Regional subduction zone kinematics and slab geometry parameters. We assume that all intra-plate deformation is localized in the back-arc deformation rate, V_B (positive for extension), and all velocities are trench-normal components. Plate (V_P), trench (V_T, positive for rollback) and overriding plate (V_{OP}, positive toward the trench) velocities are all computed in an absolute reference frame. The convergence rate at the trench, V_C, is given by $V_P + V_{OP}$ with this sign convention, and the sinking velocity of the slab, V_S, is partitioned between V_P and V_T. (We assume $V_T = V_{OP} + V_B$, so $V_S = V_P + V_{OP} + V_B$).We also indicate the width of the slab and plate, h, and the approximate bending radius at the trench, R, as well as the slab dip angle, δ. For more elaborate dip angle descriptions, see Jarrard (1986) and Lallemand et al. (2005)

2.3 Regional Subduction Kinematics and Trench Migration

Modeling subduction requires a choice of absolute reference frame because a layered mantle may produce slab anchoring at depth. Typically, absolute plate velocities relative to a hypothetical stagnant lower mantle are considered, for which hotspot reference-frames may be used. The kinematic parameters necessary to describe subduction also need to take into account the motion of the fore-arc sliver, which yields information on the kinematics of the trench. There is a long tradition of analyzing the dependence of the various kinematic (e.g. V_P) and geometrical (e.g. δ) quantities for different subduction zones to test how those are affected by parameters such as slab pull, parametrized by subducting plate age (see Fig. 3 for the definitions).

A recent review of the role of the overriding plate, the subducting plate, and mantle flow in controlling subduction is given by Heuret and Lallemand (2005). It is now clear that strong correlations between dynamic and kinematic, or geometrical, quantities only exist for a few cases (e.g. Carlson and Melia, 1984; Jarrard, 1986; Cruciani et al., 2005; Heuret and Lallemand, 2005; Lallemand et al., 2005; Sdrolias and Müller, 2006). An extreme view is to take such complications as sufficient evidence to abandon slab pull as a plate driving force (Doglioni et al., 2007). Alternatively, we think that the complications in deriving simple rules for slab behavior indicate that real subduction zones are more complicated than simplified considerations (e.g. McKenzie, 1969; Stevenson and Turner, 1972; Forsyth and Uyeda, 1975), and assumptions about steady-state subduction rates, would lead us to expect. As we show below, more realistic models of subduction provide physical explanations for many of the general trends.

Trenches are not stationary features, but are often found to retreat toward the subducting plate with respect to the lower mantle (V_T>0; Spence, 1977; Chase, 1978; Garfunkel et al., 1986). However, it has been recognized that in several regions, e.g. Marianas-Izu Bonin, trenches also appear to advance toward the subducting plates in all reference frames (V_T < 0; Carlson and Melia, 1984; Jarrard, 1986; Heuret and Lallemand, 2005; Faccenna et al., 2007). Any trench migration has important consequences for tectonics such as back-arc spreading, but also for large-scale upper mantle dynamics. Modifications in the oceanic plate areas and trench locations will affect long-term heat transport and mixing efficiency of mantle convection. Regionally, the motion of trenches is also of importance since the details of mantle flow will be affected. The migration of the slab inside the mantle induces a toroidal component of flow, with significant effects on back-arc temperatures, volcanism, and the interpretations of seismic anisotropy observations (Sect. 4.4).

As illustrated in Fig. 3, V_T can be estimated by subtracting the back-arc deformation rate, V_B, from the velocity of the upper plate, V_{OP} (Carlson and Melia, 1984; Heuret and Lallemand, 2005), assuming that erosion and accretion at trenches are negligible (Lallemand, 1995). Errors in V_T may then result if this assumption is not justified (Clift and Vannucchi, 2004), or from uncertain V_B which is only straightforward to determine for spreading oceanic seafloor (Sdrolias and Müller, 2006). The Andes are a classical example where shortening rates vary dramatically as a function of balancing technique (Kley, 1999). Given such geological complications, V_B can also be approximated by geodetic studies (Heuret and Lallemand, 2005; Doglioni et al., 2007) although the relevant time interval (compared to the seismic cycle) is not the same, and V_T might fluctuate over several Ma (Sdrolias and Müller, 2006; Schellart et al., 2007).

The most straightforward effect on trench motions, however, is that of the reference frame (Fig. 4). Funiciello et al. (2008) present an analysis of trench motions and find that almost all Pacific trenches are strongly affected by the NR component, as expected from Fig. 2. Generalizations on subduction zone behavior in terms of V_T trends should thus be treated with caution. A lower mantle reference-frame, presumably approximated by the hotspot models with significant NR components, is probably the most appropriate choice for subduction rollback studies, and the NNR reference-frame provides an endmember description for reduced NR.

Global V_T mean and standard deviation vary between 6 ± 40 mm per year for HS-3 and 11 ± 30 mm per year for NNR (Fig. 4b, predominantly rollback). It is apparent, however, that the distributions are spread out, and the value of the mean might not be significant (note the bimodality/skewness in Fig. 4b). The geometric mean of the ratio of trench motion to subducting plate velocity is 0.48 for HS-3 and 0.3 for NNR (using data from Heuret and Lallemand, 2005), i.e. trench migration rates are typically not more than ~50% of the convergence rates.

3 Global Mantle Flow and Subduction Dynamics

What are the physical models of mantle convection that can explain the kinematic surface observations? Even if plates and subduction are simply aspects of the mantle system, there are still important lessons to be learned from subdividing the forces driving the plates into different components ("force models," e.g. Solomon and Sleep, 1974; Forsyth and Uyeda, 1975; Chapple and Tullis, 1977); a short review may be found in Becker and O'Connell (2001).

Another approach is bottom up ("velocity model") where one solves for mantle flow given specified density distributions and mantle rheology, and then evaluates the tractions that apply at the base of the plates. Density anomalies may be inferred for slabs from seismicity (Hager, 1984), subduction histories (Ricard et al., 1993; Lithgow-Bertelloni and Richards, 1995; Steinberger, 2000), or seismic tomography (Hager and

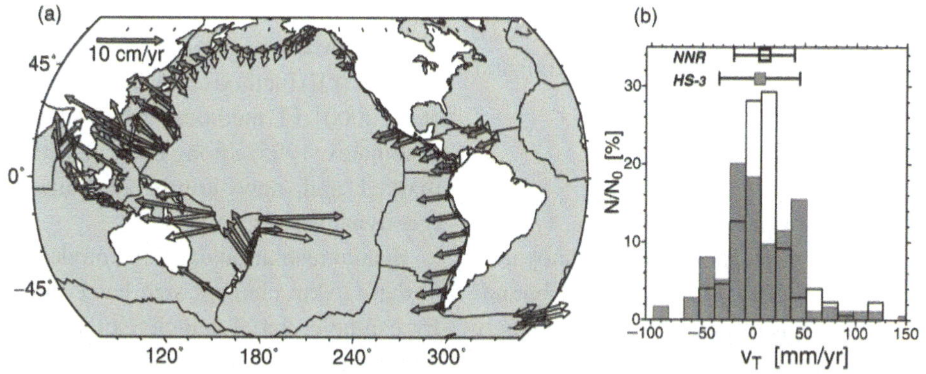

Fig. 4 (**a**) Map of trench migration, V_T, for the present-day in the HS-3 (filled) and NNR (open vectors) reference frame (averaged on 7.5 × 7.5° blocks) and, (**b**), histograms of V_T. We show mean ± one standard deviation on top of the histograms. Both figures are based on the even-length trench sampling by Heuret and Lallemand (2005) (*cf.* Funiciello et al., 2008)

Clayton, 1989; Forte and Peltier, 1987). We start with a short introduction of mantle fluid dynamics before discussing global flow models and their implications for the role of slabs and subduction for plate tectonics.

3.1 Stokes Flow and Circulation Models

Inertial forces are negligible for the mantle (infinite Prandtl number limit), and continuity of momentum simplifies to the Stokes equation. For an incompressible fluid, force balance can be written as

$$\vec{\nabla} p - \vec{\nabla} \cdot \tau = \vec{g} \Delta \rho. \quad (1)$$

An important laminar flow problem with an analytical solution is that of a viscous sphere with η_s embedded in an infinite medium with viscosity η_m. The sinking, Stokes, velocity is then given by

$$V_{\text{Stokes}} = C \frac{\Delta \rho g a^2}{\eta_m} \quad (2)$$

with a viscosity-ratio dependent

$$C = \frac{2 + 2\eta'}{6 + 9\eta'} \quad \text{where} \quad \eta' = \frac{\eta_s}{\eta_m}. \quad (3)$$

For a rigid sphere ($\eta' \to \infty$), $C = 2/9$ and for a weak bubble ($\eta' \to 0$), $C = 1/3$, indicating that V_{Stokes} is mostly sensitive to η_m and not η'. Different orientations of an elongated sinker also do not affect C by more than a factor of ~2 (Batchelor, 1967). The $\eta_m/(\Delta \rho g a)$ term in V_{Stokes} sets a timescale for any buoyancy-dominated subduction problem, and scalings of $V \propto \Delta \rho a^2/\eta_m$, and $\tau \propto \Delta \rho a$ are expected. For typical values of $\Delta \rho = 50\,\text{kg/m}^3$, $a = 100\,\text{km}$, and $\eta_m = 10^{21}\,\text{Pas}$ (e.g. Turcotte and Schubert, 2002), $V_{\text{Stokes}} \sim 4\,\text{cm/yr}$. For a Newtonian fluid with constant η, (1) further simplifies to

$$\vec{\nabla} p - \eta \nabla^2 \vec{v} = \vec{g} \Delta \rho. \quad (4)$$

where it can be seen that pressure gradients, $\vec{\nabla} p$, and viscous drag, $\eta \nabla^2 \vec{v}$ are balanced by the buoyancy forces, $\vec{g} \Delta \rho$. Without inertia, the mechanical boundary conditions plus density distributions uniquely determine an instantaneous velocity solution ("circulation model"). Without LVVs, the Stokes flow problem is linear and this allows the superposition of solutions for the flow field of point-like buoyancy sources (e.g. Batchelor, 1967). For super-critical Rayleigh numbers, subduction velocities can also be estimated from a fluid loop analysis of sinking plumes in an isoviscous convective cell (Turcotte and Oxburgh, 1967; Turcotte and Schubert, 2002). Energetically, any convective system will maintain a balance between the rates of potential energy change and the viscous dissipation (stress times strain-rates) in the fluid. How the latter partitions between the mantle and the plate may affect plate velocities (Sect. 4.3.3).

3.1.1 Numerical Solution Methods

If surface velocities and internal density anomalies are prescribed, velocities and tractions can be computed easily in spherical geometry if mantle viscosity is assumed to be only radially variable (e.g. Hager and O'Connell, 1981; Forte and Peltier, 1987; Hager and Clayton, 1989). Such semianalytical models can be solved in seconds on a modern computer, and their spatial resolution is limited mostly by the input density model, at present well resolved up to $l \sim 20$ for global seismic tomography (Becker and Boschi, 2002).

Earlier attempts to evaluate the role of LVVs in mantle flow were affected by modeling limitations (Čadeket al., 1993; Zhang and Christensen, 1993; Wen and Anderson, 1997). Iterative schemes such as that of Zhang and Christensen (1993) are usually restricted to small ($\lesssim$ three orders of magnitude) LVVs, and the resolution of early direct solution methods was limited by available computer memory (Forte and Peltier, 1994). However, the ongoing increase in computational resources has brought partial remedy. Parallelization and improved algorithms have allowed computations of more realistic global models with spectral (Moucha et al., 2007) and finite volume or element (FE) methods (e.g. Tackley et al., 1994; Zhong et al., 2000). FE methods such as CitcomS (Moresi and Solomatov, 1995; Zhong et al., 2000) are well benchmarked and now jointly developed through CIG (geodynamics.org).

Spatial resolution of global models is now routinely at the ~25 km element size level and using CitcomS, for example, initial velocity solutions can be obtained within minutes to hours depending on the resolution, numbers of CPUs, the magnitude of LVVs, and the solution algorithms used. However, strong LVVs, such as for thin weak zones, can cause slow convergence of iterative solutions for $\vec{v}$. In such cases, care has to be taken to avoid intermediate and erroneous, but not obviously wrong-looking, solutions. Such issues should be solvable using improved multigrid schemes and variable mesh refinement in the future.

3.2 Role of Slabs for Plate Motions

An important advance in our modeling capabilities to test the consistency of tractions from circulation models with observed plate velocities was provided by Ricard and Vigny (1989). In their formulation, plates are assumed rigid, thin shells with no applied forces along their boundaries whose geometry is prescribed. One may then first compute the interaction of unit motions of one plate via viscous drag on another (Gable et al., 1991), and then solve for "free" plate motions so that the torques due to the body force loading are balanced.

Several workers have applied such velocity models to test force partitioning (Deparis et al., 1995; Lithgow-Bertelloni and Richards, 1995; Čížková et al., 1998), and plate motions can be fit well (linear correlation coefficients between observed and predicted Euler vectors ~0.9). Typically, ~70% of the driving forces are found to be caused by slabs, and the rest is mainly due to gravitational sliding (ridge push) (Lithgow-Bertelloni and Richards, 1998). Becker and O'Connell (2001) found that ~40% of the total slab contribution originates in the lower mantle for typical viscosity profiles such as those from Hager and Clayton (1989).

Traction scale with the density anomalies, which may be affected by both chemical and thermal effects (e.g. Forte and Mitrovica, 2001) and depend on inversion choices when inferred from tomography. Another issue with velocity models is that the magnitude of force transmission from mantle flow to the lithospheric layer ("plate coupling") depends somewhat on the radial viscosity structure. A moderate reduction in sublithospheric viscosity in the transition zone ($\eta \sim 10^{19} \ldots 10^{20}$ Pas for depths $100 < z \leq 410$ km compared to $\sim 10^{21}$ Pas for $410 < z \leq 660$ km) is indeed preferred by inversions of geopotential and postglacial rebound data (Hager and Clayton, 1989; Mitrovica and Forte, 2004), and also expected given laboratory constraints on upper mantle rheology (Hirth and Kohlstedt, 2004). However, Lithgow-Bertelloni and Richards (1998) and Becker and O'Connell (2001) showed that even for low transition zone viscosities of order $10^{18} \ldots 10^{19}$ Pas, a large fraction of the forces driving the plates originates in the mantle. Put differently, a mantle contribution is required and lithospheric, gravitational potential energy variations alone cannot drive the plates. Becker et al. (2003) further explored the role of an asthenospheric low viscosity channel and found that seismic anisotropy prefers relatively strong coupling with global asthenospheric viscosities being no less than $\sim 10^{-3}$ of the upper mantle reference viscosity.

3.3 Edge Forces vs Mantle Drag

Can we use velocity models to make refined assessments for plate-boundary dynamics? Becker and O'Connell (2001) combined the force model approach of prescribing shear and normal forces along the plate boundaries with the velocity model of computing mantle tractions. These authors explored various combinations of forces and showed that velocity correlations were not very sensitive to edge forces. In the absence of LVVs, adding asymmetric pull by slabs on plates did not improve the model fit much over models where tractions due to sinking slabs apply symmetrically at both overriding and subducting plate ("slab suction" in the terminology of Conrad and Lithgow-Bertelloni, 2002). Figure 5 shows the reason for this force ambiguity: several of the plate driving and resisting forces are highly (anti-)correlated (Forsyth and Uyeda, 1975). This is because of plate geometry and motions from spreading centers to subduction zones: any edge-force derived torque that is based on plate boundary segments which are roughly perpendicular to plate motion will then yield similar forcing vectors. An exception is a plate-motion drag that applies underneath cratonic roots only ("plate motions at cratons" in Fig. 5).

Conrad and Lithgow-Bertelloni (2002) presented a similar, joint force analysis and focused on the partitioning between slab suction and slab pull. In contrast to Becker and O'Connell (2001), these authors found that incorporating slab pull did improve plate motions significantly. Conrad and Lithgow-Bertelloni inferred a yield stress of ~500 MPa via the slab's stress-guide behavior, and expanded their analysis to past convec-

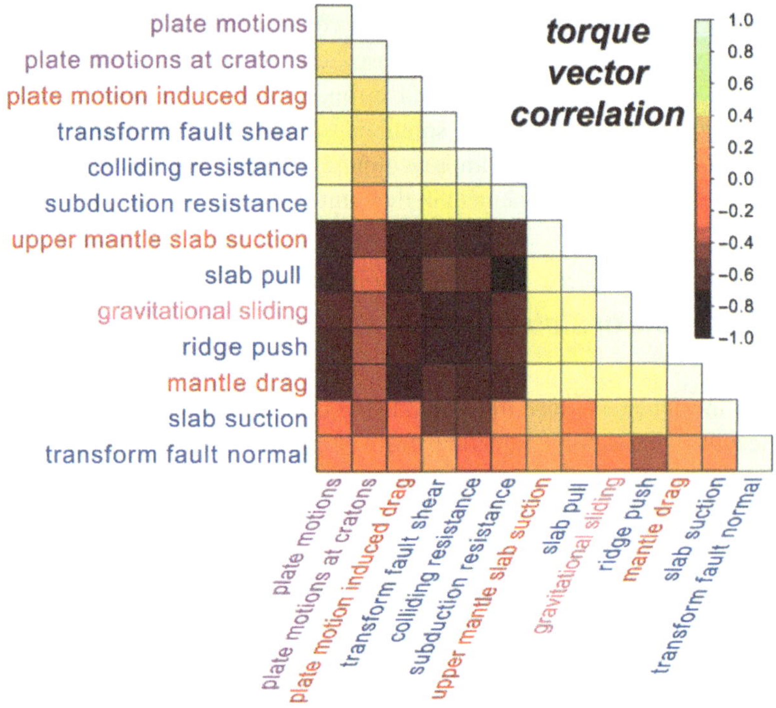

Fig. 5 Correlation of torque vectors for all major lithospheric plates and several driving and resisting forces, modified from Becker and O'Connell (2001). We show edge forces in *blue* following the classification of Forsyth and Uyeda (1975), and forces computed from global circulation models in *red* ("slab suction" refers to the slab-density induced downwelling tractions, "mantle drag" to a tomography-inferred model which includes upwellings). "Gravitational sliding" is a gravitational potential energy model for the lithosphere that includes the distributed half-space cooling force, as opposed to "ridge push" which just applies at the spreading centers. All plate velocities were computed in the NNR reference-frame, for details see Becker and O'Connell (2001)

tive settings (Conrad and Lithgow-Bertelloni, 2004) and thrust interface coupling (Conrad et al., 2004). It is likely that different assumptions about the details of plate boundary geometry and the resulting force integration may be responsible for the diverging conclusions. More importantly, Conrad and Lithgow-Bertelloni (2002) chose to not only consider correlations of plate motions, but also the ratio between oceanic and continental plate velocities, which is indeed increased if one-sided slab pull is added to the oceanic plates.

3.4 Lateral Viscosity Variations and Variable Plate Coupling

Lateral variations in the plate-mantle coupling may be important for such velocity ratios, particularly if one considers the relatively stiff cratonic keel regions compared with the hotter and compositionally distinct suboceanic asthenosphere at ~200 km depth (Ricard et al., 1991; Zhong, 2001; Čadek and Fleitout, 2003). We can infer plate motions for LVV models by prescribing weak zones along plate boundaries (Zhong and Davies, 1999). Such computations are conceptually similar to the approach of Ricard and Vigny (1989), and have some limitations (e.g. effect of weak zone viscosity on plate speeds). Those are, however, fairly well understood (King and Hager, 1990; King et al., 1992; Han and Gurnis, 1999; Yoshida et al., 2001).

Following the work by Zhong and Davies (1999) several recent computations have advanced our understanding of global plate dynamics. Conrad and Lithgow-Bertelloni (2006) evaluated the role of 3D LVVs including continental roots on coupling and found that traction amplitudes are strongly affected by LVVs whereas directions are not. Becker (2006) computed global flow using a "realistic" rheology as expected for joint diffusion/dislocation creep for dry olivine in the upper mantle (Karato, 1998; Hirth and Kohlstedt, 2004). Becker found that mantle flow directions for models with LVVs were roughly similar to estimates with only radially varying rheology on global scales, but deviations exist regionally, especially underneath the oceanic plates. Importantly, the low viscosity asthenosphere underneath the oceans increases the plate velocities there, and so affects strongly the oceanic/continental velocity ratio (Fig. 6).

Ricard and Vigny (1989)-type models with prescribed plate boundaries and plates that move freely as

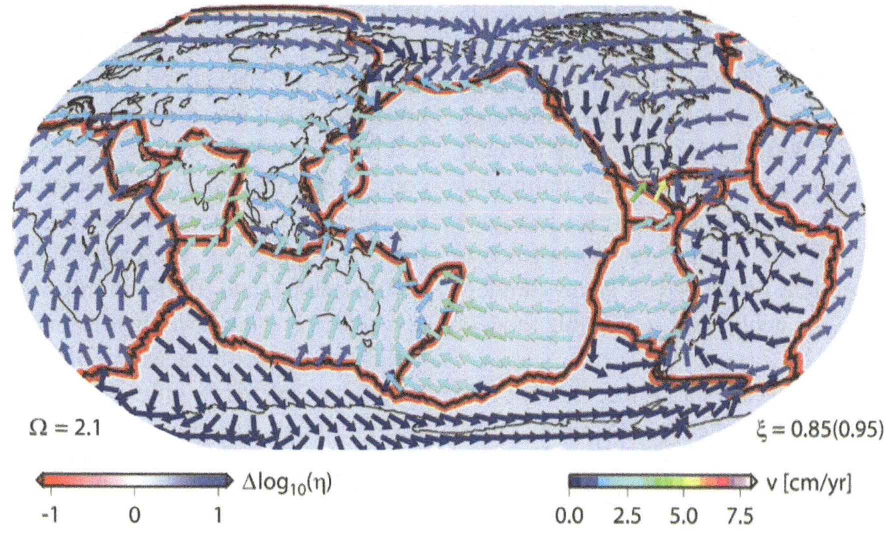

Fig. 6 Surface velocities in the NNR reference-frame as predicted by an example computation following Zhong (2001). The rheology and model setup is similar to the η_{eff} model of Becker (2006) (B06b in Fig. 2) but does not consider stiff cratonic keels, and only density anomalies from the slab model of Steinberger (2000) are used to drive flow. We show plate motions as vectors (fixed length, shading indicating amplitude) and lithospheric viscosity (weak zone geometry prescribed, global mean of viscosity, $\langle\eta\rangle$, is ~280 times the reference, 10^{21} Pas). The flow computation predicts only a small amount of NR flow (~10% of HS-3 when corrected for RMS velocities, BSK in Figure 2). Velocities correlate with NUVEL-1 at the ξ = 0.85 (0.95 velocity weighted) level. The RMS ratio between oceanic and continental plates (Ω = 2.1) is very similar to observations, while the overall RMS velocities (1.8 cm/year) are too low; this could be adjusted by tuning the weak zone formulation or the density model

driven by density anomalies explain the observed plate motions well (Lithgow-Bertelloni and Richards, 1998). This means that computations where surface motions are prescribed at the surface and density anomalies drive additional flow (Hager and O'Connell, 1981) are dynamically consistent for adequate parameters. Such forward models can, of course, not answer the question why the plate boundaries formed, but they can be used to explore how LVVs interact with plate tectonics. It is in fact possible to match several global plate tectonic scores (velocity ratios, correlations, plateness, TPR ratios) better with LVV models than with purely radially varying viscosity (Zhong, 2001; Becker, 2006). The lower viscosity asthenosphere facilitates oceanic plate motions and the temperature-dependent rheology reduces the intraplate deformation such that it is similar to that found in GSRM as shown in Fig. 1b (*cf.* Moresi and Solomatov, 1998). Even simple models such as those of Becker (2006), which were not optimized to fit plate motions, are able to match both directions (weighted correlations of ~ 0.9), and amplitudes of relative plate motions in continental and oceanic regions, without the addition of edge forces (Fig. 6).

These conflicting findings as to the role of slabs for plate driving forces imply that global plate motions may be only moderately sensitive to the variation in plate boundary strength, though this clearly needs to be explored further with models of higher regional realism (*cf.* Billen and Gurnis, 2003). Other global datasets such as the crustal stress field show similar nonuniqueness with regard to the role of deep slabs (Steinberger et al., 2001; Lithgow-Bertelloni and Guynn, 2004), while regional models provide a clearer detection of slab-induced flow for seismic anisotropy and stress (Becker et al., 2006; Humphreys and Coblentz, 2007).

3.5 Net Rotations Caused by Keels and Slabs

Another subduction-related prediction of LVV circulation models are net rotations which are relevant to determining regional kinematics (Figs. 2 and 4). Toroidal flow can be excited by LVVs within the lithosphere (e.g. Tackley, 2000b, c), and ℓ = 1 NR motion by continent–ocean asthenosphere viscosity differences (Ricard et al., 1991). The role of stiff keels was first studied with full flow solutions by Zhong (2001). The net rotations that are excited for power-law rheologies and large-scale tomographic anomalies as used by Becker (2006) are somewhat higher than those found by Zhong, even for shallower continental roots (Fig. 2), and the sense of motion (pole location) is close to that of the hotspot reference-frames. However, the amplitude of the NR is always under-predicted for any of the published geodynamic models compared to HS-3.

Given the uncertainties in establishing hotspot reference frames, the NR component of HS-3 may be an overestimate. Moreover, geodynamic forward modeling of flow and azimuthal seismic anisotropy indicates that only a moderate amount of NR shear flow (as in GJ86 or SB04 of Fig. 2) is consistent with seismology (Becker, 2008). However, accepting an under-prediction of NR from keel-based flow models for the sake of argument, Becker (2006) suggested that it might be the joint effect of regional slab dynamics, buoyant upwellings, and continental keels that leads to higher NR motion than in the simplified flow models. Slab-induced NR flow was studied first on a global scale by Zhong (2001), with relatively low resolution computations. Slab interaction with the viscosity contrast at 660 km as a source for NR was discussed by Enns et al. (2005) for regional, two-dimensional (2D) models, and is also implicit in the success of the Faccenna et al. (2007) predictions we discuss in Sect. 4.4.2. We thus conducted further tests of this suggestion by computing LVV power-law flow for models that include slab structures.

Figure 6 shows velocities from an example circulation model that is based on converting the advected slablet model of Steinberger (2000) to temperature anomalies (*cf.* Zhong and Davies, 1999). We use the η_{eff} power-law rheology as in Becker (2006) (B06b in Fig, 2), but did not prescribe any continental keels. The NR amplitudes generated by slabs alone are relatively small (BSK in Fig. 2), but the Euler pole location of HS-3 is matched well. (All NR amplitudes were normalized by the predicted NNR velocity RMS because other parameters were not optimized for these tests.) The induced NR roughly doubles if the additional effect of continental keels is considered, and different whole mantle slab models (Lithgow-Bertelloni and Richards, 1998) lead to similar results, confirming Zhong's (2001) conclusions. We also conducted preliminary tests where we replaced upper mantle tomography by slab anomalies as inferred from seismicity (*cf.* Becker and O'Connell, 2001). For such upper mantle slab models, the role of keels was reversed, i.e. flow without keels yielded higher NR components than those which included keels. However, the NR component for any of these slab circulation models is $\lesssim$40% of the original models by Becker (2006) (B06a and B06b in Fig. 2) which include upwellings and downwellings as inferred from seismic tomography. These findings may imply that keel-deflected upwellings are more efficient than downwellings in NR excitation, but this needs to be confirmed with different slab and weak zone representations.

3.6 Flow Confinement and Geopotential Fields

Lateral viscosity variations also need to be explored further in terms of the match of slab- and plate-induced flow to geopotential observations (Moresi and Gurnis, 1996; Chen and King, 1998; Zhong and Davies, 1999; Billen and Gurnis, 2001). Most global flow models without LVVs that employ a free-slip (zero shear stress) surface boundary condition (without weak zones along plate boundaries) match the geoid well, and plate motions poorly (because there is no toroidal flow). Alternatively, if plate motions are prescribed ("no slip" boundary condition with all velocity components fixed), these models lead to poor geoid fits (Thoraval and Richards, 1997). Studying slab-induced flow with LVVs, Zhong and Davies (1999) found that the fit to the geoid actually deteriorated when slabs were made stiffer than the mantle, and Čadek and Fleitout (2003) argued that impedance to flow at 660 km may be required by LVV flow models. Yet, Moucha et al. (2007) conclude that LVVs have a minor effect for geoid modeling based on tomography compared to uncertainties in the seismological models.

Such disagreement points to the requirement to incorporate diverse data into our geodynamical inversions, as velocity-related quantities are more sensitive to LVVs than stress-related measures, and the need to improve the implementation of plate boundaries in spherical flow computations. Improved models would also allow better evaluation of suggestions of dynamic topography and pressure gradients (Phipps Morgan et al., 1995; Husson and Ricard, 2004) that may be induced by closing off parts of the mantle by "slab curtains", such as around the Pacific at present. Husson et al. (2008), in particular, explored the influence of the Americas motion on the asymmetry of plate velocities in the Pacific. Such slab-imposed barriers to flow may also temporarily degrade thermal mixing with consequences for the coupling between oceanic plate convection and subcontinental heat budgets (*cf.* Lenardic et al., 2005). The next few years will likely bring new insights into how a consistent description of geopotential, anisotropy, and plate motion data may be achieved with weak spreading centers and thermo–chemically strong, but plastically yielding, convergent margins in a global model. It is also desirable to understand regional dynamics fully before models are made more comprehensive. However, we start with a cautionary note for the interpretation of global quantities based on regional models.

3.7 Large-Scale Mantle Wind and Regional Slab Circulation

Figure 7 shows two numerically computed subduction zone "streamlines" (*cf.* Hager and O'Connell, 1978). Those examples are meant for illustration only and are not supposed to be particularly realistic. Figure 7a is for a regional box with a free-slip surface boundary condition; it shows typical, small-scale circular motion that is induced by the slab buoyancy and similar to a Stokes sinker. Moreover, it is evident that the box is not large enough to exclude the effect of the side boundaries. Importantly, the streamlines indicate that the slab dip that is input (inferred from seismicity) has substantially different alignment. Clearly, the reference frame of the overriding plate and trench motions needs to be considered for proper comparisons (e.g. Olbertz et al., 1997) but the mismatch between slab structure and flow in Fig. 7a is typical even for more careful regional models.

If we prescribe plate motions on the surface in a global circulation model as in Fig. 6, the regional circulation patterns are considerably simpler. Figure 7b shows a cut out of a global computation, but using an embedded, regional, high-resolution model yields similar results. Such nested models have recently been used in a unidirectional mode, where the large-scale flow determines the velocities on the sides of a smaller box (Mihálffy et al., 2007), and in a more sophisticated fashion where communication goes both ways (Tan et al., 2006). (While the NR motions are not completely understood (Sect. 3.5), such large-scale flow that is unique to LVVs should be accounted for in nested modeling approaches but has been neglected so far.) For the example of Fig. 7b, the subduction vortices are suppressed and the flow much simplified

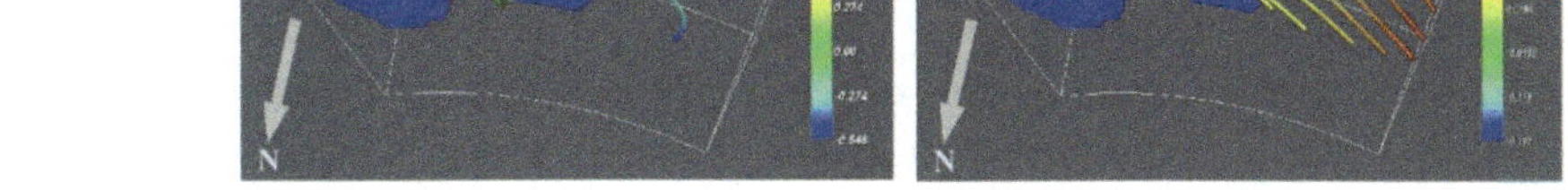

Fig. 7 Illustration of the potential importance of large-scale flow for subduction zone observables. Circulation models include upper mantle slabs as inferred from seismicity (from Gudmundsson and Sambridge, 1998) within the Japan–Kurile system (blue surfaces) which are converted to temperature anomalies, and are also stiffer than the mantle by a factor of $\eta' \sim 200$. Flow is visualized by assuming velocities are steady-state and following tracers forward and backward in time (coloring is integration time). Part (**a**) is from a regional computation with free-slip boundary conditions. Part (**b**) is a cut out of a global computation that has the same regional density contribution as (**a**) but prescribed plate motions everywhere on the surface

compared to Fig. 7a. (This is not to say that actual mantle currents need to be simple; smaller-scale effects may be averaged out in global computations.) Moreover, the dip from seismicity matches with the flow lines, as was shown globally without LVVs by Hager et al. (1983). The computation of Fig. 7b is not suited for quantitative comparisons strictly speaking, as a global subduction model with free surface conditions and weak zones should be used for consistency. However, since plate velocities of such models match the observed ones well, results would be very similar.

Subduction is likely a highly time-dependent process (e.g. Ita and King, 1998; Becker et al., 1999; Faccenna et al., 2001a; Billen and Hirth, 2007), and one may thus rightly argue that slab dips should not necessarily line up with instantaneous flow at all times (*cf.* Garfunkel et al., 1986; Lallemand et al., 2005). However, as earlier work on global flow and Fig. 7 illustrate, it is important to at least evaluate the role of large-scale currents if insights from regional models are transferred to global observables such as slab dip. With this caveat, we turn to the modeling of isolated slabs.

4 Regional Slab Dynamics

Regional models of subduction, implemented mostly in 2D, have been reviewed elsewhere with focus on deep dynamics and slab ponding at 660 km (King, 2001; Christensen, 2001), thermal structure (van Keken, 2003; King, 2007), and rheology (Billen, 2008). Recent advances include the incorporation of laboratory-derived rheologies to test how temperature-dependent viscosities plus plastic yielding may explain slab morphology (e.g. Čížková et al., 2002; Billen and Hirth, 2007), and what processes might assist subduction initiation (Toth and Gurnis, 1998; Faccenna et al., 1999; Regenauer-Lieb et al., 2001; Hall et al., 2003). However, as for the plate generation problem (Tackley, 2000a), questions remain about the appropriate weakening mechanism, the degree of strain history-dependence (e.g. Gurnis et al., 2000), and the effect of 3D plate reorganizations and continental cover.

We proceed to comment on slab rheology, discuss some of the common assumptions inherent in regional, isolated slab models, and how assumptions are sometimes intertwined with numerical issues by necessity. We will focus on fluid subduction models that are concerned with exploring the question of what actually controls the plate and plate boundary motions. In particular, we discuss those models that allow the trench to move self-consistently ("free trench") and strive to understand the causes rather than the effects of trench motion.

4.1 Free and Prescribed Trench Motions

It was seismic tomography that renewed interest in the role of trench migration in modifying the shape of subducting slabs, and several workers showed how variable motion of the trench can explain the observed "ponding" of fluid-like slabs at the 660 km phase transition (Griffiths et al., 1995; Gouillou-Frottier et al., 1995; Christensen, 1996). There, a negative Clapeyron slope of the olivine–perovskite/magnesiowüstite transition may be expected to inhibit subduction temporarily (Christensen and Yuen, 1984; Kincaid and Olson, 1987), though the likely viscosity increase between upper and lower mantle alone will also lead to changes in slab dip, flattening, or folding (e.g. Gurnis and Hager, 1988; Zhong and Gurnis, 1995; Enns et al., 2005). Weaker slabs and large rollback will promote ponding, whereas stationary, strong slabs are more likely to penetrate into the lower mantle (Davies, 1995; Christensen, 1996).

The migration of the slab inside the mantle also has an important impact on return mantle circulation and the way the resisting forces are partitioned between slab and mantle (Garfunkel et al., 1986; Conrad and Hager, 1999a). On that premise, several recent studies consider the trench motion as a proxy to infer subduction dynamics (Bellahsen et al., 2005; Enns et al., 2005; Royden and Husson, 2006), rather than prescribing V_T. Such free trench models, where V_T evolves dynamically, were pioneered in laboratory analog modeling by Jacoby and Schmeling (1981). Kincaid and Olson (1987) focused on an isolated subducting slab and showed that rates of subduction and rollback could be strongly affected by the interaction of the slab with a viscosity or density contrast at depth. Shemenda (1994) presented a wide range of laboratory experiments for a elasto–plastic slab in an inviscid mantle, with particular focus on lithospheric deformation and back-arc spreading. Zhong and Gurnis (1995) discussed 2D cylindrical numerical experiments that included a mobile, faulted margin implemented using FE with slippery-nodes (Melosh and Williams, 1989). The numerical results confirmed that plate kinematics may depend strongly on deep slab dynamics. Faccenna et al. (1996) introduced a different kind of laboratory analog model to the subduction research community. Silicone putty and glucose syrup (both viscous at the strain-rates considered) fill the role of the lithosphere and mantle, respectively.

4.2 Impact of Modeling Assumptions

Rocks deform elasto–visco–plastically, depending on the timescales of loading and ambient conditions, and both numerical and laboratory subduction experiments have been performed with elasto–plastic, visco–elasto–plastic, visco–plastic, and purely viscous rheologies. We will here focus on the latter two, but why would we expect that viscous slabs provide a sufficient description?

4.2.1 Rheology and Viscous Slabs

The role of elasticity in controlling slab-related deformation is in fact still somewhat controversial. Several large-scale features of subduction zones can be explained by elasto- or visco–plastic behavior, including the forebulge (Melosh and Raefsky, 1980; McAdoo et al., 1985; Zhong and Gurnis, 1994; Hall and Gurnis, 2005), the trench geometry (Morra et al., 2006; Schellart et al., 2007), and the shape of the slab (Hassani et al., 1997; Funiciello et al., 2003b). These results indicate that macroscopic observations, e.g. trench geometry, alone cannot be used to infer slab rheology. Large-scale kinematics such as rollback trade-off with intrinsic parameters such as slab stiffness (*cf.* Billen and Hirth, 2007).

In terms of subduction dynamics, elasticity may play an important role in initiation of subduction (Kemp and Stevenson, 1996), forming instabilities or shear localization (Muhlhaus and Regenauer-Lieb, 2005; Kaus and Podladchikov, 2006), and might lead to enhanced slab rollback (Moresi et al., 2002). However, often different rheological laws are active in such models at the same time, and the specific role of elasticity is unclear. Kaus and Becker (2007) therefore evaluated an idealized problem, the development of Rayleigh–Taylor instabilities for viscous and visco–elastic rheologies.

These authors found that the instability is sped up by the inclusion of elasticity, though only for parameter values that are likely not applicable to the Earth. For typical elasticity values as in PREM (Dziewoński and Anderson, 1981), stress fields are predicted to be different, but the temporal behavior of lithospheric instabilities is very similar for viscous and visco–elastic cases. While the effect of a supposed elastic core may need to be further explored, results by Schmeling et al. (2007) also indicate that elasticity does not significantly affect subduction dynamics.

Here, we shall therefore make the common assumption that the overall slab behavior over long time-scales can be well described by a visco–plastic fluid. Such an approach is supported by several lines of evidence: inferred intraslab deformation from seismicity in Wadati–Benioff zones are on the same order of those expected for the fluid mantle (Bevis, 1986, 1988; Holt, 1995), the transition between in-slab extension and compression (Isacks and Molnar, 1971) can be explained by a fluid slab encountering a viscosity jump (Vassiliou and Hager, 1988; Tao and O'Connell, 1993), and both seismicity (e.g. Giardini and Woodhouse, 1984, 1986; Fischer and Jordan, 1991) and tomography (e.g. van der Hilst and Seno, 1993; Widiyantoro and van der Hilst, 1997) show strong slab contortions reminiscent of fluid behavior (*cf.* Christensen, 1996; Tan et al., 2002; Ribe et al., 2007).

4.2.2 Numerical Models and Analog Experiments

To study fluid slabs, regional, 3D numerical computations can now routinely be performed at relatively high resolution (e.g. Billen and Gurnis, 2003; Piromallo et al., 2006; Stegman et al., 2006). Figure 8 shows snapshots from typical visco–plastic, numerical models. Using CitcomCU (Zhong et al., 1998) with a standard resolution of ~20 km (~4,000,000 elements) and ~80 advected material tracers per element, such computations take a few hours (using 54 CPUs) for a typical model evolving over tens of Myrs with thousands of time steps. However, the large LVVs that are inferred from laboratory creep laws still pose some problems numerically, particularly in the global,

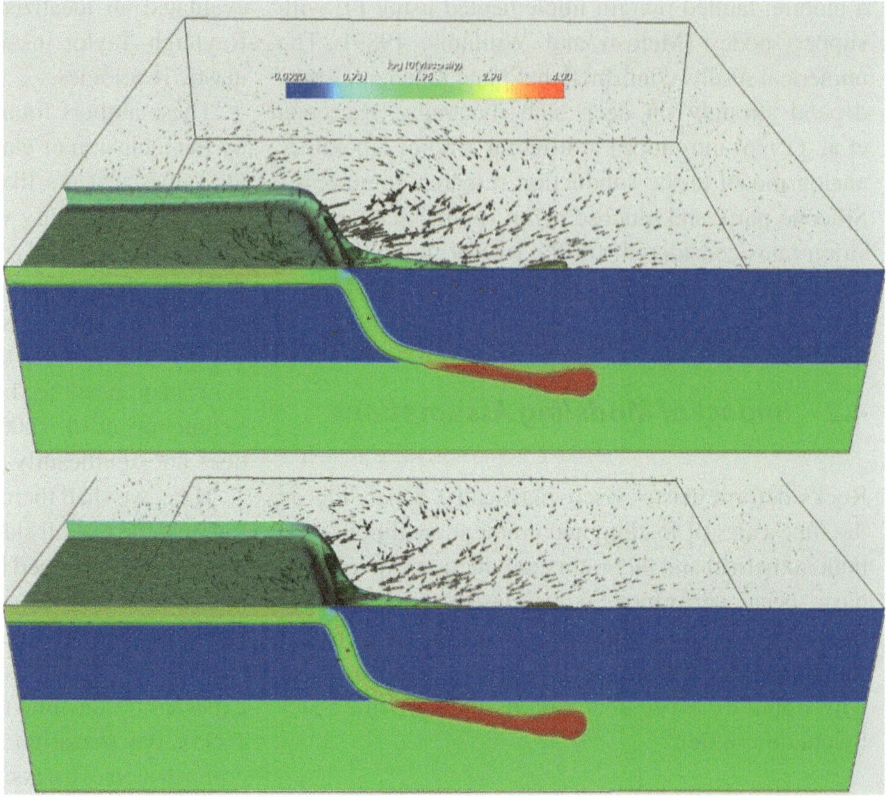

Fig. 8 Example of regional, free trench computations focusing on understanding trench kinematics. A visco–plastic slab (*green*, compositional iso-surface) sinks into a mantle with viscosity stratification (see shading for $\log_{10}$ of viscosity η^{eff}); *black vectors* denote flow velocity. Slab stiffness is $\eta' = 500$, lower/upper mantle viscosity ratio $\eta_l = 100$, and the subducting plate is attached to the side of the box on the left ("fixed ridge", *cf.* Enns et al., 2005). *Bottom figure* is for the minimum implementation of visco–plasticity (6), top is for the same model setup and time step but using the average viscosity formulation, (5). See Figure 10a for temporal evolution

spherical case where thermo–chemical problems such as entrainment remain challenging. It is therefore very useful to also consider laboratory models.

Numerous tectonic processes may be studied with analog experiments, provided that adequate scaling of material parameters to nature is ensured (e.g. Weijermars and Schmeling, 1986). One of the motivations for modeling subduction with both analog and numerical models is to complement each method's weaknesses and strengths. For the lab approach, pressure and temperature-dependence of rheology, reproducibility, and extraction of quantitative information are challenges, while physical realism and resolution can be strengths. The advantage of numerics is that all relevant quantitative measures can be easily extracted, and computer experiments can explore parameter space with perfect reproducibility.

The joint study of laboratory experiments and analytical or numerical methods has a long history in fluid dynamics (e.g. Ribe, 2003, for a related study). Applications to the subduction problem include the work of Kincaid and Olson (1987) who compared lab results on slab ponding with computer experiments by Christensen and Yuen (1984). Hassani et al. (1997) evaluated numerical models of elasto–plastic slabs following Shemenda (1994), and Becker et al. (1999) compared the analog models of Faccenna et al. (1999) with numerics. In an effort to further reduce the complexities and to understand the role of the subducting lithosphere in isolation (Christensen, 1996), many subsequent analog experiments focused on a single (stiff, dense, fluid) plate and got rid of the overriding plate (e.g. Funiciello et al., 2003a, 2004, 2006; Schellart, 2004a, b). Assuming Moore's (1965) law continues to hold, one may expect that increased computing power may soon render analog experiments obsolete. However, the lab has been catching up with improved analysis techniques (e.g. particle velocimetry, temperature visualization via polymers). Besides methodological issues, there are also often implicit assumptions, e.g. with respect to boundary conditions, whose relevance may only become evident when different approaches are compared. This is further motivation to use different tools to tackle the same problem, and we next discuss an example of such issues which, ironically, stem from trying to reproduce the simplified, single-slab experiments.

4.2.3 The Surface Boundary Condition and Visco–Plasticity

Most computations are conducted in an Eulerian framework with free-slip top surface boundary conditions. If an isolated, dense body is introduced on the top of the fluid domain, it is hard to detach the dense fluid from the top boundary in a slab-like shape, as opposed to a drip-like downwelling. This is because only horizontal motions are allowed at the top, and the corner flow geometry leads to a stress singularity in the wedge region (*cf.* Batchelor, 1967; Budiansky and Carrier, 1973). For kinematic subduction models, a similar wedge issue exists, and temperature solutions depend sensitively on discretization and choices on slab coupling (van Keken, 2003).

Several tricks are used to circumvent complications due to the wedge, including the addition of a soft, buoyant surface layer. Artifacts due to soft layer entrainment may result, however, and different methods yield convergent solutions only under relatively high resolution (Schmeling et al., 2007). Schmeling et al. also show that the averaging method for the effective rheology close to material interfaces leads to extremely different results. Eulerian codes only match results from Lagrangian, free-surface numerics, or laboratory results, if the (weakest) harmonic averaging is chosen. Such averaging of two fluid viscosities corresponds to two dashpots in series, as would be appropriate for simple shear; for pure shear, the dashpots would act in parallel (arithmetic mean). Which averaging scheme applies to subduction in general is unclear (Schmeling et al., 2007).

Another way to detach the slab in free-slip computations is the use of "Byerlee (1978)" type plasticity to locally yield a strong slab (as in Fig. 8). Taking this (pseudo-) plastic implementation as an example, Enns et al. (2005) argued that the introduction of a plastic viscosity, η_p, which is computed by dividing an appropriately adjusted, depth-dependent yield stress by the local strain-rates, allows detaching the slab without affecting its fluid behavior strongly. Such an approximate plastic behavior had been used before for subduction (Christensen, 1996; Tetzlaff and Schmeling, 2000), and is often also applied in plate generation studies (e.g. Moresi and Solomatov, 1998). This leads to another, secondary, issue: If one should treat plastic "viscosity" and the regular, creeping viscosity, η_c, as an effective viscosity (harmonic mean)

$$\eta_1^{\text{eff}} = \frac{\eta_p \eta_c}{\eta_p + \eta_c} \tag{5}$$

(e.g. Tetzlaff and Schmeling, 2000; Stein et al., 2004; Enns et al., 2005), or by using the minimum of the two viscosities

$$\eta_2^{\text{eff}} = \min(\eta_p, \eta_c) \tag{6}$$

(e.g. Moresi and Solomatov, 1998; Tackley, 2000b; Stegman et al., 2006). On the one hand, material should microscopically either behave plastically or viscously, which favors (6). On the other hand, numerical resolution will typically be coarser than plastic shear zones, which favors (5). Figure 8 shows two subduction models which only differ by the way this visco–plasticity is implemented. Differences are noticeable, but not large, for short time-histories, if sufficient resolution is used. However, subtleties such as different choices as to η^{eff} might partly explain, for example, why Stegman et al. (2006) were not able to closely reproduce Enns et al.'s (2005) results.

The extent to which viscous behavior dominates over plastic yielding in the models of Enns et al. (2005) is not entirely clear, and we are testing truly viscous and visco–plastic slab behavior for free-slip and free-surface boundary conditions at present. These problems are emphasized by Stegman et al.'s (2006) visco–plastic models where the entire trench region of the subducting plate is weakened such that no in-slab forces are transmitted; bending dissipation is, consequentially, not important. Schellart et al.'s (2007) approach limits yielding to a crustal layer, which improves slab coherence.

In nature, a reduction of shallow lithospheric strength by brittle and ductile failure is expected from rock rheology (e.g. Brace and Kohlstedt, 1980; Burov and Diament, 1995). Moreover, an increase in shallow weakening and faulting within the plate toward the trench is observed in plate flexure (McAdoo et al., 1985; Billen and Gurnis, 2005) and seismic (Ranero et al., 2003) studies, and deep slab deformation also requires departure from pure temperature-dependent viscosity (Čížková et al., 2002). However, we need to understand the rheological controls of the numerical and laboratory results better before confidently translating model results to nature. The importance of the trench formulation is also emphasized by the work of

Royden and Husson (2006). These authors use a semi-analytical formulation to treat the effect of large-scale mantle flow and find that trench migration is less affected by deep dynamics, such as mantle flow, but more sensitive to the overriding plate density structure (*cf.* Shemenda, 1994).

4.3 Dynamic Inferences from Regional Trench Kinematics

Most of the previous caveats on methodology apply to intermodel comparisons. If parameters are varied within one fixed setup, useful insights on the overall dynamics can still be gained. We proceed to discuss explanations for the regional kinematics derived from fluid slab experiments and return to slab strength in Sect. 4.3.3.

4.3.1 Back Arc Deformation and Sinking Velocities

Close to the trench, the overriding plate often deforms either in a strongly extensional sense (accommodated by back-arc spreading, "Marianas type," $V_B > 0$) or in a compressional fashion (sometimes associated with orogeny, "Chilean type," $V_B < 0$) (Uyeda and Kanamori, 1979). An overview of back-arc behavior and correlations of kinematic parameters with the style of deformation can be found in Lallemand et al. (2005).

Chase (1978) had argued that most slabs are either stationary or roll back in a hotspot reference-frame. The absolute overriding-plate motion would then control the type of back-arc deformation, and we would expect a strong correlation between rollback (V_T) and back-arc extension (V_B). While multiparametric descriptions can be found (Jarrard, 1986), real subduction zones appear to be more complicated (e.g. Carlson and Melia, 1984), and both trench motion and back-arc deformation are observed to vary in time (e.g. Sdrolias and Müller, 2006; Doglioni et al., 2007). Discrepancies between datasets built on geologic or geodetic observations may thus be expected. Newer data confirm the existence of correlation between V_B and overriding plate motion using back-arc deformation inferred from GPS geodesy (Heuret and Lallemand, 2005), although with large scatter. As for the causes of back-arc deformation, several processes are candidates. Large scale mantle currents that couple differently to slabs and via the thrust interface to the plate are among them (e.g. Dvorkin et al., 1993; Russo and Silver, 1994; Iaffaldano et al., 2006; Schellart et al., 2007), though it is unlikely that the diversity of deformation patterns can be explained by the effects of a single NR flow component as suggested by Doglioni (1990).

Heuret et al. (2007) provide a new analysis of global plate kinematics, back-arc deformation and deep slab shapes from tomography and seismicity (Heuret and Lallemand, 2005). In a plot of subducting (V_P) to overriding plate (V_{OP}) velocities, all of the Earth's subduction zones fall along a negative trend that corresponds to overall convergence rates of roughly $V_C \sim 5$ cm/yr (with large ~±50% fluctuations). Back-arc spreading and steep dips are found toward large $V_P > 0$ and $V_{OP} < 0$, while shortening and shallow dips are observed for $V_P > 0$ and $V_{OP} > 0$.

Choices in the reference frame will affect the exact values of these kinematic parameters on Earth (Fig. 4), e.g. if less NR than the HS-3 frame used in Heuret and Lallemand (2005) is applied. It is therefore more useful for broader conclusions to consider the overall slope of V_C. A general interpretation then emerges: if plates subduct with $V_C \gtrsim 5$ cm per year, the overriding plate is shortened; it is extended for convergence rates $\lesssim 5$ cm per year. It therefore appears that there is a preferred, local subduction rate V' which may be determined by regional slab dynamics and sinking into the mantle. If large-scale plate motions lead to values of V_{OP} and V_P that add up to deviations from V', then back-arc deformation results. This "anchoring" effect of the slab is confirmed in laboratory experiments with piston-driven plates and stiff slabs ($\eta' \sim 1{,}000$) where V_{OP} and V_P can be varied at will (Heuret et al., 2007).

If slabs were to consistently roll back, Heuret et al.'s (2007) observation would be in line with Chase's (1978) original argument that the overriding plate motion controls back-arc deformation. However, slabs are found to advance as well as retreat (Fig. 4b), and such, more complex, behavior is expected for stiff slabs (Bellahsen et al., 2005). Sdrolias and Müller (2006) analyze the temporal evolution of back-arc spreading using O'Neill et al.'s (2005) hotspot reference-frame and age reconstructions. These authors find that back-arc extension is limited to old oceanic lithosphere, and always preceded by overriding plate retreat away from the trench. After extension is initiated, rollback and regional slab dynamics appear to play a role, consistent with the arguments by Heuret et al. (2007) based on laboratory experiments.

4.3.2 Stokes Sinkers and Tomography

If back-arc deformation is controlled by a sinking velocity V', what controls this preferred subduction speed that would lead to deformation-neutral behavior? The first candidate for any velocity scale is V_{Stokes}, $\propto \Delta\rho/\eta_m$ (2) and this is reflected in the findings of Capitanio et al. (2007) for density- or age-dependent slab velocities. However, disagreements arise over the origin and the degree of variations from such general Stokes-scaling. Processes such as thrust interface coupling at the margin, large-scale mantle flow, and lithospheric bending may additionally affect the actual subduction velocity (e.g. Conrad and Hager, 1999a; Becker et al., 1999; Buffett and Rowley, 2006), and deep slab interactions, such as at the 660 km phase transition will lead to further complications (e.g. Zhong and Gurnis, 1996; Faccenna et al., 2001b). However, assuming that mantle drag is the important control ($V' \approx V_{\mathrm{Stokes}}$) and that slabs are weak, it is instructive to explore which other aspects of global subduction dynamics may be explained with Stokes behavior. Seismic tomography images fast anomalies in the lower mantle which can be interpreted as continuous, cold slabs at least down to ~1,200 km (e.g. Grand et al., 1997; Kárason and van der Hilst, 2000), and relative trench motion has a strong control on slab morphologies. Ricard et al. (1993) and Lithgow-Bertelloni and Richards (1998) therefore constructed a forward model of mantle structure by dropping "slablets" (superpositions of numerous Stokes sinkers) into the mantle at the locations where trenches are reconstructed to have been within the last 120 Ma. Lithgow-Bertelloni and Richards found that large-scale patterns in tomography can be matched well with such a model if the sinking velocity of the slablets is reduced in the lower mantle by a factor of $s \sim 4$. We expect $s \sim \ln(\eta_l)$, where η_l is the viscosity ratio between lower and upper mantle, because reduced sinking velocities (due to eq. 2) are accompanied by slab thickening in the lower mantle (Gurnis and Davies, 1986; Richards, 1991). The best-fit slowing factor $s \sim 4$ is compatible with estimates of $\eta_l \sim 50$ from geoid modeling (e.g. Hager and Clayton, 1989).

The correlations between slablet model and tomography are mostly limited to the longest spatial wavelengths and only ~0.4 when computed up to $\ell = 8$, but the match holds up for comparisons with more recent tomographic maps (Becker and Boschi, 2002). Using the same subduction history as Ricard et al. (1993) but allowing for lateral advection of slablets by mantle flow as induced by changing plate motions and density anomalies (Steinberger, 2000) has so far not led to a significantly better match with tomography (Becker and Boschi, 2002). This implies that more realistic global, forward subduction models may be needed which include LVVs such as weakening in the mantle wedge.

Regionally, the match of mapped slab structure and predictions from slablets can be improved by more realistic solutions for the slab sinkers, or by using more detailed plate reconstructions (Bunge and Grand, 2000; Tan et al., 2002). Kárason (2002) showed that complex slab images as imaged in several subduction zones, such as the along-strike variations in the Sunda arc (Kárason and van der Hilst, 2000), may be explained by slablet modeling and prescribed trench motions. In such models, the slab is assumed to have the same viscosity as the mantle, for simplicity, $\eta' = 1$. If a dense downwelling encounters a viscosity increase, such as at 660 km, an inverted mushroom-like shape and broadening results. However, relatively large viscosity jumps of $\eta_l \gtrsim 200$ were required to reproduce the imaged broadening of the slab anomaly in the lower mantle for the Sunda arc (Kárason, 2002).

An alternative explanation for the tomographically mapped structure at reduced η_l viscosity contrasts is the folding instability of a relatively stiff slab (Ribe et al., 2007). In this case, the mantle is treated as inviscid ($\eta' \to \infty$), and only the fluid deformation of the slab is considered (e.g. Houseman and Gubbins, 1997; Ribe, 2003). Reality is somewhere between these η'-rheological endmembers, and the joint approach of Morra et al. (2006) and Capitanio et al. (2007), where the subducting lithosphere is modeled numerically by FE, and the mantle component is dealt with semianalytically, may lead to improvements in regional models that account for the global flow contributions.

4.3.3 Lithospheric Bending and Partitioning of Viscous Dissipation

Conrad and Hager (1999a) improved on previous fluid loop analysis for subduction velocities by taking the detailed partitioning of viscous dissipation between faulted margin, bending lithosphere, and mantle into account and tested their scaling relationships with FE computations. Such models allow exploration of the parameter space between weak and strong slab dynamics. For experiments with fixed slab bending geometry, Conrad and Hager found that a large fraction of viscous dissipation was accommodated by the bending of a viscous slab, and that a scaling relationship could be derived to predict plate velocities as a function of h, R, and η'. Becker et al. (1999) were able to fit the (exponential) rate of subduction development for different slab strengths in their dynamically evolving models as a function of η' assuming that all viscous dissipation took place in the slab. However, these authors also pointed out that this only worked if the time-dependence of R was accounted for.

Also, the type of trench motion (advancing vs. retreating) was shown to be affected by the buoyancy number, which measures the magnitude of slab pull relative to slab strength (Houseman and Gubbins, 1997). Becker et al. found that subducting slabs failed to develop for a system with two plates and prescribed, far-field convergence velocities for $\eta' \gtrsim 750$, consistent with the plate velocity analysis of Conrad and Hager (1999a). A tectonic application of the subduction velocity rules from Becker et al. (1999) for oceanic and continental material entering a trench was presented in Faccenna et al. (2001a). The authors argue that the subduction rates at different trench segments in the central Mediterranean can be fit with a single, simple scaling law if buoyancy variations in subducting lithosphere due to a mix of continental and oceanic lithosphere are accounted for (*cf.* Royden and Husson, 2006).

Bellahsen et al. (2005) tested the applicability of bending scaling laws further, and presented a comprehensive analysis of rollback and subduction velocities for laboratory experiments with a single, isolated slab and stiffness ratios of $\eta' \gtrsim 1{,}000$. The dynamic behavior of such strong slabs when sinking into a layered mantle can be divided into three types: Type one always shows rollback, type two shows alternating episodes of trench rollback and advance (*cf.* Enns et al., 2005), and type three always advances (Funiciello et al., 2004). Especially the type two behavior is controlled by the interaction of the slab with the 660 km equivalent; this stage is associated with rebending of the lithosphere at depth (*cf.* Hassani et al., 1997; Čížková et al., 2002). Such deep dynamics may be recorded geologically by means of episodic back-arc basin opening, possibly associated with different amounts of lateral flow confinement (Faccenna et al., 2001b, 2004).

Subduction is time-dependent, especially in the transient phase where the initial downwelling has not yet reached the 660 km discontinuity. However, Bellahsen et al. (2005) were able to fit approximate steady-state behavior for fluid slabs once they have reached the ponding stage after interactions with 660 km (stage III in Fig. 10a). These authors used a simplified force balance based on Conrad and Hager (1999a), neglecting any overriding-plate or fault zone terms. If parameters are measured for a certain stage of subduction, the subduction velocity

$$V' \propto \frac{\Delta\rho g h L}{(2\eta'(h/R)^3 + 3A)\eta_m} \tag{7}$$

is found to scale linearly with V_c measured from experiments (Table 1 for symbols). It is clear that the strong $(h/R)^3$ dependence will complicate application of (7) to nature, and the bending geometry may be expected to vary in time and along strike.

By comparing V_{Stokes} (2) and the subduction velocity V' (7), it can be seen that the role of lithospheric viscosity (η') is modified, and typically much increased from the minor effect η' has on the Stokes velocity. In fact, V' may be normalized by a typical Stokes velocity to evaluate the role of bending and induced shear by the plate at the surface (Faccenna et al., 2007), as opposed to the vertical sinking of an isolated slab sinker. We will call V' from (7) a "modified Stokes" velocity to emphasize the role of the (age-dependent) slab pull ($\Delta\rho$ type terms). Bellahsen et al. (2005) and Funiciello et al. (2008) analyze how forces are partitioned between the lithospheric bending and mantle flow components ($(h/R)^3\eta'$ and $3A$ terms in (7), respectively). In accord with the earlier work, they find that the dominant control is the bending of the lithosphere for most settings. This view of subduction dynamics where relatively strong slabs are an important control on plate velocities also finds support in the analysis of

Buffett and Rowley (2006) who show that the direction of certain plate motions may be a consequence of the bending force balance. However, the Bellahsen et al. (2005) results are different from the analysis by Stegman et al. (2006), and at odds with conclusions by Schellart (2004b). Most of the discrepancies of published studies are likely due to different parameter choices with respect to slab stiffness. The division of viscous dissipation is expected to depend strongly on the rheology of the slab at the trench, which is poorly known in nature.

Dynamically evolving slabs may adjust into a minimum viscous dissipation configuration by modifying the bending geometry and the relative rate of trench motion (Enns et al., 2005). Recent work on the partitioning of viscous dissipation in dynamically evolving subduction models based on 3D modeling as in Fig. 8 indicates that between 40 and 50% (retreating or advancing slab, respectively) may be due to bending (Di Giuseppe et al., 2008). We conclude that bending may not be the sole, dominant source of dissipation, but it is likely of great importance. The broader implication is that plate tectonics may be more strongly affected by lithospheric strength as would be expected from an isoviscous convecting system. However, the heat flow scalings suggested by Conrad and Hager (1999b) and Korenaga (2003) may overestimate the role of the lithosphere, as they are based on constant bending geometries.

4.4 Different Modes of Trench Migration: Retreat and Advance

The other component of viscous dissipation, due to the induced flow in the mantle, is divided into a poloidal part similar to that of a Stokes sinker, and toroidal flow associated with trench migration (Garfunkel et al., 1986) and vortex-like motion around the slab (Buttles and Olson, 1998; Kincaid and Griffith, 2003;

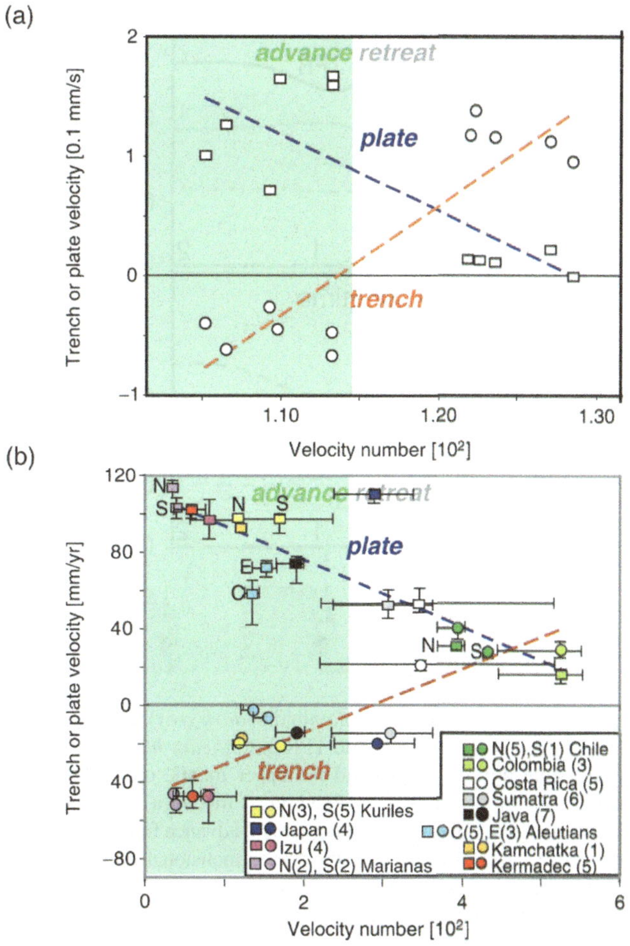

Fig. 9 (**a**) Dependence of subducting plate velocity V_p (*squares*) and trench motion V_T (*circles, cf.* Fig. 3) on "subduction velocity number," i.e., (7) normalized by a Stokes velocity (2 with $C = 1$ and $a^2 = Lh$) for the stiff ($\eta' \sim 1{,}000$) slab laboratory experiments of Bellahsen et al. (2005). (**b**) Subduction zone parameters for different trenches in nature (from Heuret and Lallemand, 2005). Legend lists color-coded trench regions with geographic subdivisions, error bars indicate the standard deviation of the mean velocities for each trench. (Figure modified from Faccenna et al., 2007, see this paper for details)

Funiciello et al., 2004, 2006; Schellart, 2004a; Piromallo et al., 2006).

4.4.1 Toroidal Flow and the Role of Slab Width

Like lithospheric bending, toroidal flow depends on rheology, and the toroidal to poloidal flow ratio may be used to judge the character of theoretical convection models compared to observed surface velocities. For regional subduction models, the decomposition of the velocity fields can be useful to understand the temporal evolution of slab sinking into the upper mantle. While large values of toroidal flow are reported ("95–100% of the mantle flux"; Schellart et al., 2007), such statements apply only to restricted layers of the upper mantle during strong rollback. It is also best to distinguish between amplitudes of divergence and vorticity, as opposed to poloidal and toroidal velocity decompositions (e.g. Tackley, 2000b). The latter are preferred, and reported here.

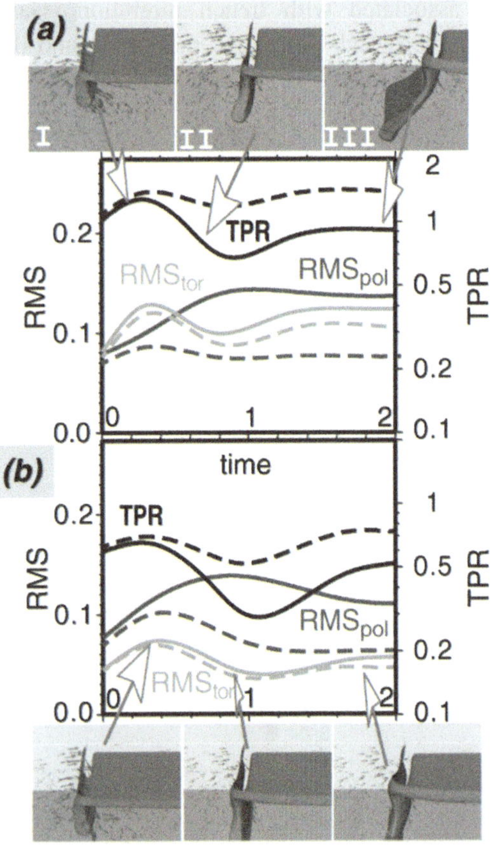

Fig. 10 Temporal evolution of poloidal (*dark gray*) and toroidal (light gray) velocity RMS, and TPR ratio (*heavy black line*) for the whole domain (solid) and the upper mantle only (*dashed lines*). We show results for a "fixed ridge" model (**a**), as in Fig. 8 and for a "free ridge" where the plate can advance freely ((**b**), as in Fig. 12; *cf.* Enns et al., 2005). Nondimensional units, unity time corresponds to ~15 Ma, and small plots show snapshots at the indicated times as in Fig. 8

From analysis of subduction models such as those shown in Fig. 8, we find that TPRs are highly time-dependent, particularly for "free ridge" models that lack a pronounced, steady-state rollback phase (Fig. 10). Mean TPR ratios are of the order of 0.5–0.9 for $\eta' = 500$ when spatially averaged over the model domain, where the high values result from fixed ridge setups where rollback is required (*cf.* Enns et al., 2005). Such overall TPR values can be compared with detailed tests for single slab snapshots by Piromallo et al. (2006). For their example geometry (similar to phase II of Fig. 10a), Piromallo et al. showed that toroidal flow itself increases with slab width W. However the TPR does not depend strongly on W, but increases with η', from 0 at $\eta' = 1$ to ~0.5 for $\eta' = 100$, saturating at ~0.7 for $\eta' \gtrsim 1{,}000$. If mantle circulation could be constrained for subduction settings independently, e.g. from seismic anisotropy, this relationship may be used to arrive at an indirect constraint for η'.

While TPR ratios are a kinematic construct, the toroidal flow component does allow evaluating the dynamic pressure components induced by rollback, and Royden and Husson (2006) present an analytical estimate of this flow component. Rollback pressure is expected to increase $\propto W$. This would imply that rollback rates, V_T, may scale inversely with W, as suggested by Dvorkin et al. (1993) and found in the laboratory experiments of Bellahsen et al. (2005). Consistently, Stegman et al. (2006) were able to parametrize V_T in their numerical models with a completely yielded trench region by poloidal and toroidal viscous dissipation in the mantle as a function of W, and also found that $V_T \propto 1/W$. In an application of their experiments, Schellart et al. (2007) point out that both rollback velocities and trench curvature in nature may be controlled by slab width. Narrow slabs such as the Scotia arc show fast rollback and convex trench shapes. Wide slabs such as in the Chilean roll back slowly or are stationary (Fig. 4), while their trenches form concave geometries with a possible stagnation point in the center (*cf.* Russo and Silver, 1994). However, both Stegman et al.'s (2006) and Schellart et al.'s (2007) results are out of scale for V_P, which is ~ one order of magnitude lower than the observed plate velocities. This may overestimate the role of V_T for subduction zone dynamics.

4.4.2 Regional Subduction Dynamics and Plate Advance

The work by Bellahsen et al. (2005) shows that certain parameters can be used to roughly predict overall convergence V_C based on V' (7), while the partitioning into V_T and V_P remains somewhat elusive. It is possible that rollback settings follow a minimum viscous dissipation principle where a balance between the Stokes-sinker flow, the shear of the plate at the surface, and the rollback-induced shear is found (Enns et al., 2005). While several groups are working on tests of this suggestion at present, a consensus or comprehensive theoretical description has yet to emerge.

However, from an empirical point of view, Faccenna et al. (2007) showed that trends exist in the subduction velocity scaling of Bellahsen et al. (2005). Trench migration, V_T, scales inversely with plate velocity, V_P. Since $V_S = V_T + V_P$ (Fig. 3), this implies that there is an anchoring effect with externally controlled V_S, similar to the results by Heuret et al. (2007) for back-arc deformation (Sect. 4.3.1). Expressed as a function of predicted subduction velocity V', trench motion, V_T, is found to increase, and plate velocity V_P to decrease, with increasing V' in the laboratory experiments (Fig. 9a). Thus, an increased subduction tendency as measured by V' (or its normalized version, the subduction velocity number of Fig. 9) is converted into increased rollback V_T because of anchoring. Interestingly, similar scalings for V_T, V_P, and V' are also found for subduction zones in nature (Fig. 9b). While there is large scatter, Faccenna et al. (2007) show that regional subduction zone parameters (from Heuret and Lallemand, 2005, based on HS-3) and the empirical relationships of Fig. 9 can adequately predict the trench-normal components of V_P and V_T.

Issues with applying the simplified Faccenna et al. (2007) results to nature were discussed extensively in their paper. For example, we would expect regions without complications from mantle wind, and those without continental material entering the trench, to behave most closely to what is expected based on models with a single subducting plate. The Sunda arc, and as a consequence the Australian plate, are accordingly not matched well, while other subduction zones are matched better. However, Fig. 11 provides a global test of the regional dynamics analysis of Faccenna et al. We computed rigid plate motions for the Australian, Pacific, and Nazca plate based on least-squaresestimates of the Euler vector that best matches the trench-normal, local velocities from Faccenna et al. (2007). It is apparent that the overall motions of the major oceanic plates that have subduction zones attached can be fit by the scaling relations that were derived from isolated, highly idealized subduction models without an overriding plate.

Part of this perhaps surprising result is due to the prescribed plate geometries and the correspondingly highly correlated driving torques (Fig. 5). While global

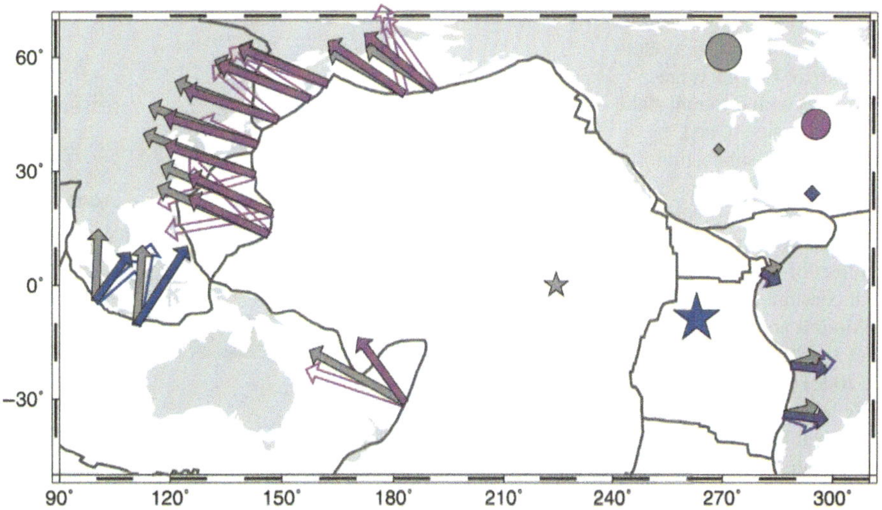

	PAC	NAZ	AUS
χ^2:	1.95	0.74	0.77
$(\Omega_m \cdot \Omega_n)/(\|\Omega_m\|\|\Omega_n\|)$:	0.913	0.909	0.775
$\|\Omega_m\|/\|\Omega_n\|$:	81%	141%	195%

Fig. 11 Best-fit plate motions (*solid colored vectors*) for rigid-plate Euler vectors based on matching the trench-normal velocity predictions of Faccenna et al. (2007) (*open solid vectors*), compared with HS-3 NUVEL (*gray vectors*, in the hotspot reference-frame). We allow for an azimuth error of the estimated trench segment normals of 30° before computing Euler poles (*star*: Australian plate; *circle*: Pacific; *diamond*: Nazca). Numbers below plot indicate χ^2 misfit, normalized correlation of Euler vectors $((\Omega_m \cdot \Omega_n)/(|\Omega_m||\Omega_n|))$ and amplitude ratio ($|\Omega_m|/|\Omega_n|$, values larger than unity meaning over-prediction of amplitudes)

models with viscous coupling at the base can predict plate motions well (Fig. 6), pulling at the right plate boundaries can also, expectedly, provide a good match. However, it is intriguing that roughly the right amplitudes of HS-3 V_p values are predicted, given that this plate model includes net rotations. This implies that regional slab dynamics may indeed be invoked to excite net rotations. At first this may seem counter intuitive, but given the possible strong effect of the slabs that are attached to the plates forming the Pacific basin, this finding deserves further study beyond the preliminary global tests that were discussed in Sect. 3.5.

4.4.3 Roll of Slab Strength for Roll Back

We had emphasized earlier that slab behavior in regional subduction experiments depends strongly on the viscosity of the slab η', particularly in the trench region. The strong and weak slab experiments by Bellahsen et al. (2005) and Stegman et al. (2006), respectively, are two endmember cases. Funiciello et al. (2008) therefore conducted additional experiments and were able to map out subduction behavior as a function of η' and V'. Figure 12 shows that models in the strong slab regime ($\eta' \gtrsim 10{,}000$, phase space "C"), on the one hand, are observed to always roll back. This is because the viscous bending is dominant, very little in-slab deformation and stretching occurs, and subduction prefers to minimize the deformation at the trench. In the C regime, dynamics are plate, or lithosphere, dominated. Weak slabs ($\eta' \lesssim 100$, phase space "A"), on the other hand, display little trench motion, deform internally, and form a Rayleigh–Taylor like viscous downwelling instead. In this regime, the dynamics are stronglymantle controlled.

In between the two extreme cases A and C is the more varied subduction behavior of phase space "B" where increasing subduction velocity decreases V_p and increases V_T; it is the high η' end of this range that was used by Faccenna et al. (2007) and discussed for Fig. 9. Only for this intermediate η' field are models with larger subduction velocity observed to be retreating, and models with smaller subduction velocity are found to be advancing (Fig. 11). While the intermediate η' regime might be interesting, it is not clear if it is relevant for nature. However, the η' ratios appear reasonable given other constraints on effective slab viscosity. Moreover, both advancing and retreating slabs are observed in nature (Fig. 4b), and back-arc deformation (Heuret et al., 2007) as well as V_P and V_T subduction velocities (Faccenna et al., 2007) follow trends that can be explained based on isolated slab experiments in this stiffness regime. The joint match of back-arc deformation and V_T is encouraging given that plate velocities (V_p) with moderate NR can be explained with global flow models (Figs. 2 and 6) without a refined representation of regional slab dynamics.

We therefore assume that viscous slab models are a useful means to understand global subduction

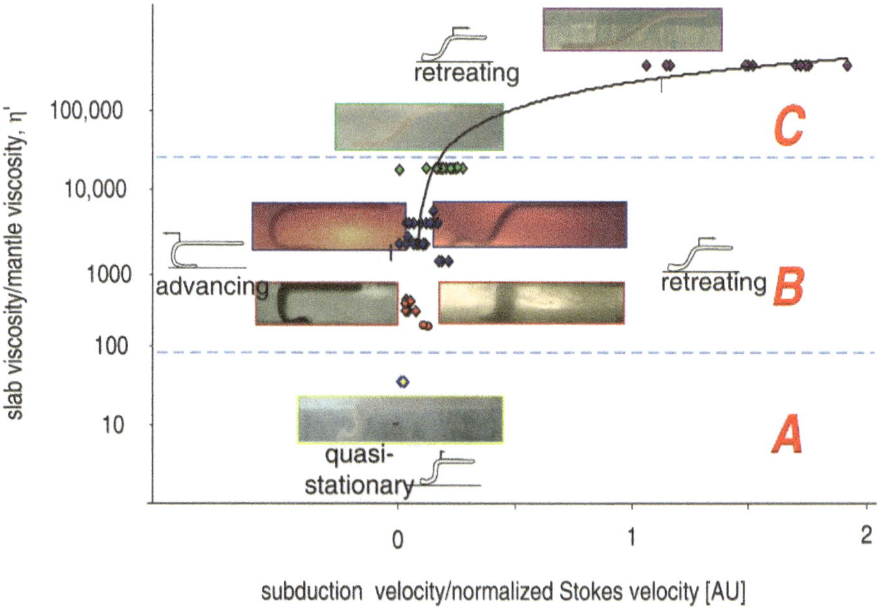

Fig. 12 Laboratory model behavior for free ridge models with different slab/mantle ratios η' and normalized subduction velocities (V' of eq. 7, divided by a normalized Stokes velocity to remove the effect of buoyancy changes at constant mantle viscosity). Models are from Bellahsen et al. (2005), Funiciello et al. (2008) (diamonds), and Schellart (2004b) (circles). Dashed lines are approximate divisions of model behavior; figure is modified from Funiciello et al. (2008)

kinematics, and that V_T is mainly controlled by deep slab dynamics. One can then turn the argument around and explore the inverse correlation between V_T and V_P as a function of η' and reference frames. Funiciello et al. (2008) summarize existing, and conduct new, free trailing-edge subduction models with an isolated viscous slab (Fig. 12). They show that the correlation between the two kinematic parameters holds for experiments with a range of η' within the intermediate field of Fig. 12. The best-fit slope $\alpha = V_T/V_P$ is found to increase from $\alpha \sim 0.5$ at $\eta' \sim 100$ to $\alpha \sim 1.75$ at $\eta' \sim 15{,}000$. In nature, the strength of the anticorrelation between V_P and V_T and the amplitude of α depend on the reference frame (Fig. 4); α varies from ~0.25 for NNR to ~0.75 for HS-3 (Funiciello et al., 2008).

Therefore, none of the α slopes for largest η' are observed in nature, as trench motion is usually smaller than plate velocities. Instead, the observed α values may indicate that slabs are relatively weak, with compatible η' values between ~150 and ~350 (Funiciello et al., 2008). Using this line of reasoning, the experiments by Bellahsen et al. (2005) were probablyconducted at η' values that are too high, and the inviscid fluid models of, e.g., Shemenda (1994) and Houseman and Gubbins (1997) with $\eta' \to \infty$ might make too drastic an assumption with regard to the neglect of mantle flow. Schellart's (2004a) experiments are closer to the η' values that are implied by Funiciello et al.'s study, but on the low end. The experiments of Schellart et al. (2007) are, however, inconsistent in that their α ratios are significantly higher than those observed in nature.

Viscosity ratios η' of the order of 100, as expected given the arguments above, are also consistent with arguments about subductability of fluid slabs (Conrad and Hager, 1999a; Becker et al., 1999), the geopotential and in-slab force transmission (Moresi and Gurnis, 1996; Zhong and Davies, 1999), as well as the flexure study of Billen and Gurnis (2005). A range of observations therefore indicate that subducting oceanic lithosphere behaves as a viscous fluid that is effectively ~500 times stiffer than the upper mantle. Is an effective viscosity a meaningful concept? In reality, temperature- and pressure-dependent diffusion and dislocation creep mechanisms control slab deformation, along with plastic yielding mechanisms that will limit maximum deviatoric stresses (Kirby and Kronenberg, 1987; Riedel and Karato, 1997; Hirth and Kohlstedt, 2004). Even for olivine, several creep law parameters are not well measured (e.g. activation volume), constrained (e.g. volatile content), or may dynamically evolve (e.g. grain size) in nature. However, several authors have studied which visco–plastic olivine-type rheologies lead to "slab-like" subduction in the upper mantle. Geodynamic models for the present-day potential temperatures (e.g. Tetzlaff and Schmeling, 2000; van Hunen et al., 2000; Billen and Hirth, 2007), and the Earth's past (van Hunen and van den Berg, 2007) have been explored. From these experiments, it is apparent that both slab shapes and average viscosities are dynamically evolving. However, η' values of $\sim 10^4 \ldots 10^5$ for the slab outside the trench are plausible, along with significant weakening toward the trench, down to $\eta' \sim 10^2 \ldots 10^3$. Lab-derived rheologies in thermo–mechanical models are thus consistent with the range of η' used in the simplified models here.

5 Conclusions

Plate kinematics may be understood by the dynamical behavior of fluid slabs sinking into a mantle that is viscously stratified. Regional modeling indicates that rollback, trench curvature, and back-arc deformation are consistent with slabs that are ~250–500 times stiffer than the mantle, in terms of the effective viscosity in the trench region. The dynamics are controlled by bending, mantle drag, and interactions with the higher viscosity lower mantle.

The net rotation of the lithosphere may be caused jointly by regional slab dynamics, buoyant upwellings, and the effect of continental keels, although global models based on slabs alone were not found to be effective in inducing net rotations as large as those of recent hotspot models. However, such global models with weak zones and other lateral viscosity variations lead to a good fit with regard to other plate tectonic scores such as intraplate deformation, and oceanic to continental plate velocity ratios.

While much progress has been made, several important issues such as the physics of the partitioning between rollback and subducting plate velocity over time are still not fully resolved. It is also clear that conclusions about subduction dynamics are somewhat dependent on the choice of the plate kinematic model and assumptions about slab rheology. This is a problem, as the strength of the slab is poorly constrained from either large-scale modeling inferences or laboratory measurements.

It appears crucial to improve the weak zone and plate boundary implementation in global subduction models. A more sophisticated rheological treatment of slabs in global flow may also improve the match between geodynamic forward models and seismic tomography. Moreover, advances in the plate boundary treatment will help to evaluate the role of the upper mechanical boundary condition for numerical models, and to perform truly quantitative comparisons between laboratory analog models and numerics.

Most of the results discussed in this review indicate that the role of the overriding plate may be small for controlling subduction dynamics. However, this also needs to be evaluated with joint lithospheric-mantle models, particularly in the context of a thick, continental overriding plate.

Acknowledgments Discussions during the SUBCO meeting in Montpellier in June 2007 helped to distill some of the ideas summarized here. We thank our reviewers Saskia Goes and Carolina Lithgow–Bertelloni as well as F. Funiciello, M. I. Billen, B. J. P. Kaus, and S. D. King for comments on an earlier version of this manuscript. In addition, we thank A. Heuret and F. Funiciello for sharing their data, and S. Zhong and L. Moresi for their computer software. Computations were conducted at the University of Southern California Center for High Performance Computing and Communications (www.usc.edu/hpcc), and we used software from CIG (geodynamics.org). Most figures were produced with the GMT software by Wessel and Smith (1991). This research was supported by NSF grants EAR-0643365 and 0451952.

Note This manuscript reflects the state of the literature during the time of original submission, the Fall of 2007.

References

Batchelor GK (1967) An introduction to fluid dynamics. Cambridge University Press, Cambridge UK

Becker TW (2006) On the effect of temperature and strain-rate dependent viscosity on global mantle flow, net rotation, and plate-driving forces. Geophys J Int 167:943–957

Becker TW (2008) Azimuthal seismic anisotropy constrains net rotation of the lithosphere. Geophys Res Lett 35, doi:10.1029/2007GL032928,correction: doi:10.1029/2008GL033946

Becker TW, Boschi L (2002) A comparison of tomographic and geodynamic mantle models. Geochem Geophys Geosyst 3:2001GC000168

Becker TW, O'Connell RJ (2001) Predicting plate velocities with geodynamic models. Geochem Geophys Geosyst 2:2001GC000171

Becker TW, Faccenna C, O'Connell RJ, Giardini D (1999) The development of slabs in the upper mantle: insight from numerical and laboratory experiments. J Geophys Res 104:15207–15225

Becker TW, Kellogg JB, Ekström G, O'Connell RJ (2003) Comparison of azimuthal seismic anisotropy from surface waves and finite-strain from global mantle-circulation models. Geophys J Int 155:696–714

Becker TW, Schulte-Pelkum V, Blackman DK, Kellogg JB, O'Connell RJ (2006) Mantle flow under the western United States from shear wave splitting. Earth Planet Sci Lett 247:235–251

Bellahsen N, Faccenna C, Funiciello F (2005) Dynamics of subduction and plate motion in laboratory experiments: insights into the plate tectonics behavior of the Earth. J Geophys Res 110, doi:10.1029/2004JB002999

Bercovici D (2003) The generation of plate tectonics from mantle convection. Earth Planet Sci Lett 205:107–121

Bercovici D, Ricard Y, Richards M (2000) The relationship between mantle dynamics and plate tectonics: a primer. In: Richards M, Gordon R, van der Hilst RD (eds.), *The History and Dynamics of Global Plate Motions*, American Geophysical Union, Washington, DC, vol. 121 of *Geophysical Monograph*, 5–46

Bevis M (1986) The curvature of Wadati-Benioff zones and the torsional rigidity of subducting plates. Nature 323:52–53

Bevis M (1988) Seismic slip and down dip strain rate in Wadati-Benioff zones. Science 240:1317–1319

Billen MI (2008) Modeling the dynamics of subducting slabs. Annu Rev Earth Planet Sci 36:325–356

Billen MI, Gurnis M (2001) A low viscosity wedge in subduction zones. Earth Planet Sci Lett 193:227–236

Billen MI, Gurnis M (2003) Multiscale dynamics of the Tonga-Kermadec subduction zone. Geophys J Int 153:359–388

Billen MI, Gurnis M (2005) Constraints on subducting plate strength within the Kermadec trench. J Geophys Res 110, doi:10.1029/2004JB003308

Billen MI, Hirth G (2007) Rheologic controls on slab dynamics. Geochem Geophys Geosyst 8, doi:10.1029/2007GC001597: Q08012

Bird P (2003) An updated digital model of plate boundaries. Geochem Geophys Geosyst 4, doi:10.1029/2001GC000252

Boschi L, Becker TW, Steinberger B (2007) Mantle plumes: dynamic models and seismic images. Geochem Geophys Geosyst 8, Q10006, doi:10.1029/2007GC001733

Brace WF, Kohlstedt DL (1980) Limits on lithospheric stress imposed by laboratory experiments. J Geophys Res 85:6248–6252

Budiansky R, Carrier GF (1973) The pointless wedge. SIAM J Appl Mech 25:378–387

Buffett BA, Rowley DB (2006) Plate bending at subduction zones: Consequences for the direction of plate motions. Earth Planet Sci Lett 245:359–364

Bunge HP, Grand SP (2000) Mesozoic plate-motion history below the northeast Pacific Ocean from seismic images of the subducted Farallon slab. Nature 405:337–340

Burov EB, Diament M (1995) The effective elastic thickness (T_e) of continental lithosphere: What does it really mean? J Geophys Res 100:3905–3927

Buttles J, Olson P (1998) A laboratory model of subduction zone anisotropy. Earth Planet Sci Lett 164:245–262

Byerlee J (1978) Friction of rock. Pure Appl Geophys 116:615–626
Čadek O, Fleitout L (2003) Effect of lateral viscosity variations in the top 300 km of the mantle on the geoid and dynamic topography. Geophys J Int 152:566–580
Čadek O, Ricard Y, Martinec Z, Matyska C (1993) Comparison between Newtonian and non-Newtonian flow driven by internal loads. Geophys J Int 112:103–114
Capitanio FA, Morra G, Goes S (2007) Dynamic models of downgoing plate-buoyancy driven subduction: Subduction motions and energy dissipation. Earth Planet Sci Lett 262:284–297
Carlson RL, Melia PJ (1984) Subduction hinge migration. Tectonophysics 102:1–16
Chapple WM, Tullis TE (1977) Evaluation of the forces that drive the plates. J Geophys Res 82:1967–1984
Chase CG (1978) Extension behind island arcs and motion relative to hot spots. J Geophys Res 83:5385–5387
Chen J, King SD (1998) The influence of temperature and depth dependent viscosity on geoid and topography profiles from models of mantle convection. Phys Earth Planet Inter 106:75–91
Christensen U (2001) Geodynamic models of deep subduction. Phys Earth Planet Inter 127:25–34
Christensen UR (1996) The influence of trench migration on slab penetration into the lower mantle. Earth Planet Sci Lett 140:27–39
Christensen UR, Yuen DA (1984) The interaction of a subducting lithospheric slab with a chemical or phase boundary. J Geophys Res 89:4389–4402
Čížková H, adek O, Slancová A (1998) Regional correlation analysis between seismic heterogeneity in the lower mantle and subduction in the last 180 Myr: implications for mantle dynamics and rheology. Pure Appl Geophys 151:527–537
Čížková H, van Hunen J, van den Berg AP, Vlaar NJ (2002) The influence of rheological weakening and yield stress on the interaction of slabs with the 670-km discontinuity. Earth Planet Sci Lett 199:447–457
Clift P, Vannucchi P (2004) Controls on tectonic accretion versus erosion in subduction zones: Implications for the origin and recycling of the continental crust. Rev Geophys 42, RG2001:1–31
Conrad CP, Hager BH (1999a) The effects of plate, bending and fault strength at subduction zones on plate dynamics. J Geophys Res 104:17551–17571
Conrad CP, Hager BH (1999b) The thermal evolution of an Earth with strong subduction zones. Geophys Res Lett 26:3041–3044
Conrad CP, Lithgow-Bertelloni C (2002) How mantle slabs drive plate tectonics. Science 298:207–209
Conrad CP, Lithgow-Bertelloni C (2004) The temporal evolution of plate driving forces: Importance of "slab suction" versus "slab pull" during the Cenozoic. J Geophys Res 109, doi:10.1029/2004JB002991
Conrad CP, Lithgow-Bertelloni C (2006) Influence of continental roots and asthenosphere on plate-mantle coupling. Geophys Res Lett 33, doi:10.1029/2005GL02562
Conrad CP, Bilek S, Lithgow-Bertelloni C (2004) Great earthquakes and slab-pull: Interaction between seismic coupling and plate-slab coupling. Earth Planet Sci Lett 218:109–122
Cruciani C, Carminati E, Doglioni C (2005) Slab dip vs. lithosphere age: no direct function. Earth Planet Sci Lett 238:298–310
Davies GF (1995) Penetration of plates and plumes through the mantle transition zone. Earth Planet Sci Lett 133:507–516
DeMets C, Gordon RG, Argus DF, Stein S (1994) Effect of recent revisions to the geomagnetic reversal time scale on estimates of current plate motions. Geophys Res Lett 21:2191–2194
Deparis V, Legros H, Ricard Y (1995) Mass anomalies due to subducted slabs and simulations of plate motion since 200 My. Earth Planet Sci Lett 89:271–280
Di Giuseppe E, van Hunen J, Funiciello F, Faccenna C, Giardini D (2008) Slab stiffness controls trench motion: insights from numerical models. Geochem Geophys Geosyst 9, doi:10.1029/2007GC001776:Q02014
Doglioni C (1990) The global tectonic pattern. J Geodynamics 12:21–38
Doglioni C, Carminati E, Cuffaro M, Scrocca D (2007) Subduction kinematics and dynamic constraints. Earth Sci Rev 83:125–175
Dumoulin C, Bercovici D, Wessel P (1998) A continuous plate-tectonic model using geophysical data to estimate plate-margin widths, with a seismicity-based example. Geophys J Int 133:379–389
Dvorkin J, Nur A, Mavko G, Ben-Avraham Z (1993) Narrow subducting slabs and the origin of backarc basins. Tectonophysics 227:63–79
Dziewoński AM, Anderson DL (1981) Preliminary reference Earth model. Phys Earth Planet Inter 25:297–356
England PC, Molnar P (1997) Active deformation of Asia: from kinematics to dynamics. Science 278:647–650
Enns A, Becker TW, Schmeling H (2005) The dynamics of subduction and trench migration for viscosity stratification. Geophys J Int 160:761–775
Faccenna C, Davy P, Brun JP, Funiciello R, Giardini D, Mattei M, Nalpas T (1996) The dynamics of back-arc extension: an experimental approach to the opening of the Tyrrhenian Sea. Geophys J Int 126:781–795
Faccenna C, Giardini D, Davy P, Argentieri A (1999) Initiation of subduction at Atlantic type margins: Insights from laboratory experiments. J Geophys Res 104:2749–2766
Faccenna C, Becker TW, Lucente FP, Jolivet L, Rossetti F (2001a) History of subduction and back-arc extension in the central Mediterranean. Geophys J Int 145:809–820
Faccenna C, Funiciello F, Giardini D, Lucente P (2001b) Episodic back-arc extension during restricted mantle convection in the Central Mediterranean. Earth Planet Sci Lett 187:105–116
Faccenna C, Piromallo C, Crespo Blanc A, Jolivet L, Rossetti F (2004) Lateral slab deformation and the origin of the arcs of the western Mediterranean. Tectonics 23, doi:10.1029/2002TC001488:TC1012
Faccenna C, Heuret A, Funiciello F, Lallemand S, Becker TW (2007) Predicting trench and plate motion from the dynamics of a strong slab. Earth Planet Sci Lett 257:29–36
Fischer KM, Jordan TH (1991) Seismic strain rate and deep slab deformation in Tonga. J Geophys Res 96:14429–14444
Forsyth DW, Uyeda S (1975) On the relative importance of the driving forces of plate motion. Geophys J R Astr Soc 43:163–200

Forte AM, Mitrovica JX (2001) Deep-mantle high-viscosity flow and thermochemical structure inferred from seismic and geodynamic data. Nature 410:1049–1056
Forte AM, Peltier WR (1987) Plate tectonics and aspherical earth structure: the importance of poloidal-toroidal coupling. J Geophys Res 92:3645–3679
Forte A, Peltier WR (1994) The kinematics and dynamics of poloidal-toroidal coupling in mantle flow: the importance of surface plates and lateral viscosity variations. Adv Geophys 36:1–119
Funiciello F, Faccenna C, Giardini D, Regenauer-Lieb K (2003a) Dynamics of retreating slabs (part 2): insights from 3D laboratory experiments. J Geophys Res 108, doi:10.1029/2001JB000896
Funiciello F, Morra G, Regenauer-Lieb K, Giardini D (2003b) Dynamics of retreating slabs (part 1): insights from numerical experiments. J Geophys Res
Funiciello F, Faccenna C, Giardini D (2004) Flow in the evolution of subduction system: Insights from 3-D laboratory experiments. Geophys J Int 157:1393–1407
Funiciello F, Moroni M, Piromallo C, Faccenna C, Cenedese A, Bui HA (2006) Mapping flow during retreating subduction: laboratory models analyzed by Feature Tracking. J Geophys Res 111, doi:10.1029/2005JB003792
Funiciello F, Faccenna C, Heuret A, Di Giuseppe E, Lallemand S, Becker TW (2008) Trench migration, net rotation and slab-mantle coupling. Earth Planet Sci Lett 271:233–240
Gable CW, O'Connell RJ, Travis BJ (1991) Convection in three dimensions with surface plates: generation of toroidal flow. J Geophys Res 96:8391–8405
Garfunkel Z, Anderson CA, Schubert G (1986) Mantle circulation and the lateral migration of subducted slabs. J Geophys Res 91:7205–7223
Giardini D, Woodhouse JH (1984) Deep seismicity and modes of deformation in Tonga subduction zone. Nature 307:505–509
Giardini D, Woodhouse JH (1986) Horizontal shear flow in the mantle beneath the Tonga arc. Nature 319:551–555
Gordon RG (2000) Diffuse oceanic plate boundaries: Strain rates, vertically averaged rheology, and comparisons with narrow plate boundaries and stable plate interiors. In: Richards MA, Gordon RG, van der Hilst RD (eds.), *The History and Dynamics of Global Plate Motion*, American Geophysical Union, Washington DC, vol. 121 of *Geophysical Monograph*, 143–159
Gordon RG, Jurdy DM (1986) Cenozoic global plate motions. J Geophys Res 91:12389–12406
Gouillou-Frottier L, Buttles J, Olson P (1995) Laboratory experiments on the structure of subducted lithosphere. Earth Planet Sci Lett 133:19–34
Grand SP, van der Hilst RD, Widiyantoro S (1997) Global seismic tomography; a snapshot of convection in the Earth. GSA Today 7:1–7
Griffiths RW, Hackney RI, van der Hilst RD (1995) A laboratory investigation of effects of trench migration on the descent of subducted slabs. Earth Planet Sci Lett 133:1–17
Gripp AE, Gordon RG (1990) Current plate velocities relative to the hotspots incorporating the NUVEL-1 global plate motion model. Geophys Res Lett 17:1109–1112
Gripp AE, Gordon RG (2002) Young tracks of hotspots and current plate velocities. Geophys J Int 150:321–361
Gudmundsson O, Sambridge M (1998) A regionalized upper mantle (RUM) seismic model. J Geophys Res 103: 7121–7136
Gurnis M, Davies GF (1986) The effect of depth-dependent viscosity on convective mixing in the mantle and the possible survival of primitive mantle. Geophys Res Lett 13: 541–544
Gurnis M, Hager BH (1988) Controls of the structure of subducted slabs. Nature 335:317–321
Gurnis M, Zhong S, Toth J (2000) On the competing roles of fault reactivation and brittle failure in generating plate tectonics from mantle convection. In: Richards MA, Gordon RG, van der Hilst RD (eds.), *The History and Dynamics of Global Plate Motions*, AGU, Washington DC, vol. 121 of *Geophysical Monograph*, 73–94
Hager BH (1984) Subducted slabs and the geoid: constraints on mantle rheology and flow. J Geophys Res 89:6003–6015
Hager BH, Clayton RW (1989) Constraints on the structure of mantle convection using seismic observations, flow models, and the geoid. In: Peltier WR (ed.), *Mantle convection: Plate tectonics and global dynamics*, Gordon and Breach Science Publishers, New York, NY, vol. 4 of *The Fluid Mechanics of Astrophysics and Geophysics*, 657–763
Hager BH, O'Connell RJ (1978) Subduction zone dip angles and flow derived by plate motion. Tectonophysics 50:111–133
Hager BH, O'Connell RJ (1981) A simple global model of plate dynamics and mantle convection. J Geophys Res 86:4843–4867
Hager BH, O'Connell RJ, Raefsky A (1983) Subduction, back-arc spreading and global mantle flow. Tectonophysics 99:165–189
Hall CE, Gurnis M (2005) Strength of fracture zones from their barymetric and gravitational evolution. J Geophys Res 110, doi:10.1029/2004JB003312
Hall CE, Gurnis M, Sdrolias M, Lavier LL, Muller RD (2003) Catastrophic initiation of subduction following forced convergence at transform boundaries. Earth Planet Sci Lett 212:15–30
Han L, Gurnis M (1999) How valid are dynamical models of subduction and convection when plate motions are prescribed? Phys Earth Planet Inter 110:235–246
Hassani R, Jongmans D, Chéry J (1997) Study of plate deformation and stress in subduction processes using two-dimensional numerical models. J Geophys Res 102:17951–17965
Heuret A, Lallemand S (2005) Slab dynamics and back-arc deformation. Phys Earth Planet Inter 149:31–51
Heuret A, Funiciello F, Faccenna C, Lallemand S (2007) Plate kinematics, slab shape and back-arc stress: A comparison between laboratory models and current subduction zones. Earth Planet Sci Lett 256:473–483
Hirth G, Kohlstedt DL (2004) Rheology of the upper mantle and the mantle wedge: A view from the experimentalists. In: Eiler J (ed.), *Inside the Subduction Factory*, American Geophysical Union, Washington DC, vol. 138 of *Geophysical Monograph*, 83–105
Holt WE (1995) Flow fields within the Tonga slab determined from the moment tensors of deep earthquakes. Geophys Res Lett 22:989–992

Houseman GA, Gubbins D (1997) Deformation of subducted oceanic lithosphere. Geophys J Int 131:535–551

Humphreys ED, Coblentz D (2007) North American dynamics and western U.S. tectonics. Rev Geophys 45, RG3001:doi:10.1029/2005RG000181

Husson L, Ricard Y (2004) Stress balance above subduction: application to the Andes. Earth Planet Sci Lett 222:1037–1050

Husson L, Conrad CP, Faccenna C (2008) Tethyan closure, Andean orogeny, and westward drift of the Pacific basin, Earth Planet Sci Lett 271:303–318.

Iaffaldano G, Bunge HP, Dixon TH (2006) Feedback between mountain belt growth and plate convergence. Geology 34:893–896

Isacks B, Molnar P (1971) Distribution of stresses in the descending lithosphere from a global survey of focal-mechanism solutions of mantle earthquakes. Rev Geophys Space Phys 9:103–175

Ita J, King SD (1998) The influence of thermodynamic formulation on simulations of subduction zone geometry and history. Geophys Res Lett 25:1463–1466

Jacoby WR, Schmeling H (1981) Convection experiments and driving mechanism. Geol Rundschau 24:217–284

Jarrard RD (1986) Relations among subduction parameters. Rev Geophys 24:217–284

Jordan TH (1978) Composition and development of the continental tectosphere. Nature 274:544–548

Kárason H (2002) Constraints on mantle convection from seismic tomography and flow modeling. Ph.D. thesis, Massachusetts Institute of Technology, Cambridge MA

Kárason H, van der Hilst RD (2000) Constraints on mantle convection from seismic tomography. In: Richards MA, Gordon RG, van der Hilst RD (eds.), *The History and Dynamics of Global Plate Motion*, American Geophysical Union, Washington DC, vol. 121 of *Geophysical Monograph*, 277–288

Karato Si (1998) Seismic anisotropy in the deep mantle, boundary layers and the geometry of convection. Pure Appl Geophys 151:565–587

Kaus BJP, Becker TW (2007) Effects of elasticity on the Rayleigh-Taylor instability: implications for large-scale geodynamics. Geophys J Int 168:843–862

Kaus BJP, Podladchikov YY (2006) Initiation of localized shear in visco-elasto-plastic rocks. J Geophys Res 111, doi:10.1029/2005JB003652

Kemp DV, Stevenson DJ (1996) A tensile, flexural model for the initiation of subduction. Geophys J Int 125:73–94

Kincaid C, Griffith RW (2003) Laboratory models of the thermal evolution of the mantle during rollback subduction. Nature 425:58–62

Kincaid C, Olson P (1987) An experimental study of subduction and slab migration. J Geophys Res 92:13832–13840

King SD (2001) Subduction: Observations and geodynamic models. Phys Earth Planet Inter 127:9–24

King SD (2007) Mantle downwellings and the fate of subducting slabs: constraints from seismology, geoid, topography, geochemistry, and pretrology. In: Schubert G, Bercovici D (eds.), *Treatise on Geophysics*, Elsevier. In press

King SD, Hager BH (1990) The relationship between plate velocity and trench viscosity in Newtonian and power-law subduction calculations. Geophys Res Lett 17: 2409–2412

King SD, Gable CW, Weinstein SA (1992) Models of convection-driven tectonic plates: a comparison of methods and results. Geophys J Int 109:481–487

Kirby SH, Kronenberg AK (1987) Rheology of the lithosphere: Selected topics. Rev Geophys 25:1219–1244

Kley J (1999) Geologic and geometric constraints on a kinematic model of the Bolivian orocline. J South Am Earth Sci 12:221–235

Korenaga J (2003) Energetics of mantle convection and the fate of fossile heat. Geophys Res Lett 30, 8:doi:10.1029/2003GL016

Kreemer C, Holt WE, Haines AJ (2003) An integrated global model of present-day plate motions and plate boundary deformation. Geophys J Int 154:5–34

Lallemand S (1995) High rates of arc consumption by subduction processes: some consequences. Geology 23:551–554

Lallemand S, Heuret A, Boutelier D (2005) On the relationships between slab dip, back-arc stress, upper plate absolute motion, and crustal nature in subduction zones. Geochem Geophys Geosyst 6, doi:10.1029/2005GC000917

Lee CTA, Lenardic A, Cooper CM, Niu F, Levander A (2005) The role of chemical boundary layers in regulating the thickness of continental and oceanic thermal boundary layers. Earth Planet Sci Lett 230:379–395

Lenardic A, Moresi LN, Jellinek AM, Manga M (2005) Continental insulation, mantle cooling, and the surface area of oceans and continents. Earth Planet Sci Lett 234: 317–333

Lithgow-Bertelloni C, Guynn JH (2004) Origin of the lithospheric stress field. J Geophys Res 109, doi:10.1029/2003JB002467

Lithgow-Bertelloni C, Richards MA (1995) Cenozoic plate driving forces. Geophys Res Lett 22:1317–1320

Lithgow-Bertelloni C, Richards MA (1998) The dynamics of Cenozoic and Mesozoic plate motions. Rev Geophys 36:27–78

Lithgow-Bertelloni C, Richards MA, Ricard Y, O'Connell RJ, Engebretson DC (1993) Toroidal-poloidal partitioning of plate motions since 120 Ma. Geophys Res Lett 20: 375–378

McAdoo DC, Martin CF, Polouse P (1985) Seasat observations of flexure: Evidence for a strong lithosphere. Tectonophysics 116:209–222

McClusky S, Bassanian S, Barka A, Demir C, Ergintav S, Georgiev I, Gurkan O, Hamburger M, Hurst K, Kahle H, Kastens K, Kekelidze G, King R, Kotzev V, Lenk O, Mahmoud S, Mishin A, Nadariya M, Ouzounis A, Paradissis D, Peter Y, Preilepin M, Reilinger R, Sanli I, Seeger H, Tealeb A, Toksoz MN, Veis G (2000) Global Positioning System constraints on plate kinematics and dynamics in the eastern Mediterranean and Caucasus. J Geophys Res 105:5695–5719

McKenzie DP (1969) Speculations on the consequences and causes of plate motions. Geophys J R Astr Soc 18:1–32

McKenzie DP, Parker RL (1967) The North Pacific; an example of tectonics on a sphere. Nature 216:1276–1280

Melosh HJ, Raefsky A (1980) The dynamical origin of subduction zone topography. Geophys J R Astr Soc 60: 333–354

Melosh HJ, Williams CA (1989) Mechanics of graben formation in crustal rocks: A finite element analysis. J Geophys Res 94:13961–13973
Mihálffy P, Steinberger B, Schmeling H (2007) The effect of the large-scale mantle flow field on the Iceland hotspot track. Tectonophysics in press
Mihálffy P, Steinberger S, Schmeling H (2008) The effect of the large-scale mantle flow field on the Iceland hotspot trade. Technophysics 447:5–18
Minster JB, Jordan TH (1978) Present-day plate motions. J Geophys Res 83:5331–5354
Mitrovica JX, Forte AM (2004) A new inference of mantle viscosity based upon joint inversion of convection and glacial isostatic adjustment data. Earth Planet Sci Lett 225:177–189
Molnar P, Stock J (1987) Relative motions of hotspots in the Pacific, Atlantic, and Indian Oceans since Late Cretaceous time. Nature 327:587–591
Moore GE (1965) Cramming more components onto integrated circuits. Electronics 38:8
Moresi LN, Gurnis M (1996) Constraints on the lateral strength of slabs from three-dimensional dynamic flow models. Earth Planet Sci Lett 138:15–28
Moresi LN, Solomatov V (1998) Mantle convection with a brittle lithosphere: thoughts on the global tectonic styles of the Earth and Venus. Geophys J Int 133:669–682
Moresi LN, Solomatov VS (1995) Numerical investigations of 2D convection with extremely large viscosity variations. Phys Fluids 7:2154–2162
Moresi LN, Dufour F, Muehlhaus HB (2002) Mantle convection modeling with viscoelastic/brittle lithosphere: numerical modeling and plate tectonic modeling. Pure Appl Geophys 159:2335–2356
Morgan WJ (1968) Rises, trenches, great faults, and crustal blocks. J Geophys Res 73:1959–1982
Morgan JP (1971) Convection plumes in the lower mantle. Nature 230:42–43
Morra G, Regenauer-Lieb K, Giardini D (2006) Curvature of oceanic arcs. Geology 34:877–880
Moucha R, Forte AM, Mitrovica JX, Daradich A (2007) Lateral variations in mantle rheology: implications for convection related surface observables and inferred viscosity models. Geophys J Int 169:113–135
Muhlhaus HB, Regenauer-Lieb K (2005) Towards a self-consistent plate mantle model that includes elasticity: simple benchmarks and application to basic modes of convection. Geophys J Int 163:788–800
O'Connell RJ, Gable CW, Hager BH (1991) Toroidal-poloidal partitioning of lithospheric plate motions. In: Sabadini R, Lambeck K (eds.), *Glacial Isostasy, Sea-Level and Mantle Rheology*, Kluwer Academic Publishers, Norwell MA, 535–551
Olbertz D, Wortel MJR, Hansen U (1997) Trench migration and subduction zone geometry. Geophys Res Lett 24:221–224
Olson P, Bercovici D (1991) On the equipartitioning of kinematic energy in plate tectonics. Geophys Res Lett 18:1751–1754
O'Neill C, Müller D, Steinberger B (2005) On the uncertainties in hot spot reconstructions and the significance of moving hot spot reference frames. Geochemistry Geophysics Geosystems 6
Phipps Morgan J, Morgan WJ, Zhang YS, Smith WHF (1995) Observational hints for a plume-fed, suboceanic asthenosphere and its role in mantle convection. J Geophys Res 100:12753–12767
Piromallo P, Becker TW, Funiciello F, Faccenna C (2006)Three-dimensional instantaneous mantle flow induced by subduction. Geophys Res Lett 33, doi:10.1029/2005GL025390
Ranero C, Phipps Morgan J, McIntosh K, Reichert C (2003) Bending-related faulting and mantle serpentinization at the Middle America trench. Nature 425:367–373
Regenauer-Lieb K, Yuen DA, Branlund J (2001) The initiation of subduction; criticality by addition of water? Science 294:578–580
Ribe NM (1992) The dynamics of thin shells with variable viscosity and the origin of toroidal flow in the mantle. Geophys J Int 110:537–552
Ribe NM (2003) Periodic folding of viscous sheets. Phys Rev E 86, 036305
Ribe NM, Stutzmann E, Ren Y, van der Hilst R (2007) Buckling instabilities of subducted lithosphere beneath the transition zone. Earth Planet Sci Lett 254:173–179
Ricard Y, Vigny C (1989) Mantle dynamics with induced plate tectonics. J Geophys Res 94:17543–17559
Ricard Y, Doglioni C, Sabadini R (1991) Differential rotation between lithosphere and mantle: A consequence of lateral mantle viscosity variations. J Geophys Res 96:8407–8415
Ricard Y, Richards MA, Lithgow-Bertelloni C, Le Stunff Y (1993) A geodynamic model of mantle density heterogeneity. J Geophys Res 98:21895–21909
Richards MA (1991) Hotspots and the case for a high-viscosity lower mantle. In: Sabadini R, Lambeck K (eds.), *Glacial Isostasy, Sea-Level and Mantle Rheology*, Kluwer Academic Publishers, Norwell MA, 571–588
Riedel MR, Karato Si (1997) Grain-size evolution in subducted oceanic lithosphere associated with the olivine-spinel transformation and its effects on rheology. Earth Planet Sci Lett 148:27–43
Royden LH, Husson L (2006) Trench motion, slab geometry and viscous stresses in subduction systems. Geophys J Int 167:881–905
Russo RM, Silver PG (1994) Trench-parallel flow beneath the Nazca plate from seismic anisotropy. Science 263:1105–1111
Schellart WP (2004a) Kinematics of subduction and subduction-induced flow in the upper mantle. J Geophys Res 109, doi:10.1029/2004JB002970
Schellart WP (2004b) Quantifying the net slab pull force as a driving mechanism for plate tectonics. Geophys Res Lett 31, 5
Schellart WP, Freeman J, Stegman DR, Moresi LN (2007) Evolution and diversity of subduction zones controlled by slab width. Nature 446:308–311
Schmeling H, Babeyko A, Enns A, Faccenna C, Funiciello F, Gerya T, Golabek G, Grigull S, Kaus BJP, Morra G, van Hunen J (2007) A benchmark comparison of subduction models. Phys Earth Planet Inter 171:198–223
Sdrolias M, Müller RD (2006) Controls on back-arc basin formation. Geochemistry Geophysics Geosystems 7, doi:10.1029/2005GC001090

Sella GF, Dixon TH, Mao A (2002) REVEL: A model for recent plate velocities from space geodesy. J Geophys Res 107, doi:10.1029/2000JB000033

Shemenda AI (1994) Subduction: Insights from Physical Modelling. Modern Approaches in Geophysics. Kluwer Academic Publishers, Dordrecht

Solomon SC, Sleep NH (1974) Some simple physical models for absolute plate motions. J Geophys Res 79:2557–2567

Spence W (1977) Aleutian arc-tectonic blocks, episodic subduction, strain diffusion, and magma generation. J Geophys Res 82:213–230

Stegman DR, Freeman J, Schellart WP, Moresi L, May D (2006) Influence of trench width on subduction hinge retreat rates in 3-D models of slab rollback. Geochemistry Geophysics Geosystems 7, doi:10.1029/2005GC001056

Stein C, Schmalz J, Hansen U (2004) The effect of rheological parameters on plate behaviour in a self-consistent model of mantle convection. Phys Earth Planet Inter 142: 225–255

Steinberger B (2000) Slabs in the lower mantle – results of dynamic modelling compared with tomographic images and the geoid. Phys Earth Planet Inter 118:241–257

Steinberger B, Schmeling H, Marquart G (2001) Large-scale lithospheric stress field and topography induced by global mantle circulation. Earth Planet Sci Lett 186:75–91

Steinberger B, Sutherland R, O'Connell RJ (2004) Prediction of Emperor-Hawaii seamount locations from a revised model of global plate motion and mantle flow. Nature 430: 167–173

Stevenson DJ, Turner JS (1972) Angle of subduction. Nature 270:334–336

Tackley PJ (2000a) The quest for self-consistent incorporation of plate tectonics in mantle convection. In: Richards M, Gordon R, van der Hilst RD (eds.), *The History and Dynamics of Global Plate Motions*, American Geophysical Union, Washington, DC, vol. 121 of *Geophysical Monograph*

Tackley PJ (2000b) Self-consistent generation of tectonic plates in time-dependent, three-dimensional mantle convection simulations 1. Pseudoplastic yielding. Geochem Geophys Geosyst 1:2000GC000036

Tackley PJ (2000c) Self-consistent generation of tectonic plates in time-dependent, three-dimensional mantle convection simulations 2. Strain weakening and asthenosphere. Geochemistry Geophysics Geosystems 1, 2000GC000043

Tackley PJ, Stevenson DJ, Glatzmaier GA, Schubert G (1994) Effects of multiple phase transitions in a three-dimensional spherical model of convection in Earth's mantle. J Geophys Res 99:15877–15901

Tan E, Gurnis M, Han L (2002) Slabs in the lower mantle and their modulation of plume formation. Geochem Geophys Geosyst 3:2001GC000238

Tan E, Choi E, Thoutireddy P, Gurnis M, Aivazis M (2006) GeoFramework: Coupling multiple models of mantle convection within a computational framework. Geochem Geophys Geosyst 7, doi:10.1029/2005GC001155

Tao WC, O'Connell RJ (1993) Deformation of a weak subducted slab and variation of seismicity with depth. Nature 361:626–628

Tarduno JA, Duncan RA, Scholl DW, Cottrell RD, Steinberger B, Thordarson T, Kerr B. C. and. Neal CR, Frey FA, Torii M, Carvallo C (2003) The Emperor Seamounts: Southward motion of the Hawaiian hotspot plume in Earth's mantle. Science 301:1064–1069

Tetzlaff M, Schmeling H (2000) The influence of olivine metastability on deep subduction of oceanic lithosphere. Phys Earth Planet Inter 120:29–38

Thoraval C, Richards MA (1997) The geoid constraint in global geodynamics: viscosity structure, mantle heterogeneity models and boundary conditions. Geophys J Int 131:1–8

Toth J, Gurnis M (1998) Dynamics of subduction initiation at pre-existing fault zones. J Geophys Res 103: 18053–18067

Turcotte DL, Oxburgh ER (1967) Finite amplitude convective cells and continental drift. J Fluid Mech 28:29–42

Turcotte DL, Schubert G (2002) Geodynamics. Cambridge University Press, Cambridge, 2nd edn.

Uyeda S, Kanamori HJ (1979) Back-arc opening and the mode of subduction. J Geophys Res 84:1049–1061

van der Hilst RD, Seno T (1993) Effects of relative plate motion on the deep structure and penetration depth of slabs below the Izu-Bonin and Mariana island arcs. Earth Planet Sci Lett 120:395–407

van Hunen J, van den Berg AP (2007) Plate tectonics on the early Earth: limitations imposed by strength and buoyancy of subducted lithosphere. Lithos 103: 217–235

van Hunen J, van den Berg AP, Vlaar NJ (2000) A thermomechanical model of horizontal subduction below an overriding plate. Earth Planet Sci Lett 182:157–169

van Keken PE (2003) The structure and dynamics of the mantle wedge. Earth Planet Sci Lett 215:323–338

Vassiliou MS, Hager BH (1988) Subduction zone earthquakes and stress in slabs. Pure Appl Geophys 128:547–624

Weijermars R, Schmeling H (1986) Scaling of Newtonian and non-Newtonian fluid dynamics without inertia for quantitative modelling of rock flow due to gravity (including the concept of rheological similarity). Phys Earth Planet Inter 43:316–330

Wen L, Anderson DL (1997) Present-day plate motion constraint on mantle rheology and convection. J Geophys Res 102: 24639–24653

Wessel P, Smith WHF (1991) Free software helps map and display data. EOS Trans AGU 72:445–446

Widiyantoro S, van der Hilst RD (1997) Mantle structure beneath Indonesia inferred from high-resolution tomographic imaging. Geophys J Int 130:167–182

Wilson JT (1973) Mantle plumes and plate motions. Tectonophysics 19:149–164

Yoshida M, Honda S, Kido M, Iwase Y (2001) Numerical simulation for the prediction of the plate motions: effects of lateral viscosity variations in the lithosphere. Earth Planets Space 53:709–721

Zhang S, Christensen U (1993) Some effects of lateral viscosity variations on geoid and surface velocities induced by density anomalies in the mantle. Geophys J Int 114: 531–547

Zhong S (2001) Role of ocean-continent contrast and continental keels on plate motion, net rotation of lithosphere, and the geoid. J Geophys Res 106:703–712

Zhong S, Davies GF (1999) Effects of plate and slab viscosities on geoid. Earth Planet Sci Lett 170:487–496
Zhong S, Gurnis M (1994) Controls on trench topography from dynamic models of subducted slabs. J Geophys Res 99:15683–15695
Zhong S, Gurnis M (1995) Mantle convection with plates and mobile, faulted plate margins. Science 267:838–842
Zhong S, Gurnis M (1996) Interaction of weak faults and non-newtonian rheology produces plate tectonics in a 3D model of mantle flow. Nature 383:245–247
Zhong S, Gurnis M, Moresi L (1998) Role of faults, nonlinear rheology, and viscosity structure in generating plates from instantaneous mantle flow models. J Geophys Res 103:15255–15268
Zhong S, Zuber MT, Moresi L, Gurnis M (2000) Role of temperature-dependent viscosity and surface plates in spherical shell models of mantle convection. J Geophys Res 105:11063–11082
Zhong S, Zhang N, Li ZX, Roberts JH (2007) Supercontinent cycles, true polar wander, and very long wavelength mantle convection. Earth Planet Sci Lett 261:551–564

Subduction with Variations in Slab Buoyancy: Models and Application to the Banda and Apennine Systems

Leigh H. Royden and Laurent Husson

Abstract Temporal variations in the buoyancy of subducting lithosphere exert a first-order control on subduction rate, slab dip and the position of the associated volcanic arc. We use a semi-analytic, three-dimensional subduction model to simulate "unforced" subduction, in which trench motion is driven solely by slab buoyancy. Model rates of subduction and model slab dip respond almost immediately to changes in the buoyancy of the subducting lithosphere entering the trench; as more buoyant slab segments correlate with slower subduction rates and steeper slab dip. The results are largely consistent with observations from the Banda and southern Apennine subduction systems, where subduction slowed and ended shortly after the entry of continental lithosphere into the trench. Over a 2 m.y. period, model subduction rates decrease from ~70 mm/year to ~30 mm/year for the Banda Arc, and from ~40 mm/year to ~20 mm/year for the Apennine Arc. Increases in model slab dip and decreases in arc-trench distance are likewise consistent with hypocenter locations and volcanic arc position along the Banda and Sunda arcs. In contrast, a time period of ~10 m.y. is needed for model subduction rates to slow to near zero, much longer than the ~3 m.y. upper bound on the observed slowing and cessation of trench motion in the Apennine and Banda systems. One possible explanation is that slab break-off, or the formation of large slab windows, occurred during the last stages of subduction, eliminating toroidal flow around the slab and allowing the slab to steepen rapidly into its final position.

1 Introduction

Slab buoyancy provides the primary driving force for subduction (e.g., Forsyth and Uyeda, 1975; Chapple and Tullis 1977; Conrad and Lithgow-Bertelloni, 2004). However, several studies have revealed a poor correlation between observed slab buoyancy and subduction kinematics (Jarrard, 1986; Doglioni et al., 1999; Lallemand et al., 2005). This may reflect the importance of other factors that affect subduction rate and has led to a number of studies that explore a wide range of factors that might affect subduction (e.g., Kincaid and Olson, 1987; Funiciello et al., 2003; Schellart, 2004, Bellahsen et al., 2005; Royden and Husson, 2006; Capitanio et al., 2007; Billen and Hirth, 2007, Stegman et al., 2006). Alternatively, it may be that temporal variations in slab buoyancy exert a short-term effect on subduction that is as important as the mean buoyancy.

L.H. Royden
Department of Earth, Atmospheric and Planetary Sciences, M.I.T., Cambridge, MA 02139, USA, lhroyden@mit.edu

S. Lallemand and F. Funiciello (eds.), *Subduction Zone Geodynamics*,
DOI 10.1007/978-3-540-87974-9,

Temporal variations in slab buoyancy occur commonly during subduction, for example when oceanic plateaus, seamounts or fragments of continental crust enter a subduction system. Thus by examining the relationship between slab density, slab geometry and subduction rate from theoretical models, and comparing these results to the time-history of subduction and magmatism in natural systems, we can begin to understand how variations in slab density may control a host of geological phenomenon. Yet, despite the extensive literature on subduction process, only a handful of studies have examined the role of variable slab buoyancy in the subduction process. These have generally aimed at examining the role of slab buoyancy in controlling slab shape (i.e., van Hunen et al., 2004) and arc curvature (i.e., Morra et al., 2006). In contrast, Martinod et al. (2005) use an analog approach to show that buoyant slab material entering the subduction system both steepens the slab angle and reduces the velocity of the trench. In this paper we investigate more systematically the length and timescales over which variations in slab buoyancy control on subduction velocity, slab geometry and subduction dynamics, with a brief comparison to two natural subduction systems.

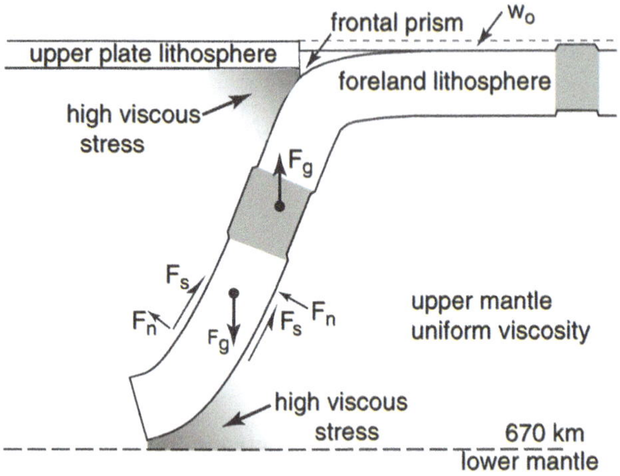

Fig. 1 Schematic diagram of the model subduction zone. *Shaded areas* within the slab denote low-density (high buoyancy) slab segments. F_g, F_n and F_s are gravitational force, and normal and tangential components of viscous forces, respectively. w_0 is pre-subduction water depth of the foreland. *Shaded areas* in the upper mantle outside of the slab denote areas of high viscous stress on the slab due to the flux of mantle in or out of this constricted region

2 Subduction Model

We use a semi-analytical model updated from Royden and Husson (2006) that includes a Newtonian viscous mantle and a slab of finite width, as measured parallel to the trench. Figure 1 shows the geometry and primary sources of stress on the slab. This model uses approximate solutions for toroidal flow around the slab and for flow above and beneath the slab. This provide a reasonable approximation of stress on the slab at distances of more than ~100 km from its side boundaries (Royden and Husson, 2006). We have neglected possible curvature of the trench and slab along strike, which also affects stress on the slab near the side boundaries.

A uniform viscosity is assigned to the slab, making up a viscously competent zone that is 50 km thick and localized within the upper (cold) portion of the slab (Table 1). Subduction is assumed to be "unforced," in that the rate of subduction and trench migration are determined only by the negative buoyancy of the subducted lithosphere and viscous stresses induced in the slab and surrounding mantle. In the model, far-field motions of the plates do not contribute to the subduction process nor play a role in trench migration. Unless otherwise specified, the foreland (unsubducted lithosphere in front of the trench) is assumed to be stationary relative to the underlying lower mantle; therefore the rate of trench motion is equal to the rate of subduction (see later portions of this paper and Royden and Husson, 2006, for alternative assumptions). This is equivalent to the analog models that fix the end of the slab to the sidewall of the tank (e.g., Kincaid and Olson, 1987).

Table 1 Model parameters for Figs. 2 and 3

w_a	Mid-ocean ridge depth	2.5 km
l	Thickness of slab	100 km
	Trench length	600 km
	Thickness of upper plate	50 km
ρ_a	Density of the asthenosphere	3,300 kg m^{-3}
ρ_c	Density of the crust/frontal prism	2,800 kg m^{-3}
ρ_w	Density of water	1,000 kg m^{-3}
	Viscosity of the upper mantle (excluding slab and lithosphere)	2.5 × 10^{20} Pa
	Viscosity of slab "core"	10^{23} Pa s
	Thickness of slab "core"	50 km

Subduction is "mature" in that the end (or "tail") of the slab extends to the top of the lower mantle, treated in this paper as a rigid boundary. This condition prevents mantle flow beneath the slab tail (Garfunkel et al., 1986). Note, however, that the position of the slab tail is determined only by viscous forces within the slab and upper mantle. The slab position is not *a priori* fixed with respect to the top of the lower mantle. This differs from tank models where the deep slab becomes fixed onto the base of the tank (see Funiciello et al., 2003). Because the process of slab deformation within the transition zone is poorly known, it is not clear which, if either, of these assumptions is more correct. There should be little difference between them when the foreland lithosphere, and thus the slab tail, is relatively stationary with respect to the top of the lower mantle. However, differences in predicted behavior could become large when there is rapid relative motion between them.

Prior to subduction, slab buoyancy can, to first order, be linearly equated with the bathymetry of the soon-to-be-subducted lithosphere (its "pre-subduction" water depth, w_0). For a slab thickness of 100 km and an asthenospheric density of 3300 kg/m^3, each kilometer change in pre-subduction water depth translates to a change in mean slab density of ~23 kg/m^3. Slab buoyancy is neutral if w_0 equals ~2.5 km, very negative if w_0 is ~6.5 km (e.g., old oceanic lithosphere), and very positive if w_0 is ~0 km (e.g., typical continental lithosphere). (See Royden and Husson, 2006, for further discussion of buoyancy and bathymetry.) Note, however, that slab buoyancy is a function only of the lithospheric material that is actually subducted and the density of that material as it is subducted. If sediments or crust are stripped from the slab as it enters the subduction zone, or if phase changes occur during subduction, the slab buoyancy changes accordingly. In this paper we ignore both of these effects, although they could easily be included with appropriate changes in slab density with depth.

3 Subduction of a Model Foreland

In Figs. 2 and 3 we show results for subduction of a model foreland consisting of a deep oceanic domain (w_0 = 6.5 km) embedded with several contrasting buoyancy domains, including a continental island (250 km wide, w_0 = 0 km), an oceanic plateau (800 km wide, w_0 = 3.5 km), and a continental margin (w_0 = 0 km) (Fig. 2a). Slab buoyancy is varied in a direction normal to the trench while all other parameters are held constant.

After subduction of oceanic lithosphere has achieved a steady-state geometry and rate (at ~70 mm/year), the contrasting buoyancy domains reach the trench and are subducted sequentially. When the continental island

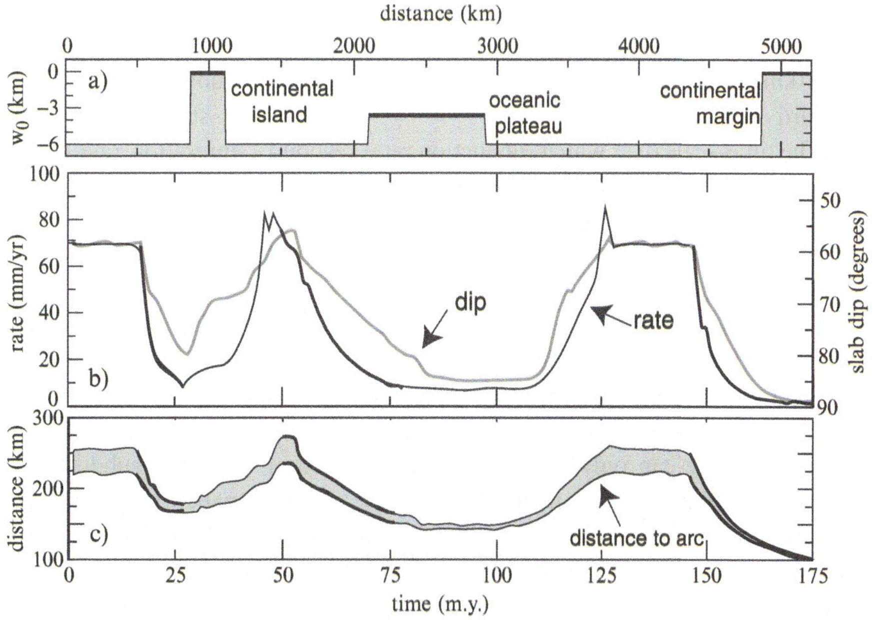

Fig. 2 (**a**) Pre-subduction water depth for the model foreland to be subducted, as a function of horizontal distance. Large water depths are equated with large negative slab buoyancy, see text for details. *Thick solid lines* represent more buoyant segments of the foreland, consistent in all panels. (**b**) Subduction rate (*black line*) and slab dip at 100 km depth (*gray line*) as a function of time. (**c**) Distance from the trench to the volcanic arc, taken as the horizontal distance over which slab deflection increases from 2 to 100 km (lower value) or to 150 km (higher value)

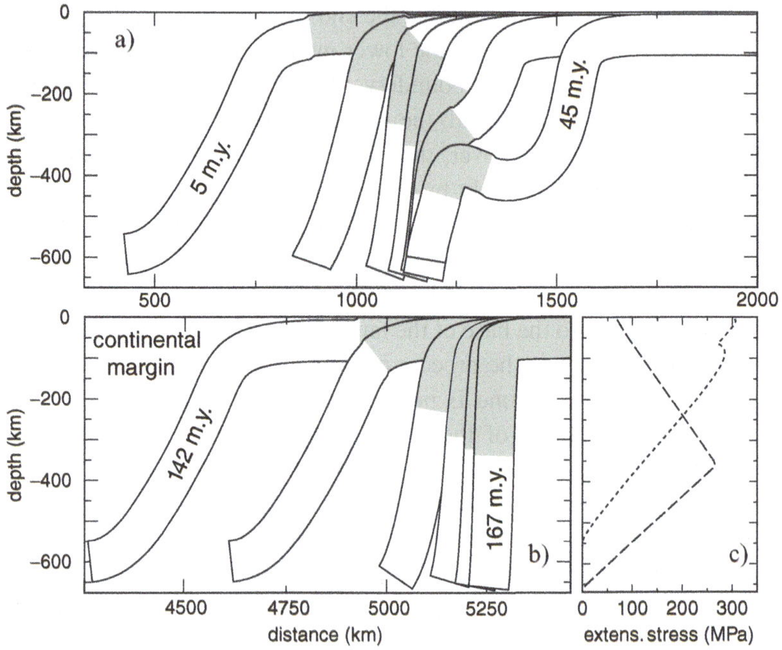

Fig. 3 Slab geometries corresponding to two subduction events of Fig. 2: (**a**) subduction of the continental island and (**b**) subduction the continental margin. *Shaded slab* units denote continental lithosphere. The time interval between slab positions is 5 m.y. with beginning and ending times as indicated. (**c**) Extensional stress along the subducted slab for steady-state subduction of oceanic lithosphere and after subduction of a continental margin

enters the trench and reaches the uppermost asthenosphere (~65 km depth), the rate of subduction drops abruptly and trench retreat slows to 5–10 mm/year (Fig. 2b). Subsequently, as the leading edge of the next oceanic segment enters the trench and reaches the uppermost asthenosphere, the subduction rate increases again to ~70 mm/year. A similar pattern of decreasing and increasing rate occurs with subduction of the oceanic plateau. Finally, rates slow and subduction is terminated when the continental margin enters the system; its leading edge is ultimately subducted to ~300 km depth km (Fig. 3, see also Ranalli et al., 2000; Regard et al., 2003).

The time-scale over which subduction rates respond to changes in slab buoyancy is generally on the order of just a few million years. When subduction rates change in response to slab buoyancy, the rates of change can be rapid only when the subduction rate is rapid. This is because each new lithospheric segment, with new buoyancy, enters the system at the rate of subduction, and the subduction system adjusts to the new slab buoyancy accordingly. Thus rapid changes in subduction rate can only occur when the subduction velocity is high; when subduction velocity is low, changes in subduction rate must occur slowly, but the rate of change can become rapid as the subduction rate speeds up.

Model steady-state subduction of the dense oceanic slab occurs at an average dip of ~60–65° (Figs. 2b and 3). When continental or plateau lithosphere enter the system, the slab steepens to more than 80° because the buoyant part of the slab at shallow depth resists sinking and slows the rate of trench retreat; meanwhile the deeper, denser part of the slab continues to sink at a faster rate. Flattening of the shallow slab occurs when dense lithosphere follows buoyant lithosphere into the trench because the dense slab at shallow depth sinks more rapidly than the buoyant material at greater depth. Thus there is a strong correlation between slab dip and trench migration rate (Fig. 2b). For narrow, highly buoyant slab segments, the dip of the slab may become inverted at depth, with a local depth-minimum coinciding with the buoyant slab segment (Fig. 3a).

Equating the distance from the subduction boundary to the volcanic arc with the horizontal distance over which slab deflection increases from 2 to 100 and 150 km, the distance from the subduction boundary to the volcanic arc is ~220–250 km during steady-state oceanic subduction for the parameters used in this paper (Fig. 2c). This decreases to ~150–170 km during subduction of the continental island and oceanic plateau areas. The greatest change in the location of the volcanic arc occurs when subduction terminates against the continental margin and the distance from the thrust belt to the volcanic arc is decreased to ~100 km. The precise distance to the arc is highly sensitive to factors like the density of material in the frontal prism (Fig. 1), but the significant decrease in arc-trench distance during slowing of subduction is a robust feature of the model results.

4 Case Studies: Banda and Apennine Systems

The subduction model used here supposes that subduction occurs solely as a result of negative slab buoyancy, so the model results are not applicable to orogenic belts like the Himalaya, which accommodate rapid inter-plate convergence driven largely by far-field forces (e.g., Conrad and Lithgow-Bertelloni, 2004). It is best applied to unforced subduction systems that display upper-plate extension concurrent with subduction, such as those of the Mediterranean region (e.g., Malinverno and Ryan, 1986; Wortel and Spakman, 2000; Royden, 1993a, b; Jolivet and Faccenna, 2000; Faccenna et al., 2003), and the region extending eastward from Indonesia to the Pacific plate (Hall, 1996; Charlton, 2000; Daly et al., 1991; Rangin et al., 1990; Hinschberger et al., 2000). The Apennine and Banda Sea systems offer excellent opportunities to compare model results with observations in systems where subduction has terminated following entry of continental lithosphere into the subduction zone. Conversely, there are also subduction systems where an increase in subduction rate can be correlated with entry of denser lithosphere into the trench, but the amount of geologic information needed to document rates and timing renders such comparisons beyond the scope of this paper.

4.1 Geological Settings

Subduction of Australian lithosphere northward beneath the Banda Arc occurred prior to Pliocene time at a rate of ~70 mm/year (Fig. 4a, after Hinschberger et al., 2000). The pre-subduction depth, w_o, of the subducted oceanic lithosphere was probably similar to that observed today in the adjacent Indian Ocean, ~5.7 km. During Pliocene time, the eastern end of the subduction system encountered the continental margin of north-western Australia and arc-continent collision occurred diachronously from west to east. During subduction of the margin, w_o decreased to its current value of ~0.05 km. The rate of subduction dropped to zero by 0.5–1.0 Ma while to the west (Sunda arc), subduction of oceanic lithosphere continues at ~70 mm/year (Kreemer et al., 2000; Richardson and Blundell, 1996; Hughes et al., 1996). Thus the Banda Sea offers an excellent opportunity to observe the reaction of subduction system to the entry of continental material into the trench, and at the same time to observe steady-state subduction of oceanic lithosphere (Sunda arc) that is probably similar to that of the Banda arc prior to collision.

The Late Cenozoic Apennine system consumed mixed oceanic and continental lithosphere along a west-dipping subduction system (Fig. 4b). Reconstruction of the southern Apennines suggests a Late Miocene subduction rate of ~30–50 mm/year or perhaps higher (e.g., Patacca et al., 1990; Fig. 5b, detailed rates taken from the unpublished work of P. Scandone, pers. com., 2006, average rate from Faccenna et al., 1997). The apparent oscillations in rate are probably artifacts of an incomplete geological record, and the maximum rates are probably the more significant. In any case, we choose a rate of 40 mm/year as the pre-collisional rate of subduction for the Apennine system. In Pliocene time, the southern Apennine trench encountered the continental Adriatic lithosphere and thrusting and subduction had

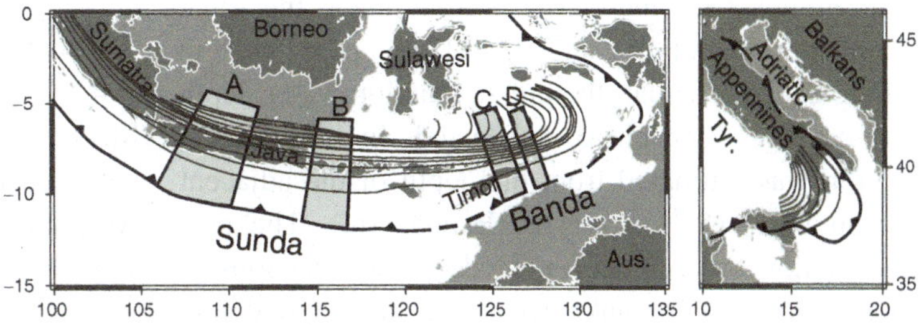

Fig. 4 Location maps for the Banda Arc and Apennine subduction systems. *Boxes* show locations of swath profiles for earthquake hypocenters in Fig. 6. Slab contours are 50 km isodepths, adapted from Sambridge and Gudmundsson (1998). Light gray domains are shallow water domains. *Solid and dashed bold curves* indicate active and fossil trench locations

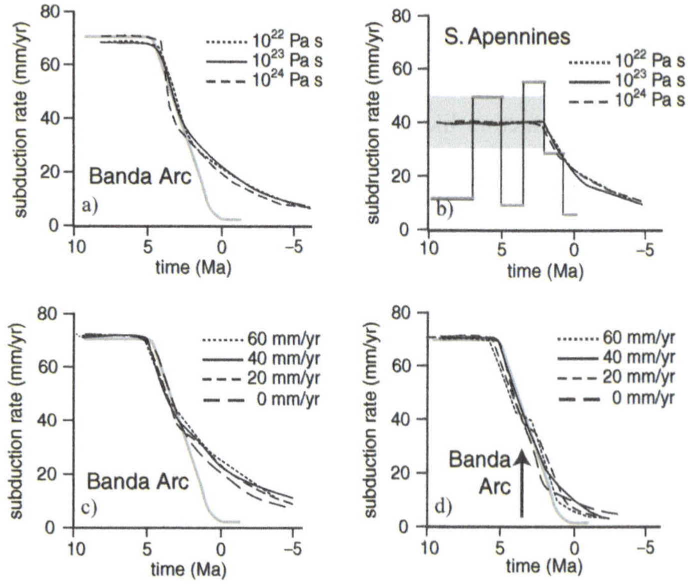

Fig. 5 Observed (*gray curves*) and model (*black curves*) rates of subduction for the Banda arc the Southern Apennines. Parameter values are given in Table 2. (**a**), (**b**) Subduction rate versus time following entry of continental lithosphere into the trench at 5 Ma. Foreland velocity relative to the top of the lower mantle is zero. *Labeled curves* correspond to different values of slab viscosity. (**c**) Subduction rate versus time for the Banda Arc following entry of continental lithosphere into the trench at ~5 Ma. *Labeled curves* correspond to different imposed values foreland velocity (from foreland toward trench) relative to the top of the lower mantle. (**d**) Same as (**c**) except that slab detachment occurs at 470 km depth at the time indicated by the *vertical arrow*

Table 2 Parameters for Figs. 5 and 7

	Pre-subduction water depth	Northward foreland velocity (relative to lower mantle)	Slab viscosity	Mantle viscosity
Figure 5a (Banda)	5.7 km	0 mm/year	10^{22} Pa s	1.9×10^{20} Pa s
	5.7 km	0 mm/year	10^{23} Pa s	1.8×10^{20} Pa s
	5.7 km	0 mm/year	10^{24} Pa s	1.5×10^{20} Pa s
Figure 5b (Apennines)	5.0 km	0 mm/year	10^{22} Pa s	2.4×10^{20} Pa s
	5.0 km	0 mm/year	10^{23} Pa s	2.2×10^{20} Pa s
	5.0 km	0 mm/year	10^{24} Pa s	1.8×10^{20} Pa s
Figure 5c,d (Banda)	5.7 km	0 mm/year	10^{22} Pa s	1.9×10^{20} Pa s
	5.7 km	20 mm/year	10^{22} Pa s	2.1×10^{20} Pa s
	5.7 km	40 mm/year	10^{22} Pa s	2.4×10^{20} Pa s
	5.7 km	60 mm/year	10^{22} Pa s	2.7×10^{20} Pa s
Figure 7	5.7 km	0 mm/year	10^{23} Pa s	1.8×10^{20} Pa s
	5.7 km	0 mm/year	10^{24} Pa s	1.5×10^{20} Pa s

ended by Quaternary time (see e.g., Patacca et al., 1990; Faccenna et al., 2001; Jolivet and Faccenna, 2000).

4.2 Model Parameters

The transition from oceanic to continental foreland was modeled for these two systems using the same parameters as in Fig. 2 and Table 1, except as noted in Table 2. In particular, the pre-subduction water depth of oceanic lithosphere subducted in the Banda region was estimated from the depth of the adjacent ocean basin (5.7 km for the Banda Arc). The pre-subduction water depth for the deepwater portions of the Apennine slab is unknown, so we chose a moderate depth of 5 km. In both cases the pre-subduction water depth of continental portion of the lithosphere is set to sea-level. A range of slab viscosities, from 10^{22} to 10^{24} Pa s, were used in the model runs, with the viscosity of the surrounding mantle chosen in each case to provide a subduction velocity of ~70 mm/year for the Banda system and ~40 mm/year for the Apennine system.

4.3 Subduction Rates

Model rates of subduction are in excellent agreement with the rapid termination of subduction within the Apennine and Banda systems for the initial period of slow-down, with velocities slowing down from ~70 to ~30 mm/year over a 2 m.y. period for the Banda Sea model, and from ~40 to ~20 mm/year over a 2 m.y. period for the Apennine system (Fig. 5a and b). This indicates that the timescale over which subduction initially responds to changes in slab buoyancy is similarly rapid in the natural systems and in the model, and is largely independent of slab viscosity. However, for all slab viscosities, the latter stages of subduction termination are significantly faster in the observed systems than in the model results. In particular, the slowing of model subduction from 20 to 0 mm/year occurs over at least 10 m.y., with a gradual slowing of rate, while subduction in the Apennine and Banda systems terminates abruptly.

One possible source of the mismatch between observed and model results is the imposed velocity of the foreland relative to the top of the lower mantle (see e.g., Schellart, 2005). The foreland lithosphere for the Apennine subduction system is approximately stationary, or at least not moving very rapidly, in the hot spot reference frame (e.g., Gordon and Jurdy, 1986). Its motion with respect to the top of the lower mantle is, presumably, similar. However, the Indian-Australian plate (the foreland for the Banda subduction system) is moving northwards in the hotspot reference frame at ~60 mm/year, bringing into question the importance of the absolute velocity of the foreland with respect to the top of the lower mantle.

Figure 5c shows the model decrease in subduction rate expected for imposed foreland velocities, relative to the top of the lower mantle, ranging from 0 to 60 mm/year, in a direction orthogonal to and toward the trench. In order to match the initial subduction velocity of ~70 mm/year and to keep the initial slab buoyancy consistent with a pre-subduction water depth of 5.7 km, the viscosity of the mantle had to be adjusted for each individual case (Table 2). With these constraints, there is little predicted variation in the rate at which subduction slows as a function of imposed foreland velocity, indicating that this is probably not the cause of the misfit between observations and model in the latter stages of subduction termination. This result is consistent with the Apennine and Banda systems exhibiting the same rapid termination in subduction regardless of foreland velocity. It appears that some process, not included in our subduction model, must be responsible for the rapid termination of subduction in the latter stages of continental margin subduction, as we discuss further below.

4.4 Slab Dip and Migration of the Volcanic Arc

A robust result of the model is that the slab dip steepens as buoyant material enters and descends into the subduction system. For example, Fig. 3 shows the model slab dip steepening from about 60–65° during steady-state subduction of oceanic lithosphere to 85° or more after buoyant lithosphere enters the trench. This compares favorably to the slab geometries indicated by earthquake hypocenters from the Sunda and Banda arcs (Fig. 6); the average slab dip in the Sunda arc, where oceanic lithosphere is being subducted in steady-state is ~55–60° between 200 and 400 km depth. The slab dip along the Banda arc, where subduction has ceased, is ~75–80°. Both model and observations show an increase in slab dip of about 20° during entry of the continental margin into the subduction system. The modern slab dip beneath the Apennines is also very steep, perhaps 80° or more (Lucente et al., 1999; Piromallo and Morelli, 2002).

Another robust result from the model is the migration of the volcanic arc (equated in the model with a depth of 100–150 km at the top of the subducted slab) toward the trench during subduction of buoyant lithosphere. Thus there is a strong correlation between subduction rate and the location of the volcanic arc. The modern position of the volcanic arc within the Southern Apennines is ~100 km from the thrust front, similar to

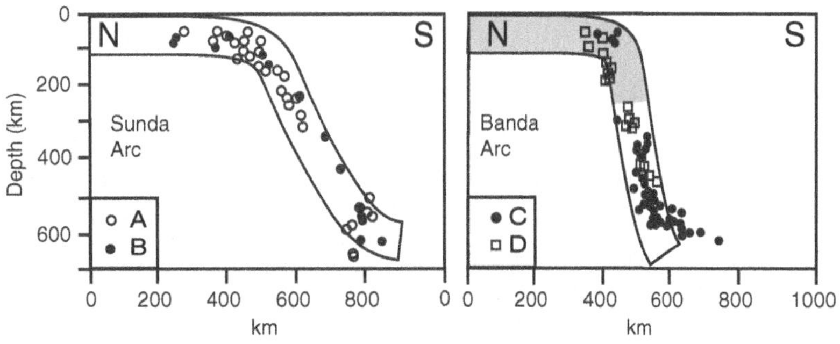

Fig. 6 Observed earthquake hypocenters (*dots*) from the Banda and Sunda arcs as given by Schoffel and Das (1999) and Das (2004). Swath locations shown in Fig. 4. *Solid lines* show model slab profiles from Fig. 3, for steady-state subduction of oceanic lithosphere (*left panel*) and following entry of the continental lithosphere (*shaded region*) into the subduction boundary (*right panel*, geometry corresponding to 152 m.y. in Fig. 2)

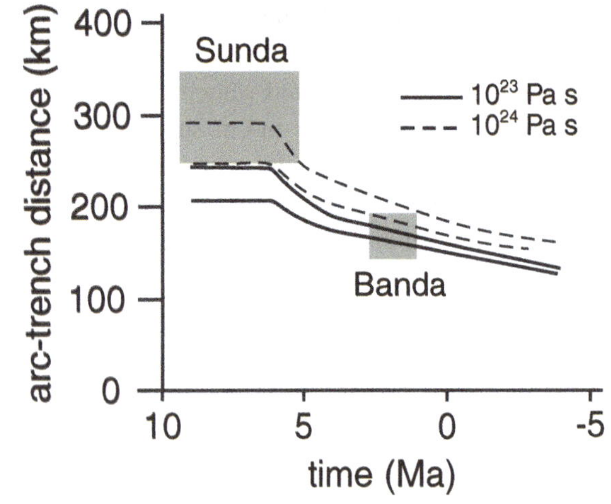

Fig. 7 Observed distance from the Banda and Sunda volcanic arcs to the adjacent trench or frontal thrust belt (*shaded squares*) as a function of time following entry of the continental lithosphere into the subduction boundary at ~5 Ma. *Lines* show model distances from trench to the front and back of the volcanic arc for the viscosities shown. Front and back of the model volcanic arc are equated with the positions where the upper slab boundary descends below 100 and 150 km depth, respectively

the model results of Fig. 2c after subduction of the continental margin. Within the Banda-Sunda arc system, the volcanic arc is ~250–350 km from the Sunda trench but only ~150–200 km from the inactive Banda thrust front. Using the Sunda arc as a proxy for the Banda arc prior to its interaction with the Australian continental margin, there appears to have been a 100 km decrease in the distance from the thrust front to the volcanic arc that is associated with subduction of the margin (Fig. 7). While we cannot rule out that some of the decrease in arc-trench distance might be due to shortening in the fore-arc region after the eruption of the last volcanic arc rocks at ~3 Ma, the fact that the slab dip is steeper slab the Banda region than in the Sunda region is also consistent with a decrease in the arc-trench distance during slowing and cessation of subduction.

5 Slab Detachment?

A number of authors (e.g., Buiter et al., 2002; Wortel and Spakman, 2000) have suggested that during the entry of continental lithosphere into a subduction zone, the lower part of the slab detaches from the upper part, in a process commonly referred to as "slab break-off" or "slab detachment." This process has been specifically suggested for both the Apennine (Buiter et al., 1998) and Banda arc (Elburg et al., 2004) systems. While a detailed investigation of this process is beyond the scope of this paper, we can make a preliminary assessment of whether this process could lead to a sufficiently rapid termination of the last stages of subduction, as indicated in Fig. 5.

Figure 5d shows the results for a simulated slab break-off at the time indicated by the vertical arrow. Although our model does not really allow for proper slab break-off, we simulate instantaneous slab break-off by truncating the slab at 470 km depth and setting the viscous stresses on the bottom of the slab to zero. This somewhat *ad hoc* modification of the forces on the slab eliminates the high stress area beneath the lower part of the slab (Fig. 1) and also eliminates the viscous stresses that result from toroidal flow around the slab (because mantle can flow through the region formerly occupied by the deep part of the slab). The results indicate a very rapid transition from slow subduction to no subduction, in good agreement with observations from the Banda Sea (Fig. 5d) and the Apennines (not shown but similar to Fig. 5d). Although not a precise simulation of slab break-off, these results suggest that slab break-off may provide a mechanism whereby the last stages of subduction occur very quickly.

Importantly, the reduction in weight of the deep slab is not the most important factor in causing rapid cessation of subduction. Rather, it is the creation of a shortcut for mantle flow that is critical in enabling rapid steepening of the slab into its final, stationary position. In particular, the creation of gaps in the slab enables rapid flow out of the geometrically confined, high-stress region beneath the deep slab (Fig. 1), thereby allowing for rapid motion of the deep slab.

6 Discussion and Conclusions

In general, rates of subduction and trench motion respond rapidly to changes in slab buoyancy. Subduction rates respond more rapidly to changes in slab buoyancy when the subduction rate is fast, and less rapidly when subduction rates are slow. Because subduction rates are faster for more negatively buoyant slabs, there is a discrepancy between the amount of time that various buoyancy domains spend in the shallow part of the subduction system and their original extent within the foreland. In Fig. 2, continental lithosphere and oceanic plateau regions initially make up ~25% of the foreland, but account for ~55% of the subduction time. Thus the duration of events within a thrust belt cannot be used as a direct measure of the original extent of various buoyancy or facies domains because the low buoyancy regions will be over-represented in the duration of thrusting.

The dip of subducted slabs at shallow depth (~100 km) tracks closely with the rate of trench migration. For the parameters used in this paper, very dense slabs (old ocean) subduct at angles of ~60–65° whereas more buoyant lithospheres, like the oceanic plateau modeled in Fig. 2, subduct at steeper angles. (The precise slab dip depends on a variety of parameters and may vary from system to system even if slab buoyancy is identical). When all other parameters are held constant, the correlation between subduction rate and slab dip at 100 km depth is surprisingly good. This is counter to the common misconception that high-density slabs dip more steeply than low-density slabs (e.g., Luyendyk, 1970). The latter is correct if the trench retreat rate is held constant while slab buoyancy is varied, but not when both subduction rate and trench-migration rate are controlled by the negative buoyancy of the slab. This agrees with the fact that little correlation is found between observed slab dip and slab density (e.g., Jarrard, 1986; Heuret and Lallemand, 2005).

Because arc-magmas are generated above subducted slabs at depths of 100–150 km, there is also a correlation between the slab buoyancy, slab dip and location of the volcanic arc. As slab dips steepen, the volcanic arc should move closer to the trench (or front of the thrust belt). This occurs most markedly during subduction of a continental margin or other large buoyant object and is illustrated by the Banda/Sunda subduction system (Fig. 2). Equating the distance from the trench to the volcanic arc with the horizontal distance over which the slab depth increases from 10 to 100 km, the model distance from the trench to the volcanic arc decreases by ~100 km during subduction of the continental margin.

When subduction terminates, the viscous stresses on the slab disappear and the stresses on the slab arise only from its buoyancy. If subduction is terminated by subduction of a continental margin or other buoyant object, the depth to which the leading edge of the buoyant lithosphere can be subducted depends on its buoyancy and the buoyancy of deeper, denser portions of the slab. This is reflected in the extensional stresses transmitted along the slab (Fig. 3c). For a uniformly buoyant slab subducting in steady-state, the extensional stress along the slab increases toward the surface. (Note that extensional stress is not necessarily zero at the surface; only the vertical component of stress, integrated along the slab, must be zero.) When subduction terminates after subducting a continental margin, the extensional stress within the subducted slab is greatest at the former ocean-continent transition. This result is intuitive because at that stage there are no longer stresses on the slab from viscous flow in the surrounding mantle. Thus the extensional stress on the slab increases upwards to the point where the density of the slab is less that the density of the surrounding mantle. This suggests slab attenuation and possible slab detachment may be most likely to occur at this site (see Fig. 3c and Wortel and Spakman, 2000; Regard et al., 2003).

Our results apply largely to unforced subduction systems, where trench motion is controlled by slab buoyancy and not by far-field constraints on relative plate velocities. Modeling studies (e.g., Hassani et al., 1997) and observations (e.g., on the Nazca subduction zone, Gutscher et al., 1999) indicate that velocities imposed on the upper plate can play a large role in determining slab geometry and kinematics. For this reason, we have chosen natural examples where the upper plate lithosphere experienced extension during the subduction process, suggesting relatively little stress coupling between the over-riding and down-going plates.

Nevertheless, some of the general concepts developed here may be extrapolated more broadly. We suggest that, for large plates like the Pacific, local and short-term variations in upper plate deformation, such as in the Japan or Kurile regions, may be more closely related to temporal and spatial variations in slab buoyancy than to the mean buoyancy of the slab. Even small variations in slab buoyancy result in speeding or slowing of trench motion. Although the change in rate may be but a minor fraction of the total trench velocity, even a change in rate of 5 mm/year over 10 m.y. can result in significant stretching or shortening of the upper plate. If this interpretation is correct, it helps to explain the apparent contradiction that slab buoyancy is the most important driver of subduction, yet there appears to be a lack of correlation between local slab buoyancy and associated upper plate deformation.

References

Bellahsen, N., Faccenna, C., and Funiciello, F., 2005, Dynamics of subduction and plate motion in laboratory experiments: Insights into the "plate tectonics" behavior of the Earth, J. Geophys. Res., 110, doi:10.1029/2004JB002999.

Billen, M.I. and Hirth, G., 2007, Rheologic Controls on Slab Dynamics, Geochem. Geophys. Geosyst. (G3), 8, Q08012, doi:101029/2007GC001597.

Buiter, S.J.H., Wortel, M.J.R., and Govers, R., 1998, The role of subduction in the evolution of the Apennines foreland basin, Tectonophysics, 296, 249–268.

Buiter, S.J.H., Govers, R., and Wortel, M.J.R., 2002, Two-dimensional simulations of surface deformation caused by slab detachment, Tectonophysics, 354, 195–210.

Capitanio, F.A., Morra, G., Goes, S., 2007, Dynamic models of downgoing-plate buoyancy driven subduction: Subduction motions and energy dissipation, Earth Planet. Sci. Lett., 262, 284–297.

Chapple, W.M., and Tullis, T.E., 1977, Evaluation of the forces that drive the plates, J. Geophys. Res., 82, 1967–1984.

Charlton, T.R., 2000, Tertiary evolution of the Eastern Indonesia collision complex, J. Asian Earth Sci., 18, 603–631.

Conrad, C.P., and Lithgow-Bertelloni, C., 2004, The temporal evolution of plate driving forces: Importance of "slab suction" versus "slab pull" during the Cenozoic, J. Geophys. Res., 109, doi:10.1029/2004JB002991.

Daly, M.C., Cooper, M.A., Wilson, I., Smith, D.G. and Hooper, B.G.D., 1991, Cenozoic plate tectonics and basin evolution in Indonesia, Mar. Pet. Geol., 8, 2–21.

Das, S., 2004, Seismicity gaps and the shape of the seismic zone in the Banda Sea region from relocated hypocenters, J. Geophys. Res., 109, doi:10.1029/2004JB003192.

Doglioni, C., Harabaglia, P., Merlini, S., Mongelli, F., Peccerillo, A., and Piromallo, C., 1999, Orogens and slabs versus their direction of subduction, Earth Sci. Rev., 45, 167–208.

Elburg, M.A., van Bergen, M.J., and Foden, J.D., 2004, Subducted upper and lower continental crust contributes to magmatism in the collision sector of the Sunda-Banda Arc, Indonesia, Geol. Soc. Am., 32, 41–44.

Faccenna, C., Mattei, N., Funiciello, R., and Jolivet, L., 1997, Styles of back-arc extension in the Central Mediterranean, Terra Nova, 9, 126–130.

Faccenna, C., Becker, T.W., Lucente, F.P., Jolivet, L., and Rossetti, F., 2001, History of subduction and back-arc extension in the central Mediterranean, Geophys. J. Int., 145, 809–820.

Faccenna, C., Jolivet, L., Piromallo, C., and Morelli, A., 2003, Subduction and the depth of convection in the Mediterranean mantle, J. Geophys Res., 108, doi:10.1029/2001JB001690.

Forsyth, D. and Uyeda, S., 1975, On the relative importance of the driving forces of plate motion, Geophys. J.R. Astr. Soc., 43, 163–200.

Funiciello, F., Faccenna, C., Giardini, D., and Regenauer-Lieb, K., 2003, Dynamics of retreating slabs: 2. Insights from three-dimensional laboratory experiments, J. Geophys. Res., 108, 2207, doi:10.1029/2001JB000896.

Garfunkel, C., Anderson, A., and Schubert, G., 1986, Mantle circulation and the lateral migration of subducted slab, J. Geophys. Res., 91, 7205–7223.

Gordon, R.G., and Jurdy, D.M., 1986, Cenozoic global plate motions, J. Geophys. Res., 91, 12389–12406.

Gutscher, M.-A., Olivet, J.-L., Aslanian, D., Eissen, J.-P., and Maury, R., 1999, The "lost Inca plateau": Cause of flat subduction beneath Peru?, Earth Planet. Sci. Lett., 171, 335–341.

Hall, R., 1996, Reconstructing Cenozoic SE Asia, in Hall, R., and Blundell, D., eds., Tectonic Evolution of SE Asia: Geol. Soc. Lond. Spec. Pub., 106, p. 203–224.

Hassani, R., Jongmans, D., and Chéry, J., 1997, Study of plate deformation and stress in subduction processes using two-dimensional numerical models. J. Geophys. Res., 102, 17951–17965.

Heuret, A., and Lallemand, S., 2005, Plate motions, slab dynamics and back-arc deformation, Phys. Earth Planet. Int., 149, 31–51.

Hinschberger, F., Malod, J.A., Réhault, J.P., Villeneuve, M., Royer, J.Y., Burhanuddin, S., 2000, Late Cenozoic geodynamic evolution of eastern Indonesia, Tectonophysics, 404, 91–118.

Hughes, B.D., Baxter, K., Clark, R.A., and Snyder, D.B., 1996, Detailed processing of seismic reflection data from the frontal part of the Timor trough accretionary wedge, eastern Indonesia, in Hall, R., and Blundell, D., eds., Tectonic Evolution of SE Asia: Geol. Soc. Lond. Spec. Pub., 106, 75–83.

Jarrard, R.D., 1986, Relations among subduction parameters, Rev. Geophys., 24, 217–284.

Jolivet, L. and Faccenna, F., 2000, Mediteranean extension and the Africa-Eurasia collision, Tectonics, 19, 1095–1106.

Kincaid, C., and Olson, P., 1987, An experimental study of subduction and slab migration, J. Geophys. Res., 92, 13, 832–13,840.

Kreemer, C., Holt, W.E., Goes, S., and Govers, R., 2000, Active deformation in eastern Indonesia and the Philippines from GPS and seismicity data, J. Geophys. Res., 105, 663–680.

Lallemand, S., Heuret, A., and Boutelier, D., 2005, On the relationship between slab dip, back-arc stress, upper plate absolute motion, and crustal nature in subduction zones, Geochem. Geophys. Geosyst., 6, Q09006, doi:10.1029/2005GC000917.

Lucente, F.P., Hiarabba, C.C., and Imini, G.C., 1999, Tomographic constraints on the geodynamic evolution of the Italian region, J. Geophys. Res., 104(B9), 20307–20327.

Luyendyk, B.P., 1970, Dips of downgoing lithospheric plates beneath island arcs, Geol. Soc. Am. Bull., 81, 3411–3416.

Malinverno, A., and Ryan, W., 1986, Extension in the Tyrrhenian sea and shortening in the Apennines as result of arc migration driven by sinking of the lithosphere, Tectonics, 5, 227–245.

Martinod, J., Funiciello, F., Faccenna, C., Labanieh, S., and Regard, V., 2005, Dynamical effects of subducting ridges: Insights from 3-D laboratory models, Geophys. J. Int., 163, 1137–1150.

Morra, G., Regenauer-Lieb, K., and Giradini, D., 2006, Curvature of oceanic arcs, Geology, 34, 877–880.

Patacca, E., Sartori, R. and Scandone, P., 1990, Tyrhennian basin and Apenninic arcs: Kinematic relations since late Tortonian times, Mem. Soc. Geol. Ital., 45, 425–451.

Piromallo, C., and Morelli, P., 2002, P-wave tomography of the mantle under the Alpine-Mediterranean area, J. Geophys. Res., 108, doi:10.1029/2002JB001757.

Ranalli, G., Pellegrini, R., and d'Affizi, S., 2000, Time-dependence of negative buoyancy and the subduction of continental lithosphere, J. Geodyn., 30, 539–555.

Rangin, C., Jolivet, L., and Pubellier, M., 1990, A simple model for the tectonic evolution of southeast Asia and Indonesian region for the past 43 My, Bull. Geol. Soc. France 8, 889–905.

Regard, V., Faccenna, C., Martinod, J., Bellier, O., and Thomas, J.C., 2003, From subduction to collision: Control of deep processes on the evolution of convergent plate boundary, J. Geophys. Res., 108, doi:10.1029/2002JB001943.

Richardson, A.N., and Blundell, D.J. 1996, Continental collision in the Banda Arc. in Hall, R., and Blundell, D., eds., Tectonic Evolution of SE Asia: Geol. Soc. Lond. Spec. Pub., 106, p. 47–60.

Royden, L.H., and Husson, L., 2006, Trench motion, slab geometry and viscous stresses in subduction systems, Geoph. J. Int., doi: 10.1111/j.1365-246X.2006.03079.x.

Royden, L.H., 1993a, The tectonic expression of slab pull at continental convergent boundaries, Tectonics, 12, 303–325.

Royden, L.H., 1993b, Evolution of retreating subduction boundaries formed during continental collision, Tectonics, 12, 629–638.

Sambridge, M., and Gudmundsson, O., 1998, Tomography with irregular cells, J. Geophys. Res., 103, 773–781.

Schellart, W.P., 2004, Kinematics of subduction and subduction-induced flow in the upper mantle, J. Geophys. Res., 109, B07401, doi:10.1029/2004JB002970.

Schellart, W.P., 2005. Influence of the subducting plate velocity on the geometry of the slab and migration of the subduction hinge, Earth Planet. Sci. Lett., 231, 197–219.

Schoffel, H.-J. and Das, S., 1999, Fine details of the Wadati-Benioff zone under Indonesia and its geodynamic implications, J. Geophys. Res., 104, 13,101–13,114.

Stegman, D.R., Freeman, J., Schellart, W.P., Moresi, L., and May, D., 2006, Influence of trench width on subduction hinge retreat rates in 3-D models of slab rollback, Geochem. Geophys. Geosyst., 7, Q03012, doi:10.1029/2005GC001056.

van Hunen, J., van den Berg, A.P., and Vlaar, N.J., 2004, Various mechanisms to induce shallow flat subduction: A numerical parameter study, Phys. Earth Planet. Int., 146, 179–194.

Wortel, M.J.R., and Spakman, W., 2000, Subduction and slab detachment in the Mediterranean-Carpathian region, Science, 290, 1910–1917.

Continental Collision and the STEP-wise Evolution of Convergent Plate Boundaries: From Structure to Dynamics

Rinus Wortel, Rob Govers and Wim Spakman

Abstract Particularly interesting stages in the evolution of subduction zones are the two main transient stages: initiation and termination. In this contribution the focus is on the second of these: terminal stage subduction, often triggered by continental collision or arc-continent collision. The landlocked basin setting of the Mediterranean region, in particular the western-central Mediterranean, provides unique opportunities to study terminal stage subduction and its consequences.

We use seismic tomography results on lithosphere and upper mantle structure as a source of information on plate boundary structure, and concentrate on the lithospheric scale aspects. Combining this structural information with process-oriented numerical modelling studies and regional observations, we present a 3D model for convergent plate boundary evolution after collision, in which slab detachment and the formation of tear or STEP (Subduction-Transform-Edge-Propagator) faults are key elements. A STEP fault laterally decouples subducting lithosphere from non-subducting lithosphere in a scissor type of fashion. It enhances the ability of a slab to retreat through the mantle flow around the edge of the subducted slab. In this way collision and back-arc extension may occur in close proximity. In our study area this specifically pertains to collision along the north African margin, STEP formation in easterly direction, CCW rotation of the southern Apennines slab and the opening of the Tyrrhenian Sea. Vertical tearing of subducted lithosphere may play an important role as well, but is probably not crucial. On the basis of the good agreement between the Mediterranean-based model and the evolution of the Tonga-Fiji region we expect that the model may shed light on other complex convergent plate boundary regions, as well.

In summary: Upon continental (or arc-continent) collision, along-trench variations in lithospheric properties of the subducting lithosphere may lead to disruption and segmentation of the subduction system. Following slab detachment along limited segments of a convergent plate boundary, the development of STEP faults is expected. These faults contribute to an increase in arc curvature within plate boundary segments. This contributes to the sinuous geometry of long subduction systems such as in the western and southwest Pacific.

Keywords Subduction • Seismic tomography • Continental collision • Slab detachment • STEP fault • Roll-back

R. Wortel
Department of Earth Sciences, Faculty of Geosciences, Utrecht University, Budapestlaan 4, 3584 CD Utrecht, The Netherlands,
wortel@geo.uu.nl

S. Lallemand and F. Funiciello (eds.), *Subduction Zone Geodynamics*,
DOI 10.1007/978-3-540-87974-9, © Springer-Verlag Berlin Heidelberg 2009

1 Introduction

Among the stages that may be distinguished in subduction zone evolution, the most intriguing ones probably are the transient stages of initiation and termination. Transient stages typically result in a wider spectrum of expressions than approximately steady state conditions. In this contribution the focus is on terminal stage subduction, often triggered by continent-continent or arc-continent collision (both hereafter referred to as continental collision). In this terminal stage the topic of back-arc extension deserves special attention, in view of Royden's (1993) finding of a peculiar combination of collisional events and escape type of arc migration and back-arc extension.

Since Karig (1971), the relation between subduction and back-arc extension (in combination with migration of the arc involved) is well-known. Identifying hinge migration as resulting from the gravitational instability of dense oceanic lithosphere (Elsasser, 1971) was a major step in understanding the physical cause of the connection. The process became commonly referred to as (slab) roll-back or slab retreat. Subsequent progress was hampered by the intrinsic short-comings of 2D modelling. This led Dvorkin et al. (1993) to investigate the significance of 3D aspects of subduction zones, particularly a finite width slab allowing mantle flow around the slab edges. Several studies expanding on this topic and other 3D aspects of subduction have further elucidated their role in subduction zone evolution and back-arc extension (Funiciello et al., 2003; Schellart, 2004; Bellahsen et al., 2005; Stegman et al., 2006; Schellart et al., 2007).

Because continental collision plays a key role in terminating subduction we choose the Mediterranean region, the evolution of which is characterized by continental collision and terminal stage subduction (Le Pichon, 1982; Jolivet and Faccenna, 2000; Wortel and Spakman, 2000), as our study area. We start from the regional geodynamic setting. Deploying information on lithosphere and upper mantle structure as obtained from seismic tomography we concentrate on the geodynamic process causing and allowing roll-back. Adopting a lithospheric scale approach, we combine the structural information with physical guidance from modelling studies to arrive at an evolutionary model for terminal stage subduction and accompanying back-arc extension.

2 Terminal Stage Subduction in Mediterranean Region: North Africa and the Sicily-Tyrrhenian Region

Collision has been part of geodynamic analyses of the Mediterranean for many decades (e.g., Argand, 1924), albeit originally predominantly in the context of the Africa-Europe collision in the Alps. In a plate tectonic framework, the role of collision in the evolution of the wider Mediterranean region has been highlighted in different ways by Le Pichon (1982) and Royden (1993).

Le Pichon (1982) focused on the role of Africa-Europe collision in the westernmost Mediterranean (Gibraltar arc region) and the Arabia-Anatolia collision in the east (Bitlis region) in establishing a landlocked basin setting. In such a setting the gravitational instability of the African lithosphere of the Mediterranean promoted roll-back of subduction zones and associated back-arc extension. Jolivet and Faccenna (2000) further advanced our understanding of the regional evolution by pointing out the contemporaneity of the start of extension in the Mediterranean and the role of the decrease in absolute velocity of the African plate in causing an acceleration in slab retreat. On a somewhat smaller scale, Royden (1993) concentrated on the kinematic relationship between collisional processes in parts of the Alpine-Mediterranean region and escape type of arc migrations in adjacent regions, such as the Carpathians, the Hellenic and Apennine arcs and the Betics-Rif thrust belt.

In this paper we use tomographic results to delineate 3D aspects of subduction zones and process-based understanding to investigate the physical processes giving rise to a kinematic pattern very similar to the type described by Royden (1993).

2.1 Deep Structure and Regional Evolution

Seismic tomography has played a major role in producing new information concerning the deep structure of subduction zones, in particular where the subducted lithosphere's geometry is not or only partially outlined by earthquake hypcentres. This has very clearly been the case in the Mediterranean region where indeed earthquake hypocentres only provided very limited windows on the deep structure of the convergent plate boundary zone (Giardini and Velonà, 1991; Spakman et al., 1993; Wortel and Spakman, 1992, 2000; Spakman and Wortel, 2004; Piromallo and Morelli, 2003; Lucente et al., 1999). Our early studies of the geodynamic implications of seismic tomography results (Spakman et al., 1988; Wortel and Spakman, 1992) showed the significance of the new source of information on structure of the lithosphere and upper mantle.

The Mediterranean region shows arc migration and back-arc extension in both the western Mediterranean (Liguro-Provencal Basin and Tyrrhenian Sea) and the Aegean region. Since in the Aegean case back-arc extension is fully active (McClusky et al., 2000), terminal stage subduction can be studied best in the western Mediterranean. We refer to Spakman and Wortel (2004) for a detailed overview of the tomographic results available for the western-central Mediterranean, in combination with resolution tests, a discussion of other tomographic results and a regional scale interpretation.

There is general consensus on the first order kinematic evolution of the western Mediterranean. Characteristic feature is the Late Oligocene to Recent arc migration starting from a location along the eastern margin of Iberia and southern France (e.g., Dewey et al., 1989; Auzende et al., 1973; Rehault et al., 1984; Doglioni et al., 1997; Faccenna et al., 2001a; Schettino and Turco, 2006). The migrating trench system made contact with the north African continental margin in the Langhian (~15 Ma). In a second episode of extension starting at about 10 Ma, the extension continued with an eastward shift of activity (Fig. 1), towards the Tyrrhenian realm (Carminati et al., 1998; Faccenna et al., 2001b). Malinverno and Ryan (1986) illustrated the relevance of Elsasser's (1971) gravitational sinking and arc migration model for the Tyrrhenian region.

The present-day plate boundary between Africa/Adria and Eurasia is often drawn from the Apennines, through the highly curved Calabrian Arc towards the Maghrebides in northern Africa. As shown by Carminati et al. (1998) on the basis of seismic tomography results, however, this apparently continuous arc does not correspond to an equally continuous subduction zone. Here we display the regional seismological structure in a series of horizontal slices through the BS2000 model of Bijwaard and Spakman (2000) (Fig. 2). The sections show a high velocity anomaly of the Calabrian slab and clearly indicate the absence of a subducted slab underneath the plate boundary segment in and to the west of western Sicily.

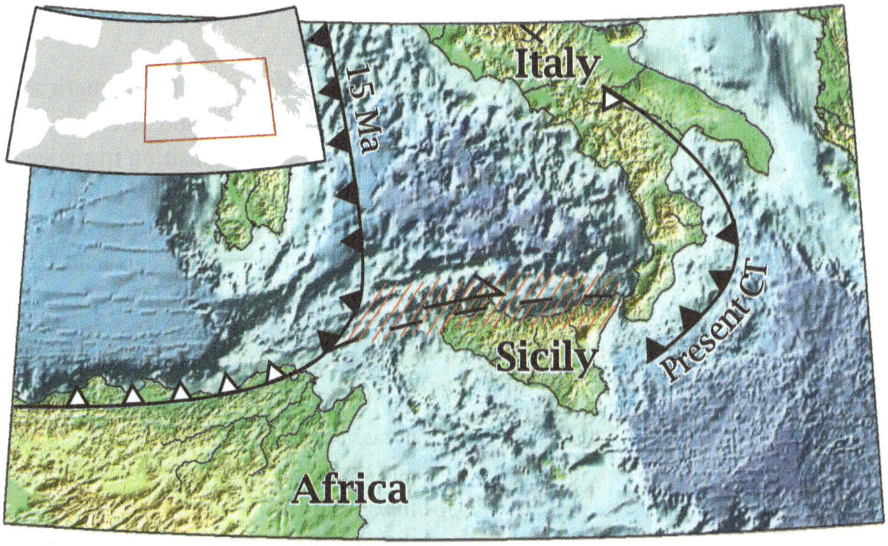

Fig. 1 Regional topographic/bathymetric map of the Sicily-Tyrrhenian Sea region (see inset for location of the main figure within the larger western-central Mediterranean region). It shows the location of the convergent plate boundary in the western-central Mediterranean in the Langhian (~15 Ma) and at Present. *CT* = Calabria Trench. White dents indicate detachment of subducted slab, black dents indicate continuous slabs. The exact location of the transition from detached slab towards continuous slab in Southern Italy is uncertain. The STEP fault is shown as a deformation zone, with an overall dextral sense of shear

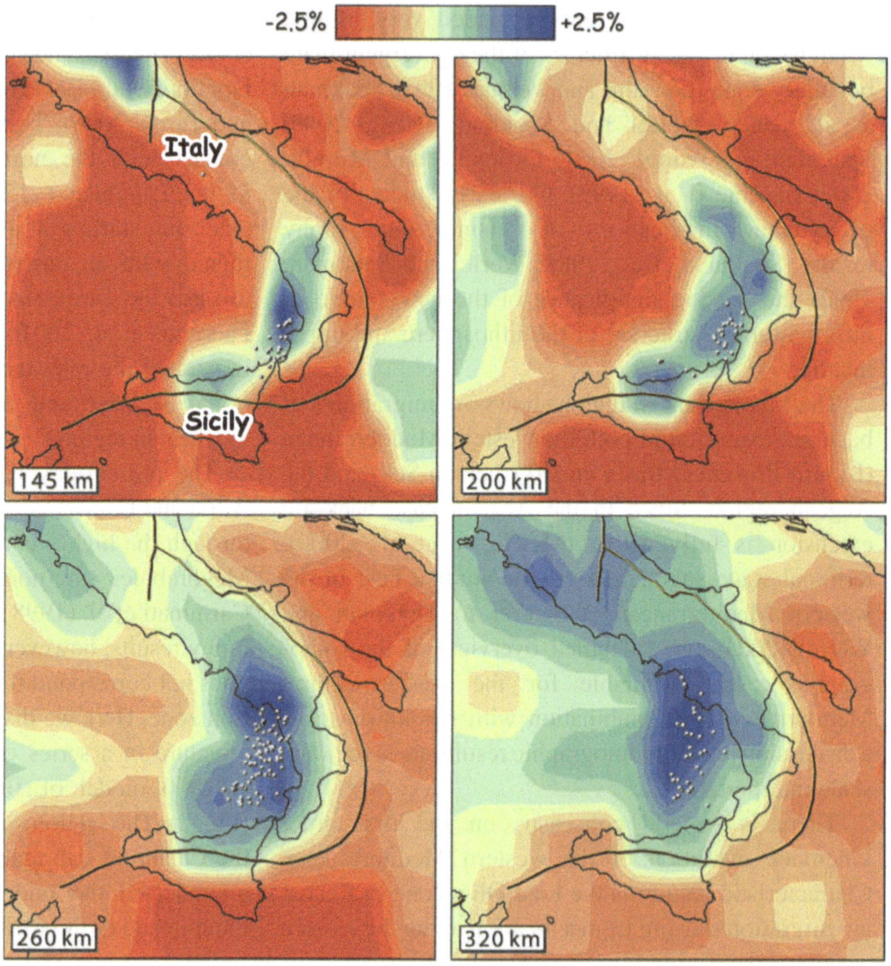

Fig. 2 Horizontal sections through the BS2000 seismic tomography model of Bijwaard and Spakman (2000) for the Sicily-Tyrrhenian region, at the depths indicated. P-wave velocity anomalies are given in percentages relative to the ambient depth-dependent velocity of the one-dimensional reference model ak135 (Kennett et al., 1995). White dots indicate earthquake hypocentres

2.2 Plate Boundary Evolution

The seismological structure in Fig. 2 pertains to the present-day situation. In order to develop a model for the evolution of the plate boundary we use results from modelling studies and follow the plate boundary evolution in a process-oriented way. We start from the stage of collision between the outwardly migrating (in S, SW and E direction) trench system and the north African continental margin (Fig. 1). In doing so we summarize earlier work and put it in a regional perspective.

2.2.1 Continental Collision and Slab Detachment

The migrating trench in the western Mediterranean reached the North African margins at ~15 Ma. Van de Zedde and Wortel (2001) numerically investigated the evolution of stress and strength in subducting lithosphere in case of transition from oceanic to continental lithosphere subduction. They showed that continental collision (in particular the soft type of collision, i.e., upon slab retreat) can lead to very shallow slab detachment (breakoff) of the subducted slab. This is schematically shown in Fig. 3a.

The seismic velocity distribution (Spakman and Wortel, 2004; Wortel and Spakman, 2000) shows evidence for the absence of high velocity slab material, and hence for slab detachment, below the north African margin in the ~100–300 km depth range, almost along the margin's entire length. It does not show clear evidence for a distinct pattern of lateral migration of the slab detachment process, neither to the west nor to the east. In analyses of convergent margin evolution it is important that slab detachment occurs with a considerable delay time (several Myr) relative to the time of

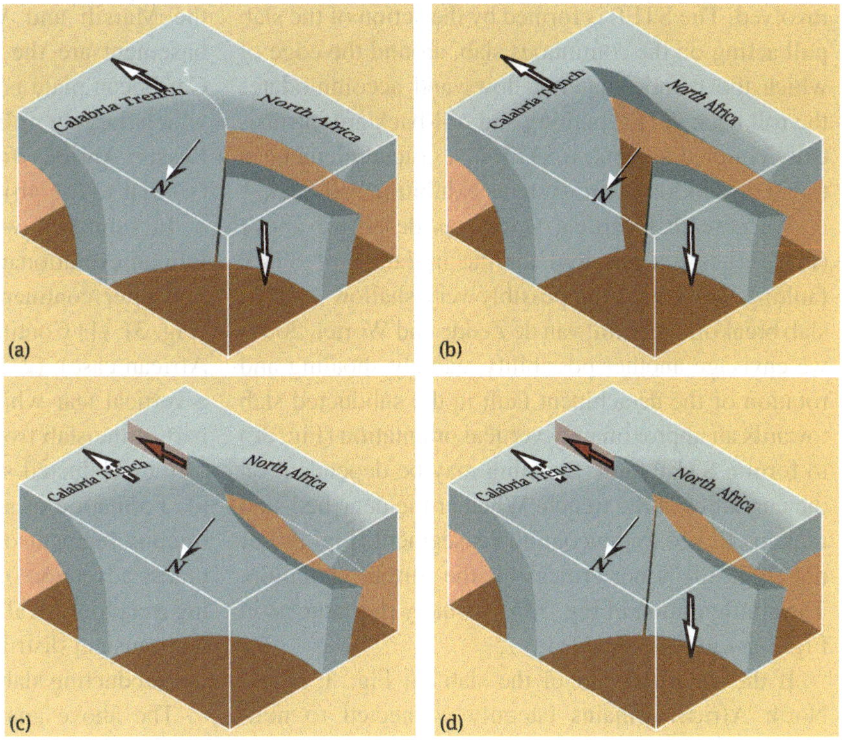

Fig. 3 Schematic representation of the subduction zone geometry in the Sicily-Tyrrhenian region. The region approximately corresponds with that in Fig. 1. The vertical scale of the boxes is 300 km. (**a**) Detached slab after collision of the African continent with the migrating trench system in the western-central Mediterranean. A vertical tear separating the detached part from the continuous part is indicated. Dotted line, labelled S, indicates future STEP fault. (**b**) Formation of a STEP fault and subsequent trench retreat (roll-back). (**c**) and (**d**) An alternative scenario for STEP fault generation: The detachment fault in the subducted slab may shoal and rotate towards an approximately vertical orientation. The detached slab segment (in c) may tear away from the continuous segment (see **d**). In this case the configuration closely resembles that in (**a**)

first arrival of the continental lithosphere at the trench (van de Zedde and Wortel, 2001).

The lateral variation in contact properties along the trench system in the western-central Mediterranean implied that slab detachment was probably confined to the north African segment, whereas the Calabrian segment, in which the oceanic Ionian lithosphere was consumed, remained continuous. This leads to the next phase.

2.2.2 Formation of a Vertical Slab Tear

A detached slab along a major part of the plate boundary gives rise to in-plate bending stresses (Yoshioka and Wortel, 1995), tending to tear away the detached part from the continuous part. We assume a vertical tear to develop as in Fig. 3a (this is not a critical assumption; an alternative is presented below and in Fig. 3c and d). The vertical tear leads to the creation of a free slab edge. This is an important evolutionary step in the process, since both analogue and numerical experiments have convincingly shown the prominence of transport of asthenospheric and deeper mantle material around the edge of the slab, allowing the slab to retreat further and faster (see references given in Introduction).

2.2.3 Formation of a STEP Fault

Carminati et al. (1998) and Carminati and Wortel (2000) proposed a tear fault as the connection between the north African and the Calabrian plate boundary segments (Fig. 1). The physical need for this type of fault was recognized from the beginning of plate tectonic theory (Isacks et al., 1968: "scissors type of faulting"; also regularly referred to as tear fault or transfer fault). Govers and Wortel (2005) drew attention to the non-rigid body nature of the relative motions along such a fault, and named it STEP fault (**S**ubduction-**T**ransform-**E**dge-**P**ropagator). They investigated the potential for propagation and the stress field and vertical motions associated with its activity. The motion along a STEP fault, in general, and the dextral motion indicated in Fig. 1, in particular, are to be understood as the overall sense of motion of a possibly broad deformation zone, in which also block rotations can be involved. The STEP is formed by the action of the slab pull acting on the continuous slab, around the edge of which the mantle material flows and accommodates the roll back. STEP faulting and roll-back are intrinsically coupled aspects; a STEP is a crucial element in the 3D evolution of arc systems exhibiting roll-back.

We note that vertical tearing as described above (Fig. 3a) is not required for the initiation of STEP faulting. In view of the possibly very shallow level of slab breakoff (~35 km; van de Zedde and Wortel, 2001) we envisage another possibility, namely shoaling and rotation of the detachment fault in the subducted slab towards an approximately vertical orientation (Fig. 3c) to form a STEP fault. Shoaling may be dependent on the curvature of the trench. Whether the detached slab remains linked to the continuous segment (Fig. 3c) or not (Fig. 3d) is not critical for the further evolution. The configuration in Fig. 3d is virtually the same as in Fig. 3a, with a vertical tear.

If the detached part of the slab (in Fig. 3: below North Africa) remains laterally connected to non-detached subducted lithosphere (in Fig. 3c: below Calabria) the slab pull acting on the non-subducted lithosphere at the surface, is concentrated in the trench segment where the subducting slab is still continuous in the downdip direction (in Fig. 3c: along the Calabria Trench segment). STEP fault formation, indicated by the red arrow, is to be expected in a very similar way as in Fig. 3a and b.

2.2.4 Slab Retreat with or without Back-Arc Extension

Once a STEP fault has been initiated the adjacent continuous slab segment can roll-back. Depending on the regional tectonic setting, the slab retreat can be accompanied by either advective motion as a rigid body or deformation of the overriding plate (Fig. 4; see also Govers and Wortel, 2005). Figure 4 illustrates this by two situations with different boundary conditions on the left side of the model: (a) no "east-west" motion between left and right side of the model, and (b) east-west motion is possible. For the Tyrrhenian situation eastward rigid body type advection (relative to the African lithosphere) of the overriding lithosphere was/is not possible (situation as in Fig. 4a) and deformation developed, giving rise to the back-arc extension in the Tyrrhenian Sea, of which the Marsili and Vavilov basins with their oceanic basement are the most prominent expressions. The Caribbean plate is an example of an overriding plate which largely follows the eastward motion of the Lesser Antilles trench through rigid body motion (similar to Fig 4b).

In summary, we connected a series of processes into an evolutionary model for plate boundary evolution after continental collision. The components are (Fig. 3): (1) Continental collision ("soft", in the north African case); (2) Slab detachment; (3) Formation of a vertical tear which laterally separates the detached part of the slab from the adjacent still continuous part of the subducted slab. This step is not a crucial one; (4) Formation of a STEP fault and retreat of the continuous segment of the slab; (5) Back-arc extension, unless advection of the overriding follows the retreating trench. Basically, the entire process is governed by the transient distribution of slab pull forces acting on the subducting slab.

The above analysis was aimed at abstracting an essential aspect of collision and subsequent plate boundary evolution, which – as was realized by Royden (1993) – can be studied very well in the Mediterranean context. For other analyses of back-arc extension and arc migration in the western-central Mediterranean we refer to Faccenna et al. (2001a, b, 2004), Wortel and Spakman (2000) and Spakman and Wortel (2004), and other references given in Sect. 2.1.

3 Test of Evolutionary Model

3.1 Magmatism

Although the scenario developed for the Sicily-Tyrrhenian region seems physically sound, we would like to subject the above model scenario to new tests, preferably against data which have not been used in the development of the scenario. This could be done by using observations from the same region but of a different type. Magmatic data are valuable in this context, since they bear information on transient process, including their timing. Maury et al. (2000) and Coulon et al. (2002) concluded that alkaline magmatism in northern Africa is accounted for by slab breakoff, as was proposed by Carminati et al. (1998). In the Tyrrhenian realm, data on spatial and temporal varia-

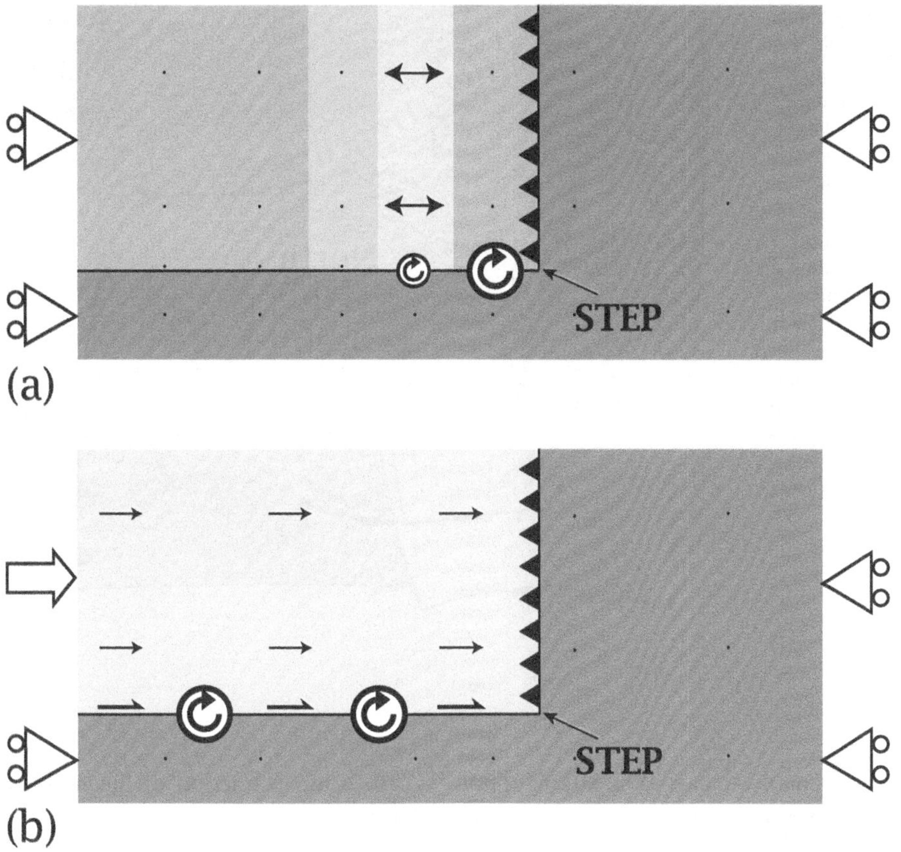

Fig. 4 Schematic representation (map view) of two situations showing a STEP fault near the edge of a subducting slab, in combination with the motion or deformation of the overriding plate. Subduction and subduction direction are indicated by black dents. Left and right are taken to correspond with west and east, respectively. The NW part represents the overriding plate. (**a**) The left side boundary condition is: no "eastward" motion (indicated by rollers). Subduction of (oceanic) lithosphere is accompanied by eastward trench retreat (roll-back) and back-arc extension (light areas in overriding plate). The region of extension may migrate with the retreating trench, towards the right/east. The STEP fault shows dextral motion, not uniform along the STEP fault, possibly in combination with CW rotations. (**b**) At the left boundary a velocity boundary condition applies, implying that the overriding plate moves relative to the subducting plate. If the relative motion equals the trench retreat rate, no intraplate deformation in the overriding plate occurs. This is rigid body advection of the overriding plate, and the STEP fault acts like a regular transform fault, with dextral sense of motion

tions in magma composition (Peccerillo, 2005) support the scenario developed. However, we consider this as not yet convincing in view of the lack of spatial resolution.

3.2 Tonga-Vitiaz Trench Region

Alternatively, we can test the scenario in a completely different tectonic environment, with a similar starting situation. Collisional (arc-continent) events must have occurred in many regions in the geological history. Preferably, such a test is to be carried out in a region which experienced collision recently enough to show observational evidence concerning the evolution of the process. There are several reasons for selecting the Tonga-Fiji region as an excellent candidate for such a test: among these are (1) The starting situation is very similar to the one in the western-central Mediterranean; a long uninterrupted trench system, north-northeast of Australia (Hall, 2002; Schellart et al., 2006); (2) Magnetic anomaly data allow reconstruction of the complicated recent history of back-arc extension, and last but not least, (3) In this region the presence of STEP faults is demonstrated in a uniquely convincing way by a detailed seismicity and focal mechanism study of Millen and

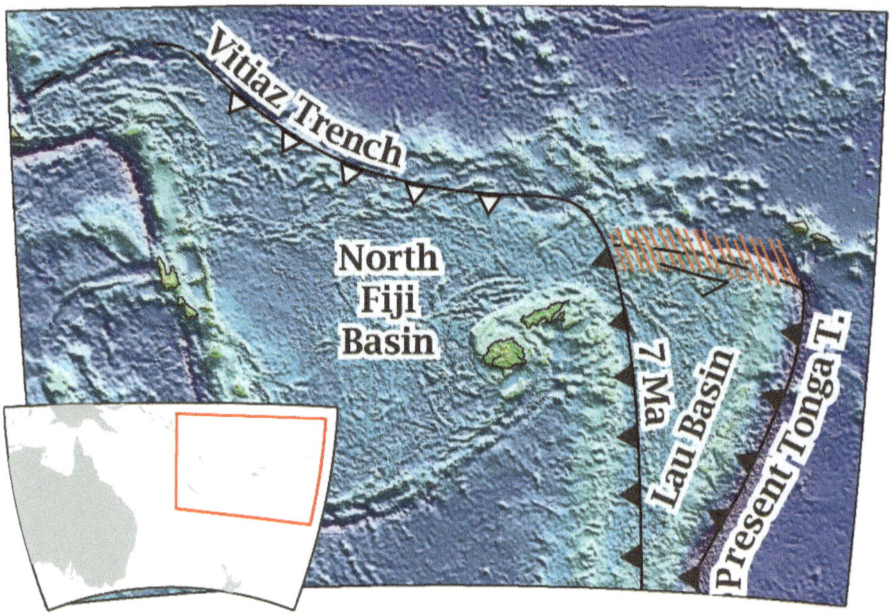

Fig. 5 Regional topographic/bathymetric map for the Tonga-Vitiaz Trench region (see inset for location of the main figure within the larger southwest Pacific region). It shows the approximate location of the convergent plate boundary in this part of the SW Pacific in the Middle–Late Miocene and the present location of the Tonga Trench. White dents indicate presumed detachment of subducted slab, black dents indicate continuous slabs. The STEP fault is shown as a deformation zone, with an overall sinistral sense of shear

Hamburger (1998): A narrow band (~40 × 80 km) of normal faulting events ("hinge faulting", along vertical fault planes) with focal depths in the range 18–88 km provides direct evidence for tearing of the Pacific plate.

The region shown in Fig. 5 encompassing the Tonga-Fiji, the New Hebrides and the Vitiaz trench is a complicated segment of the plate boundary between the Pacific plate and the (Indo)-Australian plate. The sum of the back-arc spreading rates and the motion of the Pacific plate produces the world's highest subduction rate (~24 cm/year, Bevis et al., 1995; Doglioni et al., 2007), in the Tonga trench.

Figure 6 shows horizontal sections from the seismic tomography model BS2000 (Bijwaard and Spakman, 2000). In view of the high subduction velocity at the Tonga Trench effects of recent slab segmentation are to be found at greater depths than in the Calabrian subduction zone. A distinct slab edge is visible at the northern limit of the Tonga trench. Note the difference in orientation between the slab anomaly at 870 km depth and the Tonga slab anomalies in the upper mantle at 420 and 330 km depth. Another conspicuous feature is the distribution of deep earthquake hypocentres, not only in the very active Tonga subduction zone, but also outside the Tonga slab, in the northwestern part of the region displayed (see Fig. 6, section at 630 km depth).

In a reconstruction of the regional evolution Hall (2002) shows a long trench system, the Melanesian Arc, which was subject to major reorganization, probably in Miocene time. Hall (2002) identified the collision of continental type of lithosphere (the Ontong Java Plateau) with the Melanesian Arc as the likely cause of the plate boundary reorganization. Whether this collision affected the region of the present-day Vitiaz trench (Fig. 5) or rather the Solomon Islands region more to the west is uncertain. A similar analysis was made by Schellart et al. (2006) who distinguished a smaller continental type of block to the east of the Ontong-Java Plateau, the Melanesian Borderland, which in his reconstruction did collide with the trench system at the location of the present-day Vitiaz trench, in the Middle Miocene (~10 Ma).

In both reconstructions active motion of the Pacific plate brought a continental block (Melanesian Borderland, Ontong Java Plateau) into collisional contact with part of the trench system. Presumably, continent-arc collision of this type led to termination of the convergence and subsequent slab detachment in that particular part of the trench, now called Vitiaz Trench (Fig. 7a). Several studies have proposed a fossil subduction origin for the bathymetric feature of the Vitiaz trench (Pelletier and Auzende, 1996, and references therein). Subsequent development of a

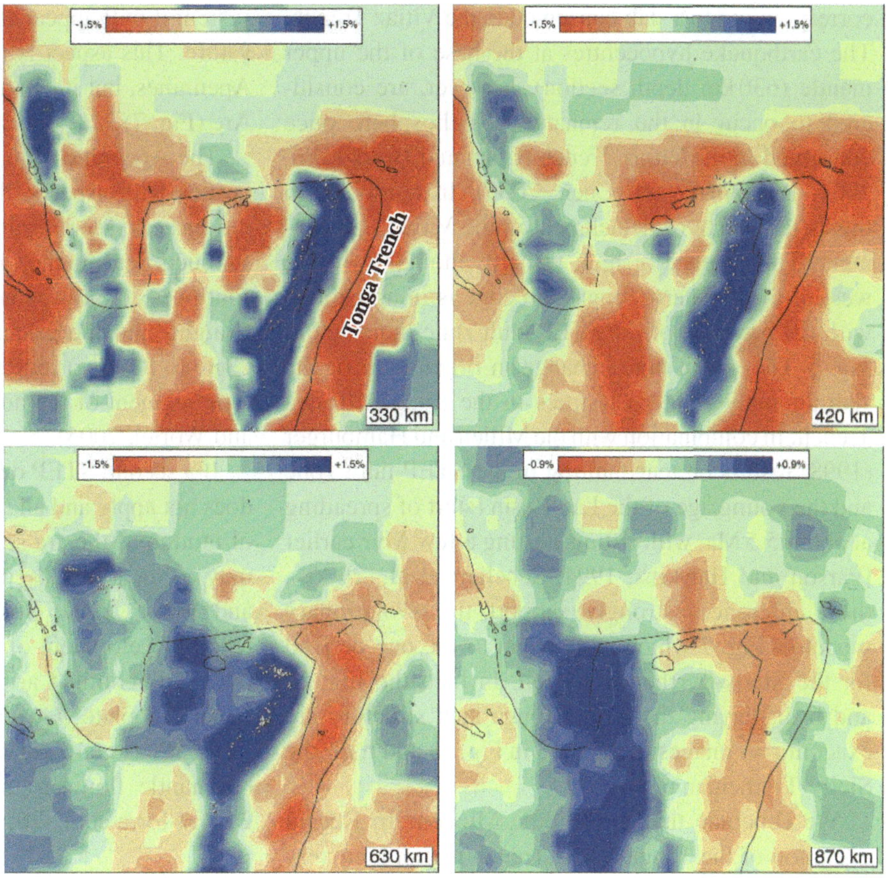

Fig. 6 Horizontal sections through the BS2000 seismic tomography model of Bijwaard and Spakman (2000) for the Tonga-Vitiaz Trench region, at the depths indicated. P-wave velocity anomalies are given in percentages relative to the ambient depth-dependent velocity of the one-dimensional reference model ak135 (Kennett et al., 1995). White dots indicate earthquake hypocentres

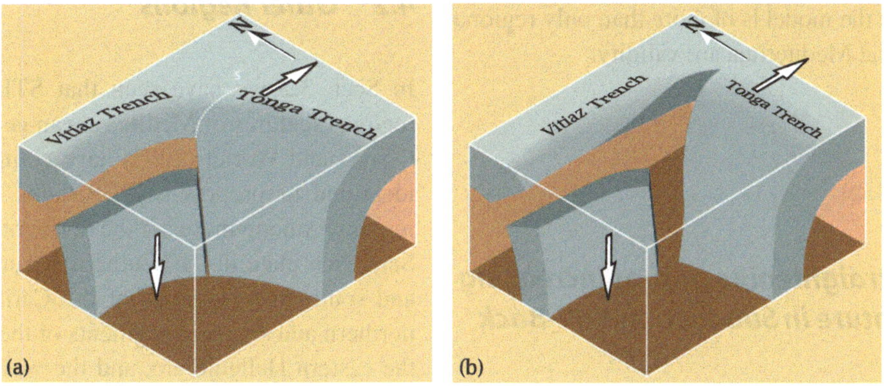

Fig. 7 Schematic representation of the subduction zone geometry in the Tonga-Vitiaz Trench region. The model applies to the region approximately corresponding with that in Fig. 5. Vertical scale of the box is 300 km. a Detached slab after collision of continental type of plateau (Melanesian Borderland, Schellart et al., 2006) with trench system in the SW Pacific (the present-day Vitiaz Trench). A vertical tear separating the detached part from the continuous part is indicated. *Dotted line*, labelled S, indicates future STEP fault. b Formation of a STEP fault and subsequent trench retreat (roll-back) of the Tonga trench

free slab edge set the stage for STEP faulting to be initiated. This in turn allowed for the rapid eastward retreat of the Tonga slab (Fig. 7b), and the opening of the Lau Basin (see Fig. 5 for location). Recognizing that the resolution in the northern part of the 330 km depth section (Fig. 6) is not sufficient to draw direct conclusions from the tomographic images it appears that there is no anomaly in the uppermost mantle corresponding to a slab consumed in the Vitiaz Trench. The earthquake hypocentres at the base of the upper mantle (630 km depth section), however, are considered to occur in the remnants of lithosphere once consumed in the Vitiaz Trench (Chen and Brudzinski, 2001; Brudzinski and Chen, 2003). In combination this is indicative of detachment in the Vitiaz slab segment. The absence of high velocity anomalies to the west of the northern Tonga slab edge is in agreement with the presence of the STEP fault zone indicated in Fig. 5 The above noted change in orientation of the slab anomaly (Fig. 6) testifies to the rotation Tonga Trench. In combination with the Millen and Hamburger (1998) results on tear faulting in the STEP fault zone and the young age of the Lau Basin (start of spreading at about 5.5 Ma, with rifting starting a few Myr earlier [Parson and Hawkins, 1994; Schellart et al., 2006]), these inferences strongly support the evolutionary model in Fig. 7 which is identical to the one derived for the Sicily-Tyrrhenian region. In fact, the Fig. 7a and b displaying the Vitiaz-Tonga slab structure are exact mirror images of the ones in Fig. 3a and b for the Sicily-Tyrrhenian region.

We conclude that the Middle Miocene to Recent evolution of the Tonga-Vitiaz Trench region is in very good agreement with the evolutionary model derived from the Sicily-Tyrrhenian region. This is taken as an indication that the model is of more than only regional (western-central Mediterranean) validity.

4 Discussion

4.1 Arc Straightening and/or Increasing Curvature in Subduction Roll-Back

The free slab edge allows for sideways migration of asthenospheric and deeper mantle material around it, and consequently higher roll-back velocities relative to adjacent segments of the same trench system. In case there is only one STEP fault in the system the retreat of the slab will lead to a rotation of the plate boundary segment. If the starting configuration was arc-shaped, such a rotation leads to straightening of the convergent arc segment involved. On the other hand, if we consider the combination of the convergent segment and the STEP segment(s) of the evolving arc system it will appear to lead to an increase in arc curvature. This aspect is particularly clear in case of the Apennines, but it can also be recognized in the Tonga Arc (Fig. 7).

The Apennines-Calabrian arc evolution bears similarities with aspects of the Hellenic arc evolution in the eastern Mediterranean, where the African plate subducts beneath the Crete and the Aegean region. The southwestern arc segment is convergent and relatively straight, and the eastern part (from south of central Crete to Rhodes) represents the expression of a STEP type segment of the arc (Huchon et al., 1982; Govers and Wortel, 2005).

If a second STEP develops, the above straightening does not apply and an approximately translational type of motion of the arc segment between the two STEPS may occur, as proposed for the Gibraltar arc (Lonergan and White, 1997; Spakman and Wortel, 2004), the recent development of the Calabrian arc and the relative motion of the Caribbean plate relative to the South American plate (Govers and Wortel, 2005). For this 2-STEP case, arc curvature may further be increased by mantle flow around both slab edges (Schellart, 2004; Morra et al., 2006).

4.2 Other Regions

In Sect. 3.2 we have seen that STEP fault are not regional features of Mediterranean geodynamics only. Govers and Wortel (2005) have compiled a series of identified or proposed STEP faults in several other regions. Among these are the northern boundary of the Sandwich plate in the southern Atlantic, the northern and southern boundaries of the Caribbean plate, the northern and southern segments of the Carpathian arc, the eastern Hellenic arc, and the western edge of the north Sulawesi trench in Indonesia. In addition, Wortel et al. (2006) identified a STEP fault along the eastern part of the Cyprus arc.

In the Pacific realm, apart from the Tonga region, Schellart et al. (2003) identified roll-back in combination with STEP type of faulting in the Kuriles arc and back-arc region and investigated the STEP's near-surface expression by analogue modelling. In fact, the western Pacific Ocean provides a setting where STEP formation is to be expected because a basic condition for slab roll-back, gravitational instability of the oceanic lithosphere involved in subduction, is amply fulfilled in view of the predominantly Mesozoic lithospheric age. In this light, the tear in the subducting Izu-Bonin slab identified by Miller et al. (2004) may lead to a new arc segmentation phase in the region.

We expect STEP faults to be intrinsic tectonic elements in highly curved arcs, and anticipate that the number of identified STEP faults will further increase.

4.3 Arc Segmentation as a Repetitive Process?

Both in the Western Mediterranean and in the Tonga-Vitiaz Trench region we started from long continuous arcs, as postulated in regional studies. One may wonder whether perceptions of that type are the consequence of incomplete knowledge about early stages of arc development and subduction zone evolution. The postulated long arcs and/or the associated subducted slabs were possibly already segmented in early stages of the regional evolution, of which direct evidence was lost. Along these lines we hypothesize that the above evolutionary model applies to long arc segments which, upon growing by outward arc migration, become prone to collision and segmentation along the lines of the model considered. The new segments may again grow in length (as expressed at the surface) till new continental collisional events cause a repetition of the segmentation process. This may happen to a long trench system such as the Tonga-Kermadec trench, in the future, by the interaction with continental plateaus or ridges. The process we have considered in this paper, then leads to a new complication added to the earlier arc structure, and as such it may be a modular process within a repeating scheme.

4.4 Other Aspects of Terminal Stage Subduction Involving Slab Detachment

In this contribution we concentrated on continental collision and arc migration and back-arc extension. If continental collision results in slab detachment other expressions or accompanying processes are to be expected, notably in the field of vertical motions and magmatism. We briefly mention some pertinent studies.

Buiter et al. (2002) and Gerya et al. (2004) studied vertical motions by using an elastic rheology and a viscous rheology for the descending plate, respectively. In the former study the total magnitude of vertical motion was found to possibly amount to as much as 2–6 km. A striking result in the latter study was the long time span of the uplift process after the occurrence of slab detachment (~20 Myr).

Wortel and Spakman (1992) and in particular Davies and von Blanckenburg (1995) drew attention to the implications of slab breakoff for magmatic processes, with emphasis on the generation of alkaline magmas. For several other regions slab detachment has been proposed to account for special types of (post-collisional), usually K-rich magmatism, for example, for the Himalayas by Kohn and Parkinson (2002), for Central-Southern Italy by Wortel et al. (2003), and for Mexico by Ferrari (2004). De Boorder et al. (1998) have highlighted the association of mineralization zones with regions where slab detachment has been proposed. A crucial element in both magmatism and mineralization is the possible rise of asthenospheric material to unusually shallow depths after shallow slab breakoff.

5 Conclusions

We have identified a series of sequential process components for the evolution of a plate boundary in part of which continental collision occurs and combined them into a three-dimensional evolutionary model. Lateral (i.e., along strike of the trench) variation in nature of the lithosphere arriving at the trench, from continental lithosphere to oceanic lithosphere, sets the stage for the entire process. Continental collision triggers the process by causing slab detachment, and STEP fault formation is a key element in the further evolution. The various components represent stages in a continuous process, governed by the transient distribution in slab pull forces. As such, it is an evolutionary model for arc segmentation, which leads to strongly varying patterns of back-arc spreading.

Whereas we have derived the model for the Sicily-Tyrrhenian area, the Tonga-Vitiaz Trench region shows a strikingly similar evolution, which testifies to its more general validity. Hence, we expect that it will shed light on other convergent plate boundary regions with a history involving collision.

Acknowledgements RW thanks Serge Lallemand and Francesca Funiciello for the invitation to contribute to the very stimulating Subduction Zone Geodynamics Conference, Montpellier, June 2007. Discussions with participants are much appreciated. Our research is carried out under the programme of the Vening Meinesz Research School of Geodynamics (VMSG) and contributes to the Netherlands Research Centre of Integrated Solid Earth Science (ISES), and the EUROMARGINS Program of the European Science Foundation, 01-LEC-EMA22F (WESTMED project). Constructive reviews by Eugenio Carminati and Wiki Royden are gratefully acknowledged.

References

Argand E (1924) La tectonique de l'Asie. Comptes-rendu du XIIIe Congrès géologique international: 170–372

Auzende JM, Bonnin J, Olivet JL (1973) The origin of the western Mediterranean basin. J. Geol. Soc. London 129: 607–620.

Bellahsen N, Faccenna C, Funiciello F (2005) Dynamics of subduction and plate motion in laboratory experiments: Insights into the "plate tectonics" behavior of the Earth. J. Geophys. Res. 110: (B01401): doi:10.1029/2004JB 002999.

Bevis M, et al. (1995) Geodetic observations of very rapid convergence and back-arc extension at the Tonga arc. Nature 374: 249–251.

Bijwaard H, Spakman W (2000) Nonlinear global P-wave tomography by iterated linearised inversion. Geophys. J. Int. 141: 71–82.

Brudzinski M, Chen WP (2003) A petrological anomaly accompanying outboard earthquakes beneath Fiji-Tonga: Corresponding evidence from broadband P and S waveforms. J. Geophys. Res. 108: (B6, 2299): doi:10.1029/ 2002JB002012.

Buiter SJH, Govers R, Wortel MJR (2002) Two-dimensional simulations of surface deformation caused by slab detachment. Tectonophysics 354(3–4): 195–210.

Carminati, E, Wortel MJR (2000) Subduction zone rollback in 4D: The Langhian to Recent evolution of the Africa-Eurasia plate boundary in the central Mediterranean. Abstracts from *XXV General Assembly of the European Geophysical Society*, Nice (France), April 2000.

Carminati E, Wortel MJR, Spakman W, Sabadini R (1998) The role of slab detachment processes in the opening of the western-central Mediterranean basins: Some geological and geophysical evidence. Earth Planet. Sci. Lett. 160(3–4): 651–665.

Chen WP, Brudzinski MR (2001) Evidence for a large-scale remnant of subducted lithosphere beneath Fiji. Science 292: 2475–2479.

Coulon C et al. (2002) Post-collisional transition from calc-alkaline to alkaline volcanism during the Neogene in Oranie (Algeria): Magmatic expression of a slab breakoff. Lithos 62: 87–110.

Davies JH, von Blanckenburg F (1995) Slab breakoff: A model of lithosphere detachment and its test in the magmatism and deformation of collisional orogens. Earth Planet. Sci. Lett. 129: 85–102.

de Boorder H, Spakman W, White SH, Wortel MJR (1998) Late Cenozoic mineralization, orogenic collapse and slab detachment in the European Alpine Belt. Earth Planet. Sci. Lett. 164(3–4): 569–575.

Dewey JF, Helman ML, Turco E, Hutton DHW, Knott SD (1989) Kinematics of the western Mediterranean, in Coward MP, Dietrich D, Park RG (Eds.) Alpine Tectonics, Geol. Soc. Lond. Spec. Publ. 45: 265–283.

Doglioni C, Gueguen E, Sàbat F, Fernandez M (1997) The western Mediterranean extensional basins and the Alpine orogen. Terra Nova 9: 109–112.

Doglioni C, Carminati E, Cuffaro M, Scrocca D (2007) Subduction kinematics and dynamic constraints. Earth-Sci. Rev. 83: 125–175.

Dvorkin J, Nur A, Mavko G., Ben-Avraham Z (1993) Narrow subducting slabs and the origin of backarc basins. Tectonophysics 227: 63–79.

Elsasser WM (1971) Sea-floor spreading as thermal convection. J. Geophys. Res. 76: 1101–1112.

Faccenna C, Becker TW, Lucente FP, Jolivet L, Rossetti F (2001a) History of subduction and back-arc extension in the central Mediterranean. Geophys. J. Int. 145: 809–820.

Faccenna C, Funiciello F, Giardini D, Lucente P (2001b) Episodic back-arc extension during restricted mantle convection in the Central Mediterranean. Earth Planet. Sci. Lett. 187(1–2): 105–116.

Faccenna C, Piromallo C, Crespo-Blanc A, Jolivet L, Rossetti F (2004) Lateral slab deformation and the origin of the western Mediterranean arcs. Tectonics, 23: TC1012, doi:10.1029/ 2002TC001488.

Ferrari L (2004) Slab detachment control on mafic volcanic pulse and mantle heterogeneity in central Mexico. Geology 32: 77–80.

Funiciello F, Faccenna C, Giardini D, Regenauer-Lieb K(2003) Dynamics of retreating slabs: 2. Insights from three-dimensional laboratory experiments. J. Geophys. Res. 108(B4, 2207): doi:10.1029/2001JB000896.

Gerya T, Yuen DA, Maresch WV (2004) Thermomechanical modelling of slab detachment. Earth Planet. Sci. Lett. 226: 101–116.

Giardini D, Velonà M (1991) The deep seismicity of the Tyrrhenian Sea. Terra Nova 3: 57–64.

Govers R, Wortel MJR (2005) Lithosphere tearing at STEP faults: Response to edges of subduction zones. Earth Planet. Sci. Lett. 236: 505–523.

Hall R (2002) Cenozoic geological and plate tectonic evolution of SE Asia and the SW Pacific: Computer-based reconstructions, model and animations. J. Asian Earth Sci. 20: 353–431.

Huchon P, Lyberis N, Angelier J, Le Pichon X, Renard V (1982) Tectonics of the Hellenic trench: A synthesis of sea-beam and submersible observations. Tectonophysics 86: 69–112.

Isacks B, Oliver, J, Sykes LR (1968) Seismology and the new global tectonics. J. Geophys. Res. 73: 5855–5899.

Jolivet L, Faccenna C (2000) Mediterranean extension and the African-Eurasian collision. Tectonics 19: 1095–1106.

Karig DE (1971) Origin and development of marginal basins in the Western Pacific. J. Geophys. Res. 76: 2542–2561.

Kennett BLN, Engdahl ER, Buland R (1995) Constraints on seismic velocities in the Earth from travel times. Geophys. J. Int. 122: 108–124.
Kohn MJ, Parkinson CD (2002) Petrologic case for Eocene slab breakoff during the India-Asian collision. Geology 30: 591–594.
Le Pichon X (1982) Land-locked oceanic basins and continental collision: The eastern Mediterranean as a case example. In: Hsu K (ed), Mountain Building Processes Academic Press, London, pp. 201–211.
Lonergan L, White N (1997) Origin of the Betic-Rif mountain belt. Tectonics 16: 504–522.
Lucente FP, Chiarabba C, Cimini GB, Giardini D (1999) Tomographic constraints on the geodynamic evolution of the Italian region. J. Geophys. Res. 104: 20307–20327.
Malinverno A, Ryan WBF (1986) Extension in the Tyrrhenian Sea and shortening in the Apennines as a result of arc migration driven by sinking of the lithosphere. Tectonics 5: 227–245.
Maury RC et al. (2000) Post-collisional Neogene magmatism of the Mediterranean Maghreb margin: A consequence of slab breakoff. C.R. Acad. Sci. Paris/Earth Planet. Sci. 331: 159–173.
McClusky S et al. (2000) Global Positioning System constraints on plate kinematics and dynamics in the eastern Mediterranean and Caucasus. J. Geophys. Res. 105(B3): 5695–5719.
Millen DW, Hamburger MW (1998) Seismological evidence for tearing of the Pacific plate at the northern termination of the Tonga subduction zone. Geology 26: 659–662.
Miller MS, Kennett BLN, Lister GS (2004) Imaging changes in morphology, geometry, and physical properties of the subducting Pacific plate along the Izu-Bonin-Mariana arc. Earth Planet. Sci. Lett. 224: 363–370.
Morra G, Regenauer-Lieb K, Giardini D (2006) Curvature of oceanic arcs. Geology 34: 877–880.
Parson LM, Hawkins JW (1994) Two-stage ridge propagation and the geological history of the Lau backarc basin. Proc. Ocean Drill. Program Sci. Results 135: 819–828.
Peccerillo A (2005) Plio-Quaternary volcanism in Italy. Springer, Berlin, Heidelberg, New York: 365 pp.
Pelletier B, Auzende JM (1996) Geometry and structure of the Vitiaz Trench linament (SW Pacific). Mar. Geophys. Res. 18: 305–335.
Piromallo C, Morelli A (2003) P wave tomography of the mantle under the Alpine-Medieterranean area. J. Geophys. Res. 108(B2, 2065): doi:10.1029/2002JB001757.
Rehault J-P, Boillot G, Mauffret A (1984) The Western Mediterranean basin geological evolution. Mar. Geol. 55: 447–477.
Royden LH (1993) The evolution of retreating subduction boundaries formed during continental collision. Tectonics 12: 629–638.
Schellart WP (2004) Kinematics of subduction and subduction-induced flow in the upper mantle. J. Geophys. Res. 109(B07401): doi:10.1029/2004JB002970.
Schellart WP, Jessell MW, Lister GS (2003) Asymmetric deformation in the backarc region of the Kuril arc, northwest Pacific: New insights from analogue modeling. Tectonics 22(5, 10047): doi:10.1029/2002TC001473.
Schellart WP, Lister GS, Toy VG (2006) A Late cretaceous and Cenozoic reconstruction of the southwest Pacific region: Tectonics controlled by subduction and slab rollback proocesses. Earth-Sci. Rev. 76: 191–233.
Schellart WP, Freeman J, Stegman DR, Moresi L, May D (2007). Evolution and diversity of subduction zones controlled by slab width. Nature 446: doi:10.1038/nature05615.
Schettino A, Turco E (2006) Plate tectonics of the western Mediterranean region during the Oligocene and early Miocene. Geophys J. Int. 166: 1398–1423.
Spakman W, Wortel R (2004) A tomographic view on western Mediterranean geodynamics. In: Cavazza W et al. (Eds): The TRANSMED Atlas: The Mediterranean region from crust to mantle, Springer, Berlin, Heidelberg, New York, pp. 31–52.
Spakman W, Wortel MJR, Vlaar NJ (1988) The Hellenic subduction zone: A tomographic image and its geodynamic implications. Geophys. Res. Lett. 15: 60–63.
Spakman W, van der Lee S, van der Hilst R (1993) Travel-time tomography of the European-Mediterranean mantle down to 1400 km. Phys. Earth Planet. Int. 79: 3–74.
Stegman DR, Freeman J, Schellart WP, Moresi L, May D (2006) Influence of trench width on subduction hinge retreat rates in 3-D model of rollback. Geochem. Geophys. Geosyst. 7(Q03012): doi:1029/2005GC001056.
van de Zedde DMA, Wortel MJR (2001) Shallow slab detachment as a transient source of heat at midlithospheric depths. Tectonics 20: 868–882.
Wortel MJR, Spakman W (1992) Structure and dynamics of subducted lithosphere in the Mediterranean region. Proc. K. Ned. Akad. Wet. 95: 325–347.
Wortel MJR, Spakman W (2000) Subduction and slab detachment in the Mediterranean-Carpathian region. Science 290: 1910–1917.
Wortel MJR, van Bergen MJ, van de Zedde DMA, Spakman W (2003) Continental collision, slab detachment and K-rich magmatism: Central-southern Italy as a type example. Abstracts from EGS – AGU – EUG Joint Assembly, Nice (France), April 2003.
Wortel MJR, Spakman W, Govers R (2006) Deep structure and evolution of the Cyprus Arc, Abstracts from EGU, Vienna, Austria, April 2006.
Yoshioka S, Wortel MJR (1995) Three-dimensional numerical modeling of detachment of subducted lithosphere. J. Geophys. Res. 100: 20223–20244.

Part II

Seismic Tomography and Anisotropy

Seismic Anisotropy of Subduction Zone Minerals–Contribution of Hydrous Phases

David Mainprice and Benoit Ildefonse

Abstract The seismology is the most effective method to explore the structure of subduction zones to great depth. The distinguishing feature of the mantle in the subduction regions is the presence of hydrated phases, which transport water into the Earth's interior and release it with dramatic local consequences, triggering earthquakes and melting. The seismological detection of these hydrous phases and geodynamic interpretation of flow in the hydrous mantle depend on knowledge of the anisotropic elastic properties and the characteristics of the wave propagation in anisotropic media. In this paper we briefly recall the distinguishing features of anisotropic wave propagation and the observable parameters. We suggest the Vp/Vs ratio is a physically sound parameter than can be observed by seismology in anisotropic regions of the Earth, whereas the Poisson's ratio, which is often quoted, is not directly observable and does not correspond to the characteristics of wave propagation in an anisotropic or isotropic medium. We report for the first time the ratios of Vp/Vs1 and Vp/Vs2, where Vs1 is the fastest and Vs2 slowest S-wave velocities of an anisotropic media. We present the current knowledge of the anisotropic seismic properties of hydrous minerals in the upper mantle, transition zone, and lower mantle that are stable along low temperature geotherms associated with subduction, and identify which minerals are likely to influence seismological observations because they have very high volume fractions, or very high anisotropies, or both of these.

In the upper mantle antigorite and talc are exceptionally anisotropic (Vp 71%, Vs 68% and Vp 65%, Vs 68%, respectively) and chlorite is also very anisotropic for S-waves (Vp 35%, Vs 76%). Comparatively less anisotropic are hornblende (Vp 27%, Vs 31%), used as proxy for tremolite in subduction zones, and clinohumite (21.8%, Vs 15.9%). Brucite at 4 GPa (Vp 26.5%, Vs 30.9%) and the dense hydrogen magnesium silicate (DHMS) phase A at 9 GPa (Vp 9.3%, Vs 17.6%) are the only hydrous minerals stable in the upper mantle that have had their elastic properties measured at in situ mantle pressures. Except for the phase A, all these minerals are more anisotropic than olivine for at least one parameter (Vp, Vs, Vp/Vs1, or Vp/Vs2). In the transition zone the major phases hydrous wadsleyite and ringwoodite have moderate to weak anisotropies (Vp 16.3%, Vs 16.5%, and Vp 1.9%, Vs 4.4%, respectively). The DHMS Superhydrous B has a moderate anisotropy (Vp 6.9%, Vs 11.6%). At greater depth the DHMS phase D at 24 GPa (Vp 10.8%, Vs 18.0%)

D. Mainprice
Université Montpellier 2, Géosciences Montpellier,
CNRS/INSU, UMR 5243, CC 60, Place Eugène Bataillon,
34095 Montpellier, Cedex 5, France, David.Mainprice@gm.univ-montp2.fr

S. Lallemand and F. Funiciello (eds.), *Subduction Zone Geodynamics*,
DOI 10.1007/978-3-540-87974-9,

is the only hydrous phase that can transport hydrogen from the transition zone into the lower mantle. The phase D is more anisotropic for S than P waves like many of the hydrous phases. From these data it is clear that hydrous phases are in general very anisotropic. However, pressure can play a strong role in reducing anisotropy. It is the case for brucite and talc, in which increasing pressure from ambient to 4 GPa reduces the anisotropy by about 50% for both P and S waves. In contrast, the anisotropy of the DHMS phase A does not change significantly with pressure.

Our picture of the seismic anisotropy of hydrated minerals remains incomplete; the elastic properties of many have not been measured even at ambient conditions (e.g., 10 Å phase and phase E) or not measured in their true elastic symmetry (e.g., clinochlore). The majority of hydrous minerals have not been measured at high pressure, and none related to hydrated mantle have been measured at elevated temperature. We have shown in the few cases where hydrous minerals have been measured as a function of pressure, that this variable has an important effect on the velocity distribution and in most cases reduces the degree of anisotropy, hence we would expect seismic anisotropy to play a key role in the determination of the shallow structure of subduction zones in the upper mantle.

Keywords Seismic anisotropy • Elastic properties • Hydrous minerals • Subduction

1 Introduction

Subduction zones are regions of extensive recycling of hydrated materials (sediments, hydrated oceanic crust and upper mantle) that are transformed by metamorphic processes into a series of high-pressure mineral at great depths. Many of the minerals postulated to exist at great depth in subduction zones have only been discovered in the laboratory at extreme conditions of pressure and temperature, with no natural occurrences having been reported. On the other hand minerals postulated to exist on the subduction plane (metamorphosed sediments and oceanic crust) and the hydrated forearc mantle (hydrated peridotites) are exposed on the Earth's surface by the slow rise of forearc serpentine seamounts, the relatively rapid exhumation processes of high pressure metamorphic rocks or the even more dynamic ascent of volcanic xenoliths. One group of minerals that are expected to be volumetrically important components of the hydrated mantle, which is taking part in the recycling of hydrogen at depth, are the hydrated minerals, although the contribution of anhydrous silicates is now also known to be significant (Smyth and Jacobsen, 2006; Hirschmann, 2006). Knowledge of the seismic properties of minerals at depth is obviously essential for the interpretation of seismic data in terms of petrology and the anisotropic seismic properties are vital for deduction of flow patterns in the mantle. Knowledge of the seismic properties of hydrated phases and nominally anhydrous phases containing hydrogen allows the exploration of degree of hydration of the mantle, which is clearly a dynamic process in subduction zones with implications for seismogenesis (e.g., Seno et al., 2001; Hacker et al., 2003b; Jung et al., 2004), magma generation (e.g., Ulmer and Trommsdorff, 1995) and metamorphic transformations (e.g., Hacker et al., 2003a). In this paper we will investigate the anisotropic elastic properties of hydrated phases and nominally anhydrous phases containing hydrogen and the implications for seismic properties.

Seismology measures the bulk elastic properties of the Earth, with contributions from anelastic and viscoelastic properties. The bulk elastic properties are controlled to first order by the volumetrically dominant elements and hence in this study of seismic properties of subducted rocks it is natural to consider first minerals that have volume fractions of more than 10%. For seismic anisotropy the same reasoning holds, but if a mineral is exceptionally anisotropic, the classical example is layered silicates (e.g., biotite) in crustal rocks, then even volume factions of less than 10% can have an important effect on the rock properties. The degree of crystallographic preferred orientation also has a major influence on the contribution of anisotropic mineral to overall properties of a rock. In the case of mica minerals they have very strong shape and mechanical anisotropy, and hence they tend to be strongly oriented in almost all rock types from undeformed sediments to highly deformed metamorphic rocks. In the context of the strong localised flow field in subduction zones, it is highly likely that most rocks are plastically deformed to some degree and the minerals have crystallographic preferred orientation (CPO).

From a mineral physics point of view the input to a subduction zone can be separated into minerals of the thin veneer of sediments, hydrated oceanic crust, and upper mantle. The temperature of the subducting plate, controlled essentially by its age and subduction rate, mediate the metamorphic processes that produce a sequence of transformations of hydrous minerals (e.g., Arcay et al., 2005; Fumagalli and Poli, 2005). Theses minerals subsequently release fluids as the temperature of the slab rises, that hydrate the surrounding mantle and trigger flux melting (e.g., Ulmer and Trommsdorff, 1995) (Fig. 1), and rising melt reduces seismic velocities (e.g., Vp −6%) in the mantle wedge (Nakajima et al., 2001). The process of a series of linked transformations of hydrous phases with overlapping stability fields of decreasing water content

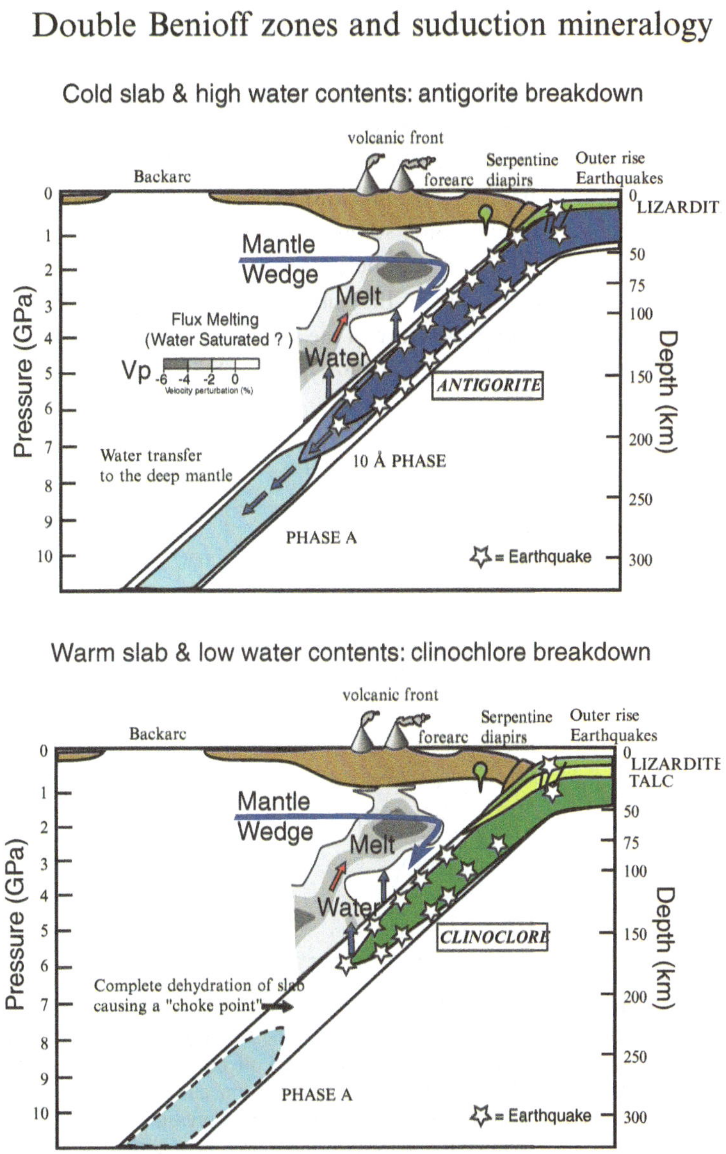

Fig. 1 Schematic cross-section of a subduction system with the double Benioff zones and their associated mineralogy in cold and warm suduction, simplified after the scenarios of Fumagalli and Poli (2005) and Poli and Schmidt (2002). The Vp seismic velocities are from Nakajima et al. (2001) in NE Japan, considered to be a cold subduction. The arrow in the Mantle Wedge represents the corner flow direction

with depth can continue to depths of the transition zone and the lower mantle (e.g., Kawamoto et al., 1996). When slab temperatures are sufficiently low (e.g., N.E. Japan), when old plates are subducted, antigorite is stable over a significant depth range in the slab (Fig. 1), and its breakdown will trigger earthquakes in the form of a double Benioff zone (e.g., Fumagalli and Poli, 2005; Brudzinski et al., 2007). At greater depth the dense hydrous magnesium silicate (DHMS) called the phase D could be preserved to 1,200-km depth, being then the only hydrated phase capable of recycling hydrogen into the lower mantle (Fig. 2). Alternatively, in warmer subduction zones the hydrous phases may be completely de-hydrated at some critical depth, called a choke point (Kawamoto et al., 1996), resulting in the breaking of the chain of hydrous phases with depth and the failure to transfer hydrogen to greater depth. Fumagalli and Poli (2005) suggest this could happen at 200-km depth if the overlap in stability fields between the 10 Å and phase A is not maintained by low temperature and water saturation. In this case it is rather the breakdown of clinochlore with a higher temperature stability that causes the double Benioff zone. Hence the presence of hydrous phases in the deeper part of the subduction zone will depend on the past temperature of the slab at shallower depths necessary to maintain the hydrous chain, and

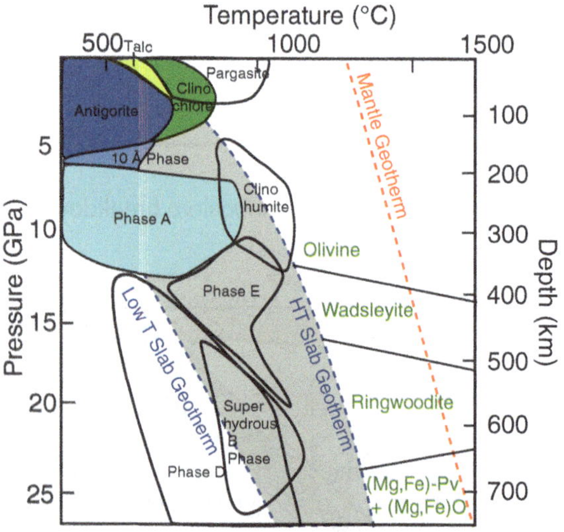

Fig. 2 Pressure-Temperature diagram with the main hydrous and anhydrous phases in subduction zones. The stability fields of hydrous phases reported by Schmidt and Poli (1998) and Fumagalli and Poli (2005) for the low-pressure region, and those of Iwamori (2004) and Ohtani et al. (2004) for high-pressures. Geotherms for low and high temperature slabs, and the mantle are taken from Peacock (1990)

of course its current temperature for the phase stability of any given hydrous mineral.

Estimates of the petrology of subducted material have been made based on rocks exposed at the surface, samples dredged from the ocean floor (e.g., Michibayashi et al., 2007) and xenoliths (e.g., Michibayashi et al., 2006), thermodynamical calculations (e.g., Hacker et al., 2003a; Bousquet et al., 2005, 1997) and laboratory experiments (e.g., Ohtani et al., 2004; Iwamori, 2004; Fumagalli and Poli, 2005). In general some combination of all three approaches is effectively incorporated in these studies to different degrees, resulting is some variations in the volume fractions reported by different authors. For the Earth's mantle we have illustrated the positions of the main anhydrous and hydrous phases in pressure-temperature space together with the mantle and subduction zone geotherms in Fig. 2 using the stability fields of hydrous phases reported by Schmidt and Poli (1998) and Fumagalli and Poli (2005) for the low pressure region, and those of Iwamori (2004) and Ohtani et al. (2004) for high pressures.

The hydrous phases can be divided into 3 main groups with increasing depth;

1. Low pressure ($P < 5$ GPa) minerals, such as antigorite (13.0 wt% H_2O), clinochlore (13 wt% H_2O), talc (4.8 wt% H_2O), Ca-amphibole tremolite or pargasite (2.3 wt% H_2O) are commonly observed in exposed metamorphic rocks.
2. Moderate pressure (5–7 GPa) minerals, such as phlogopite (4.8 wt% H_2O), 10Å phase (10–13 wt% H_2O), clinohumite (2.8 wt% H_2O) in hydrated peridotites, lawsonite (11.5 wt% H_2O) and zoisite (2.0 wt% H_2O) in hydrated metamorphosed basalts and potassium rich phengite (4.6 wt% H_2O) in metamorphosed sediments.
3. High pressure (>7 GPa) minerals such as the k-amphibole richterite (2.1 wt% H_2O), topaz-OH (10.0 wt% H_2O), phase Egg (11–18 wt% H_2O) and the DHMS or alphabet phases, phase A (12 wt% H_2O), phase B (3 wt% H_2O), phase superhydrous B (2 wt% H_2O), phase D (10 wt% H_2O), and phase E (11 wt% H_2O), most of which have never observed at the Earth's surface.

The general trend there will be a decrease in the water content in the hydrated slab with depth as the hot surrounding mantle raises the slab temperature by thermal conduction, causing the breakdown and reduction of the volume faction of hydrous minerals, and finally releasing water to the adjacent mantle. As a word of caution it should be said that the heterogeneous water distribution, which is an intrinsic feature of the subduction zone system, implies the presence or the absence of hydrated minerals, as well as the occurrence of both water saturated and water undersaturated conditions. The stability of a hydrated mineral implies neither the presence of free water phase nor water saturation (Poli and Schmidt, 2002). The relatively low temperature in subducted slabs favours the formation of hydrous minerals that contain large amounts of water as hydroxyl groups (e.g., antigorite, chlorite, phases E and D), which increases the likelihood of water under saturated assemblages. The appearance of a fluid is related to the thermodynamic definition of saturation and controlled by assemblages buffering the chemical potential of water (e.g., Bolfan-Casanova, 2005); the relationships between the amount of water and the presence of free water are not straightforward. Experiments are often conducted in water-saturated conditions, with free water present (e.g., Fumagalli and Poli, 2005; Schmidt and Poli, 1998), whereas studies of natural samples often infer water undersaturated assemblages, particularly at very high pressure (e.g., Chinner and Dixon, 1973).

In what follows we will develop tensor methods to assess the contribution of these minerals to seismic anisotropy and their potential for giving diagnostic seismic signals to determine the degree of hydration in the mantle.

2 The Relationship Between Elastic and Seismic Anisotropy

There are two types of elastic waves, which propagate in an isotropic homogeneous elastic medium, the faster compressional (or longitudinal) wave with displacements parallel to propagation direction, and the slower shear (or transverse) waves with displacements perpendicular to the propagation direction. In anisotropic elastic media there are three types, one compressional and two shear waves with in general three different velocities (Fig. 3). In order to understand the displacements associated with different waves, and their relationship to the propagation direction and elastic anisotropy, it is important to consider the equation of propagation of a mechanical disturbance in an anisotropic elastic medium. The solution of elastodynamical equation for the displacement of monochromatic plane wave is given by Christoffel equation as

$$\det |\mathbf{C}_{ijkl}\,\mathbf{n}_j\,\mathbf{n}_l - \rho V^2\,\delta_{ik}| = 0$$

where $\mathbf{C}_{ijkl}$ is elastic stiffness fourth rank tensor, $\mathbf{n}_j$ the propagation unit vector, V are the three phase velocities, ρ is density, and δ_{ik} is the Kronecker delta. We can simplify this equation by introducing the Christoffel (Kelvin-Christoffel or acoustic) tensor $\mathbf{T}_{ik} = \mathbf{C}_{ijkl}\,\mathbf{n}_j\,\mathbf{n}_l$ and three wave moduli $M = \rho V^2$. The Christoffel tensor is symmetric because of the symmetry of the elastic constants, and hence the eigenvalues of the 3 × 3 Christoffel tensor are three positive real values of the wave moduli (M) corresponding to ρVp^2, $\rho Vs1^2$, $\rho Vs2^2$ of the plane waves propagating in the direction n. The three eigenvectors of the Christoffel tensor are the polarization directions (also called vibration, particle movement or displacement vectors) of the three waves. As the Christoffel tensor is symmetric the three eigenvectors (and polarization) vectors are mutually perpendicular. In the most general case there are no particular angular relationships between polarization directions ($\mathbf{p}_k$) and the propagation direction ($\mathbf{n}_j$). However, typically the P-wave polarization direction is nearly parallel and the two S-waves polarizations are nearly perpendicular to the propagation direction, and they are termed quasi-P or quasi-S waves to distinguish them from the P and S of isotropic media (see Fig. 2).

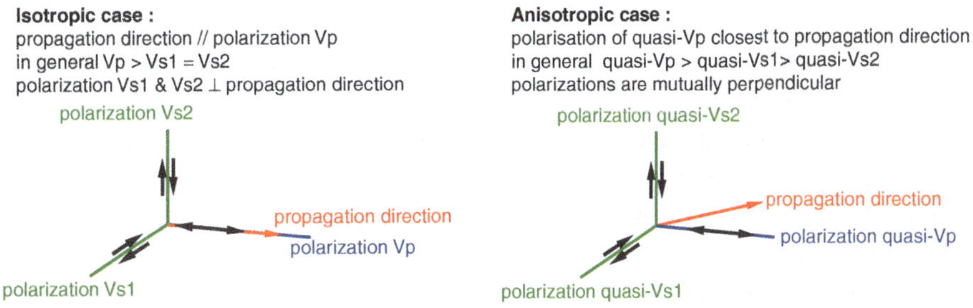

Fig. 3 Relation between polarisation and propagation directions of waves in isotropic and anisotropic elastic media

In general, the three waves have polarizations that are perpendicular to one another and propagate in the same direction with different velocities, with Vp > Vs1 > Vs2. Computer methods for calculation seismic wave velocities in anisotropic are given by Mainprice (1990).

In the case of subduction zones it has become standard practice to use the isotropic Vp/Vs ratio or Poisson's ratio to characterise the seismic response related to hydrated minerals (e.g., serpentine group minerals, Christensen, 1996, 2004), or the presence of dehydration fluids (e.g., Ito, 1990; Kono et al., 2007), or as yet undetermined minerals with low Vp/Vs (e.g., Rossi et al., 2006). Classically for an isotropic medium there is only one Vp/Vs ratio as the velocity is the same for all propagations directions and there is only one Vs velocity. In seismology the Poisson's ratio (ν) has been calculated form the classical isotropic relationship

$$\nu = \frac{1}{2}\left[\frac{(V_p/V_s)^2 - 2}{(V_p/V_s)^2 - 1}\right]$$

However, in anisotropic elastic media there are three velocities (Vp, Vs1, Vs2) and hence two ratios Vp/Vs1 and Vp/Vs2, and of course these ratios vary with direction. The anisotropic Poisson's ratio (ν_{ijkl}) varies with direction in a complex manner. Poisson's ratio is defined by the elastic strain in two orthogonal directions, the longitudinal (or axial) direction ($\mathbf{x}_i$) and transverse (or lateral) direction ($\mathbf{y}_i$). The lateral strain is defined by $-\boldsymbol{\varepsilon}_{ij}\mathbf{y}_i\mathbf{y}_j$ along $\mathbf{y}$ and the longitudinal strain by $\boldsymbol{\varepsilon}_{ij}\mathbf{x}_i\mathbf{x}_j$ along $\mathbf{x}$, where $\boldsymbol{\varepsilon}_{ij}$ is the infinitesimal strain tensor. The anisotropic Poisson's ratio ($\boldsymbol{\nu}_{ijkl}$) is given as the ratio of lateral to longitudinal strain (Sirotin and Shakolskaya, 1982) as

$$\nu_{ijkl} = -\frac{\varepsilon_{ij}\mathbf{y}_i\mathbf{y}_j}{\varepsilon_{kl}\mathbf{x}_k\mathbf{x}_l} = -\frac{S_{ijkl}\mathbf{x}_i\mathbf{x}_j\mathbf{y}_k\mathbf{y}_l}{S_{mnop}\mathbf{x}_m\mathbf{x}_n\mathbf{x}_o\mathbf{x}_p}$$

where $\mathbf{S}_{ijkl}$ is the elastic compliance fourth rank tensor in 4 index notation. To compare $\boldsymbol{\nu}_{ijkl}$ to seismic wave propagation we could choose the axial and transverse strain directions to be parallel to the orthogonal P- and S-wave polarization directions, respectively. However, this analogy is clearly not exact, as the strains associated with seismic waves in an elastic medium require shear strain components not present in the definition of Poisson's ratio. Hence, we can find no sound physical reason for using Poisson's ratio in the characteristic seismic properties of anisotropic or even isotropic media when the more direct observation of Vp/Vs ratios for an isotropic medium and Vp/Vs1 plus Vp/Vs2 for anisotropic medium have a clear physical meaning.

3 Seismic Anisotropy of Hydrated Phases

The anisotropic seismic properties of hydrous phases shown in Fig. 2 are estimated to have volume fractions well above 10% and we will use these to illustrate the anisotropic behaviour of hydrous minerals in rocks of peridotite composition in subduction zones.

3.1 Upper Mantle Phases

3.1.1 Above 180 km Depth: Low Temperature Serpentine Minerals

First, we will consider the case of serpentine polymorphs, i.e., lizardite, chrysotile, and antigorite. Although there is no direct information on the amount of water in the peridotite layer of slabs, it should contain water due to serpentinization or chloritization. Serpentine is commonly observed on the ocean floor near fracture zones, in particular at slow-spreading ridges (e.g., Lagabrielle et al., 1998; Mével, 2003; Ildefonse et al., 2007). Schmidt and Poli (1998) estimated that the upper 5 km of the peridotite layer suffers about 10% serpentinization; Ranero and Sallarès (2004) estimated about 17% of serpentinization in the upper 20 km of the peridotite layer of the Nazca plate that is subducted at the North Chile trench. Lizardite is volumetrically the most important serpentine mineral in serpentinites, especially retrograde metamorphosed (hydrated) peridotites. The low temperature serpentine minerals chrysotile and lizardite occur widely in serpentinites, commonly together, and correspond to inferred temperature conditions ranging from the surface up to perhaps as high as 400°C (Evans, 2004). The low temperature serpentine minerals are observed to form by hydration on sea floor of abyssal peridotites at mid-ocean ridges under static conditions (e.g., Mével, 2003), and are particularly abundant in the thick lithosphere of slow spreading ridges (e.g., Cannat, 1993; Lagabrielle et al., 1998). Although controversial, it appears that chrysotile is meta-stable in the laboratory and eventually transforms to lizardite (e.g., Grauby et al., 1998; Evans, 2004). The presence of the low temperature serpentine minerals in the hydrothermally altered oceanic crust and lithosphere lowers the density and seismic velocities (e.g., Carlson and Miller, 1997; Christensen, 2004).

In the subducted oceanic lithosphere the major release of water in fluid-saturated systems is caused by the breakdown of antigorite in cold slabs (Ulmer and Trommsdorff, 1995), with an associated reduction in P-wave velocity of 5.1% (Kono et al., 2007). Antigorite is expected to be the serpentine mineral stable at relatively higher temperatures and much higher pressures of the upper mantle (Evans et al., 1976), for example in the forearc upper mantle (e.g., Kamiya and Kobayashi, 2000; Bostock et al., 2002; Brocher et al., 2003; Hyndman and Peacock, 2003). In particular, the dehydration of antigorite has been associated with the formation of double seismic zones (e.g., Peacock, 2001; Hacker et al., 2003b), which are know to be global phenomena (Brudzinski et al., 2007). As pointed out by Christensen (2004) and Watanabe et al. (2007), much of the interpretation of wedge mantle seismic velocities in terms of the volume fraction of serpentine (e.g., Carlson and Miller, 2003) has been made by incorrectly using an empirical relationship between volume fraction of low temperature lizardite-chrysotile serpentinites with Vp, Vs and density established by Christensen (1966), rather than stable high-pressure phase antigorite. Christensen (1978, 2004), Kern (1993), Kern et al. (1997), and Watanabe et al. (2007) have all studied samples with clearly identified high antigorite volume fractions. These samples show that antigorite has higher Vp, Vs and lower Vp/Vs ratio than lizardite-chrysotile serpentinites. In addition Kern and Tubia (1993), Horen et al. (1996) and Song et al. (2004) also report ultrasonic measurements of serpentine samples that have transverse anisotropy, but do not identify which serpentine minerals are present. Both Christensen (2004) and Watanabe et al. (2007) have produced new empirical relationships between volume fraction of high temperature antigorite serpentinites to Vp, Vs and density. Kern (1993) and Kern et al. (1997) reports a strong crystal preferred orientation of c-axes with over three times random densities normal to the foliation. Watanabe et al. (2007) report a strong foliation, probably indicative of CPO, in 50% of their antigorite samples. There are no single crystal measurements of the elastic tensor of serpentine minerals due to the difficulty in finding natural single crystals of sufficient size and quality. Recently Auzende et al. (2006) and Pellenq et al. (2008) have used transferable semi-empirical interatomic potentials to compute the elastic tensor of single crystal lizardite and antigorite, respectively, at ambient pressure. The lizardite elastic constants of Auzende et al. (2006) predict linear compressibility for a- and c-axes of 2.3×10^{-3} and 20.5×10^{-3} GPa^{-1} are in reasonable agreement with the experimentally measured values of 2.7×10^{-3} and 9.7×10^{-3} GPa^{-1} reported by Hilairet et al. (2006), being good for the a-axis, but two times too high for the c-axis. The antigorite elastic constants of Pellenq et al. (2008) predict linear compressibility for a- and b-axes of 2.3×10^{-3} and 2.2×10^{-3} GPa^{-1}, which agree well with values of 3.7×10^{-3} and 3.3×10^{-3} GPa^{-1} reported by Hilairet et al. (2006), however the predicted value for the c-axis 17.9×10^{-3} GPa^{-1} is again twice the measured value of 8.2×10^{-3} GPa^{-1}. Reynard et al. (2007) have used *ab-initio* methods to calculate the elastic properties of pseudo-hexagonal lizardite, but comparisons between experimentally measured compressibility along the c-axis with the predicted compressibility is five times greater than measured, and hence they are not acceptable to predict seismic properties. Both the model elastic constants of Auzende et al. (2006) and Pellenq et al. (2008) predict very strong anisotropy for Vp (76.7% and 71.2%) and dVs (81.7 and 67.6%) with high Vp velocities and high shear wave splitting (dVs) anisotropy in the basal plane for lizardite and antigorite, respectively. The polarization of the fastest S-waves (S1) is parallel to the basal plane in both polymorphs. All the features in Fig. 4 are very characteristic of layer silicates. The Vp/Vs1 and Vp/Vs2 ratios for lizardite and antigorite are shown in Fig. 5, they have the general pattern of high values in the basal plane and low values around the c-axis. In detail, the Vp/Vs1 is more complex with small region of high Vp/Vs1 around the c-axis. Overall, lizardite and antigorite are characterised by high Vp/Vs1 and Vp/Vs2 for propagation directions around the basal plane. The velocity ratios have high anisotropies, 36.1–41.2% for Vp/Vs1 and 92.8–100.4% for Vp/Vs2 for antigorite and lizardite, respectively. The values of the velocity ratios vary from 1.25 to 4.30 with direction.

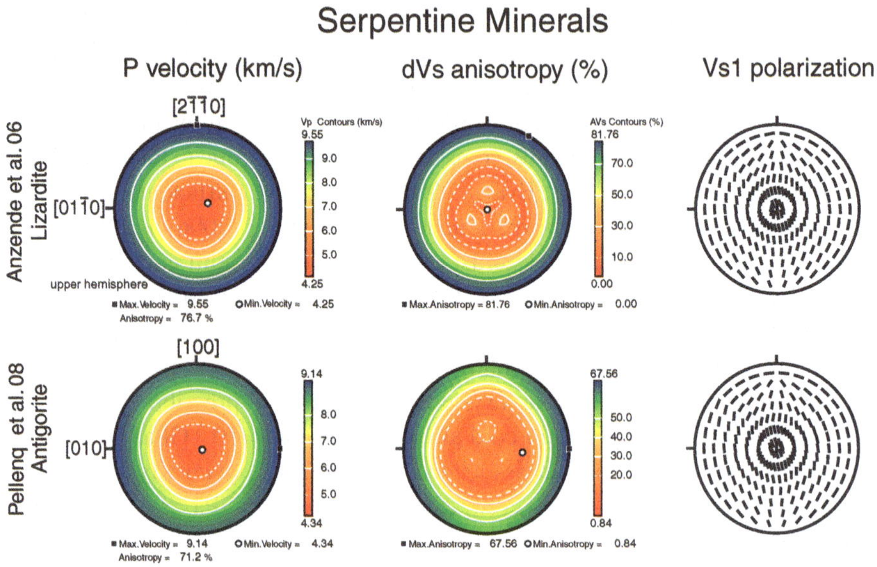

Fig. 4 Upper hemisphere pole figures of P-wave velocity, S-wave splitting anisotropy (dVs as a percentage) and the polarization direction (white lines) of the fastest S-wave (Vs1) for single crystal lizardite and antigorite, calculated using the elastic tensors reported by Auzende et al. (2006) and Pellenq et al. (2008) at ambient pressure. The c-axis is in the centre of the pole figures, other crystallographic reference directions in the basal plane are marked

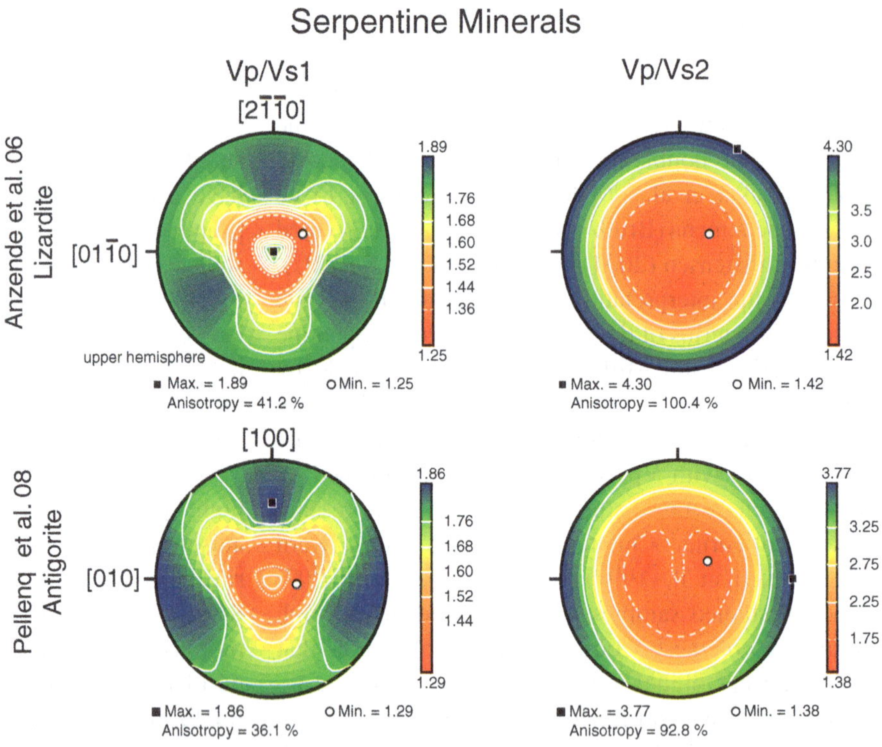

Fig. 5 Upper hemisphere pole figures of Vp/Vs1, Vp/Vs2 for single crystal lizardite and antigorite, calculated using the elastic tensors reported by Auzende et al. (2006) and Pellenq et al. (2008) at ambient pressure. The c-axis is in the centre of the pole figures, other crystallographic reference directions in the basal plane are marked

The Voigt-Reuss-Hill average isotropic values of Vp/Vs for Auzende et al. (2006) and Pellenq et al. (2008) elastic constants are 1.77 and 1.73, respectively. For reference, the experimentally measured Vp/Vs values given by Christensen (2004) at 1 GPa pressure for isotropic lizardite and antigorite are 2.17 and 1.87, respectively; the Vp/Vs ratio is higher in lizardite than antigorite in both the calculated and experimental values.

3.1.2 Above 180 km Depth: Minerals Stable at Higher Temperature

Several of the minerals present in the first 180-km depth range of subduction zones are stable to higher temperatures than antigorite. Among the more important minerals likely to be stable in warm slabs are talc, brucite, clinochlore and tremolite-pargasite.

Talc ($Mg_3Si_4O_{10}(OH)_2$) is of great geological importance having a widespread occurrence from the surface of the seafloor to depths of 150 km in subduction zones. Perhaps, it is best known to form during the serpentinization of abyssal peridotites at mid-ocean ridges (e.g., Cannat, 1993; Mével, 2003) being associated with alteration of orthopyroxene to serpentine + talc, where talc can represent up to 41% by volume of the rock (e.g., Hacker et al., 2003a). At the Mid-Atlantic Ridge, talc locally represents 100% of ODP cores drilled at 14°50′N (Shipboard Scientific Party, 2004). It also a mineral commonly sampled in detachment faults associated to oceanic core complexes (e.g., MacLeod et al., 2002; Escartín et al., 2003; Boschi et al., 2006). In subduction zones, it forms in the mantle wedge from the breakdown of antigorite to give talc + forsterite + water (e.g., Evans et al., 1976), this reaction has recently be found to explain the global pattern of seismicity that displays double Benioff zones (Brudzinski et al., 2007). At greater depths, talc releases water in the dehydration reaction to enstatite at *ca* 1.7 GPa 660°C (60-km depth) (e.g., Ulmer and Trommsdorff, 1995). Talc should accordingly be present from shallow depths to 60 km. However talc can still be present at depths of 100 km or more, where it forms in metamorphosed oceanic crust at ultra-high pressure conditions during the hydration of mafic eclogites, even at water under-saturated conditions lacking a free fluid phase (Poli and Schmidt, 1997). The elastic constants of talc have recently been determined by ab initio methods by Mainprice et al. (2008). At ambient pressure talc is perhaps the most elastically anisotropic mineral with P-wave anisotropy of 80% and S-wave anisotropy of 85%. At a representative pressure of its occurrence in a subduction zone of 0.9 GPa (40-km depth) the P-wave anisotropy is 64.9% and S-wave anisotropy is 65.1% (Fig. 6). The Vp/Vs1 and Vp/Vs2 ratios have anisotropies of 45.7% and 95.2%, respectively. The velocity ratios vary from 1.05 to 3.18 with direction.

Brucite $Mg(OH)_2$ is interesting as it is a model system for understanding dense hydrous magnesium minerals (alphabet phases) under hydrostatic compression and an important structural unit of many sheet silicates, such as chlorite, lizardite and talc. It is occasionally a major phase in low temperature and pressure oceanic serpentinites, as at Hess Deep (e.g., Mével and Stamoudi, 1996), or at the Mid-Atlantic Ridge (e.g., Bach et al., 2004, 2006). At higher pressures brucite commonly occurs in association with antigorite (Hostetler et al., 1966) at upper mantle PT conditions shown in Fig. 2, and could constitute 13% and 7% by volume of a hydrated harzburgite and lherzolite, respectively, according to Hacker et al. (2003a). Brucite is a very anisotropic phase at room pressure, however at pressures approximating to the in-situ conditions ($P = 4$ GPa) in the upper mantle the anisotropy is reduced, but still very high at 26.5% and 30.9%

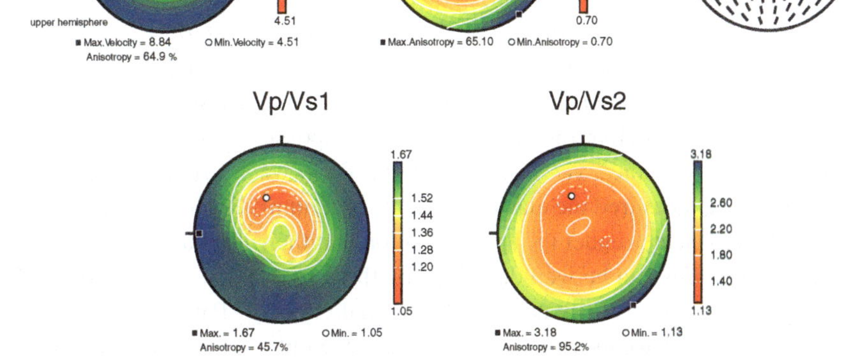

Fig. 6 Upper hemisphere pole figures of P-wave velocity, S-wave splitting anisotropy (dVs as a percentage), the polarization direction (white lines) of the fastest S-wave (Vs1), Vp/Vs1 and Vp/Vs2 for single crystal talc triclinic ($c\bar{1}$) symmetry, calculated using the elastic tensor by Mainprice et al. (2008) at a pressure of 0.9 GPa. The c-axis is in the centre of the pole figures, other crystallographic reference directions in the basal plane are marked

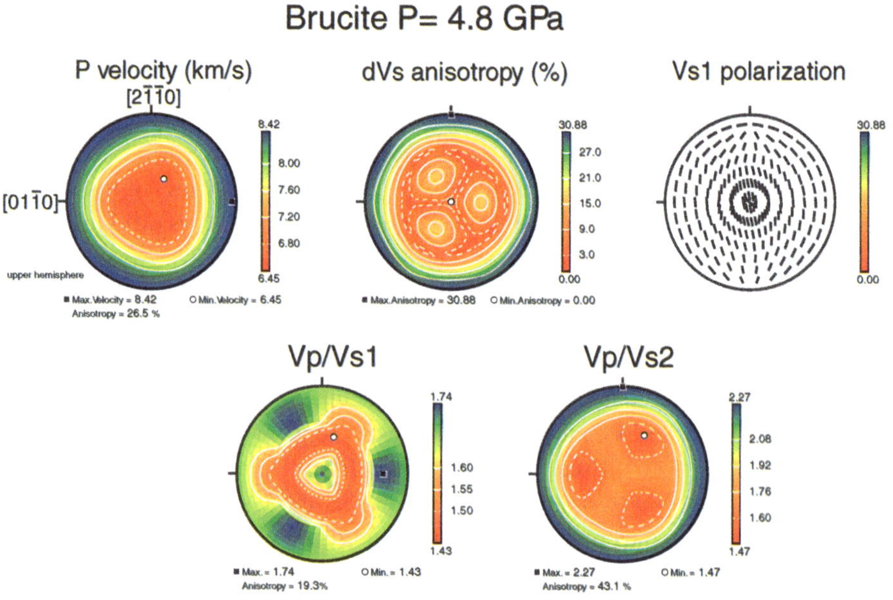

Fig. 7 Upper hemisphere pole figures of P-wave velocity, S-wave splitting anisotropy (dVs as a percentage), the polarization direction (white lines) of the fastest S-wave (Vs1), Vp/Vs1 and Vp/Vs2 for single crystal brucite (trigonal symmetry P-3m1), calculated using the elastic tensor by Jiang et al. (2006) at 4.8 GPa pressure. The c-axis is in the centre of the pole figures, other crystallographic reference directions in the basal plane are marked

for Vp and Vs, respectively (Fig. 7). Brucite like other layer structures has the highest value of Vp/Vs2 in the basal plane of 2.27 and anisotropies of Vp/Vs1 and Vs2 of 19.3% and 43.1% that is considerably less than antigorite at room pressure. For reference the Voigt-Reuss-Hill average isotropic of Vp/Vs ratio for Brucite is 1.68.

The clinochlore end member of the chlorite series is a strong candidate to be present in the hydrated upper mantle. Experimental evidence reported by Fumagalli and Poli (2005) for the presence of clinochlore in hydrated lherzolite and harzburgite at 20–40%wt fractions in the pressure range up to 6 GPa and temperatures around 700°C make clinochlore a possible alternative to antigorite to explain mantle wedge seismic anisotropy. Clinochlore is the product of breakdown of tremolite or pargasite in hydrous peridotites (Fumagalli and Poli, 2005). The temperature dependent overlap of stability fields of clinochlore and the 10 Å phase has major implications for the transfer of H_20 to higher pressures phases at depth (Fumagalli and Poli, 2005). However, the elastic constants of the 10 Å phase have not yet been measured, so that seismic interpretation of such a transfer is at present impossible in the anisotropic case. Alexandrov and Ryzhova (1961a) have measured the elastic constants of a number of layer silicates under the approximation that their elastic properties can be well described by the 5 independent constants corresponding to hexagonal crystal symmetry. They have measured phlogopite for example, however we do not illustrate the seismic properties of this mineral, as it is stable at higher temperatures than commonly associated with subduction. In the case of clinochlore, they only managed to measure ultrasonic velocities corresponding to 4 elastic constants (C_{11},C_{33},C_{44} and C_{66}), but as $C_{12} = C_{11}-2C_{66}$ we have the value for C_{12}. However, the value for C_{13} has not been measured. To estimate the values of C_{13} we used the systematics of all Cij for layer silicates reported by Alexandrov and Ryzhova (1961a). A plot of Cij/density versus Cij in order of magnitude (C_{11}, $C_{66,}$ C_{33}, C_{12}, C_{13} and C_{44}) for muscovite, phlogopite, biotite and clinochore shows that the values of C_{13} lie between C_{12} and C_{44} and simple linear interpolation gives $C_{13} = C_{44} + (C_{12}-C_{44})/2 = 40.2$ GPa for clinochlore. The elastic constant of clinochlore can be further verified by comparing the experimentally measured linear compressibility (β) along the a- and c-axes given by Pawley et al. (2002), $\beta_a = 3.0 \times 10^{-3}$ GPa^{-1} and $\beta_c = 5.1 \times 10^{-3}$ GPa^{-1}, with values calculated from Cij of clinochlore of $\beta_a = 2.7 \times 10^{-3}$ GPa^{-1} and $\beta_c = 8.1 \times 10^{-3}$ GPa^{-1}, which are very close for the a-axis and slightly too high for the c-axis. We calculated values of linear compressibility from the unit cell data of Welch and Crichton (2002) using $\beta a = (1-a/a_0)/P$ where a is the cell edge at pressure P and a_0 is the cell length at 10^{-4} GPa pressure, and obtained $\beta_a = 3.0 \times 10^{-3}$ GPa^{-1} and $\beta_c = 5.0 \times 10^{-3}$ GPa^{-1}, very similar to the results of Pawley et al. (2002). To obtain a better match between the linear compressibilities would require adjusting

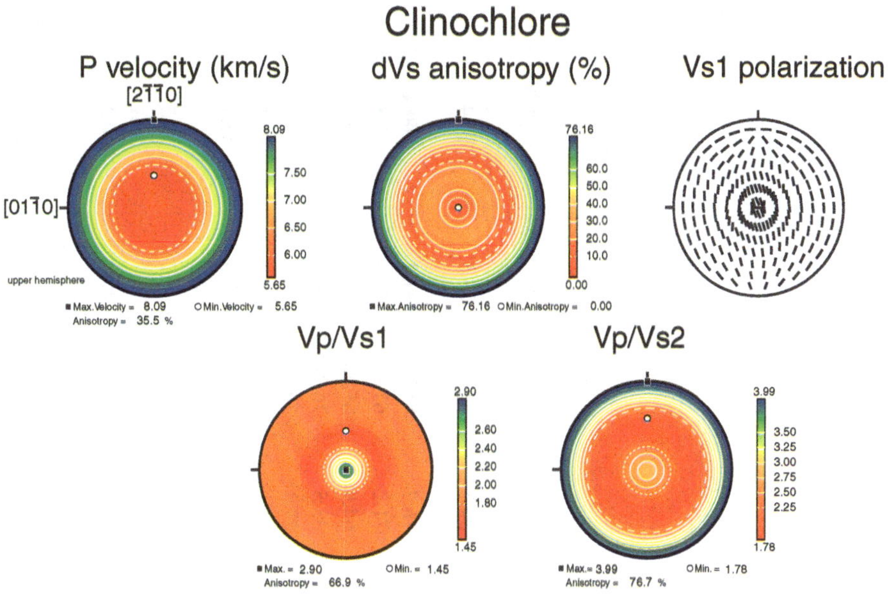

Fig. 8 Upper hemisphere pole figures of P-wave velocity, S-wave splitting anisotropy (dVs as a percentage) and the polarization direction (white lines) of the fastest S-wave (Vs1) for single crystal clinochlorite, calculated using the constants C_{11}, C_{33}, C_{44}, C_{66} and C_{12} reported by Alexandrov and Ryzhova (1961a) and C_{13} estimated by the authors, as described in the text. The elastic tensor constructed here has the hexagonal symmetry rather than the monoclinic (C2/m) symmetry of clinochlore (*see text*). The c-axis is in the centre of the pole figures, other crystallographic reference directions in the basal plane are marked

C_{13} and other Cij (see Nye, 1957) that are constrained by the ultrasonic data, but further analysis is complicated by the fact that Alexandrov and Ryzhova (1961a) do not give the composition of their crystal, hence more elaborate constraint on the elastic constants is not justified. Calculated clinochlore seismic properties (Fig. 8) have a similar pattern of anisotropy to antigorite, typical of layered silicates, with high Vp and very strong shear wave splitting in the basal plane, but with a higher symmetry due to hexagonal symmetry of the clinochore elastic constants. The P-wave anisotropy is significantly lower than antigorite at 35.5%, but the shear wave splitting is as strong at 76.2%. The Vp/Vs distribution is similar that the lizardite and antigorite in Fig. 5. The anisotropy of Vp/Vs is high at 66.9% and 76.7% for Vp/Vs1 and Vp/Vs2, respectively, and similar in magnitude to antigorite. The highest values of Vp/Vs2 are over 3.50. For comparison the Voigt-Reuss-Hill average isotropic values of Vp/Vs are 2.02, which is the highest of the all the hydrous phases in this study (Table 1).

In the upper mantle the amphiboles most likely to occur in hydrated peridotites are tremolite-actinolite or pargasite (e.g., Fumagalli and Poli, 2005). Hacker et al. (2003a) have estimated volume fractions of tremolite to be between 19–34% for hydrated lherzolite. The destabilization of tremolite produces clinochlore in water saturated peridotites (Fumagalli and Poli, 2005), hence it is restricted to about 2.5 GPa (80 km) in the shallow forearc, but it can be stable to much higher temperatures than most other hydrous phases. As the elastic constants of tremolite have never been measured we use the ‘generic’ amphibole hornblende to illustrate the probable anisotropy of this phase. The elastic properties of hornblende have been measured by Alexandrov and Ryzhova (1961b). The monoclinic symmetry of hornblende with a mirror plane normal to [010] is clearly displayed by its seismic properties in Fig. 9. Hornblende is very anisotropic for non-layer structure with a P-wave and S-wave anisotropy of 27.1% and 30.7%, respectively. The Vp/Vs anisotropy is also very strong with 24.3% and 39.2% for Vp/Vs1 and Vp/Vs2, respectively.

3.1.3 From 180 to 400 km Depth: Minerals Stable at Higher Pressure

During subduction, dehydration of antigorite from about 120 km, depending on the temperature, produces forsterite + enstatite ± chlorite + fluid. Between 120 and 180 km, hydration of forsterite produces the hydrous mineral clinohumite (Stalder and Ulmer, 2001). At greater depths it is usually assumed that clinohumite transforms to the phase A. However, clinohumite is stable in the presence of hydrogen containing forsterite (hydrous olivine) + clinoenstatite ± melt + H_2O at 12 GPa (<360 km) (Smyth et al., 2006).

Table 1 Isotropic and anisotropic parameters of selected hydrous phases

Mineral	Density (kg/m^3)	Shear Mmodulus (GPa)	Bulk Mmodulus (GPa)	Poisson's ratio	Vp (km/s)	Vs (km/s)	Vp/Vs	AVp (%)	AVs (%)	AVp/Vs1 (%)	AVp/Vs2 (%)
Olivine reference Mineral	3355	78.01	129.45	0.25	8.34	4.82	1.73	**24.3**	**17.9**	**25.5**	**16.4**
Lizardite	2580	33.63	60.90	0.27	6.30	3.56	1.77	**76.7**	**81.7**	**41.2**	**100.4**
Antigorite	2720	37.15	62.43	0.25	6.36	3.67	1.73	**71.2**	**67.6**	**36.1**	**92.8**
Talc $P = 0.9$ GPa	2858	49.21	58.57	0.17	6.56	4.13	1.59	**64.9**	**65.1**	**45.7**	**95.2**
Clinochlore	2800	29.27	79.25	0.34	6.48	3.20	2.02	**35.5**	**76.2**	**66.9**	**76.7**
Hornblende	3124	42.84	88.47	0.29	6.83	3.70	1.84	**27.1**	**30.7**	24.3	**39.2**
Clinohumite	3261	73.09	124.72	0.26	8.25	4.73	1.74	21.8	15.9	24.7	**16.7**
Brucite $P = 4$ GPa	2605	50.05	74.34	0.22	7.36	4.38	1.68	**26.5**	**30.9**	19.3	**43.1**
Phase A $P = 9$ GPa	3191	77.43	157.41	0.29	9.04	4.93	1.83	9.3	17.6	10.2	9.1
Superhydrous B	3327	96.99	153.83	0.24	9.23	5.40	1.71	6.9	11.6	18.2	10.5
Phase D $P = 24$ GPa	3934	169.86	279.28	0.25	11.33	6.57	1.72	10.8	**18.0**	20.3	**25.1**
HydroWadsleyite	3395	98.57	149.23	0.23	9.09	5.39	1.69	16.3	16.5	20.9	**20.2**
HydroRingwoodite $P = 20$ GPa	3781	137.43	250.87	0.27	10.71	6.03	1.78	1.9	4.4	4.5	5.8

Pressure is room pressure (10^{-4} GPa) except where indicated. Shear modulus, bulk modulus, Poisson's ratio, Vp, Vs and Vp/Vs are Hill averages of the Voigt and Reuss isotropic elastic bounds. AVp, AVs, AVp/Vs1, AVp/Vs2 are anisotropic parameters discussed in the text. Marked in bold the mineral names and their anisotropy parameters that are higher than olivine at ambient constants using the elastic constants of Abramson et al. (1997).

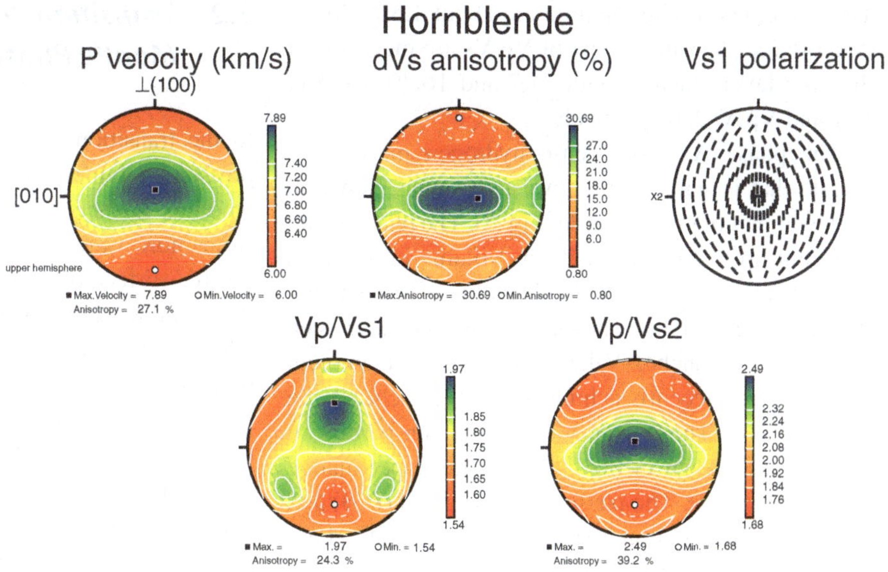

Fig. 9 Upper hemisphere pole figures of P-wave velocity, S-wave splitting anisotropy (dVs as a percentage), the polarization direction (white lines) of the fastest S-wave (Vs1), Vp/Vs1 and Vp/Vs2 for single crystal hornblende (monoclinic C2/m symmetry), calculated using the elastic tensor by Alexandrov and Ryzhova (1961b). The c-axis is in the centre of the pole figures, other crystallographic reference directions in the basal plane are marked

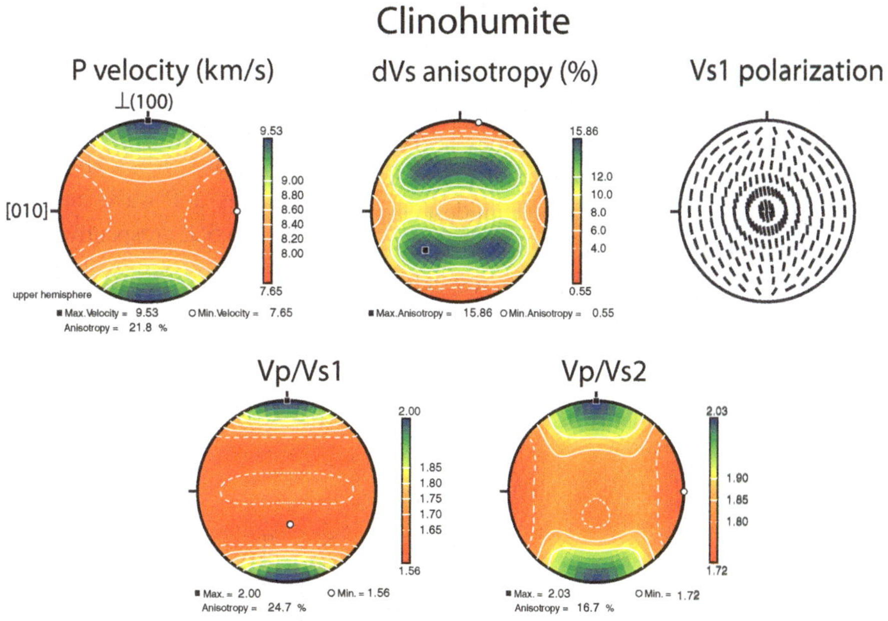

Fig. 10 Upper hemisphere pole figures of P-wave velocity, S-wave splitting anisotropy (dVs as a percentage), the polarization direction (white lines) of the fastest S-wave (Vs1), Vp/Vs1 and Vp/Vs2 for single crystal clinohumite monoclinic (C2/m symmetry), calculated using the elastic tensor by Fritzel and Bass (1997) at ambient pressure. The c-axis is in the centre of the pole figures, other crystallographic reference directions in the basal plane are marked are the pseudo-orthorhombic axes used by Fritzel and Bass (1997)

Ti-rich Clinohumite is observed in ultramafic rocks exhumed from ultrahigh-pressure conditions in Dabie Shan, China (Hermann et al., 2007). This body consists of layered mafic and ultramafic rocks ranging from harzburgites to orthopyroxenites, associated with coesite-eclogites with a peak metamorphic conditions of 4 GPa and 700°C, compatible with 120-km depth on a warm subduction geotherm. Clinohumite is monoclinic, and its elastic constants have been measured in the unusual setting ($\alpha = 100.8°$, instead of the conventional angle $\beta = 100.8°$ with $\alpha = 90°$) by Fritzel and Bass (1997) to show the similarity to olivine. In Fig. 10 one can see many features that resemble olivine, with high Vp parallel to the a*-axis and high shear wave splitting at approximately 45° to a*- and c-axes. Clinohumite has Vp and dVs anisotropies of 21.8% and 15.9%, whereas olivine has values of 24.3% and 17.9%, respectively, hence very similar to olivine.

The velocities of Clinohumite are about 1.5% slower than olivine. The anisotropy of Vp/Vs are much lower than the layer silicates with 24.7 and 16.7% for Vp/Vs1 and Vp/Vs2, respectively.

The phase A is the dense hydrogen magnesium silicate that is stable at the lowest pressures, hence in the lower part of upper mantle (Fig. 2). The phase A can represent up to 47% and 22% by volume in hydrated harzburgite and depleted lherzolite, respectively (Hacker et al., 2003a). The elastic constants have been measured by Sanchez-Valle et al. (2006) at ambient conditions and recently at high pressure (Sanchez-Valle et al., 2008). The anisotropic seismic properties of phase A are shown in Fig. 11 at in-situ upper mantle pressure of 9 GPa. The phase A is hexagonal (P63) and the velocity distribution is surprisingly different to hexagonal and trigonal symmetry elastic constants illustrated for lizardite, clinochlore and brucite. As illustrated in Mainprice (2007) the phase A has a velocity distribution similar to lizardite, clinochlore and brucite at ambient pressure, hence the role of pressure is very important in DHMS minerals, as also described for the phase D by Mainprice et al. (2007). Notice the pseudo-tetragonal symmetry, which is most obvious in dVs and Vp/Vs2 in Fig. 11. The anisotropy of the phase A at in-situ upper mantle pressure is relatively modest compared to other layered structures at 9.3%, 17.6%, 10.2%, and 9.1% for P, S, Vp/Vs1 and Vp/Vs2, respectively.

3.2 Transition Zone and Lower Mantle Phases

In the transition zone (410–670-km depth) the high-pressure polymorphs of olivine, wadsleyite and ringwoodite, in their hydrous forms have the potential to store more water as hydroxyl than the oceans on the Earth's surface (e.g., Jacobsen, 2006). Wadsleyite and ringwoodite can be synthesized with up to 3.1 wt% and 2.8 wt% H_2O, respectively. The maximum water solubility in wadsleyite at transition zone conditions (15 GPa, 1400°C) is about 0.9 wt% (Demouchy et al., 2005). In ringwoodite, water solubility decreases with increasing temperature. In subducting slabs, hydrous wadsleyite and ringwoodite can represent volume fractions of about 50–60%, resulting in the shallower region of the mantle transition zone having a larger water storage potential than the deeper region (Ohtani et al., 2004).

In the upper part of the transition zone (410–520-km depth), anhydrous wadsleyite is replaced by hydrous wadsleyite in the hydrated mantle. The elastic constants of hydrous wadsleyite have been determined at ambient conditions as a function of water content by Mao et al. (2008). Hydrous wadsleyite with 1.66 wt% H_2O has a P-wave and S-wave anisotropy of 16.3% and 16.5%, respectively (Fig. 12). This compares with anhydrous wadsleyite using the elastic constants of

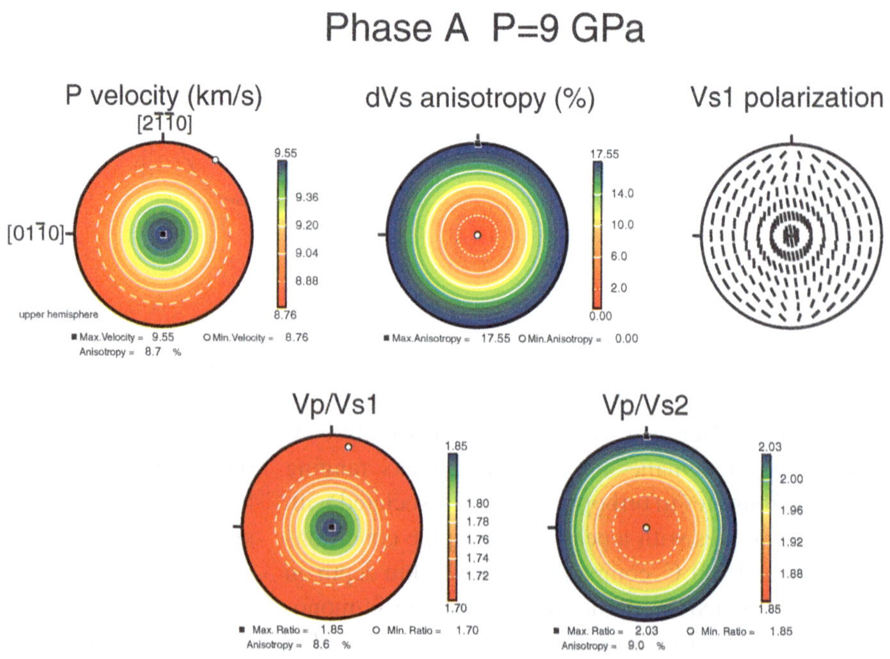

Fig. 11 Upper hemisphere pole figures of P-wave velocity, S-wave splitting anisotropy (dVs as a percentage), the polarization direction (white lines) of the fastest S-wave (Vs1), Vp/Vs1 and Vp/Vs2 for single crystal phase A (hexagonal symmetry P63), calculated using the elastic tensor Sanchez-Valle et al. (2008) at 9.0 GPa. The c- axis is in the centre of the pole figures, other crystallographic reference directions in the basal plane are marked

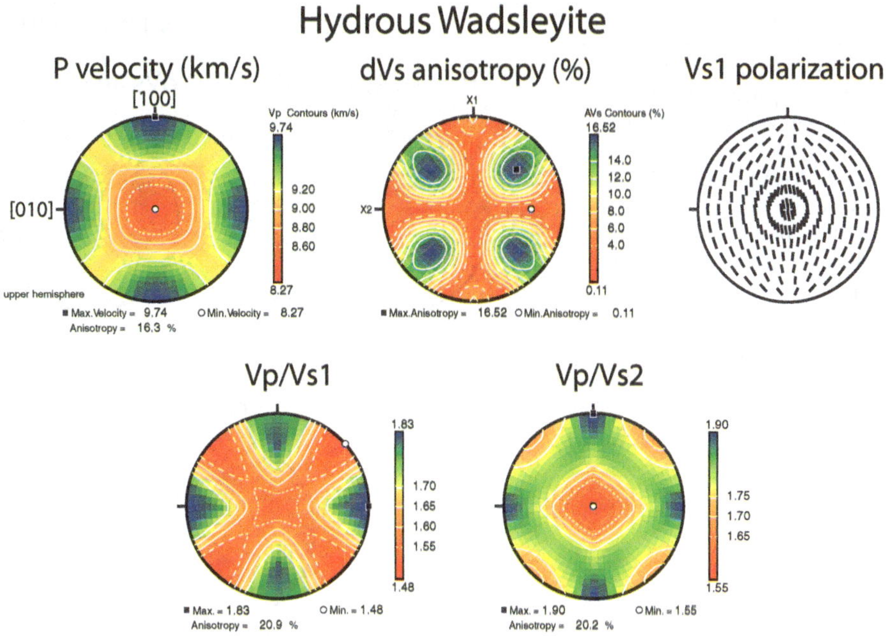

Fig. 12 Upper hemisphere pole figures of P-wave velocity, S-wave splitting anisotropy (dVs as a percentage), the polarization direction (white lines) of the fastest S-wave (Vs1), Vp/Vs1 and Vp/Vs2 for single crystal hydrous wadsleyite (orthorhombic symmetry Imma), calculated using the elastic tensor Mao et al. (2008) at ambient pressure. The c-axis is in the centre of the pole figures, other crystallographic reference directions in the basal plane are marked

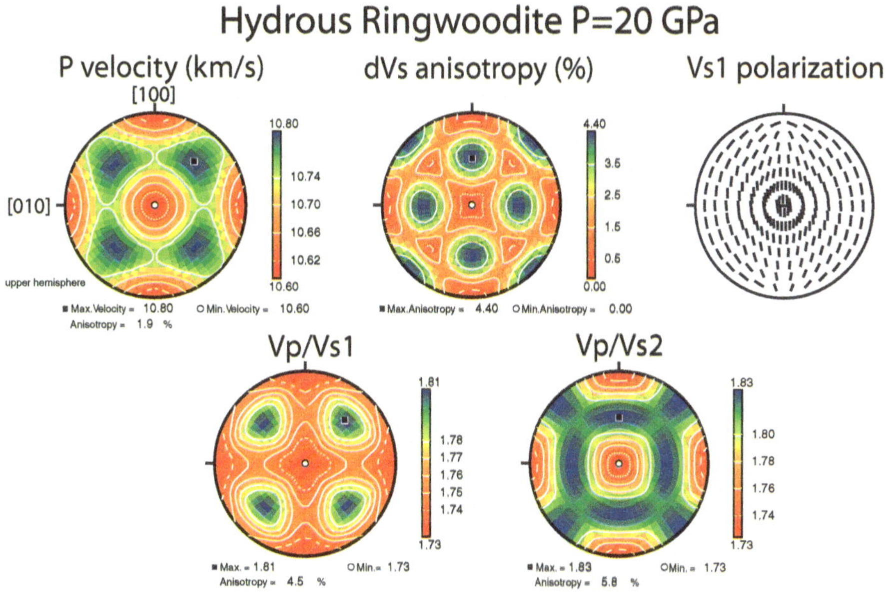

Fig. 13 Upper hemisphere pole figures of P-wave velocity, S-wave splitting anisotropy (dVs as a percentage), the polarization direction (white lines) of the fastest S-wave (Vs1), Vp/Vs1 and Vp/Vs2 for single crystal hydroRingwoodite (cubic symmetry Fd3m), calculated using the elastic tensor Wang et al. (2006) at 20 GPa. The c-axis is in the centre of the pole figures, other crystallographic reference directions in the basal plane are marked

Zha et al. (1997), with values of 15.4% and 16.8%, respectively. Hydrous wadsleyite has a Vp/Vs1 and Vp/Vs2 of 20.9% and 20.2%, respectively, whereas anhydrous wadsleyite has values of 22.0% and 19.1%. Clearly it is not possible to distinguish between anhydrous and hydrous wadsleyite on the basis of anisotropy, although there is a linear decrease of velocity with water content at room pressure (Mao et al., 2008).

In the lower part of the transition zone (520–670-km depth), anhydrous ringwoodite is replaced by its hydrated counterpart hydrous ringwoodite in the hydrated mantle. The elastic constants of hydrous ringwoodite have been measured as a function of pressure to 9 GPa by Jacobsen and Smyth (2006) and to 24 GPa by Wang et al. (2006). Hydrous ringwoodite at the in-situ transition zone pressure of 20 GPa has a very low anisotropy, 1.9% for P-waves and 4.4% S-waves (Fig. 13). Anisotropy for Vp/Vs1 and Vp/Vs2 is slightly stronger at 4.5% and 5.8%, respectively.

The superhydrous phase B (also called C by some authors) is another DHMS mineral with orthorhombic (P21 mn) symmetry is stable over the PT conditions of

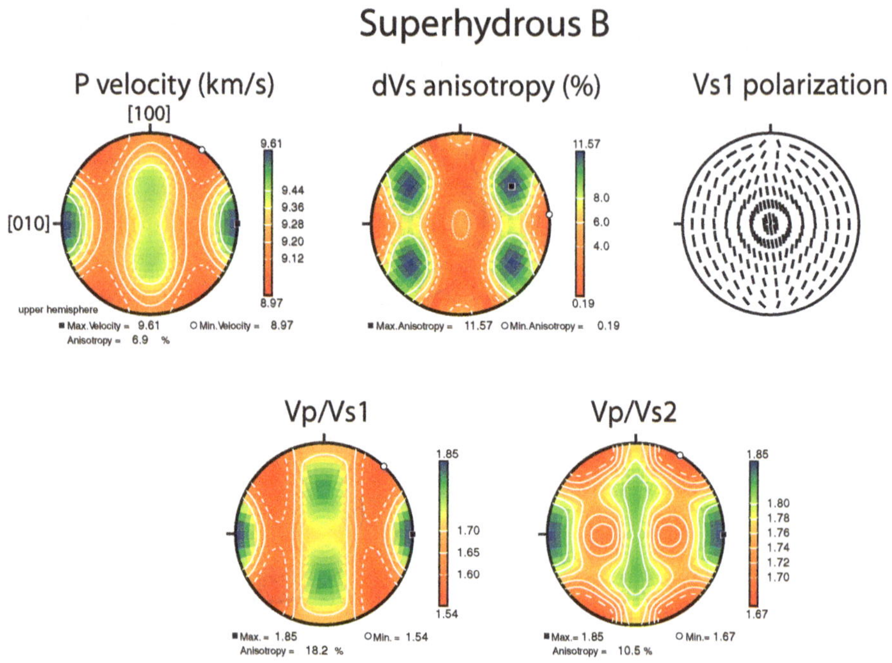

Fig. 14 Upper hemisphere pole figures of P-wave velocity, S-wave splitting anisotropy (dVs as a percentage), the polarization direction (white lines) of the fastest S-wave (Vs1), Vp/Vs1 and Vp/Vs2 for single crystal Superhydrous B phase orthorhombic (Pbnm symmetry), calculated using the elastic tensor by Pacalo and Weidner (1996) at ambient pressure. The [001] axis is in the centre of the pole figures, other crystallographic reference directions in the basal plane are marked

the transition zone (Fig. 2) constitutes up to 66.5% and 53.3% of hydrated harzburgite and lherzolite according to Iwamori (2004), whereas Ohtani et al. (2004) estimated between 10% in the upper part and 30% in the lower part of the transition zone. Pacalo and Weidner (1996) measured the single crystal elastic constants of the superhydrous phase B at ambient conditions. The anisotropic seismic properties shown in Fig. 14 reveal that the superhydrous phase B is only moderately anisotropic with P-wave anisotropy of 6.9% and S-wave anisotropy of 11.6%. The Vp/Vs1 and Vp/Vs2 anisotropies are also moderate at 18.2% and 10.5%, respectively.

The phase D is yet another DHMS mineral and the only known hydrogen bearing mineral stable to lower mantle pressures. Recently, the elastic constants of the phase D have been calculated using first principle methods by Mainprice et al. (2007) to pressures of 84 GPa. At great depth, high temperatures are expected to limit the stability field of the phase D. The percentages of the phase D present in the transition zone and possibly the lower mantle are between 63.5% and 56.3% for harzburgite and lherzolite according to Iwamori (2004) and 15% for a 2 wt% H_2O peridotite according to Ohtani et al. (2004). As with talc, brucite and phase A the velocity distribution of the phase D changes with pressure. In Fig. 15 the anisotropic seismic properties are illustrated for a pressure corresponding to the top of the lower mantle (P = 24 GPa). The velocity distributions clearly illustrate the trigonal nature of the phase D. The anisotropy is higher than the super hydrous B phase, with P-wave anisotropy of 9.2% and S-wave of 18.0%.

4 Elastic and Seismic Properties as a Function of Pressure

As already mentioned above, the elastic constants of only a few hydrous minerals has been measured, and most of these at ambient conditions. A small number of minerals have been measured as a function of pressure, and these are illustrated in Fig. 16. To our knowledge, no measurements of elastic constants of hydrous minerals as a function of temperature have been published. Pressure has a very important effect on the seismic anisotropy of brucite $Mg(OH)_2$, which is considered as an important structural component of many hydrous minerals, and hence one may suspect that most hydrous minerals will be pressure sensitive. At ambient conditions, brucite is one of the most anisotropic minerals with P wave and S wave anisotropies of 58% and 46%, respectively. Anisotropy reduces with pressure, to become about half those values at upper mantle pressures (Jiang et al., 2006). However, even at upper mantle pressures, brucite remains very

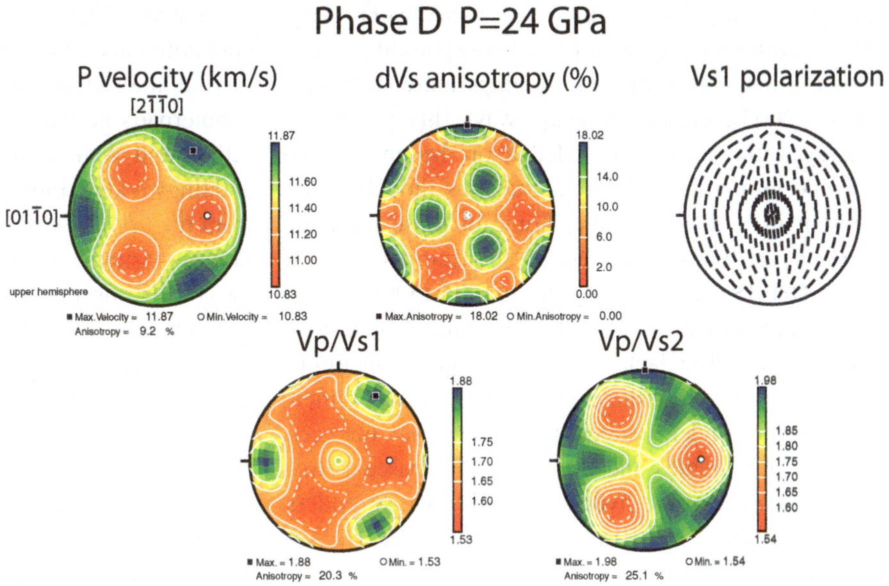

Fig. 15 Upper hemisphere pole figures of P-wave velocity, S-wave splitting anisotropy (dVs as a percentage), the polarization direction (white lines) of the fastest S-wave (Vs1), Vp/Vs1 and Vp/Vs2 for single crystal phase D (trigonal symmetry P-31 m), calculated using the elastic tensor Mainprice et al. (2007) at 24 GPa. The c-axis is in the centre of the pole figures, other crystallographic reference directions in the basal plane are marked

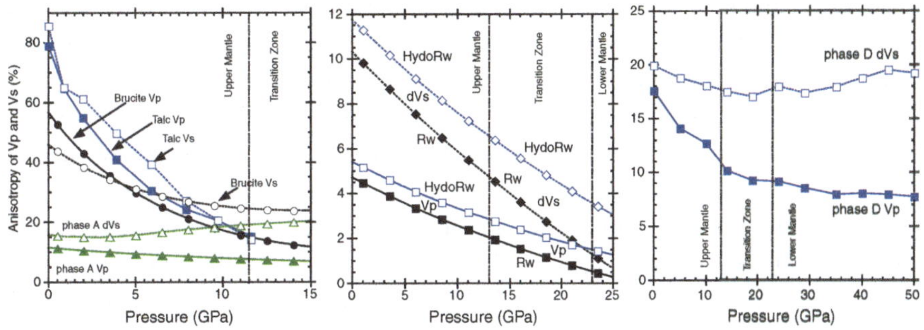

Fig. 16 Variation with pressure of P- and S-wave anisotropy for the hydrous phases talc, brucite, phase A, ringwoodite, and phase D. For reference the anhydrous ringwoodite is also plotted

anisotropic compared with most silicates. Talc ($Mg_3Si_4O_{10}(OH)_2$) is probably the most elastically anisotropic mineral reported to-date at room pressure (Mainprice et al., 2008). Like brucite it is a structural element of many hydrous phases, and also like brucite its anisotropy is reduced by 50% at 5 GPa pressure, but even so it is highly anisotropic with Vp and Vs anisotropy around 40%. In the same pressure range, the DHMS phase A has a radically different pressure sensitivity for anisotropy. In fact phase A is rather insensitive to pressure, with the S-wave anisotropy actually increasing with pressure. However, these anisotropy parameters do not reveal all the changes in velocity distribution, which are very important with pressure for brucite and phase A. In the case of hydrous ringwoodite we can also compare with anhydrous ringwoodite. From the data of Wang et al. (2006), the anisotropy of both minerals decreases with pressure, with hydrous ringwoodite being systematically 1% or 2% more anisotropic than ringwoodite. The pressure sensitivity of phase D is very different to phase A, the anisotropy decreasing significantly for P-wave anisotropy, but remaining nearly constant with pressure S-wave. Clearly it is not easy to predict the exact nature of the pressure sensitivity of hydrous minerals, as the examples presented in Fig. 16 are all different, with exception of brucite and talc that have very similar chemistries.

5 Discussion

The elastic constants of many hydrous phases have yet to be measured at ambient conditions, very few have been measured at high pressures and none, with the possible exception of lawsonite ($CaAl_2Si_2O_7(OH)_2$ H_2O, common component of metamorphosed glaucophane schists) as a function of temperature (Schilling et al., 2003). Hence, in many ways this paper has surveyed the present knowledge of the anisotropic seismic properties of hydrous minerals thought to be present in hydrated mantle. Some of these minerals have never been observed at the Earth's surface (e.g., DHMS) and can only be inferred to be present from high pressure and high temperature laboratory experiments. The lack of pressure and temperature derivatives of the elastic constants for hydrous minerals further hinders direction comparisons between seismic observables and experimentally based petrological models of the hydrated mantle, which renders inferences about the hydrogen budget of the deep Earth to be quite speculative.

The other aspect, which is almost entirely missing in this paper, is the crystal-preferred orientation of hydrated minerals. There is virtually no CPO data apart from the information cited on antigorite form natural samples and the general observation that layered structured minerals almost always form strong preferred orientations due to their anisotropic shape and anisotropic mechanical properties. Therefore, the single crystal seismic anisotropies should be taken as an upper bound for rock samples due to two majors causes; a) the statistical nature of CPO in polycrystalline samples implies that crystals are never perfectly aligned, b) the presence of other minerals will, in general, have differently aligned elastic anisotropy will further reduce the resulting seismic anisotropy of multi-phase mantle rocks at depth.

It is one of the major objectives of this paper to address a first overview of the contribution of hydrated phases to seismic anisotropy of the hydrated mantle associated with subduction, within the limitations cited above. In making this overview we hope to emphasis their importance and stimulate further work on the elastic and mechanical properties of these minerals. Perhaps the first group of minerals to consider are the ones we know the most about in terms of elasticity and CPO, the nominally anhydrous minerals that dominate the composition of the pyrolite mantle, i.e., olivine, wadsleyite, and ringwoodite, leaving aside Mg-perovskite of the lower mantle as it is known to have a very low hydrogen concentation (Bolfan-Casanova, 2005). These minerals are considered to be able to sequestrate significant volumes of water due to their high volume fractions in the mantle. The change of anisotropy in hydrogen containing anhydrous major phases (hydrous forms of olivine, wadsleyite, ringwoodite) compared to their dry counterparts is partly answered by data in Mao et al. (2008) and Wang et al. (2006) on hydrous wadsleyite and ringwoodite (Figs. 12 and 13), with a 1–2% increase from the dry phase and the same velocity distribution. Some constraint on hydrous olivine can be obtained by considering clinohumite to be a 'super saturated' analogue for olivine, as its velocity distribution is almost identical to olivine when using the crystal setting chosen by Fritzel and Bass (1997). Jacobsen et al. (2006) work on hydrous olivine only gives the isotropic elastic parameters. The comparison between dry olivine and clinohumite show 2.5% and 2.0% reduction in anisotropy in the hydrous phase. The changes in anisotropy of these major mantle phases are quite minor with the incorporation of hydrogen and it seems unlikely that seismic anisotropy could be used as a diagnostic tool to study the degree of hydration of the mantle. For these minerals the P- and S-wave seismic velocities are reduced in the hydrogen rich conditions (1 wt% H_20, Smyth and Jacobsen, 2006) by about –0.25 km/s for Vp for ringwoodite and –0.20 km/s for wadsleyite at ambient conditions. However, when the P-waves velocities are extrapolated to in-situ mantle pressure conditions, the velocity difference between wet and dry phases disappears. For S-waves the effect is slightly stronger, with a velocity reduction in ringwoodite of –0.3 km/s for hydrous phase, whereas in wadsleyite there is a velocity increase of about +0.2 km/s at ambient conditions. When extrapolated to mantle pressure, the differences in S-wave velocities between dry and wet ringwoodite and wadsleyite are reduced from –0.3 to –0.2 km/s, and +0.2 to +0.1 km/s, respectively. An alternative seismic approach to studying the hydrated mantle, suggested by Van der Meijde et al. (2003), is to measure the displacement of the '410' km seismic discontinuity that is caused by the olivine to wadsleyite phase transformation due to water content. The phase transformation depth is affected by the usual considerations of pressure and temperature, plus the effect of different equilibrium concentrations of hydrogen in the phases, where wadsleyite has been considered to have considerably higher concentrations than olivine. The temperature has a very strong effect on the equilibrium hydrogen concentration in wadsleyite. Increasing the temperature from 900°C to 1400°C reduces the concentration from 2.2 wt% to 0.9 wt% with no significant effect of pressure in the range 13–18 GPa (Demouchy et al., 2005). Wadsleyite with 0.9 wt% water has almost the same concentration as olivine at the similar conditions given by recent experiments (Mosenfelder et al., 2006; Litasov et al., 2007) at around 1200°C, raising the temperature of olivine also dramatically decreases hydrogen concentration to 0.01 wt% at 1400°C, well below that of wadsleyite. Hence, recent experimental progress supports the idea of water having a significant effect on the depth of the '410 km' discontinuity. However, the depth of the transition will also depend on temperature near 410 km. At temperatures of a subducted slab, i.e., 900–1200°C, there is almost no effect, but at 1400–1500°C of a typical mantle geotherm (Fig. 2), the effect should be present with a ratio of wadsleyite to olivine hydrogen concentration of nearly 100.

We can see that hydrous minerals have a wide variation in anisotropy (see Table 1, Fig. 16), with layered structures at ambient conditions having exceptionally high anisotropy for P- and S-waves (e.g., lizardite, antigorite, talc, brucite, clinochlore). However the lack of good quality single crystal elastic tensors at pressure is a severe handicap when discussing the seismic properties of antigorite, clinochlore, and to some extent clinohumite, even though excellent quality elastic constants are available at ambient conditions. Even for amphiboles, no measurements have been made apart from the 1961 ultrasonic measurements of hornblende at ambient conditions (Alexandrov and Ryzhova, 1961b). To give some indication of the magnitude of anisotropy in hydrous phases we can use anhydrous olivine as a reference anisotropic mineral of geodynamic importance, where its anisotropy has had a major impact on interpretation of mantle seismic anisotropy. Olivine has values for the anisotropy parameters given in Table 1 as follows; Vp of 24.3%, dVs of 17.9%, Vp/Vs1 of 25.5% and Vp/Vs2 of 16.4% using the olivine elastic constants at ambient conditions of Abramson et al. (1997). The anisotropy of olivine is not very sensitive to pressure, whereas the hydrous minerals seem to be, with exception of the phase A. The minerals that have at least one anisotropic parameter greater than olivine are lizardite, antigorite, talc, brucite, clinochore, hornblende, clinohumite, hydrous wadselyite, and phase D. Phase A is very close to olivine with its S-wave parameter. Only the transition zone minerals superhydrous B and hydroringwoodite are significantly less anisotropic than olivine. Hydrous minerals are therefore in general very anisotropic; hence the interpretation of subduction zone seismology has to take this fact into account.

6 Conclusions

We have undertaken a survey of the available data on the single crystal elastic tensors for hydrous minerals likely to occur in hydrated mantle of subduction zones. Many hydrous minerals have not had their elastic constants measured even at ambient conditions (e.g., 10Å phase, phase E), some have not been measured in their true symmetry (e.g., clinochlore, phlogopite) and most have not been measured at high pressure and none related to hydrated mantle have measured at high temperature. For serpentine minerals most of the comparison with seismic observables have been made using wave speed, Vp/Vs or Poisson's ratio. Available elastic tensors from atomic modelling shows that lizardite and antigorite are very anisotropic minerals, these minerals are often associated in serpentinites with talc and brucite, which are also very anisotropic. There are several reports of strong CPO in antigorite rocks or equivalently strong anisotropy of ultrasonic velocities. However, the universally used isotropic Vp/Vs ratio and Poisson's ratio are clearly not applicable to anisotropic rocks composed of oriented anisotropic hydrous minerals. We have presented an analysis of Poisson's ratio that is clearly not a good parameter to describe seismic anisotropy, as the axial and transverse strains that define Poisson's ratio are compressional or extensional, whereas the wave propagation involves compressional and shear strains associated with Vp, Vs1 and Vs2. For an anisotropic medium Vp/Vs has to be replaced by two ratios Vp/Vs1 and Vp/Vs1, but until now seismologists have not reported these ratios. We have shown in the few cases where hydrous minerals have been measured as function of a pressure, that this variable has an important effect on the velocity distribution and in most cases reduces the degree of anisotropy, hence we would expect seismic anisotropy to play a key role in the determination of the shallow structure of subduction zones in the upper mantle.

Acknowledgements DM thanks Carmen Sanchez-Valle (ETH Zurich) for access to her results on the phase A at high pressure prior to publication. DM also thanks the Japanese Society for the promotion of Science (JSPS) for funding a fellowship to visit Katsuyoshi Michibayashi at Shizuoka University where this research seismic properties of subduction zones was initiated in the summer of 2006, and to Wolfgang Friederich and Bernhard Stoeckhert for the invitation to present a keynote on this subject at the symposium Subduction Dynamics: Bridging the Scales in Bochum, Germany, on which this paper is partly based. We also thank Serge Lallemand and Francesca Funiciello for organising the Subduction Zone Dynamics conference in Montpellier and editing this volume of Frontiers in Earth Sciences. We thank Romain Bousquet and the other reviewer for their constructive comments on the manuscript. This contribution is also made possible thanks to the support for computer infrastructure (my_Mac) to DM from INSU-CNRS (France) and the European Science Foundation (ESF) under the EUROCORES Programme EuroMinScI, project euroSLAB through contract No. ERAS-CT-2003-980409 of the European Commission, DG Research, FP6.

References

Abramson EH, Brown JM, Slutsky LJ, Zang J (1997) The elastic constants of San Carlos olivine to 17 GPa. J Geophys Res 102:12253–12263.

Alexandrov KS, Ryzhova TV (1961a) Elastic properties of rock-forming minerals 2 layered silicates. Bull Acad Sci USSR Geophys Ser 11:871–875.

Alexandrov KS, Ryzhova TV (1961b) Elastic properties of rock-forming minerals 1 pyroxenes and amphiboles. Bull Acad Sci USSR Geophys Ser 9:1165–1168.

Arcay D, Tric E, Doin M-P (2005) Numerical simulations of subduction zones: effect of slab dehydration on the mantle wedge dynamics. Phys Earth Planet Inter 149:133–153.

Auzende A-L, Pellenq RJ-M, Devouard B, Baronnet A, Grauby O (2006) Atomistic calculations of structural and elastic properties of serpentine minerals: the case of lizardite. Phys Chem Miner 33:266–275. Doi:101007/s00269-006-0078-x.

Bach W, Garrido CJ, Paulick H, Harvey J, Rosner M (2004) Seawater-peridotite interactions: first insights from ODP Leg 209, MAR 15°N. Geochem Geophys Geosyst. Doi:10.1029/2004GC000744.

Bach W, Paulick H, Garrido CJ, Ildefonse B, Meurer WP, Humphris SE (2006) Unraveling the sequence of serpentinization reactions: petrography, mineral chemistry and petrophysics of serpeninites from MAR 15°N (ODP Leg 209, Site 1274). Geophys Res Lett 33, L13306. Doi:10.1029/2006GL025681.

Bolfan-Casanova N (2005) Water in the Eath's mantle. Miner Mag 63:299–257.

Boschi C, Früh-Green GL, Delacour A, Karson JA, Kelley DS (2006) Mass transfer and fluid flow during detachment faulting and development of an oceanic core complex, Atlantis Massif (MAR 30°N). Geochem Geophys Geosyst. Doi:10.1029/2005GC001074.

Bostock MG, Hyndman RD, Rondenay S, Peacock SM (2002) An inverted continental Moho and serpentinization of the forearc mantle. Nature 417:536–538.

Bousquet R, Goffé B, Henry P, Le Pichon X, Chopin C (1997) Kinematic thermal and petrological model of the central Alps: lepontine metamorphism in the upper crust and eclogitisation of the lower crust. Tectonophysics 273:105–127.

Bousquet R, Goffé B, Le Pichon X, de Capitani C, Chopin C, Henry P (2005) Comment on "Subduction factory: 1. Theoretical mineralogy, densities, seismic wave speeds, and H_2O contents" by Bradley R. Hacker, Geoffrey A. Abers, and Simon M. Peacock. J Geophys Res 110:B02206. Doi:10.1029/2004JB003450.

Brocher TM, Parsons T, Tréhu A-M, Snelson C, Fischer MA (2003) Seismic evidence for widespread serpentinized forearc upper mantle along the Cascadia margin. Geology 31:267–270.

Brudzinski MR, Thurber CH, Hacker BR, Engdahl ER (2007) Global prevalence of double Benioff zones. Science 316:1472–1474.

Cannat M (1993) Emplacement of mantle-rocks in the seafloor at mid-ocean ridges. J Geophys Res 98:4163–4172.

Carlson RL, Miller DJ (1997) A new assessment of the abundance of serpentinite in the oceanic crust. Geophys Res Lett 24:457–460.

Carlson RL, Miller DJ (2003) Mantle wedge water contents estimated from seismic velocities in partially serpentinized peridotites. Geophys Res Lett 30:1250. Doi:101029/2002GL016600.

Chinner GA, Dixon JE (1973) Some high-pressure parageneses of the Allalin gabbro, Valais, Switerland. J Petrol 14:185–202.

Christensen NI (1966) Elasticity of ultrabasic rocks. J Geophys Res 71:5921–5931.

Christensen NI (1978) Ophiolites seismic velocities and oceanic crustal structure. Tectonophysics 47:131–157.

Christensen NI (1996) Poisson's ratio and crustal seismology. J Geophys Res 101:3139–3156.

Christensen NI (2004) Serpentine peridotites and seismology. Int Geol Rev 46:795–816.

Demouchy S, Deloule E, Frost DJ, Keppler H (2005) Pressure and temperature-dependence of water solubility in Fe-free wadsleyite. Am Mineral 90 1084–1091.

Escartín J, Mével C, MacLeod CJ, McCaig AM (2003) Constraints on deformation conditions and the origin of oceanic detachments: the Mid-Atlantic Ridge core complex at 15. Geochem Geophys Geosys. Doi:1010.1029/2001GC000278.

Evans BW (2004) The serpentinite multisystem revisited: chrysotile is metastable. Int Geol Rev 46:479–506.

Evans BW, Johannes W, Oterdoom H, Trommsdorf V (1976) Stability of chrysotile and antigorite in the serpentine multisystem. Schweiz Mineral Petrogr Mitt 56:79–93.

Fumagalli P, Poli S (2005) Experimentally determined phase relations in hydrous peridotites to 6–5 GPa and their consequences on the dynamics of subduction zones. J Petrol 46:555–578.

Fritzel TLB, Bass JD (1997) Sound velocities of clinohumite, and implications for water in Earth's upper mantle. Geophys Res Lett 24:1023–1026.

Grauby O, Baronnet A, Devouard B, Schoumacker K, Demirdjian L (1998) The chrysotile-polygonal serpentinite suite synthesized from a $3MgO–2SiO_2$–excess H_2O gel [abs.]: Terra Nova Abstract Suppl 1, p. 24.

Hacker BR, Abers GA, Peacock SM (2003a) Subduction factory 1. Theoretical mineralogy densities seismic wave speeds and H_2O contents. J Geophys Res 108. Doi:101029/2001JB001127.

Hacker BR, Peacock SM, Abers GA, Holloway SD (2003b) Subduction factory 2. Are intermediate-depth earthquakes in subducting slabs linked to metamorphic dehydration reactions? J Geophys Res 108. Doi:101029/2001JB001129.

Hermann J, Fitz Gerald JD, Malaspina N, Berry AJ, Scmbelluri M (2007) OH bearing planar defects in olivine produced by the breakdown of Ti-rich humite minerals from Dabie Shan (China). Contrib Mineral Petrol 153:417–428.

Hilairet N, Daniel I, Reynard B (2006) P–V equations of state and the relative stabilities of serpentine varieties. Phys Chem Miner 33:629–637.

Hirschmann MM (2006) Water melting and the deep Earth H_2O cycle. Annu Rev Earth Planet Sci 34:629–653.

Horen H, Zamora M, Dubuisson G (1996) Seismic wave velocities and anisotropy in serpentinized peridotites from Xigaze ophiolite: abundance of serpentine in slow spreading ridge: Geophys Res Lett 23:9–12.

Hostetler PB, Coleman RG, Evans BW (1966) Brucite in alpine serpentinites. Am Mineral 51:75–98.

Hyndman RD, Peacock SM (2003) Serpentinization of the forearc mantle. Earth Planet Sci Lett 212:417–432.

Ildefonse B, Christie DM, Mission Moho Workshop Steering Committee (2007) Mission Moho workshop: drilling through the oceanic crust to the Mantle. Scientific Drilling 4:11–18. Doi:10.2204/iodp.sd.4.02.2007.

Ito K (1990) Effects of H_2O on elastic wave velocities in ultrabasic rocks at 900°C under 1GPa. Phys Earth Planet Inter 61:260–268.

Iwamori H (2004) Phase relations of peridotites under H_2O-saturated conditions and ability of subducting plates for transportation of H_2O. Earth Planet Sci Lett 227:57–71.

Jacobsen SD (2006) Effect of water on the equation of state of nominally anhydrous minerals In: H Keppler and JR Smyth (Eds) Water in Nominally Anhydrous Minerals. Rev Min Geochem 62:321–342.

Jacobsen SD, Smyth JR (2006) Effect of water on the sound velocities of ringwoodite in the transition zone. In: SD Jacobsen and S van der Lee (Eds) Earth's Deep Water Cycle, American Geophysical Union. Geophys Monogr 168:131–145.

Jacobsen SD, Jiang F, Smyth JR, Duffy TS, Mao Z, Holl CM, Frost DJ (2006) Sound velocities of hydrous olivine and the effects of water on nominally anhydrous minerals. Eos Trans AGU 87(52) Fall Meet Suppl Abstract V53F-03.

Jiang FM, Speziale S, Duffy TS (2006) Single-crystal elasticity of brucite, Mg(OH)(2), to 15GPa by Brillouin scattering. Am Mineral 91:1893–1900.

Jung H, Green HW II, Dobrzhinetskaya LF (2004) Intermediate-depth earthquake faulting by dehydration embrittlement with negative volume change. Nature 428:545–549.

Kamiya S, Kobayashi Y (2000) Seismological evidence for the existence of serpentinized wedge mantle. Geophys Res Lett 27:819–822.

Kawamoto T, Hervig RL, Holloway JR (1996) Experimental evidence for a hydrous transition zone of the early Earth's mantle. Earth Planet Sci Lett 142:587–592.

Kern H (1993) P- and S-wave anisotropy and shear wave splitting at pressures and temperature in possible mantle rock and their relation to the rock fabric. Phys Earth Planet Int 78: 245–256.

Kern H, Tubia JM (1993) Pressure and temperature dependence of P- and S-wave velocities seismic anisotropy and density of sheared rocks from Sierra Alpujata massif (Ronda peridotites Southern Spain). Earth Planet Sci Lett 119:191–205.

Kern H, Liu B, Popp T (1997) Relationship between anisotropy of P and S wave velocities and anisotropy attenuation in serpentinite and amphibolite. J Geophys Res 102:3051–3065.

Kono Y, Ishikawa M, Arima M (2007) Effect of H_2O released by dehydration of serpentine and chlorite on compressional wave velocities of peridotites at 1GPa and up to 1000°C Phys Earth Planet Int 161:215–223.

Lagabrielle Y, Bideau D, Cannat M, Karson JA, Mével C (1998) Ultramafic-mafic plutonic rock suites exposed along the Mid-Atlantic Ridge (10°N-30°N); symmetrical-asymmetrical distribution and implications for seafloor spreading processes. In: Buck WR, Delaney PT, Karson JA, Lagabrielle Y (Eds) Faulting and magmatism at mid-ocean ridges. Geophysical Monograph, American Geophysical Union, Washington, DC, United States, pp. 153–176.

Litasov KD, Ohtani E, Kagi H, Jacobsen SD, Ghosh S (2007) Temperature dependence and mechanism of hydrogen incorporation in olivine at 12.5–14.0GPa. Geophys Res Lett 34:L16314. Doi:10.1029/2007GL030737.

Mainprice D (1990) An efficient fortran program to calculate seismic anisotropy from the lattice preferred orientation of minerals. Comput Geosci 16:385–393.

Mainprice D (2007) Seismic anisotropy of the deep Earth from a mineral and rock physics perspective. In: Schubert G (Ed) Treatise in Geophysics, Volume 2 pp. 437–492. Oxford, UK: Elsevier.

Mainprice D, Le Page Y, Rodgers J, Jouanna P (2007) Predicted elastic properties of the hydrous D Phase at mantle pressures: implications for the seismic anisotropy of subducted slabs near 670km discontinuity and in the lower mantle. Earth Planet Sci Lett 259:283–296.

Mainprice D, Le Page Y, Rodgers J, Jouanna P (2008) *Ab initio* elastic properties of talc from 0 to 12GPa: interpretation of seismic velocities at mantle pressures and prediction of auxetic behaviour at low pressure. Earth Planet Sci Lett 274:327–338.

Mao Z, Jacobsen SD, Jiang F, Smyth JR, Holl CM, Frost DJ, Duffy TS (2008) Single-crystal elasticity of wadsleyites, b-Mg_2SiO_4,containing 0.37–1.66 wt.% H_2O. Earth and Planet Sci Lett 266:78–89.

MacLeod CJ, Escartin J, Banerji D, Banks GJ, Gleeson M, Irving DHB, Lilly RM, McCaig AM, Niu Y, Allerton S, Smith DK (2002) Direct geological evidence for oceanic detachment faulting: the Mid-Atlantic Ridge, 15 degrees 45′ N. Geology 30:879–882.

Mével C (2003) Serpentinization of abyssal peridotites at mid-ocean ridges C. R. Geoscience 335:825–852.

Mével C, Stamoudi C (1996) Hydrothermal alteration of the upper-mantle section at Hess Deep. In: Mével C, Gillis KM, Allan JF, Meyer PS (Eds), Proc. ODP, Sci. Results 147:College Station, TX (Ocean Drilling Program), pp. 293–309. Doi:10.2973/odp.proc.sr.147.017.1996.

Michibayashi K, Abe N, Okamoto A, Satsukawa T, Michikura K (2006) Seismic anisotropy in the uppermost mantle, back-arc region of the northeast Japan arc: petrophysical analyses of

Ichinomegata peridotite xenoliths. Geophys Res Lett 33:L10312. Doi:10.1029/2006GL025812.
Michibayashi K, Tasaka M, Ohara Y, Ishii T, Okamoto A, Fryer P (2007) Variable microstructure of peridotite samples from the southern Mariana Trench: evidence of a complex tectonic evolution. Tectonophysics 444:111–118. Doi:10.1016/j.tecto.2007.08.010.
Mosenfelder JL, Deligne NI, Asimow PD, Rossman GR (2006) Hydrogen incorporation in olivine from 2–12 GPa. Am Mineral 91:285–294.
Nakajima J, Matsuzawa T, Hasegawa A, Zhao D (2001), Three-dimensional structure of Vp, Vs and Vp/Vs beneath north-eastern Japan: implications for arc magmatism and fluids. J Geophys Res 106:21843–21857.
Nye JF (1957) Physical properties of crystals–their representation by tensors and matrices. University Press, Oxford.
Ohtani E, Litasov K, Hosoya T, Kubo T, Kondo T (2004) Water transport into the deep mantle and formation of a hydrous transition zone. Phys Earth Planet Int 143–144: 255–269.
Pacalo REG, Weidner DJ (1996) Elasticity of superhydrous. B Phys Chem Miner 23:520–525.
Pawley AR, Clark SM, Chinnery NJ (2002) Equation of state measurements of chlorite pyrophyllite and talc. Am Mineral 87:1172–1182.
Peacock SM (1990) Fluid processes in subduction zones. Science 248:329–337.
Peacock SM (2001) Are the lower planes of double seismic zones caused by serpentine dehydration in subducting oceanic mantle? Geology 29:299–302.
Pellenq RJ-M, Mainprice D, Ildefonse B, Devouard B, Baronnet A, Grauby O (2008) Atomistic calculations of the elastic properties of antigorite at upper mantle conditions: application to the seismic properties in subduction zones. submitted to EPSL.
Poli S, Schmidt MW (1997) The high-pressure stability of hydrous phases in orogenic belts: an experimental approach on eclogite-forming processes. Tectonophysics 273:169–184.
Poli S, Schmidt MW (2002) Petrology of subducted slabs. Annu Rev Earth Planet Sci 30:207–235.
Ranero CR, Sallarès V (2004) Geophysical evidence for hydration of the crust and mantle of the Nazca Plate during bending at the north Chile Trench. Geology 32:549–552.
Reynard B, Hilairet N, Balan E, Lazzeri M (2007) Elasticity of serpentines and extensive serpentinization in subduction zones. Geophys Res Lett 34:L13307. Doi:101029/2007GL030176.
Rossi G, Abers GA, Ronenay S, Christensen DH (2006) Unusual mantle Poisson's ratio, subduction, and crustal structure in central Alaska. J Geophys Res 111:B09311. Doi:10.1029/2005JB003956.
Sanchez-Valle C, Sinogeikin SV, Smyth JR, Bass JD (2006) Single-crystal elastic properties of dense hydrous magnesium silicate phase A. Am Mineral 91:961–964.
Sanchez-Valle C, Sinogeikin SV, Smyth JR, Bass JD (2008) Sound velocities and single-crystal elasticity of DHMS phase A to high pressure and implications for seismic velocities in subducted slabs. Phys Earth Planet Int 171:229–239.
Schilling FR, Sinogeikin SV, Bass JD (2003) Single-crystal elastic properties of lawsonite and their variation with temperature. Phys Earth Planet Int 136:107–118.
Schmidt MW, Poli S (1998) Experimentally based water budgets for dehydrating slabs and consequences for arc magma generation. Earth Planet Sci Lett163:361–379.
Seno T, Zhao D, Kobayashi Y, Nakamura M (2001) Dehydration of serpentinized slab mantle: seismic evidence from southwest Japan. Earth Planets Space 53:861–871.
Shipboard Scientific Party (2004) Site 1268. In: Kelemen PB, Kikawa E, Miller DJ, et al., Proc. ODP, Init. Repts., 209: College Station, TX (Ocean Drilling Program), pp. 1–171. Doi:10.2973/odp.proc.ir.209.103.2004.
Sirotin Y, Shakolskaya MP (1982) Fundamentals of Crystal Physics. Mir, Moscow.
Smyth JR, Frost DJ, Nestola F, Holl CM, Bromiley G (2006) Olivine hydration in the deep upper mantle: effects of temperature and silica activity. Geophys Res Lett 33:L15301. Doi:101029/2006GL026194.
Smyth JR, Jacobsen SD (2006) Nominally anhydrous minerals and Earth's deep water cycle In: SD Jacobsen and S van der Lee (Eds) Earth's Deep Water Cycle. Am Geophys Union Geophys Monogr 168:1–11.
Song I, Suh M, Woo Y-K, Haob T (2004) Determination of the elastic modulus set of foliated rocks from ultrasonic velocity measurements. Engin Geol 72:293–308.
Stalder R, Ulmer P (2001) Phase relations of a serpentine composition between 5 and 14 GPa: significance of clinohumite and phase E as water carriers into the transition zone. Contrib Mineral Petrol 140:670–79.
Ulmer P, Trommsdorff V (1995) Serpentine stability to mantle depths and subduction-related magmatism. Science 268:858–861.
Van der Meijde M, Marone F, Glardini D, van der Lee S (2003) Seismic evidence for water deep in Earth's upper mantle. Science 300:1556–1558.
Wang JS, Sinogeikin V, Inoue T, Bass JD (2006) Elastic properties of hydrous ringwoodite at high-pressure conditions. Geophys Res Lett 33:L14308. Doi:101029/2006GL026441.
Watanabe T, Kasami H, Ohshima S (2007) Compressional and shear wave velocities of serpentinized peridotites up to 200 MPa. Earth Planets Space 59:233–244.
Welch M, Crichton WA (2002) Compressibility of clinochlore to 8 GPa at 298 K and a comparison with micas. Eur J Mineral 14:561–565.
Zha CS, Duffy TS, Mao HK, Downs RT, Hemley RJ, Weidner DJ (1997) Single-crystal elasticity of beta-Mg_2SiO_4 to the pressure of the 410 km seismic discontinuity in the Earth's mantle. Earth Planet Sci Lett 147:E9–E15. Doi:10.1016/S0012-821X(97)00010-1.

Local Earthquake Tomography in the Southern Tyrrhenian Region of Italy: Geophysical and Petrological Inferences on the Subducting Lithosphere

Marco Caló, Catherine Dorbath, Dario Luzio, Silvio G. Rotolo and Giuseppe D'anna

Abstract We obtained a high-resolution seismic tomography of the Ionian lithosphere subduction using a new approach based on: (a) the Double-Difference technique for inversions and (b) the statistical post-processing of a great number of preliminary models (Weighted Average Model, WAM method); the latter was used to increase reliability and resolution. In the tomographic model, the high-velocity portion of the steeply dipping Ionian slab is well imaged, as is an underlying low-Vp ($\approx$7.0 km/s) aseismic region. We propose that the low-velocity region can be assigned to a partially hydrated (serpentinized) mantle of the subducting Ionian slab, which progressively dehydrates with depth in dense high-pressure phases. In the mantle wedge overlying the slab, large areas, characterized by low-Vp ($\leq$7.0 km/s) and high Vp/Vs ($\geq$1.85), have been imaged beneath the Stromboli and Marsili volcanoes down to a maximum depth of 180 km. We have interpreted these areas as being regions where mantle partial melts form and accumulate and which then feed the present-day Aeolian Archipelago magmatism .

Keywords Local earthquake tomography • Subduction • Ionian slab • Southern Tyrrhenian sea • Hydrous peridotite

1 Introduction

The subduction of the Ionian microplate, which dips steeply with NW polarity beneath the Tyrrhenian Basin, is controlled by the collision between Africa and Eurasia. The subduction started during the Oligo-Miocene, and the front was very wide (Jolivet and Faccenna, 2000); however it is still active, along a front less than 300 km long, beneath the Aeolian Archipelago. The Tyrrhenian back-arc basin (Fig. 1) opened during early Pliocene as a result of an extensive roll-back of the subduction front (Faccenna et al., 2005 and references therein).

The Ionian lithosphere is estimated to be 125 km thick (Gvirtzman and Nur, 2001; Pontevivo and Panza, 2006), and it is composed of a 6–8 km thick Meso-Cenozoic sedimentary cover overlying an 8–9 km thick Mesozoic oceanic crust; the latter is a remnant of the former Tethyan realm (Catalano et al., 2001). Instead, the Tyrrhenian Basin is floored by a much younger (6–2 Ma) oceanic crust as thin as 7–10 km in the Vavilov and Marsili sub-basins (Panza et al., 2007).

A number of recent tomographic studies have focused on subduction in Southern Italy. In their research the authors used various methods based almost exclusively on teleseismic data (Piromallo and Morelli, 2003; Montuori et al., 2007 and references

M. Caló
CFTA Dipartimento di Chimica e Fisica della Terra, Università di Palermo, Via Archirafi 36, 90100, Palermo, Italy, marco.calo@unipa.it/co.cal@tiscali.it

S. Lallemand and F. Funiciello (eds.), *Subduction Zone Geodynamics*,
DOI 10.1007/978-3-540-87974-9, © Springer-Verlag Berlin Heidelberg 2009

Fig. 1 Structural map of the Tyrrhenian-Apennine domain (modified from Faccenna et al., 1996)

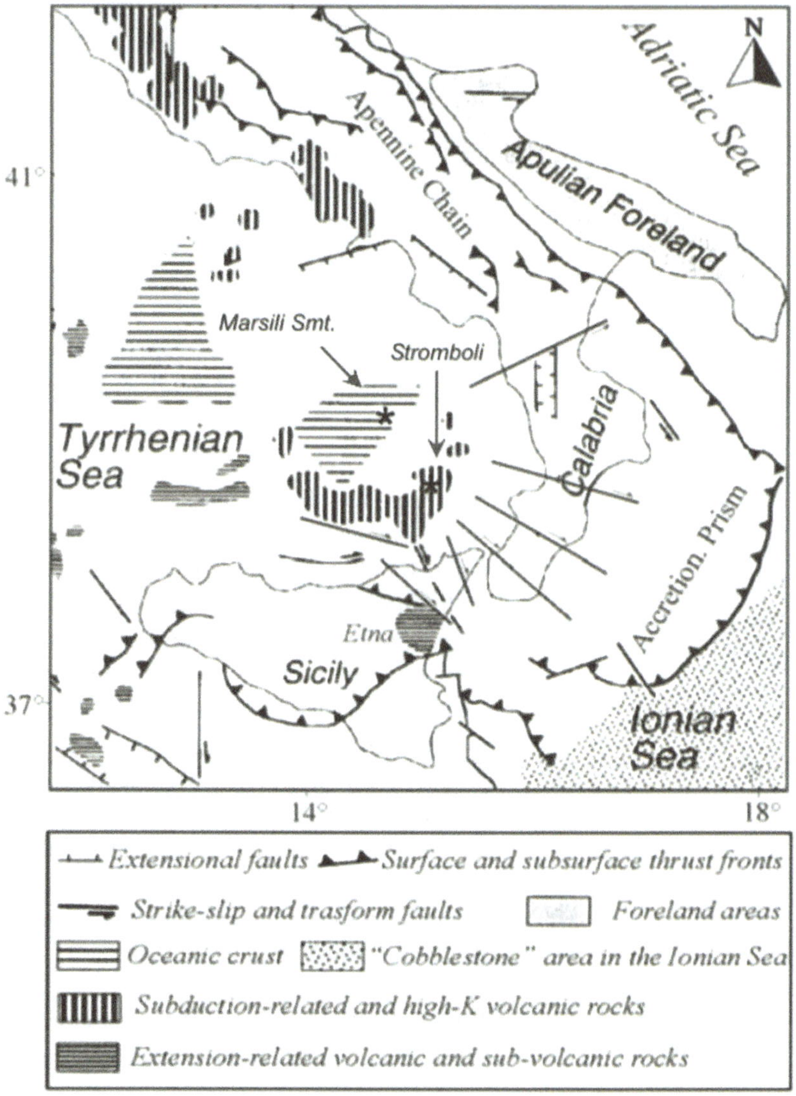

therein). A general agreement exists regarding the presence of the Ionian slab imaged as a steep high velocity body which is surrounded by slightly lower velocity volumes, generally interpreted as being hot mantle domains (African and Tyrrhenian). Nevertheless, the resolution of the existing tomographies have not allowed to investigate on heterogeneities within the subducting lithosphere.

The principal aim of this paper is to interpret the peculiar velocity distribution within the Ionian slab, obtained by high-resolution seismic tomography based on local earthquake data, in the light of mineral phase equilibria governing the progressive dehydration of H_2O-Mg-bearing silicates in the ultramafic portion of the subducted lithosphere. A secondary aim concerns the possible presence of deep regions of accumulation of partial melts in the mantle domains above the slab.

2 Data and Method

The data set used in this study contains the P and S arrival-times taken from the INGV's Italian seismicity catalogue (CSI and others) regarding earthquakes located in the window 14°30′E–17°E and 37°N–41°N and recorded during the period 1981–2005 (Fig. 2a). The data set was updated with arrival times, not included in the INGV catalogues, picked out on wave-forms recorded from some temporary arrays. The selected

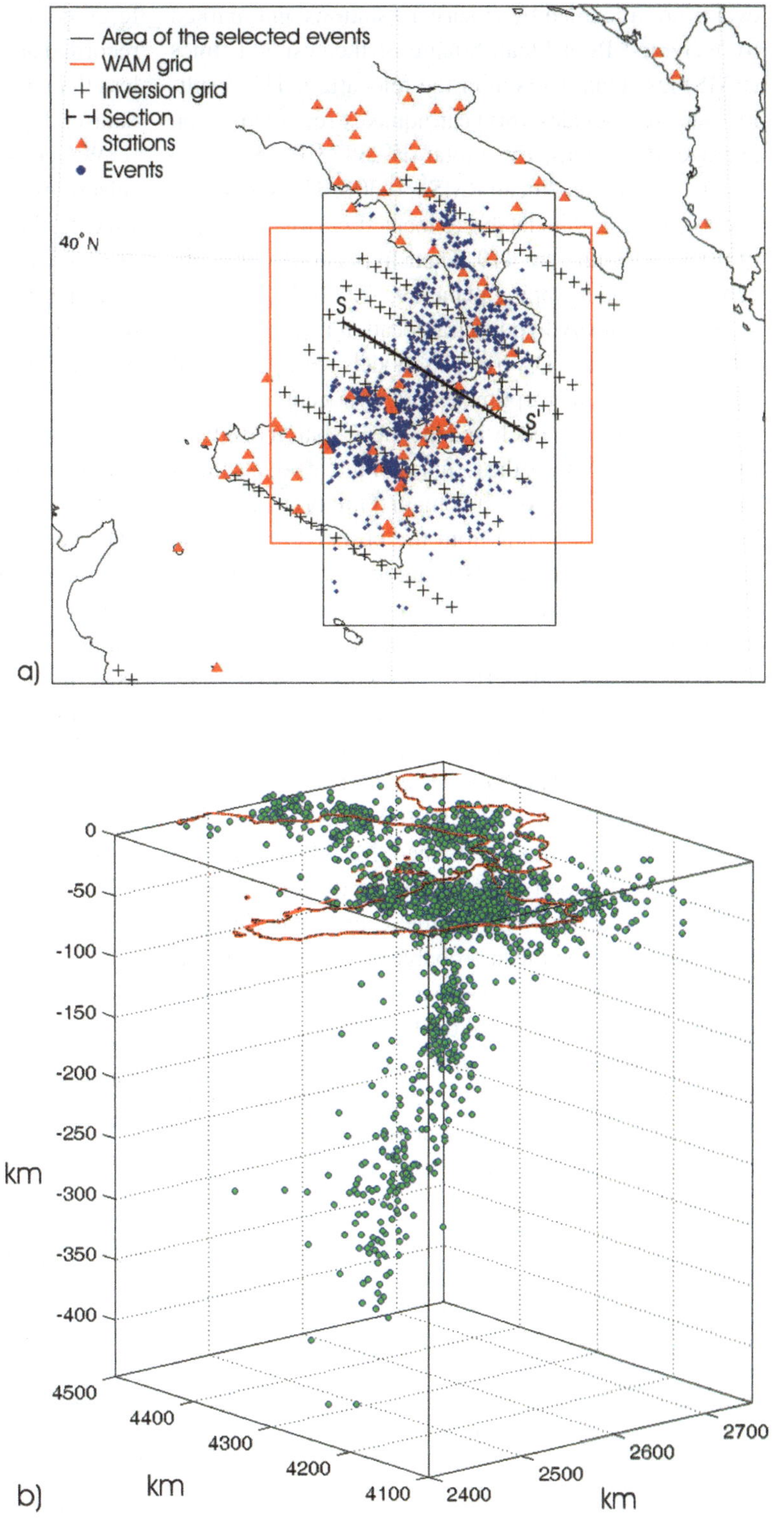

Fig. 2 (**a**) Area of study with epicentres and stations used; S-S′ is the trace of the WAM tomographic sections shown in Figs. 7–9; the small crosses mark the grid used for the synthetic tests. (**b**) Three-dimensional view of the hypocenters

events are: recorded by at least 10 stations and marked by weighted Root Mean Square of the residual times (RMS) less than 0.50 s after the relocation. The resulting data set includes 1800 earthquakes recorded by on average 16 stations, for a total of 28873 P- and 9990 S- arrival times and mean RMS = 0.30 s. The subcrustal events given in this database, located between a depth of 40 and 360 km, are 419. Their hypocentres are concentred around a dipping plane 69°–72° NW (Fig. 2b).

The preliminary hypocentral location and the optimisation of the initial 1D Vp and Vs models were performed by a procedure that optimizes the hypocentral coordinates, the velocity models and the mean station residuals by minimizing besides the L_2 norm of the residuals times their coherence in the offset domain (Giunta et al., 2004). This technique descends from the observation that the relocation of a set of events using an unsuitable model produces residual times of the rays which show large coherence intervals in the offset domain.

To produce an initial 1D model with constant vertical gradients within the layers, as required by the tomographic inversion code, this procedure, initially implemented for HYPOPC71 code (Lee and Lahr, 1985), was adapted to HYPOINVERSE-2000 code (Klein, 2002). The model obtained (MC) represents the initial vertical velocity distribution for the tomographic inversions and for some synthetic and experimental tests (Table 1). The mean probable error on the epicentral coordinates of the re-located events is ~2.5 km, whereas on the hypocentral depths it is ~4.5 km.

To carry out the 3D inversions we adopted the tomoDD code (Zhang and Thurber, 2003) in this work. This code removes some fuzziness in the final velocity model by using differential arrival times together with the absolute one (Zhang and Thurber, 2003). The differential times are calculated from pairs of earthquakes recorded at a common station with inter-event distance less than a selected threshold, therefore, the size of the dataset depends essentially on the threshold value.

Table 1 Velocities and depth of the MC model

	MC model	
Km	Vp (km/s)	Vs (km/s)
0	2.10	1.20
5	5.60	3.20
10	5.80	3.31
13	6.30	3.60
20	7.00	4.00
26	7.70	4.40
40	7.95	4.54
50	8.05	4.60
100	8.10	4.63
245	8.19	4.68
340	8.22	4.70

We performed several inversions using different selections of absolute and differential data obtained by varying the maximum accepted RMS and inter-event distance, respectively.

In particular, we used several thresholds to extract many differential datasets. The least numerous one, obtained fixing the maximum inter-event distance to 6 km, contains 23992 P- and 7492 S- differential times, while the largest, obtained with a threshold of 20 km, encloses 78352 P- and 26086 S- differential times.

The velocity model was parameterized as a regular 3D inversion grid and the values assigned to the nodes were interpolated in a generic point inside a cell using a tri-linear interpolation algorithm (Thurber, 1993). Numerous grids with different node positions were used in the inversions of experimental data. The average lengths of the horizontal grid edges, excluding the boundary nodes, were 310 and 190 km (Fig. 2a) whereas the vertical one was 340 km. The average spacing along each direction was 17, 38 and 13 km, respectively.

3 Reliability Tests

We performed tests with synthetic and experimental data to evaluate the resolution, the stability and the systematic dependence of the results on each parameter picked a priori.

3.1 Synthetic Tests

We performed synthetic tests using the real distribution of the stations and the hypocenters located with the MC model. Travel-times were calculated in the 3D model (MCS) resulting from an experimental data inversion (Fig. 3a). Starting from the MC model and using these synthetic data without adding any noise, the result of the test showed that the information contained in the calculated travel times was able to reconstruct velocity distributions closely matching the MCS model in the volumes where *DWS* (Derivative Weight Sum, Toomey and Foulger, 1989) is greater than 50 (Fig. 3b). The restoration-resolution test (Zhao

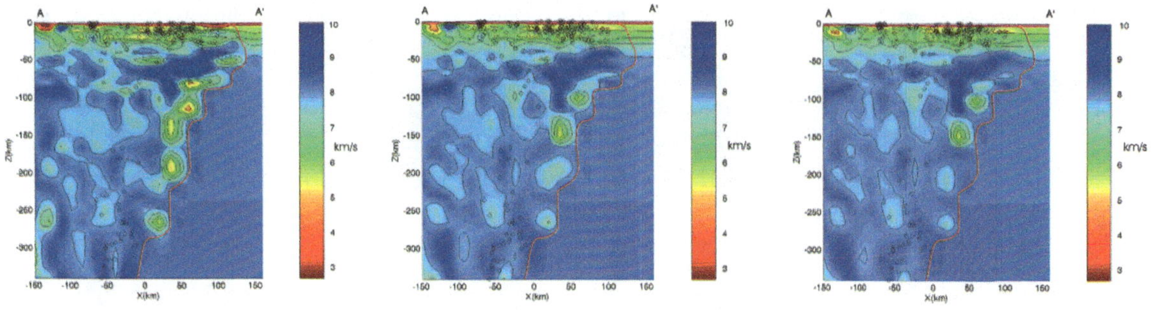

Fig. 3 (**a**) Vertical section through the MCS model obtained using the inversion grid A (Fig. 4a). (**b**) Synthetic test. (**c**) Restoration Resolution test. A plan of the grid used to perform these tests is displayed in Fig. 2a. The red lines limit the portion of the poorly resolved model (*DWS* < 50)

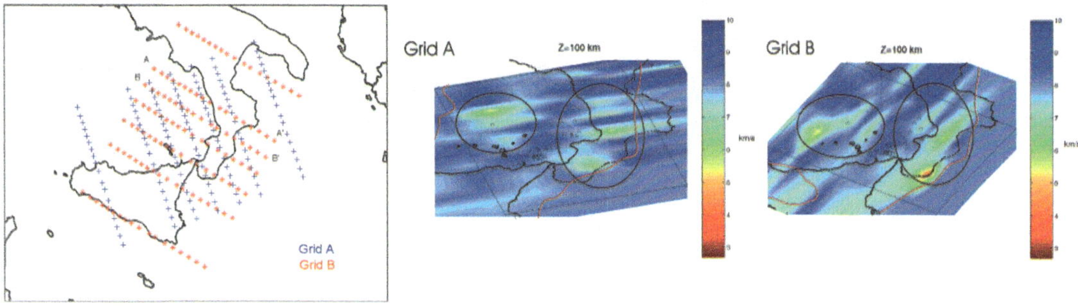

Fig. 4 Models regarding grids A and B used for the experimental tests. The two horizontal sections (centre and right) show the plane at a depth of 100 km. Compare the similarity of the velocity patterns within the circled areas The red lines limit the portion of the poorly resolved model (*DWS* < 50)

et al., 1992), consisting in the inversion of the synthetic data perturbed with a random noise having standard deviation equal to the L_2 norm of the residuals of the experimental data inversion (0.62 s), highlighted the stability of the inverse problem (Fig. 3c).

3.2 Experimental Tests

Experimental tests were carried out to evaluate the dependence of the final velocity model on the spatial parameterization, by rotating, translating and deforming a reference grid.

We used horizontally anisotropic inversion grids with inter-distance ratio between the nodes ~0.5 along the *x* and *y* grid axes. Other authors have used similar grids to investigate quasi-2D structures having strike a priori known (Zhang and Thurber, 2003; Shelly et al., 2006), while we introduced this anisotropy of the inversion grid to increase the volume of the model space sampled varying the azimuth of the grid axes.

To compare all the velocity models obtained using the various inversion grids tested we (re)sampled them at the nodes of a fixed grid by the same trilinear interpolation algorithm of tomoDD. The relative standard deviation of the velocity values were ~1% in the volumes around the earthquake hypocentres, but increase up to ~7% in the border volumes. This variability does not change the features of the velocity model for anomalies having linear dimensions greater than 15–25 km (Fig. 4).

The influence of the initial model on the final velocity distribution was tested by effecting numerous inversions with different starting 3D and 1D models. We performed inversions using models characterized by slow and/or fast bodies that simulated the investigable portion of the down-going slab. In all these tests the experimental data produced 3D models comparable to the MCS model (Fig. 5). A 1D model with velocities 1 km/s higher than the MC below a depth of 40 km, led to a velocity distribution similar to that of the MCS model (Fig. 5).

Finally, we tested the dependence of the results on the data selection procedure by inverting two data sets:

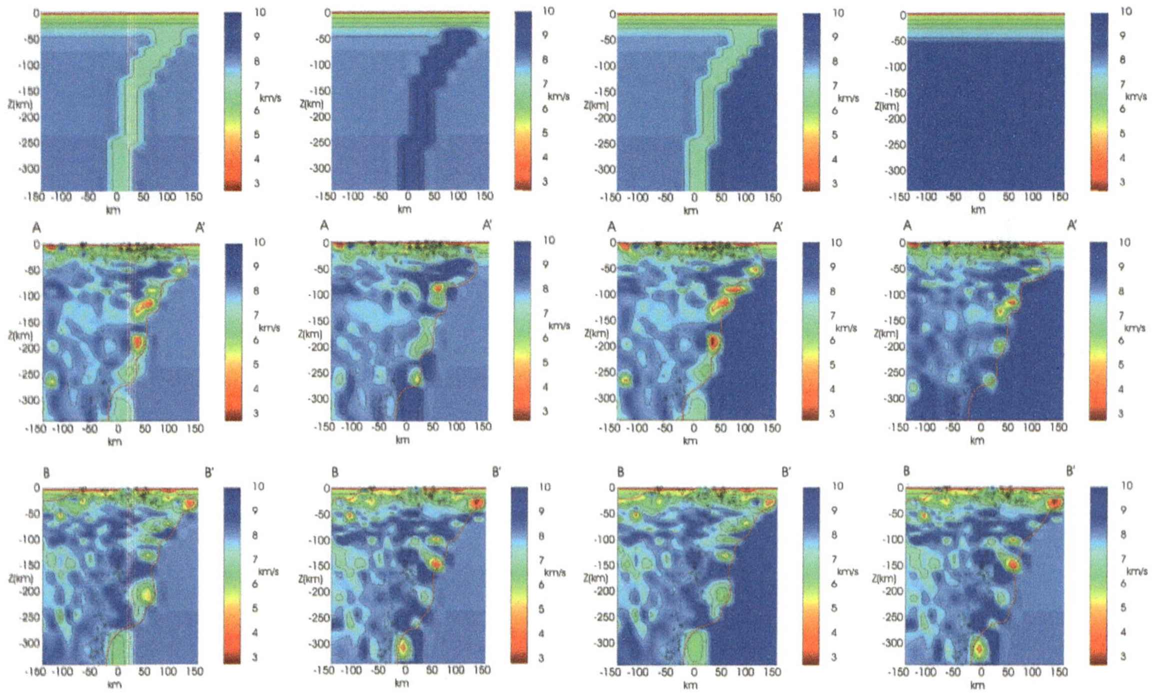

Fig. 5 Dependence of the tomographic results on the initial models. In the first row 1D and 3D initial models are displayed below which you can see the vertical sections obtained along the profiles A-A′ and B-B′. The red lines limit the portion of the poorly resolved model (*DWS* < 50)

the first included the events that occurred between 1988 and 2002, while the second encompassed the 1981–2005 period. We obtained numerous differential data for both data sets by varying the inter-event distance. All the inversions gave coherent results, however in the 1981–2005 data set the *DWS* increased consistently in many nodes that were poorly constrained by the smaller data set.

The spatial distribution of the differences between arbitrary pairs of models obtained in the experimental tests always shows, beside a low spatially incoherent component, low systematic misfits which have never skewed the interpreted velocity anomalies.

4 Tomographic Model

4.1 Construction

All the tests performed showed that the location of the anomalies is well constrained by the experimental data, yet in some inversions their extension and mean velocity value, in few but important portions of the *investigated volume*, are not sufficiently constrained. To overcome the limits of each tomographic inversion we adopted an average model for the final tomography, that is practically independent of the subjective choice of the parameters that affect the inversion process and is therefore more reliable (Calò et al., 2007, 2008). The final average distribution of all the parameters investigated will be named WAM (Weighted Average Model).

To construct this model among all velocity distributions obtained in the experimental tests, we selected $n = 27$ models, achieved with the perturbations of the input parameters that produced significantly different models.

In particular, beside some tomographies performed using various selection rules of absolute and differential data, others carried out with 1D and 3D starting models and with deformed and translated grids, we picked out several inversions relative to different rotations of the anisotropic grid to recover the spatial isotropy of the tomographic investigation.

These models, excluding the external cells, were resampled, using the same interpolation algorithm of tomoDD, on the nodes of a fixed grid (WAM grid) and the corresponding *DWS* distributions in a identical grid were interpolated by a trilinear interpolation algorithm. The WAM grid is centred at 38°40′N–15°.30′E (Fig. 2) and have horizontal spacing of 7 km and vertical spacing of 6 km.

In the WAM, the velocities of each node were calculated as weighted average by taking into account the *DWS* value of each inversion.

The equations used to determine the weights for the velocity values were:

$$\text{If: } DWS_i = 0 \Rightarrow w_i = 0$$

$$\text{If: } DWS_i \le LIM \Rightarrow w_i = \frac{LIM}{50}$$

$$\text{If:} DWS_i > LIM \Rightarrow w_i = \overline{DWS_i}\frac{(DWS_i - LIM)}{(DWS_{imax} - LIM)} + LIM$$

where $DWS_{i\,max}$ and $\overline{DWS_i}$ are the maximum and averaged *DWS* values in the *i*-th inversion respectively, *LIM* a threshold of *DWS* defined to pick out the nodes where the velocity estimates were reliable and the constant 50 was empirically optimized. The *DWS* was set to zero in the nodes where it was not possible to assign a velocity estimate. The *LIM* value was fixed at a *DWS* threshold equal to 100, as often adopted in literature to delimit well resolved areas (Zhang et al., 2004). The isosurface *DWS* = *LIM* will enclose the *investigated volume* of the final tomographic model. To determine the Vp and Vs weighted standard deviations in each node the same weights adopted to calculate the mean velocity were used.

The *DWS(i,j,k)* in the *(i,j,k)*-node of the WAM grid was determined with the following relationship:

$$DWS(i,j,k) = \frac{\sum_{i=1}^{n} w_i DWS_i(i,j,k)}{\sum_{i=1}^{n} w_i}$$

where w_i are the weights previously defined.

The average hypocentral model was determined; the mean standard deviations of each hypocentral coordinate are σ_x = 2.83 km, σ_y = 2.69 km and σ_z = 3.26 km respectively, the mean RMS is 0.11 s with standard deviation 0.05 s.

4.2 Reliability of the WAM Tomography

The effectiveness of the weighting technique that we propose in this study was verified and optimized by synthetic tests (Calò et al., 2007, 2008). The results of the tests together with the comparison between the restoration of a *classic* tomography and a *WAM tomography* are synthetically described.

To assess the reliability of the WAM method we implemented an extended version of checkerboard test that should be always performed before an experimental WAM tomography.

At first, we calculated the synthetic arrival times for a checkerboard model (Fig. 6a). Each patch is 28 × 28 × 24 km^3 large (4 nodes of the WAM grid along

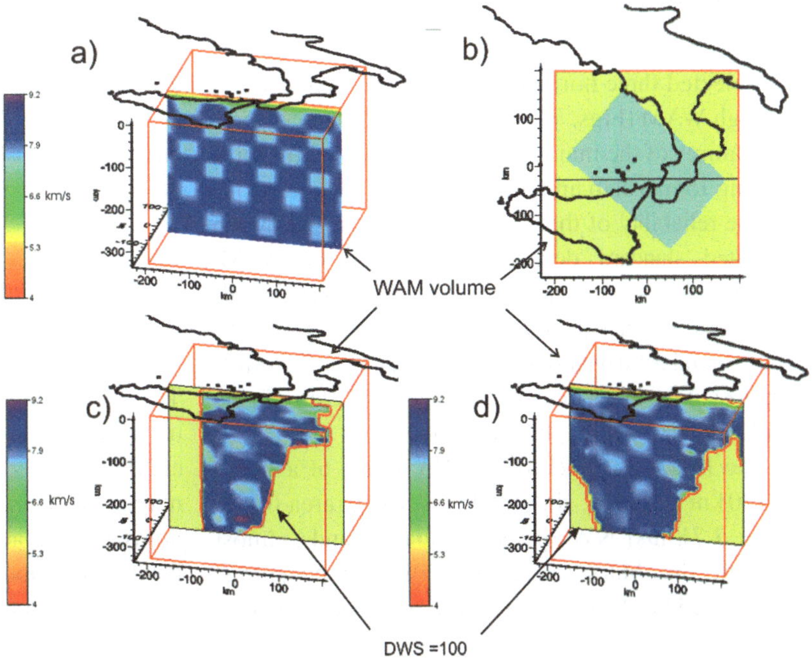

Fig. 6 (**a**) Checkerboard model used to calculate the synthetic travel-times, (**b**) example of an inversion grid (light-blue area) used to realize a single inversion, (**c**) recovered model of a single inversion relative to the inversion grid displayed in Fig. 6c, (**d**) reconstructed Vp WAM model. The red lines in the Fig. 6c, d delimit the *investigated volumes* (*DWS* > 100)

each direction), with velocities +6% and −6% alternatively with respect to the MC model.

Afterwards, we perturbed the synthetic data with increments normally distributed with null mean and standard deviation 0.6 s.

Subsequently, we performed about 60 inversions changing the input parameters and, for each one, we calculated a *restoration index* (Calò et al., 2007, 2008) to quantify the mean recovery of the synthetic model. Each inversion reconstructed approximately the true model within its *investigated volume* (*DWS* > 100) as shown in the example of Fig. 6c and d.

Finally, we constructed the velocity models (Fig. 6d), determined the standard deviation and calculated the *restoration index* for the WAM. The mean restoration for the single inversions was about 49% while for the WAM, constructed averaging n (n = 1...19) synthetic models relative to inversions characterized by very different input parameters vectors, increases up to 58% with an approximately quadratic dependence by the number n of the inversions.

We verified that there is in a good anti-correlation between the *restoration index* and the velocity standard deviation distributions of WAM (Calò et al., 2007, 2008).

The reconstructed WAM shows how the features of the true model are well restored without evident smearing effects and the *investigated volume* has been enlarged.

4.3 Results

We selected three horizontal and one vertical section through WAM (Figs. 7–10) to enhance some geometric features of the main anomalous bodies of the subducting Ionian slab and Southern Tyrrhenian mantle.

The reliability of these features was assessed by the weighted standard deviation distribution of WAM, determined as the standard deviation of the selected models divided by the square root of the number of models. These standard deviations were smaller than 0.06 km/s for most of the model and did not exceed 0.12 km/s in the border areas (Figs. 7a1, b1, c1, 8b and 9b).

The horizontal slice of the Vp model at a depth of 100 km (Fig. 7a), shows a low velocity area (6.6 < Vp < 7.7 km/s) NW of the Aeolian islands almost beneath the Marsili seamount. The hypocentres, within ±25 km from this section, are located within an SW-NE elongated high velocity body (Vp > 8.2 km/s) that images a portion of the Ionian slab. There is a remarkably low-Vp area (6.6 < Vp < 7.7 km/s) beside this body, located below the Calabrian Arc. The low velocity body seems to be sandwiched between two high-Vp bodies.

The section at a depth of 170 km (Fig. 7b) clearly shows the concave shape of the Ionian slab, here characterized by a 50–60 km large high velocity area. In this slice the low-Vp area, just beneath the Calabrian arc, is somewhat discontinuous. The low velocity regions NW of the Aeolian arc are still present.

At a depth of 190 km (Fig. 7c) the sandwich structure is even clearer, and is characterized by high Vp with a thin (20–25 km) low velocity layer. It is worth noting that at this depth the low velocity anomalies beneath the Marsili area are no longer imaged.

The vertical sections of the Vp, Vs and Vp/Vs models (Figs. 8–10) depict a portion (50–60 km thick) of the steeply dipping (69°–72° towards NW) descending Ionian slab, which is characterized by high velocities, both in Vp and Vs. Recent seismic studies using the Near-Vertical Reflection technique (CROP project; Finetti, 2005) have detected a thin, steep reflective layer down to a depth of about 140 km, which has been interpreted as being the top portion of the Ionian slab. It is portrayed in Fig. 7a within the high Vp region. Just below, that is, most likely in the Ionian lithospheric mantle, there is a 20–25 km thick area characterized by low-Vp (6.6–7.7 km/s) and -Vs (4.2–4.5 km/s). These low velocity bodies dip coherently with the overlying high velocity ones and fade out at a depth of 230–250 km. On the top of the Ionian slab vertically elongated low-velocity areas (Vp ≤ 7.0 km/s and Vs ≤ 4.5 km/s; Vp/Vs ratios > 1.85) beneath Stromboli and Marsili volcanoes reach a maximum depth of 180 km.

5 Petrological Constraints

The resolution of the tomographic results has allowed us to discuss the petrological implications of phase equilibria in the subducted lithospheric (harzburgitic) mantle with reasonable reliability. Hydration and dehydration reactions in the subducted ultramafics, by far the most abundant rock type, have relevant effects on mechanical properties, whole rock-H_2O budget and

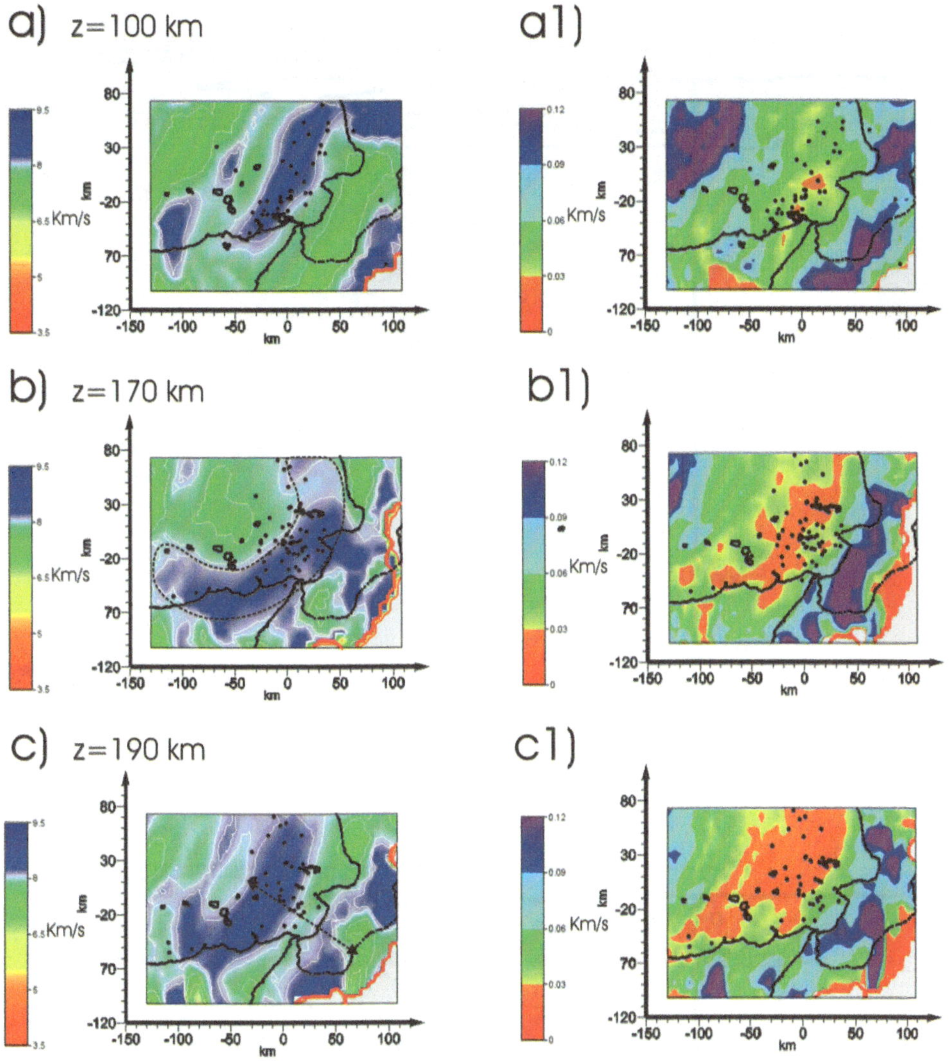

Fig. 7 Velocity planes at depths of 100, 170 and 190 km (a, b, c, respectively) and corresponding standard deviation distribution (a1, b1, c1, respectively). The red lines limit the portion of the poorly resolved model (*DWS* < 100). The black points are the projections onto the given plane of the hypocenters within ±25 km. The dashed black line in the slice at a depth of 170 km highlights the concave shape of the Ionian slab. The dashed double arrow in the slice at a depth of 190 km indicates the sandwich slab structure

consequently on P- and S-waves velocities (Hacker et al., 2003).

The P-T stability of H_2O-Mg-bearing phases in (hydrous) ultramafic systems is well constrained by experimental phase equilibria (Schmidt and Poli, 1998; Fumagalli and Poli, 2005; Ulmer and Trommsdorff, 1995 and references therein): antigorite (i.e., the high P polymorph of serpentine and also the major H_2O-carrier in the mantle; *Ulmer and Trommsdorff*, 2005) persist to a maximum pressure of 8 GPa (at 500°C), while its upper thermal stability is constrained at 600–650°C (at 5 GPa); chlorite instead breaks down for $T > 800$°C. Both phases transform through various reactions to "10 Å-phase" first and, at $P > 6$ GPa, to "phase-A" (Ulmer and Trommsdorf, 1999 a/b; Fumagalli and Poli, 2005), both belonging to the Dense Hydrous Magnesian Silicates (DHMS) group.

The placing of the maximum *T–P* bound of serpentine-chlorite stability at a specified depth into the subducted lithospheric mantle, depends on many variables that affect the thermal regime of the slab (e.g., slab

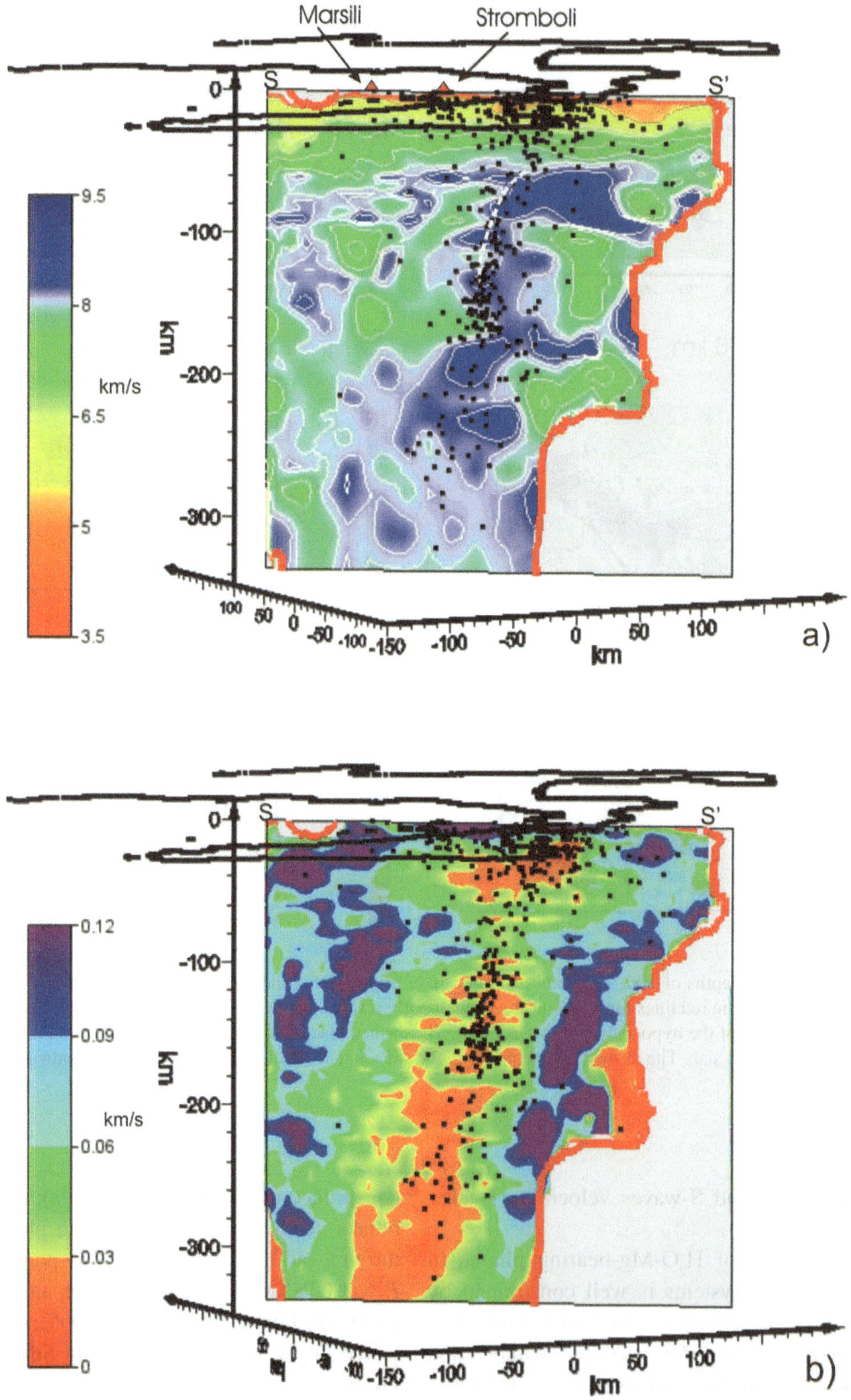

Fig. 8 (**a**) Vp section; (**b**) corresponding distribution of the standard deviation. Earthquakes within 15 km of the sections are also plotted. The best resolved area (*DWS* > 100) lies above the red line. The white dashed line in Fig. 7a is the trace of the NVR interpretative section (Finetti, 2005)

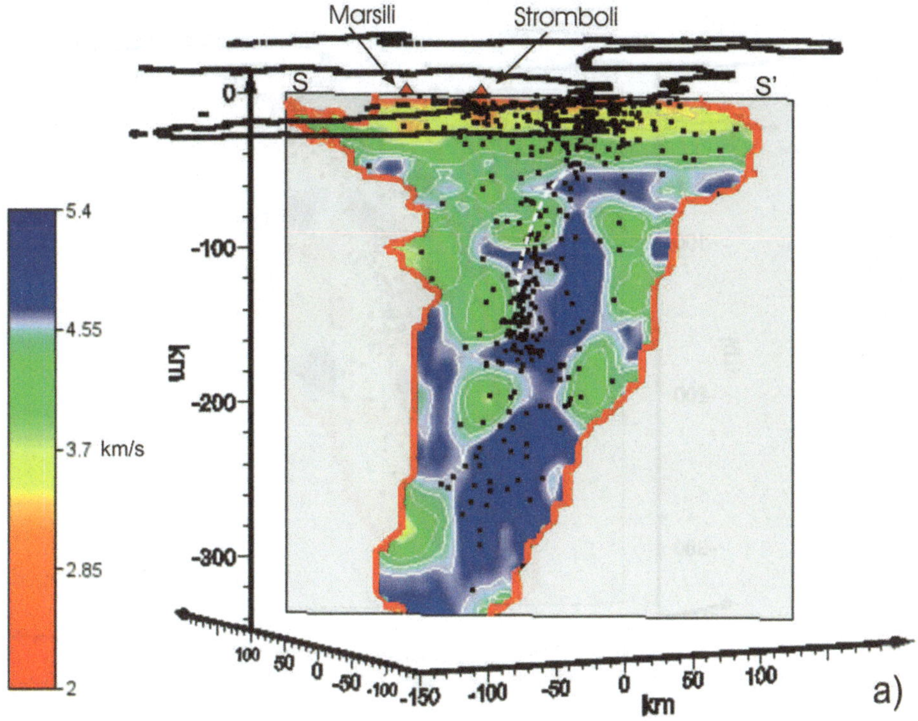

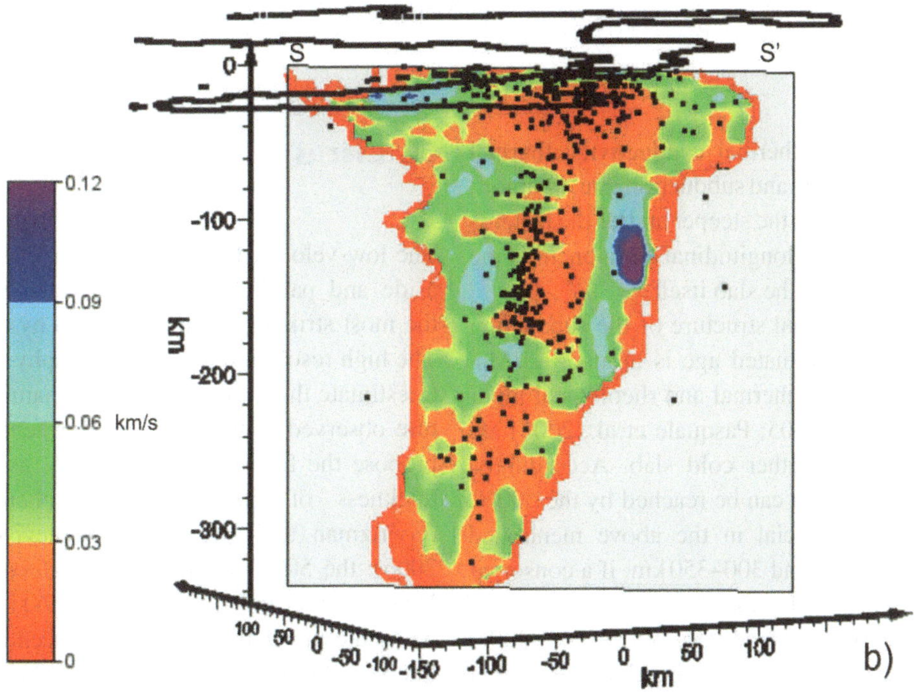

Fig. 9 (**a**) Vs section; (**b**) corresponding distribution of the standard deviation. Earthquakes within 15 km of the sections are also plotted. The best resolved area (*DWS* > 100) lies above the red line. The white dashed line in Fig. 9a is the trace of the NVR interpretative section (Finetti, 2005)

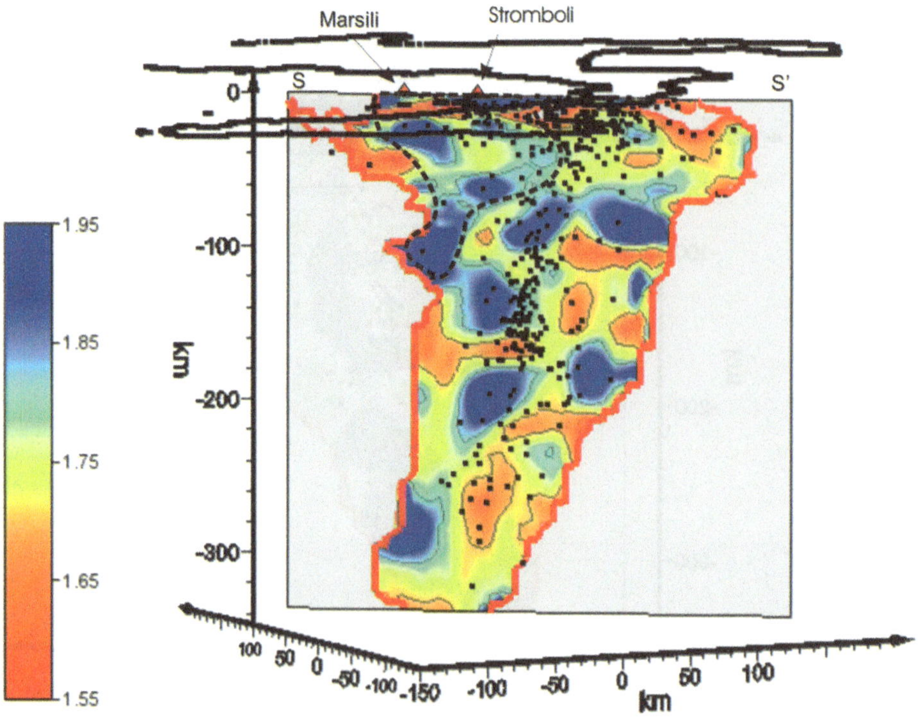

Fig. 10 Vp/Vs section. Earthquakes within 15 km of the sections are also plotted. The best resolved area (*DWS* > 100) lies above the red line. The black dashed line encloses the region of probable production/accumulation of the Tyrrhenian mantle partial melts

age, dip angle, rock's thermal conductivity, thermal distribution in the mantle and subduction rate; Iwamori, 2004). As a general rule, the steeper and/or the faster is the slab, the lower is the longitudinal component of the thermal gradient within the slab itself.

As regards the thermal structure of the subducting Ionian slab (whose estimated age is around 120 Ma, Carminati et al., 2005), thermal and rheological models (Carminati et al., 2005; Pasquale et al., 2005) are in agreement with a rather cold slab. Accordingly, the maximum depth that can be reached by the 600°C isotherm, which is crucial in the above mentioned phase equilibria, is around 300–350 km, if a conservative value of 3 cm/year for the subduction velocity of the Ionian slab is assumed (Rupke et al., 2004; Carminati et al., 2005). Must be mentioned here that the fluids released by dehydration reactions in the ultramafic portion of the slab are not univocally destined to migrate into the overlying mantle wedge (via fault-controlled pathways), but, at least for steeply subducting slabs, as the Ionian one, a partial up-dip flow (within the lower portion of the subducted slab) can be expected (Abers, 2005).

6 Discussion and Conclusions

The low-velocity bodies imaged in Figs. 8 and 9, just aside and parallel to the high-velocity volume, are the most striking feature produced by our inversions. The high resolution of the tomography has allowed us to estimate the dimensions of the main velocity structure observed in the Southern Tyrrhenian Sea and to propose the following assignations, assuming a total thickness of the Ionian lithosphere in 125 km (Gvirtzman and Nur, 2001; Pontevivo and Panza, 2006): the 50–60 km thick *high-velocity body* (8.2 < Vp < 8.8 km/s and 4.75 < Vs < 5 km/s) is interpreted as being the upper portion of the descending slab, most likely composed of: (1) ~10 km of eclogite, former oceanic crust (Vp = 8.4–8.7 km/s), and (2) ~50 km of anhydrous harzburgite (8.4 < Vp < 8.8 km/s), virtually undistinguishable one from the other.

The immediately underlying 20–25 km thick *low-velocity bodies* (7.0 < Vp < 7.7 km/s and 3.8 < Vs < 4.2 km/s) are instead attributed to a partially hydrated (serpentinized) harzburgite. These bodies are coincident with the inner (i.e., colder) portion of the slab and are separated by faster protrusions, probably corresponding to less hydrous mantle regions. The portion of the slab underlying the low velocity bodies should have a remaining thickness around 40–55 km.

Serpentinite is, in fact, characterized by low elastic wave velocities and high Poisson's ratio (Christensen, 2004). The relation between the decrease of Vp with increasing serpentinization in peridotites (Christensen, 2004) suggests that the decrease of Vp down to 7.0 km/s can be achieved with a 30–40 vol% of serpentinization. It is worth noting that this low-velocity region is aseismic, in agreement with the ductile rheology of serpentine-bearing rocks (Peacock and Hyndmann, 1999).

Interestingly, the low-velocity body disappears at a depth of 230–250 km, that is, close to the experimentally determined maximum pressure stability of antigorite-chlorite assemblages in hydrous peridotites: ca. 6.0 GPa at 600°C according to Fumagalli and Poli (2005) or even slightly higher P according to Hacker et al. (2003). The vanishing of the low-velocity region at a depth of > 230 km could thus be ascribed to the dehydration of the serpentine group minerals to H_2O-poor high pressure phases (e.g., the "phase A"). These latter are closely similar to anhydrous lherzolite minerals in their seismic properties (Hacker et al., 2003).

By comparing the tomographic results of this study with recent studies in the same region (Piromallo and Morelli, 2003; Montuori et al., 2007), there is a broad agreement regarding the position, the shape and the dimension of the main structures beneath the Southern Tyrrhenian basin, namely, the high velocity portion of the slab. The low velocity regions located SE of the high velocity body in the tomographies of Piromallo and Morelli (2003) and Montuori et al. (2007), being positioned outside the inferred slab volume are interpreted as lateral flows of hot astenospheric mantle (Montuori et al., 2007). In our model the low velocity regions occur well within the Ionian lithosphere and consequently must be viewed as being the result of hydration processes affecting the ultramafic portion of the slab.

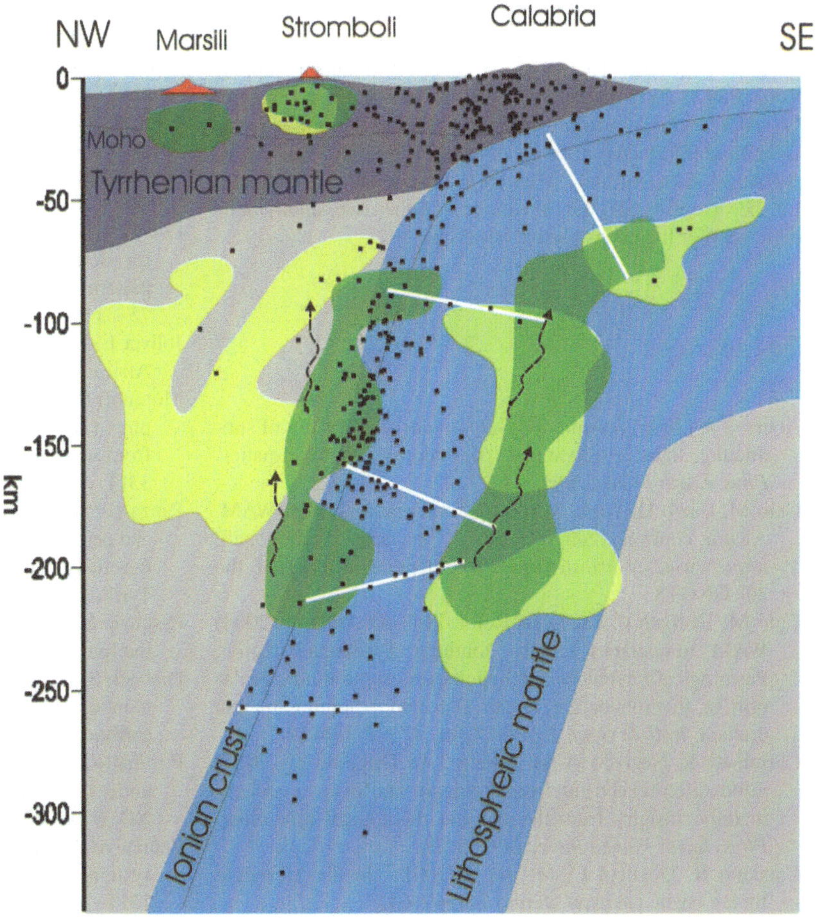

Fig. 11 Schematic sketch of the Ionian subduction scenario. The yellow and green bodies are relative to the low Vp and low Vs zones, respectively. The possible pathways (black arrows) for fluids released by dehydration of the ultramafic portion of the slab, include an up-dip migration within the subducted ultramafic lithosphere and, through *bend faults* (white lines), to the top of the subducted and the overlying mantle wedge

As regards the structure of the mantle wedge, our data suggest that partial melting, triggered by slab-released fluids, is likely to occur at a depth ≤180 km, in the large area characterized by low Vp and Vs in Figs. 8 and 9. This depth is in good agreement with experimental petrology constraints (e.g., Schmidt and Poli, 1998) and geochemical/petrological models (Tommasini et al., 2007).

The low-Vp and Vs and high Vp/Vs bodies imaged within the Tyrrhenian lithosphere beneath Stromboli, and less clearly Marsili volcanoes (Figs. 8–10), argue for the accumulation of significant amounts of mantle partial melts (i.e., some thousands of km^3) feeding the present-day volcanic activity.

The sketch in Fig. 11 summarizes the main features of the velocity models discussed above and illustrates two possible pathways for fluids released by dehydration of the ultramafic portion of the slab: (1) an up-dip migration (within the subducted ultramafic lithosphere) enhancing hydration of more superficial ultramafics and (2) through transverse channels, (e.g., *bend faults*) which deliver fluids to the top of the subducted crust and hence to the overlying mantle wedge.

Acknowledgments This research received financial support from PRIN 2004 funds from the Italian Ministry, project *"Hypotheses on the crustal structures along the Termini Imerese-Gela band (Sicily) based on the integrated analysis of geophysical and petrological data"*. We wish to acknowledge F. Funiciello for her careful editorial handling and the two anonymous reviewers for their constructive criticism.

References

Abers G A (2005) Seismic low-velocity layer at the top of subducting slabs: observations, predictions, and systematics. *Phys. Earth Planet. Inter., 149*, 7–29.

Calò M, Luzio D, Vitale M (2007) Tomografia sismica WAM. Alcuni confronti con tecniche di inversione tradizionali. *http://www2.ogs.trieste.it/gngts/*. Extended abstract of the 26° GNGTS.

Calò M, Dorbath C, Luzio D, Rotolo S G, D'Anna G (2008) WAM tomography in the southern Tyrrhenian region. Petrological inferences and hypotheses on the fluid circulation in the subducting Ionian slab and adjoining mantle domain. *Boll. di Geof. Teor. ed Appl., 42 N2*, 136–141.

Carminati S, Negredo A M, Valera J L, Doglioni C (2005) Subduction-related intermediate-depth and deep seismicity in Italy: insights from thermal and rheological modelling. *Phys. Earth Planet. Inter., 149*, 65–79.

Catalano R, Doglioni C, Merlini S (2001) On the Mesozoic Ionian basin. *Geophys. J. Int., 144*, 49–64.

Christensen N I (2004) Serpentinites, peridotites and seismology. *Intern. Geol. Rev., 46*, 795–816.

Faccenna, C, Davy P, Brun J P, Funiciello R, Giardini D, Mattei M, Nalpas T (1996) The dynamic of backarc basins: an experimental approach to the opening of the Tyrrhenian Sea. *Geophys. J. Int., 126*, 781–795.

Faccenna C, Civetta L, D' Antonio M, Funiciello F, Margheriti L, Piromallo C (2005) Contraints on mantle circulation around the deforming Calabrian slab. *Geophys. Res. Lett., 34*, L05302, doi:10.1029/2006GL028899.

Finetti I R (2005) The Calabrian arc and subducting Ionian slab from new CROP seismic data. *CROP PROJECT: Deep Seismic exploration of the Central Mediterranean and Italy*, 393–412, ISBN: 0444506934. Elsevier.

Fumagalli P, Poli S (2005) Experimentally determined phase relations in hydrous peridotites to 6.5 GPa and their consequences on the dynamics of subduction zones. *J. Petrol., 46*, 555–578.

Giunta G, Luzio D, Tondi E, De Luca L, Giorgianni A, D'Anna G, Renda P, Cello G, Nigro F, Vitale M (2004) The Palermo (Sicily) seismic cluster of September 2002, in the seismotectonic framework of the Tyrrhenian Sea-Sicily border area. *Ann. Geophys., 47*, 1775–1770.

Gvirtzman Z, Nur A (2001) Residual topography, lithospheric structure and sunken slab in the central Mediterranean. *Earth Plan. Sci. Lett., 187*, 117–130.

Hacker B R, Abers G A, Peacock S M (2003) Subduction factory 1) Theoretical mineralogy, densities, seismic wave speeds, and H_2O contents. *J. Geophys. Res., 108* (B1) 2029, 03311, doi:10.129/2001JB001127.

Iwamori H (2004) Phase relations of peridotites under H_2O-saturated conditions and ability of subducting plates for traportation of H_2O. *Earth Plan. Sci. Lett., 227*, 57–71.

Klein W F (2002) HYPOINVERSE-2000, a Fortran program to solve for earthquake locations and magnitudes. *U.S.G.S. Open-file rep.*, 02–171.

Lee W H K, Lahr J C (1985) HYPO71: a computer program for determining hypocenter, magnitude and first motion pattern of local earthquake. *U.S.G.S. Open-file rep.*, 75–311.

Jolivet L, Faccenna C (2000) Mediterranean extension and the Africa-Europe collision. *Tectonics, 19*, 1095–1106.

Montuori C, Cimini G B, Favali P (2007) Teleseismic tomography of the southern Tyrrhenian subduction zone: new results from seafloor and land recordings. *J. Geoph. Res., 112* (B0) 3311, doi:10.129/2005JB004114

Panza G F, Peccerillo A, Aoudia A, Farina B (2007) Geophysical and petrological modeling of the structure and composition of the crust and upper mantle in complex geodynamic setting. The Tyrrhenian sea and surroundings. *Earth Sci. Rev., 80*, 1–46.

Pasquale V, Verdoya M, Chiozzi P (2005) Thermal structure of the Ionian slab. *Pure Appl. Geophys., 162*, 967–986.

Peacock S M, Hyndmann R D (1999) Hydrous minerals in the mantle wedge and the maximum depth of subduction thrust earthquakes. *Geophys. Res. Lett., 26*, 2517–2520.

Piromallo C, Morelli A (2003) P wave tomography of the mantle under the Alpine-Mediterranean area. *J. Geophys. Res., 108*, NO. B2, 2065.

Pontevivo A, Panza G F (2006) The Lithosphere-Astenosphere system in the Calabrian arc. *Pure Appl. Geophys., 163*, 1617–1659.

Rupke L H, Morgan J P, Hort M, Connolly J A D (2004) Serpentine and the subduction zone water cycle. *Earth Planet. Sci. Lett., 223*, 17–34.

Schmidt M W, Poli S (1998) Experimentally based water budgets for dehydrating slabs and consequences for arc magma generation. *Earth Planet. Sci. Lett., 163*, 361–379.

Shelly D R, Beroza G C, Zhang H, Thurber C H, Ide S (2006) High-resolution subduction zone seismicity and velocity structure beneath Ibaraki Prefecture, Japan. *J. Geophys. Res., 111*, B06311, doi:10.1029/2005JB004081.

Thurber C H (1993) Local earthquake tomography: velocities V_P/V_S – theory. Seismic tomography. Theory and practice. *Chapman & Hall*, 563–580.

Tommasini S, Heumann A, Avanzinelli R, Francalanci L (2007) The fate of high-angle dipping slabs in the subduction factory: an integrated trace element and radiogenic isotope (U, Th, Sr, Nd, Pb) study of Stromboli Volcano, Aeolian Arc, Italy. *J. Petrol., 48*, 2407–2430.

Toomey D R, Foulger G R (1989) Tomography inversion of local earthquake data from the Hengill- Grensdalur central volcano complex, Iceland. *J. Geophys. Res., 94 B12*, 17497–17510.

Ulmer P, Trommsdorf V (1995) Serpentine stability to mantle depths and subduction-related magmatism. *Science, 268*, 858–861.

Ulmer P, Trommsdorf V (1999a) Phase relations of hydrous mantle subducting to 300km. In: Fei Y. W., Bertka C., & Mysen B. (eds) Mantle petrology: field observations and high pressure experimentation: a tribute to Francis R. (Joe) Boyd. *Geochemical Society Special Publications 6*, 259–281.

Ulmer P, Trommsdorf V (1999b) Serpentine stability to mantle depths and subduction-related magmatism. *Science, 268*, 858–861.

Zhang H, Thurber C H (2003) Double-difference tomography: the method and its application to the Hayward fault, California. *Bull. Seismol. Soc. Am., 93*, 1175–1189.

Zhang H, Thurber C H, Shelly D, Ide S, Beroza G C, Hasegawa A (2004) High-resolution subducting-slab structure beneath northern Honshu, Japan, revealed by double-difference tomography. *Geology, 32, 4*, 361–364.

Zhao D, Hasegawa A, Horiuchi S (1992) Tomographic imaging of P and S wave velocity structure beneath northeaster Japan. *J. Geophys. Res., 97*, 19909–19928.

Part III

Great Subduction Zone Earthquakes

Effect of Subducting Seafloor Topography on the Rupture Characteristics of Great Subduction Zone Earthquakes[1]

S. Das and A.B. Watts

Abstract Improvements in the quality and quantity of seismological data, together with technological advances in marine geophysics, mean that we are now able to examine in detail the influence of sea floor topography on the rupture process of great subduction earthquakes. Subducting seamounts were first suspected to affect the rupture process of a great earthquake in the 1986 M_w 8.0 Andreanof Islands earthquake, where large slip was seen in isolated round patches in the direction of the plate subduction. Since then, we have been able to show that a ridge and trough feature on the subducting oceanic plate stalled the rupture process of the 2001 M_w 8.4 Peru for ~30 s, and then broke, thereby resulting in the third largest earthquake worldwide since the 1960s. An important question is how much of a subducting oceanic plate bathymetric feature remains intact after subduction and how it affects earthquake rupture on the subduction plane. Recent high quality bathymetric and seismic surveys from the Middle America trench, for example, shows both the scars on the hanging wall associated with the subduction of a seamount, as well as large, clear, subducted seamounts after subduction. This paper discusses the rupture histories of four great subduction earthquakes in the Indian and Pacific oceans and examines the relationships between these histories and bathymetric features on the subducting oceanic plate.

Keywords Subduction-earthquake-rupture • seafloor-topography • South-America • Alaska • Indonesia

1 Introduction

It has long been recognised that subducting bathymetric features may influence the behaviour of subduction zone earthquakes. As far back as the mid-1970s, Kelleher and McCann (1976) considered the circum-Pacific belt and suggested that bathymetric highs such as aseismic ridges may control the location and frequency of great subduction zone earthquakes. Spence (1977) studied the Aleutian arc and suggested that the plate boundary was segmented by "distinctive bathymetry" such as submarine canyons perpendicular to the trench. For the 1965 M_w 8.7 Aleutians, Alaska earthquake, he suggested that the greatest earthquake slip occurred at zones under transverse canyons which were locked prior to the occurrence of the earthquake. This conclusion was based on the clustering of aftershocks in these regions. In today's jargon, we could say that

S. Das
Department of Earth Sciences, University of Oxford, Parks Road, Oxford, OX1 3PR, UK, das@earth.ox.ac.uk

[1]Submitted to Proceedings of SUBCO Conference, Springer; Accepted February 27, 2008.

S. Lallemand and F. Funiciello (eds.), *Subduction Zone Geodynamics*,
DOI 10.1007/978-3-540-87974-9,

he identified the major "asperities" of this earthquake. The well known segmented plate boundary off the Nankai trough, with its long documented earthquake history (Ando 1975) is remarkable, and shows how more than one such segment can break at a time, that is, how an earthquake at times propagates across the segment barrier, but not on other occasions.

Though the idea that submarine topography may affect the earthquake rupture process was implied in these very early studies, the basis on which they were made was sometimes tenuous, mainly due to the paucity of both earthquake and bathymetric data. For example, it is only in the late 1980s that the very first analysis of digital seismograms were carried out to find the history and distribution of the earthquake fault slip over the entire rupture area. Since then, the quality and reliability of such solutions has improved significantly. This improvement in seismological data has been accompanied by technological advance in swath bathymetry and satellite altimetry such that a much more complex, but reliable, picture has now started to emerge.

Many types of seafloor topography, irrespective of their origin, could affect the earthquake rupture process. This is particularly true of constructional features, such as oceanic islands and seamounts, for example, the Louisville Ridge (Scholz and Small 1997), off the Nankai region (Bangs et al. 2006; Kodaira et al. 2000) and off Costa Rica (von Huene et al. 1995, 2000; Bilek et al. 2003), which have been built up on the seafloor by volcanism.

Intuitively, all bathymetric highs, if they are still intact after subduction must locally increase the coupling across the subduction interface. In particular, subducting seamounts have now been suggested as a primary source of dense fracture networks in the forearc (Dominguez et al. 1999) and as being the region of high slip in subduction zone earthquakes (Cloos 1992; Scholz and Small 1997; Bangs et al. 2006). Scholz and Small (1997) have discussed how such seamounts could increase the recurrence time interval between earthquakes. Other features that have been proposed to influence earthquakes include fracture zone ridges and troughs, for example, off southern Peru (Robinson et al. 2006) and aseismic ridges, for example, the Nazca Ridge (Spence et al. 1999). Bilek (2007) provides a broad overall review of some aspects of the problem, and discusses some great earthquakes that occurred in the 1950s and the 1960s. Some details of the rupture process, such as rupture speed are not discussed there, due to the fact that using the earlier data many such important parameters could not be determined. The slip distributions for very old earthquakes could not be tested for 'robustness', again due to poor data quality. We emphasize the use of reliable robust features for the tectonic interpretation of the earthquake slip distributions for the recent earthquakes discussed in this paper.

Ruff (1989) considered the role played by trench sediments in controlling the size of subduction zone earthquakes. He showed that 13 of 19 large subduction earthquakes, including the three largest, occurred in zones with excess trench sediments, and speculated that "excess trench sediments are associated with the subduction of a coherent sedimentary layer, which at elevated temperature and pressure, forms a homogeneous and strong contact zone between plates". We shall return to this point later in the paper to see whether some of the recent great earthquakes fit this picture.

A fundamental question which relates to the degree of coupling across the subduction interface is, when a seamount subducts, does it get sheared off or is it preserved intact after subduction? The best example of an obducted seamount was found in the Snow Mountain volcanic complex in northern California (Macpherson 1981). A less clear example is known in Japan (Maruyama and Yamasaki 1978), where the whole structure had to be reconstructed by retro-deforming the thrust faults in order to separate out the seamount. However, whether all seamounts are obducted or if some are subducted is a question that had remained unanswered. Also, if a seamount is subducted intact, how well is the morphology of its crest preserved? For example, does a guyot retain its flat top or an atoll its fringing barrier reef? High-quality swath bathymetry data together with seismic and magnetic anomaly surveys of the forearc have started to address this question. Off the coast of New Zealand (Collot et al. 2001), along the Costa Rica subduction zone (von Huene et al. 1995, 2000), and along the Nankai trough (Kodaira et al. 2000; Bangs et al. 2006), scars can be observed in the forearc, formed by the penetration of seamounts. In the case of Costa Rica, a seismic section adjacent to a scar, clearly shows the 1.5 km high preserved seamount under the forearc which has been raised up due to it. This seamount is also seen in the bathymetric, gravity and magnetic data (Barckhausen et al. 1998). Along the Nankai trough, a 13 km thick by 50 km wide velocity anomaly was imaged in a tomographic model using wide-angle seismic refraction data (Kodaira et al. 2000). The velocity anomaly aligns with the Kinan seamount chain to its southeast, which is parallel to the plate motion vector there. The existence of this anomalous region was supported by a magnetic anomaly study, as well as by a bathymetric high above it. It was interpreted as a subducting seamount, which sits directly on the rupture plane, and within the rupture area, of the 1946 Nankaido earthquake, and is presumed to have stopped the rupture from propagating rapidly further. Even in cases where the seamount is damaged or broken up during subduction, the seismic coupling across the interface there would still be expected to be large.

In all the above discussion, we have assumed that it is the lower plate that controls the earthquake process. However, Collot et al. (2004) have suggested that earthquake ruptures in Ecuador and Colombia, on the subducting Nazca plate interface, were mainly controlled by transverse crustal faults in the overriding South America plate. Their study was based on seismic reflection rather than refraction (i.e. wide-angle) data and it is difficult on some of their profiles (e.g. SIS-32, SIS-24) to image crustal faults beneath the thick (up to 6km) sedimentary sequences of the forearc. However, as far as earthquake ruptures are concerned, both the underthrusting and the overriding plates probably play a role.

In this paper, we provide three examples of earthquakes in which the rupture process appears to have been influenced by the topography of the subducting oceanic plate. These three examples have been chosen for detailed discussions, as the distribution of slip on these faults are considered by us to be reliable, that is, the slip distributions have been tested for 'robustness'. The spatial grid size used in the inversion of the broadband body wave seismograms for these earthquakes was 20km. In addition, the 2004 and 2005 Sumatra earthquakes are mentioned, though not discussed in detail, firstly, due to their importance, and secondly, since much work is underway in the area, and rigorous discussion on them will only be possible in the future.

2 Seamounts, Asperities, and Barriers

Seamounts are, as Fig. 1 clearly shows, ubiquitous features of the ocean floor. They range in size from up to widths of 100km and heights of 5km above the surrounding seafloor depths and are found in a range of tectonic settings including mid-ocean ridges, transform faults and fracture zones, and deep-sea trench outer rises (Hillier and Watts 2007). They are also found in the plate interiors, especially at hotspots and on the crests of mid-plate topographic swells. It is perhaps not surprising therefore that when seamounts arrive on a subducting oceanic plate at a deep-sea trench, they might play some role in the earthquake cycle. The original papers of McCann and co-workers at Lamont–Doherty pointed to the role of relatively large-scale bathymetric features such as aseismic ridges, banks and rises. However, the advent of high-resolution satellite altimetry and swath mapping techniques to measure relatively short-wavelength bathymetry suggest that these features play an equally, if not more, important role.

Bilek et al. (2003) and Bilek and Engdahl (2007) have discussed how subducting seamounts can act as either asperities or barriers to rupturing, based on the work of Kodaira et al. (2000) and Abercrombie et al. (2001). Scholz and Small (1997) suggested that a hotspot generated ridge of seamounts can act as an asperity for earthquakes in Tonga–Kermedec while Robinson et al. (2007) were able to track bathymetric features-including seamounts-into a region of a fault plane that had acted as a barrier for the southern Peru earthquake.

In seismology, we generally refer to regions of high slip (and therefore high stress drop) as "asperities" (the word 'asperity' in tribology means a protuberance on one side of a surface). We generally call "barriers" regions with little or no slip. The barrier of one earthquake can therefore be the asperity of a future earthquake along the same plate boundary. Earthquakes often terminate at barriers. The asperity is like a rivet across the fault interface, which when ruptured, causes a great earthquake.

In the definition of barriers and asperities given above, we made reference to the amount of slip, but not to other rupture properties such as aftershock concentration and rupture speed. As discussed by Das and Henry (2003), with specific examples, no universal relation exists between the regions of high or low slip in an earthquake and aftershock concentration. The general pattern is that few, usually the smaller, aftershocks occur in the regions of high slip but exceptions exist. Along the Alaska subduction zone, they showed that along the same portion of the fault, different behaviour was seen in two earthquakes

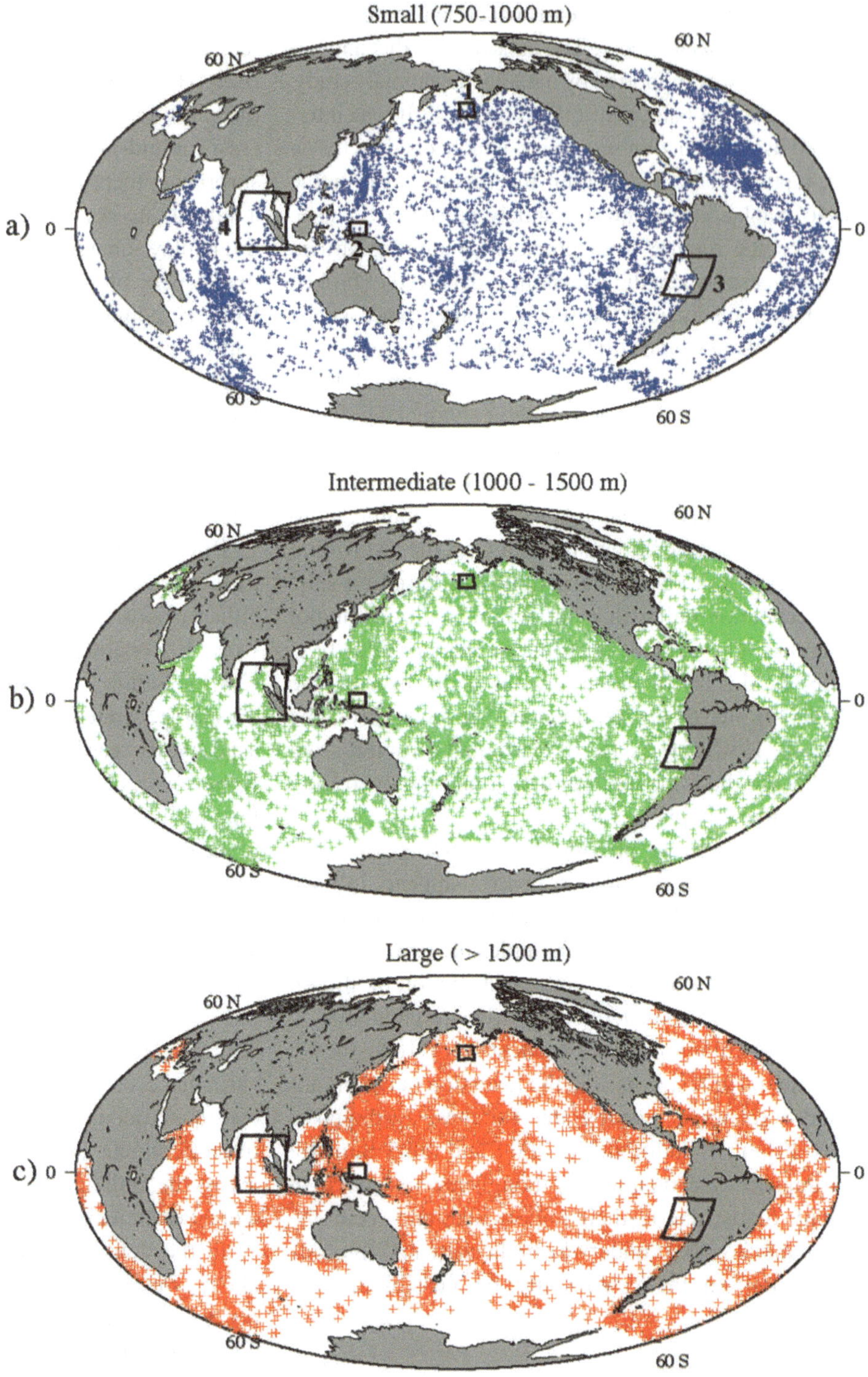

Fig. 1 The global distribution of seamounts, ridges and banks based on shipboard bathymetry data (Hillier and Watts 2007). The bathymetric features have been colour coded according to their height above the surrounding regional seafloor. The thick lines outline the focus sites in this paper: 1-Andreanof Islands, Alaska; 2-Biak, Indonesia; 3-southern Peru; 4-Sumatra

about 30 years apart, and have outlined a scenario in which this could happen. We show in this paper that while subducting plate morphology and the earthquake cycle maybe linked, there is also no universal relation between the regions of high and low slip and specific bathymetric highs and lows on the seafloor. A similar conclusion was also reached by Bilek (2007).

3 Specific Examples of Earthquake Ruptures Which Appear to be Affected by Subducting Seafloor Roughness

3.1 *The 1986 Mw 8.0 Andreanof Islands, Alaska Earthquake*

This was the first earthquake which seemed to suggest a correlation between the high slip areas in an earthquake and topography of the subducting oceanic plate. The earthquake occurred on the subduction interface where the Pacific plate is descending under the Aleutian arc. The earthquake rupture was 280 km long along strike and the fault plane had a dip of about 20°. The final slip distribution on the fault, obtained by Das and Kostrov (1990) by inversion of broadband body wave seismograms, is plotted in Fig. 2a. It is seen that most of the slip on the fault was confined to the elongated region of higher slip in the west. The complete distribution of slip time history, obtained by Das and Kostrov (1990), but not shown here, illustrates how this region continued to slip from about 25 s after rupture nucleation until the almost complete cessation of all motion on the fault, with the part of the fault to the east having little activity from 35 s on after rupture initiation. The shape and trend of the high slip region to the west was clearly developed 40 s after the time of rupture initiation and about 50% of the moment release occurred in this region, which covers only about 25% of the total fault area. A region of low slip parallel to the high slip region and to its east is also clear. The result that the major slip in this earthquake was to the west of the hypocenter agrees with other studies of this earthquake (Hwang and Kanamori 1986; Boyd and Nabelek 1988; Houston and Engdahl 1989). The average speed in the 90 km section of the fault to the east of the hypocenter (the star in Fig. 2a) was found to be ≈3.2 km s^{-1} (0.85v_s) and ≈2 km s^{-1} (0.5v_s) in the 190 km section lying to the west, where v_s is the shear wave speed of the medium. The average rupture speed through the central region of low slip was about the shear wave speed and that through the region of high slip was less than 0.5v_s. It must be kept in mind that the quality of seismic data and the station distribution in 1986 was not as good as it is today. Nevertheless, the results, which were tested for robustness by Das and Kostrov (1990), suggest that the region of higher rupture speed (but lower slip) and the region of lower rupture speeds (but higher slip) represent mechanically weaker and stronger regions of the fault interface, respectively, relative to the state of stress on the fault when the earthquake occurred. The high slip region overlapped in large part the region where quiescence was observed prior to the 1986 event (Kisslinger 1988), but we see that the earthquake did not nucleate in this region (Fig. 2a). Instead, it nucleated 90 km to the east at a region which was not the region of highest slip on the fault but was a local maximum in slip in the eastern part of the fault. Thus the observed quiescence seems to have correctly identified the high slip region of the future earthquake but not the zone of rupture initiation. Based on the relocated hypocenters obtained by Ekstrom and Engdahl (1989), Das and Henry (2003) showed that the region of high slip has fewer aftershocks than the rest of the fault, both for the 3-week and the 1-year aftershock distributions.

Figure 2a, b shows the plate convergence vector (DeMets et al. 1994), and that there are a number of bathymetric features on the subducting oceanic plate visible in the GEBCO bathymetry and satellite-derived gravity data. The most prominent is the Amlia Fracture Zone, a N–S trending ridge and trough feature that formed as a result of early Tertiary transform motion between the Kula and Pacific plates (e.g. Mortera-Gutierrez et al. (2003)). Other features include small seamounts and a number of ridges. Some of the ridges have an E–W trend (Fig. 2a, b), which is orthogonal to the Amlia Fracture Zone and, hence, parallel to the spreading direction. Others, however, appear to be isolated features with no clear trend. Irrespective of this, individual features on the subducting Aleutian oceanic plate are large, reaching heights of up to 500 m above the surrounding regional seafloor depth and widths of up to 25 km (Fig. 2a).

From the earthquake slip, it is seen that the trends of the regions of high and low slip in the earthquake generally align with the plate convergence direction. The region of high slip of the Andreanof Islands earthquake, consisting of three isolated round patches, generally align along the direction of plate subduction as well as with clusters of bathymetric features on the subducting oceanic plate. The GEBCO grid and individual ship tracks (e.g. cruises p6829 and pi4s2) show (Fig. 2a) a number of visible seamounts on the subducting oceanic plate to the south of the central region of low slip (and high rupture speed), but with a trend aligning with the western high slip (low rupture speed) region. Since no swath bathymetry surveys of the kind that have been

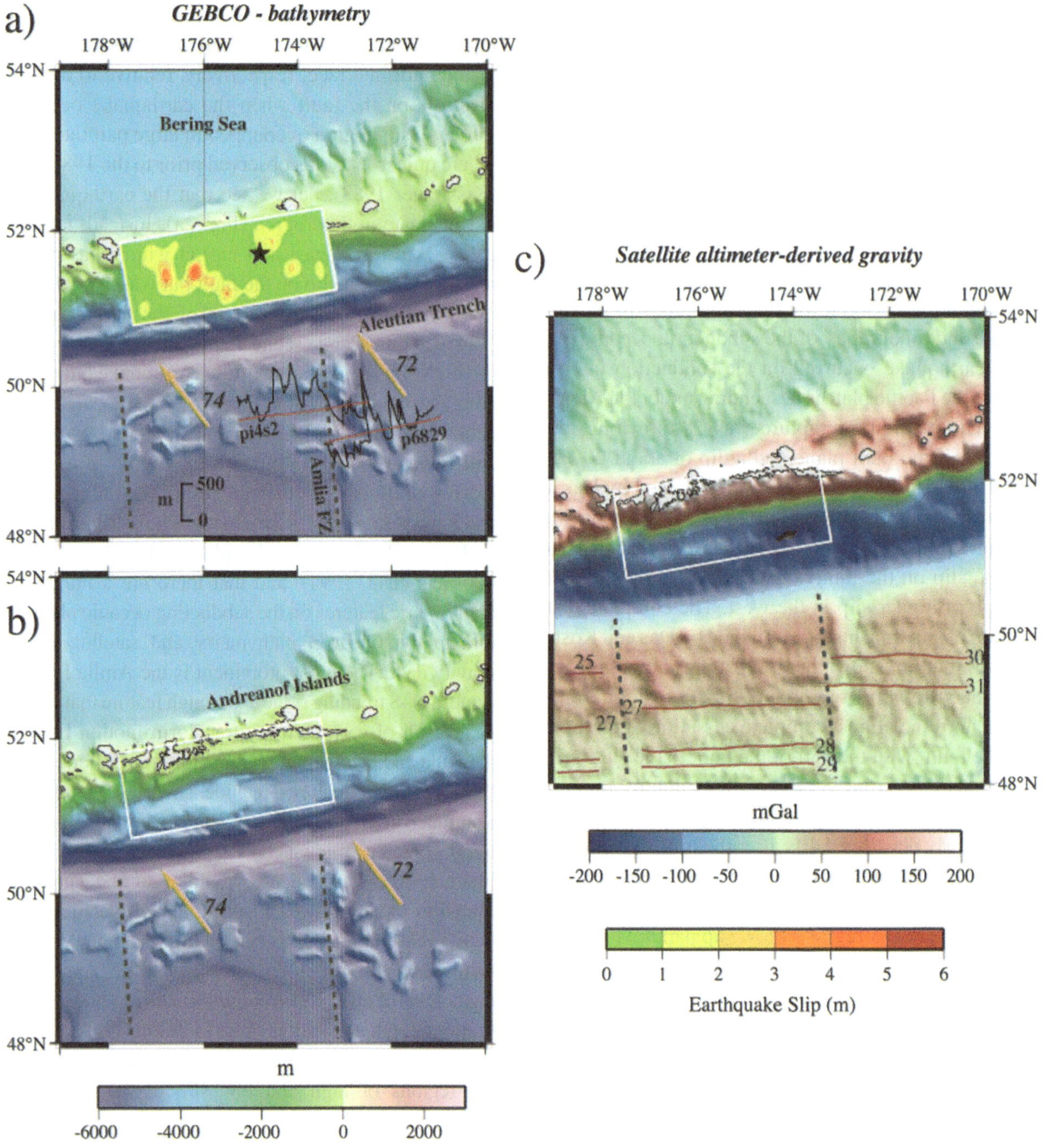

Fig. 2 (**a**) The final slip distribution for the 1986 M_w 8.0 Andreanof Islands, Alaska earthquake (Das and Kostrov 1990), superimposed on the GEBCO (2003) bathymetry data. The *star* shows the region where rupture initiated. The scale used to plot the slip distribution (in metres) is shown. The plate convergence direction is shown as brown arrows, with the plate velocity (mm per year) marked next to them (from DeMets et al. (1994)). *Black lines* show the bathymetry along ship tracks (indicated by the *dark red lines*) acquired during cruises p6829 and pi4cs. The bathymetry scale (m) is shown. Dark grey dashed lines show fracture zones based on Cande et al. (1989). (**b**) The *rectangular* slip area for this earthquake superimposed on the GEBCO bathymetry data, to reveal the indentation of the forearc in the region of the slip zone. (**c**) The *rectangular* slip area for the Andreanof Islands earthquake, superimposed on the satellite-derived gravity field of Sandwell and Smith (1997)-Version 15.2. *Red lines* show magnetic lineations based on Cande et al. (1989). The numbers next to the red lines are magnetic polarity chrons (chrons 25–31 correspond (approximately) to 57–70 Ma). The scale key for the gravity field (mGal) is shown

carried out in Costa Rica or the Nankai trough exist in this region yet, we cannot tell if some seamounts have been subducted here. However, the indentation of the forearc slope between longitudes 174° W and 177° W (Fig. 2b), together with the slip distribution, is very suggestive.

Figure 2b, c also show that the bathymetric indentation in the Aleutian fore-arc correlates with a strongly negative free-air gravity anomaly. Song and Simons (2003) and Llenos and McGuire (2007) constructed an average trench-island arc gravity anomaly and subtracted it from the observed anomaly. They noted that the resulting residual anomaly correlated with variations in the seismogenic behaviour of the subduction zone and suggested that both observations might be linked to the structure of the fore-arc. We believe, however, that the indentation of the fore-arc in the region of the Andreanof island earthquake and its associated negative free-air anomaly is the consequence, at least in part, of seamount subduction which in turn is the main control on the seismicity.

The Amlia Fracture Zone and a second fracture zone to its west appear to be related to the ends of the fracture. Extrapolating these, we see that the earthquake has terminated close to the western fracture zone. The earthquake slip is small (remember that very small slip may not be resolvable (Das and Kostrov 1990)) beyond the extrapolation of the Amlia Fracture Zone, in the eastern direction. Thus, both ends of the earthquake rupture appear to have been terminated by bathymetric features on the subducting oceanic plate.

3.2 The 1996 Mw 8.2 Biak, Indonesia Earthquake

At the time of its occurrence, the February 17, 1996 Biak earthquake was the largest thrust earthquake worldwide since 1977. It occurred where the Pacific plate is subducting under New Guinea. Henry and Das (2002) inverted body wave seismograms to determine its rupture history. They found that the earthquake occurred on a shallow dipping thrust fault (9° dip). The rupture propagated bilaterally at an average speed of ~90% of the shear wave speed, on a fault extending 180 km west and 50 km east of the hypocenter, with an extremely variable width ranging from 30 to 100 km at different locations along strike (Fig. 3a). The mean slip over the $230 \times 100\,km^2$ fault area was 4 m and the mean stress drop 1.9 MPa. The slip distribution is highly non-uniform over the fault, with the largest slip of ~12 m being near the hypocentral depth (~10 km). The rupture process is very complex, propagating first to the west and then, after a about 15-s delay, to the east. This delay was interpreted by Henry and Das (2002) as being due to the existence of a inhomogeneous barrier just east of the hypocenter, which initially acted as a barrier to rupture propagation towards the east, but subsequently failed due to stress increase on it generated by the rupture to its west, and with the rupture then continuing to propagate eastwards. This barrier also acted as a stress concentrator before the earthquake, giving rise to foreshocks and was also the initiator of rupture for this earthquake. The aftershock zone corresponds closely to the region in which rupture occurred, the area with greatest aftershock density lying entirely within the area of highest moment release, an exception to the usual behaviour of regions of high slip having fewer afrershocks (Das and Henry 2003).

Figure 3a shows that the main region of high slip, and the postulated barrier to its east, are both generally aligned in the direction of the plate motion vector. Moreover, there is an arcuate gravity high in the region of high slip which might represent one or more subducted seamounts. The GEBCO bathymetry data suggest that compared to the Aleutians there are relatively few bathymetric features on the subducting oceanic plate. However, the satellite-derived gravity data shows some structure (Fig. 3b), and magnetic anomaly studies (Weissel and Anderson 1978; Hegarty et al. 1988) suggest one or more fracture zones may intersect the New Guinea trench at a high angle. This suggests that the "inhomogeneous barrier" (that is a barrier due to higher strength rocks) identified by Henry and Das (2002) could have actually been a "geometrical barrier" caused by the non-planar ascept of the fault plane. A second concentrated gravity high to the west of this is close to the second patch of high slip to the west.

3.3 The 2001 Mw 8.4 Peru Earthquake

This earthquake is a remarkably good example where bathymetric features on the subducting oceanic plate initially acted a "barrier", stalling the rupture, and then

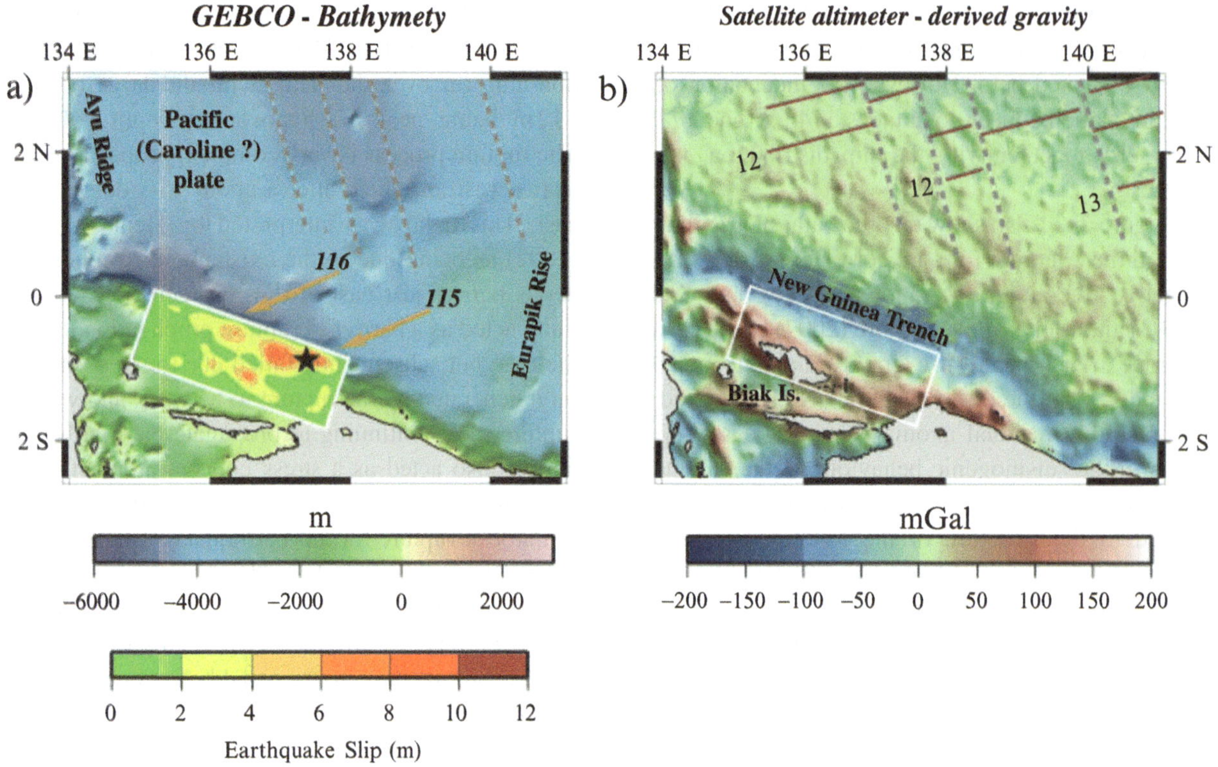

Fig. 3 (**a**) The slip distribution for the 1996 M_w 8.2 Biak, Indonesia earthquake (Henry and Das 2002) superimposed on the GEBCO (2003) bathymetry data. (**b**) Satellite-derived gravity field of Sandwell and Smith (1997) - Version 15.2, with the *rectangular* zone of (**a**) marked. *Grey dashed lines* and *red lines* show fracture zones and magnetic lineations (Hegarty and Weissel 1988). Symbols, scales, etc. are as in Fig. 2

was overcome, thereby acting as an "asperity", and producing the world's third largest earthquake since 1965.

The 2001 southern Peru earthquake occurred on a 400 km long fault dipping 15°, within a ~1,000 km long seismic gap on the plate boundary where the Nazca plate subducts under the South American plate. Figure 4 illustrates the tectonic setting of the earthquake. The history of known earthquakes in the region (Fig. 5) shows that earthquakes in the southern Peru region has been followed within about a decade by one in northern Chile, more than once in known history.

By analyzing broad band seismograms, Robinson et al. (2006) found that the earthquake had a very complex rupture history, which is shown in Fig. 6. A schematic of the final slip, together with the bathymetry, is shown in Fig. 7a. The earthquake unilaterally ruptured southeastwards, initially propagating ~70 km, then propagated around a ~6,000 km^2 region (the barrier), after which it continued propagating further southeastwards. At ~48 s after initiation, though the earthquake rupture was 180 km long along strike, it had accumulated less than 15% of its final moment. About 54 s after initiation, the initially unbroken barrier (shown by the dotted line in Fig. 7a) started breaking, and caused large slip in its vicinity, particularly, in the shallow regions updip of it. This asperity (once the barrier is completed surrounded by slipping regions, it becomes an asperity) ruptures completely in ~36 s, at an average speed of ~2.8 km s^{-1}, which is lower than the average rupture speed of ~3.5 km s^{-1} (this is the same as the local shear wave speed in the medium) over the rest of the fault. The term asperity is generally associated with regions of high slip, but in this case, the feature we call an asperity is a region of relatively lower slip. However, we still refer to it as an asperity, as it was the rupture of this "rivet" that transformed the earthquake from a M_w 7.8 at 48 s after initiation, into an M_w 8.4 earthquake eventually. This barrier is a very robust feature of Robinson et al.'s (2006) solution which fits the data very well.

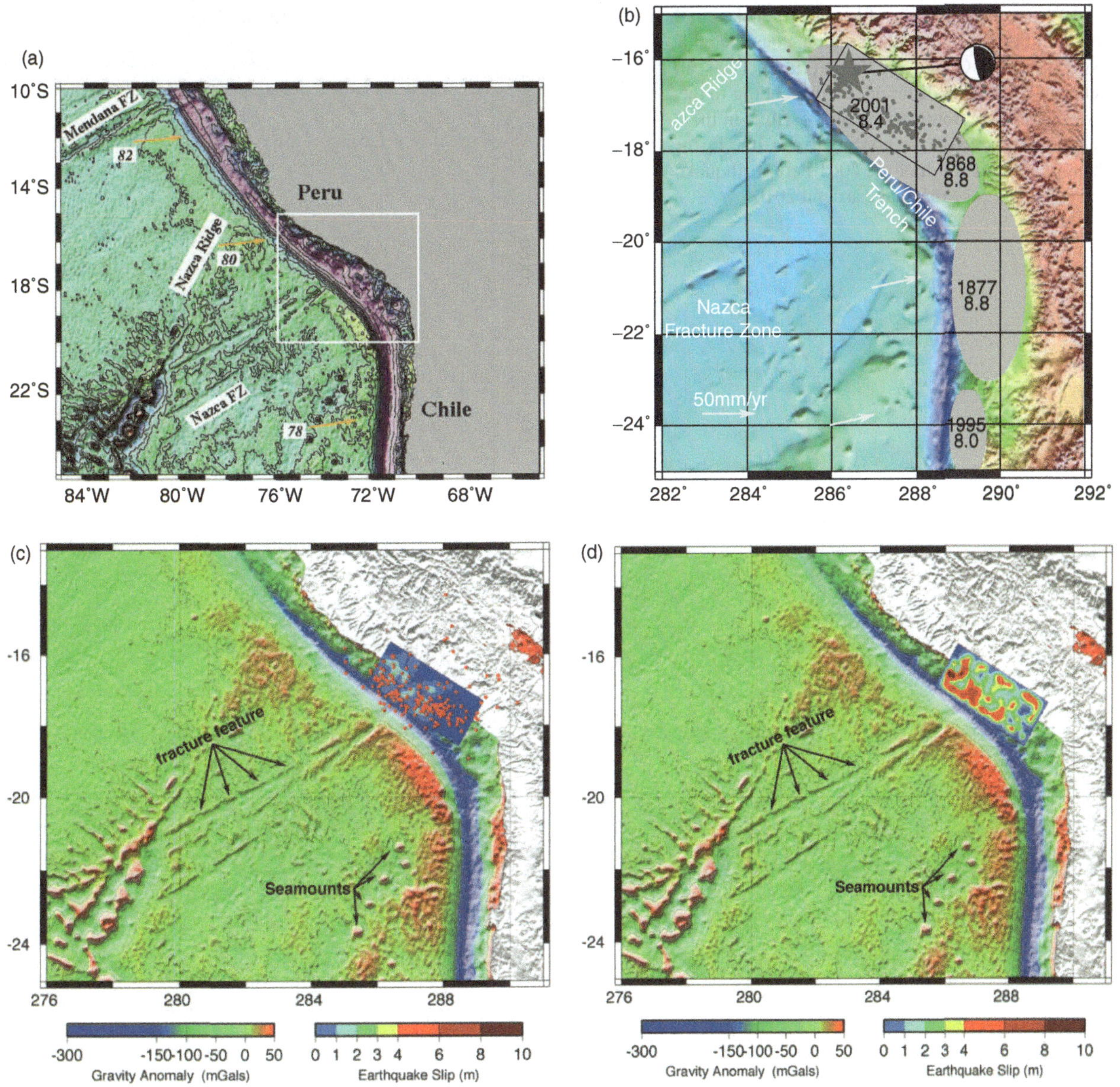

Fig. 4 (**a**) Tectonic setting of the M_w 8.4 2001 southern Peru earthquake. (**b**) The 1-day aftershocks of the earthquake, together with the rupture areas of the previous earthquake in the region outlined in *grey*. *Arrows* show the direction of plate convergence (from DeMets et al. (1994)), its length showing the plate velocity. (**c**) Contoured satellite-derived gravity field of Sandwell and Smith (1997)-Version 15.2-off-shore of the Peru earthquake, and the slip distribution about 42 s after rupture initiation shown. (**d**) Same as (**c**) but with the final co-seismic slip in the earthquake plotted. Symbols, etc. are as in Fig. 2

Shipboard bathymetry data (Fig. 7a, b) show a 25–50 km wide ridge that rises up to 700 m above the expected depth of the oceanic crust for its age (thin red line in Fig. 7a, b) and flanking 20–25 km wide trough that is up to 300 m deeper than the expected depth, which can be traced for a distance of 275 km from the trench seaward wall to the Nazca Fracture Zone in the southwest. The ridge, scarp, and trough feature resemble the morphology of an oceanic fracture zone and may be an eastern continuation of the Nazca Fracture Zone. However, the ridge, scarp and trough have a different azimuthal trend than the fracture zone, and magnetic lineation compilations (e.g. Cande et al. 1989) suggest there is little or no age offset across it. Irrespective of its origin, the feature aligns remarkably well with the first barrier and, we

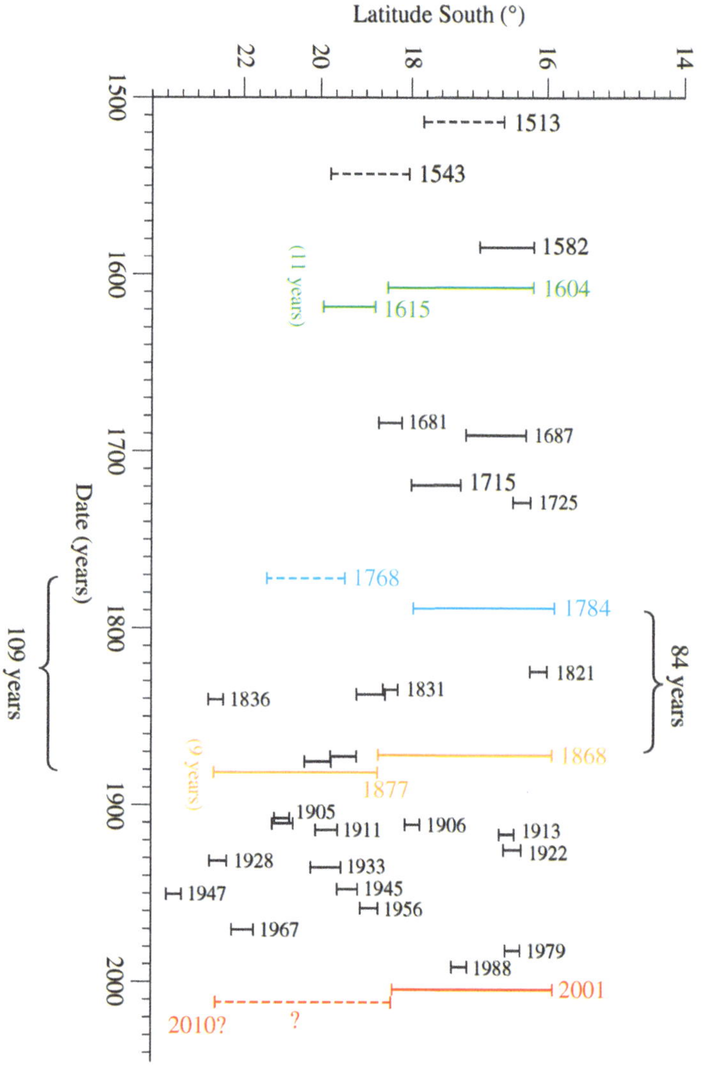

Fig. 5 Space-time history of large earthquakes in the southern Peru-northern Chile region, constructed using data from Comte and Pardo (1991). The scale between latitudes is different, corresponding to the projection of the South American coastline in the north–south direction. According to the time-history, the region to the south of the 2001 Peru earthquake is a candidate for earthquake around 2010

suggest, increases the coupling between the two sides of the fault, resulting in a highly heterogeneous earthquake rupture history. The aftershock distribution (Fig. 4) was found to be very non-uniform over the fault, and the trapezoidal shape of the first asperity coincides with a trapezoidal region of lower aftershock density, visible both in the 24-h and the 6-month aftershock distributions (Robinson et al. 2006). This is in contrast to the usual observation that regions of higher slip have fewer aftershocks (Das and Henry 2003).

That such complexity can arise in reality is reinforced from sandbox experiments in the laboratory, where it is found that nucleation of rupture can occur in front of a subducting seamount on a ramp, and then propagate in its wake subhorizontally (Dominguez et al. 2000).

3.4 The 2004 and 2005 Great Sumatra Earthquakes

The M_w 9.3 December 26, 2004 Sumatra earthquake is the second largest earthquake of the last 100 years worldwide. It was followed 3 months later, on March 28, 2005 by a M_w 8.7 immediately to its south, on the continuation of the same plate boundary (Fig. 8). Larger aftershocks of the two earthquakes are shown in this figure. In the first earthquake, the rupture

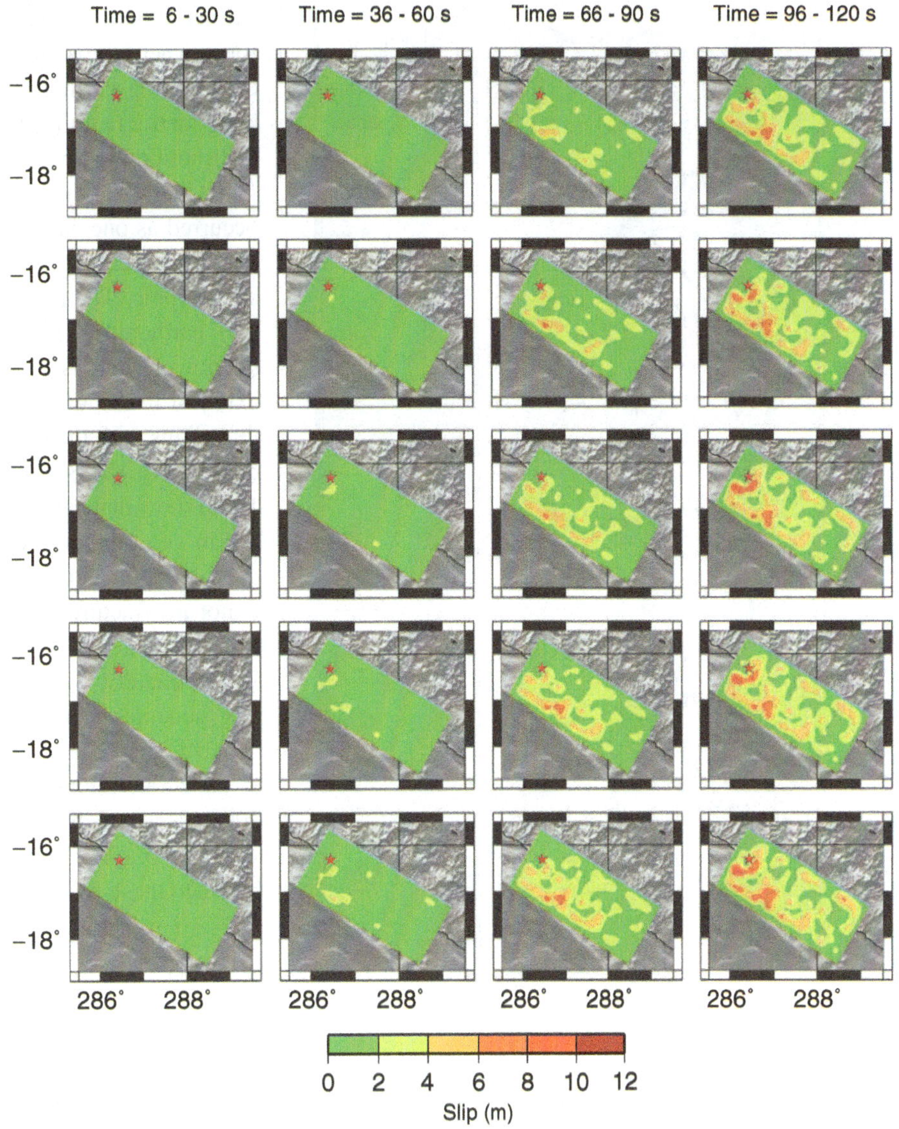

Fig. 6 Complete rupture and slip history for the 2001 Peru earthquake (based on Robinson et al. 2005). The snapshots are shown at 6 s intervals. The scale used for the earthquake slip is shown

initiated at its southern end and unilaterally propagated northwards. Three months later, the second earthquake commenced some distance south of the termination of the first one and propagated primarily southwards.

Based on our previous studies, we would expect that the two Sumatra earthquakes, the largest and second largest earthquakes worldwide since 1965, to have been influenced by bathymetric features on the subducting oceanic plate. However, Fig. 8 shows that the NW part of the subducting Indo–Australian plate is remarkably smooth due to the influence of the Bengal deep-sea fan which developed following uplift of the Himalaya during the Eocene (Curray et al. 2003). Even the northern segment of the 90° East aseismic ridge is subducted and obscured by sediment at its intersection with the trench west of the Andaman Islands. The southwestern part, however, is underlain by thinner sediments and is associated with a number of linear ridges, troughs and small isolated seamounts. The most prominent feature on the GEBCO bathymetry map (Fig. 8a) is the N–S trending Investigator Fracture Zone which, interestingly, intersects the trench at a point where there is an abrupt change in the distribution of aftershocks associated with the 2005 Sumatra earthquakes, namely, there are no aftershocks here. The fracture zone, which formed following a major rotation of the spreading direction as Greater India cleared the Wallaby Plateau ~100 Ma, is marked by a sharp bathymetric offset that is small in the south and reaches 700 m in the north (Larson et al. 1978). The satellite-derived gravity map

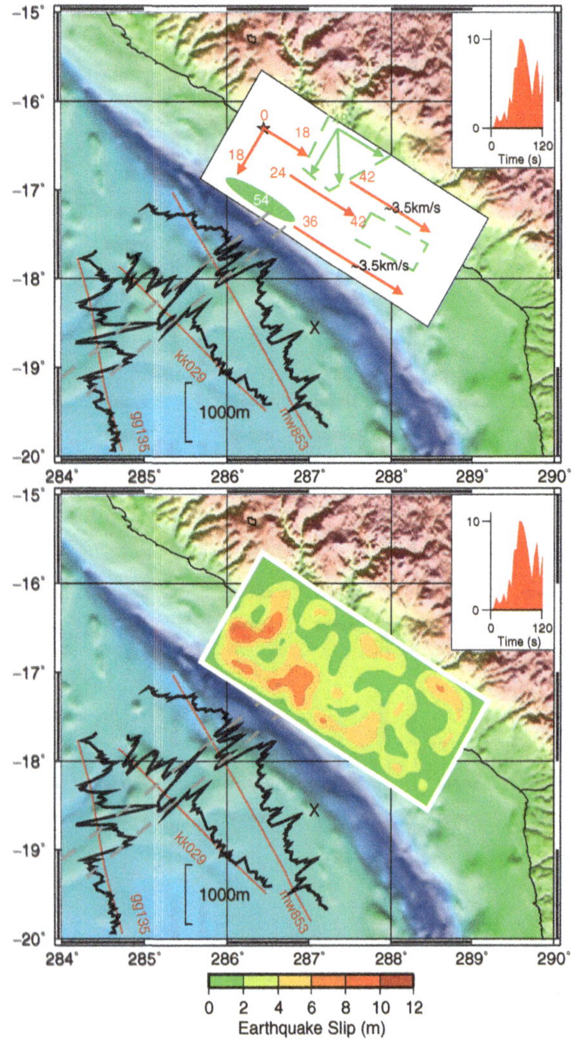

Fig. 7 (**a**) Schematic rupture process for the 2001 Peru earthquake superimposed on the GEBCO (2003) bathymetry data. *Arrows* show the earthquake rupture directivity, labeled with start and selected end times of segments. *Red* indicates the primary rupture and green the rupture of asperities. The two asperities are shown by *green dashed lines*. Approximate rupture speeds are indicated in *black*. *Red lines* on the ocean show selected ship tracks mw853, kk029, gg135, which have been labeled according to their cruise identification in the GEODAS data base. *Thick black lines* show the residual bathymetry (i.e. the observed bathymetry minus the bathymetry expected for the age of the oceanic crust) projected orthogonally onto the ship track. *Thick grey dashed lines* show the ridge and trough feature that can be correlated on each ship track. The 'moment-rate function' for the earthquake is shown in the inset, the ordinate being in 10^{19}N m s^{-1}. (from Robinson et al. 2006). (**b**) Same as (**a**) but showing the final slip distribution. (based on Robinson et al. 2006)

(Fig. 8b) suggest, however, a number of additional fractures zones between Investigator and the 90° East Ridge that are now obscured by sediments of the Bengal fan. Moreover, Deplus et al. (1998) have shown that the extinct Wharton basin spreading centre intersects the forearc near the southern extremity of the aftershocks of the 2005 earthquake. Irrespective, it is likely that bathymetric features have had a major influence on the distribution of slip and aftershocks.

If the 2004 and 2005 Sumatra earthquakes had occurred as one earthquake, resulting in an even larger one, its effects would obviously have been even more catastrophic. The physical origin of what caused these two earthquakes to occur 3 months apart remains unknown. There is, however, an inflection of the trench axis just to the west of the Investigator Fracture Zone that aligns with the region between the two great earthquakes (Fig. 8a). Seismic studies (e.g. Moore and Curray 1980) show that the forearc in this region is underlain by ~3 km of sediment. Unfortunately, this obscures the deep structure and so we do not currently know whether or not any fracture zone related feature has actually have been subducted at the Sumatra trench.

Fortunately, several marine geophysical surveys (e.g. seismic, heat flow, swath bathymetry) are being, or have very recently been, conducted (e.g. Henstock et al. 2006; Ladage et al. 2006; Delisle and Zeibig 2007, etc.), which will significantly increase our understanding of the deep structure of the forearc region between these two great earthquakes.

4 Conclusions

With the improvement in both seismic and bathymetric and satellite-derived gravity data, we are now starting to see the influence of sea floor topography on the rupture process of great subduction earthquakes. It is interesting to note that its effect on the earthquakes discussed in detail here are somewhat different. For the 1986 Andreanof Islands earthquake, the region which appears as a line of subducting seamounts had high slip. But for the 2001 Peru earthquake, the barrier region we identified as being due to subducting seafloor topography, had low slip. In both cases, these regions were identified as being due to subducting bathymetric highs, and had slower rupture speed and fewer aftershocks than the areas surrounding it. We suggest that the difference in behaviour may be due to the part of the seismic cycle the earthquake occurs in. Since eventually the entire plate boundary has to "catch-up", the region of low slip in one earthquake, must have higher

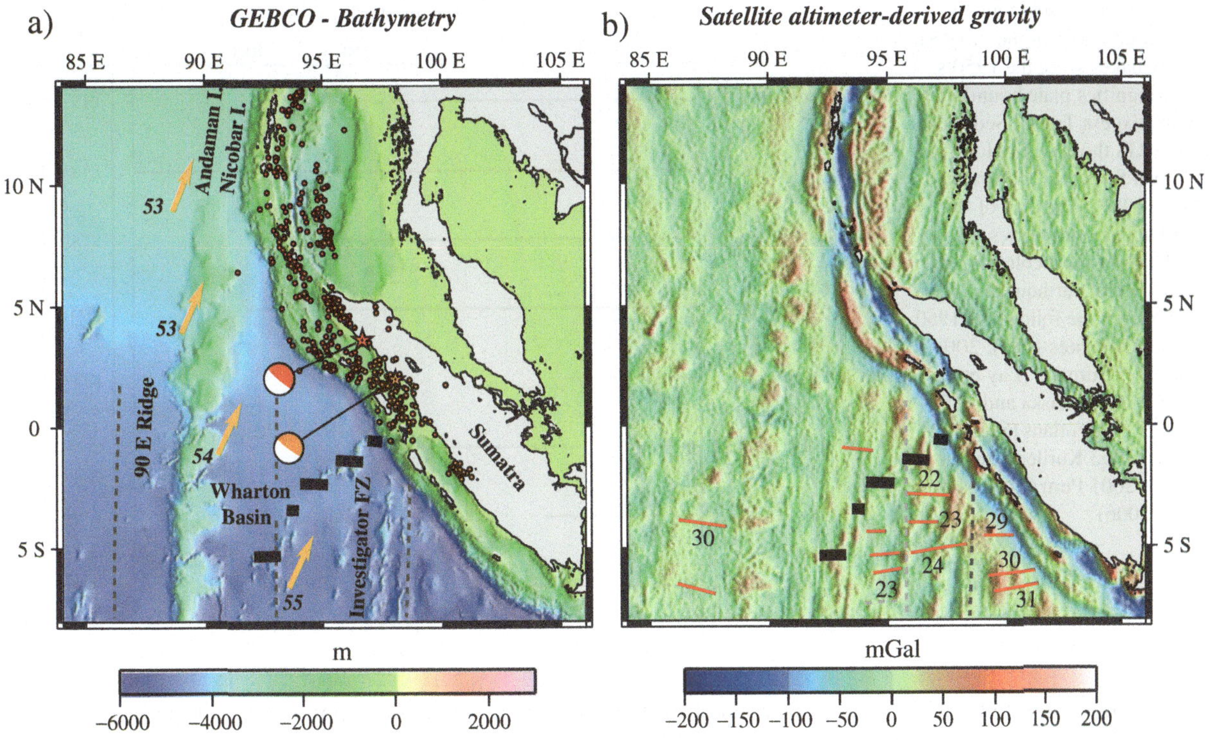

Fig. 8 Tectonic setting of the 2004 and 2005 Sumatra earthquakes. (**a**) The larger aftershocks of the M_w 9.3, 26 December, 2004 (*red*) and the M_w 8.7 March 28, 2005 earthquakes (*orange*), superimposed on the the GEBCO (2003) bathymetry data, are shown. *Thick black lines* show the location of the extinct spreading centre that according to Deplus et al. (1998) generated oceanic crust in the Wharton basin. (**b**) Satellite-derived gravity anomaly based on Sandwell and Smith (1997)-Version 15.2. *Grey dashed lines* and *red lines* show fracture zones and magnetic lineations respectively (Larson et al. 1978)

slip in some future earthquake. Das and Henry (2003) suggested that the high slip region for the 1986 Andreanof Islands earthquake was a low slip region in the preceding 1957 M_w 8.8 Rat Islands, Alaska earthquake. By this reasoning, we can speculate that the next large earthquake which occurs within the rupture zone of the 2001 Peru earthquake will either have large slip in the region of the barrier that delayed rupture, or, that the barrier region will slip aseismically.

We have not considered in our discussion so far the role played by the trench sediment fill in controlling the rupture history of great subduction earthquakes. Ruff (1989) pointed out that the truly great earthquakes such as the M_w 9.5 May 22, 1960 Chile earthquake, the M_w 9.2 March 28, 1964 Alaska earthquake and the M_w 9.0 November 4, 1952 Kamchatka earthquake all occurred in subduction zones with excess trench sediments. An updated figure to include recent great earthquakes and their relationship to the nature of the trench fill is given in Fig. 9. The table shows that both the 2004 and the 2005 Sumatra earthquakes are also associated with excess trench fill. The 2001 Peru earthquake, on the other hand, occurred in region which Ruff (1989) identified as having a "horst and graben structure" rather than "excess sediments". If we look only at the truly great earthquakes ($M_w > 9$), we can say that the December 26, 2004 Sumatra earthquake supports Ruff's hypothesis, so that all known $M_w \geq 9$ earthquakes occur in subduction zones with excess trench sediments.

It is interesting to note in Fig. 9 that with the exception of the 1957 Aleutian and the 1965 Kuril events, all the other subduction earthquakes have dips of less than 15°, the dip for the 1960 earthquake not being reliably known. Thus, most of the megathrust earthquakes seem to occur on nearly horizontal faults. The dips of the 1957 and 1965 earthquakes may be less reliable, and perhaps the dips were actually shallower than believed now. Future earthquakes in this region should resolve this issue.

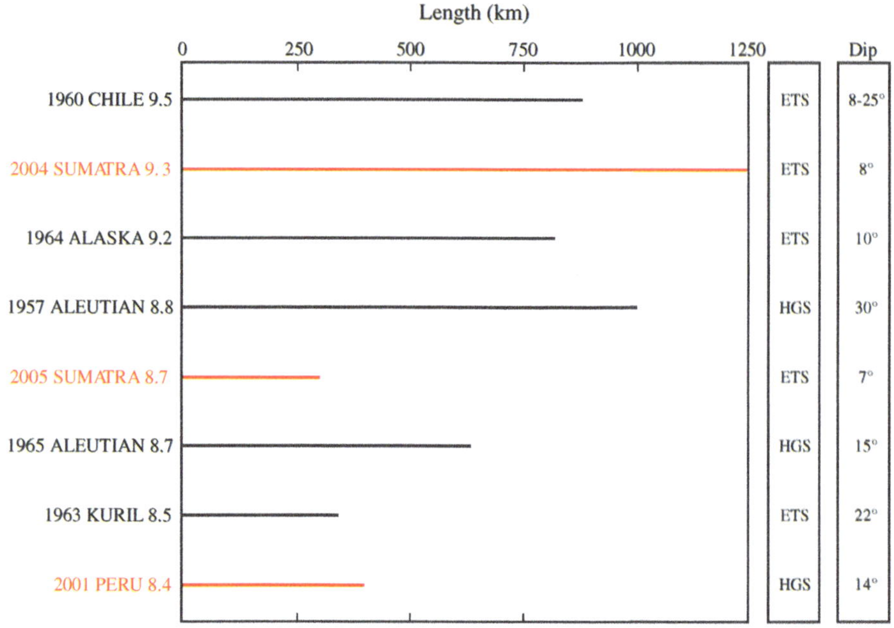

Fig. 9 Length and moment-magnitude (M_w) of the greatest earthquakes since the 1950's. ETS signifies plate boundaries with 'Excess Trench Sediments' and HGS those with 'Horst and Graben Structure', following Ruff (1989). The dips of the earthquake fault planes, shown in the right-most column for the different earthquakes, were taken from the following: 1960 Chile (Cifuentes 1989); 2004 and 2005 Sumatra (Lay et al. 2005); 1964 Alaska and 1957 and 1965 Aleutians (Sykes 1971); 1963 Kurile (Kanamori 1970); 2001 Peru (Robinson et al. 2006)

One question that still remains is how much of the seamount exists after it is subducted to depths of 20 or 30km, to be able to affect the earthquake rupture on the lower parts of the subduction plane? Recent swath bathymetry data from the Middle America trench shows subducted seamounts ~15–20km from the trench (von Huene et al. 2000). If the shape of the subducting seamount is preserved as it subducts deeper and deeper, it is clear that the fault would not be planar and the rupture would either have to shear the base of the seamount or, it would have to rupture over the uplifted surface of the seamount. Such a scenario has been suggested for a recent earthquake near Java (Abercrombie et al. 2001).

A related question is, what determines whether a seamount is obducted on to the forearc or subducted at a trench? One factor that may play a crucial role is the strength of the coupling of a volcanic seamount with the underlying oceanic crust at the boundary that separates them (Macpherson 1981). Therefore, when a seamount does subduct, it suggests that this coupling is very strong and the integrity of the seamount is more likely to be better preserved as further subduction proceeds. Another factor might be the mode of isostatic compensation of the seamount on the subducting plate. Oceanic flexure studies show that, in general, seamounts that form off-ridge are regionally supported while on-ridge have a more local compensating 'root' (Watts et al. 1980). A consequence of this is that as an off-ridge seamount approaches a trench, the leading edge of its broad compensating 'root' will be removed, so that when it arrives at the trench it is only partially supported and will be carried down with the subducting plate. An on-ridge seamount, in contrast, will have little or none of its support removed before it arrives at a trench, and so will tend to remain buoyant and be obducted rather than subducted.

We suggest that that the subduction of bathymetric features that scallop the forearc, leaving scars and long run-out debris flows may be more common than we know now and will be identified with increasing frequency with improving bathymetric, gravity, magnetic and seismic surveys. We hope that our work will encourage such studies for major subduction zones around the world.

Acknowledgements We would like to thank Dave Sandwell of IGPP, SIO, UCSD for providing his satellite-derived gravity data V15.1. We would also like to thank the Editor, Serge Lallemand, who handled our paper, as well as Sue Bilek and an anonymous reviewer for very thoughtful comments, which helped to improve and clarify our paper.

References

Abercrombie RE, Antolik M, Felzer K, Ekström G (2001) The 1994 Java tsunami earthquake: slip over a subducting seamount. J Geophys Res 106:6595–6607.

Ando M (1975) Source mechanisms and tectonic significance of historical earthquakes along the Nankai trough. Tectonophysics 27:119–140.

Bangs NLB, Gulick SPS, Shipley TH (2006) Seamount subduction erosion in the Nankai Trough and its potential impact on the seismogenic zone. Geol Soc Am 34:701–704.

Barckhausen U, Roeser, HA, von Huene, R (1998) Magnetic signature of upper plate structures and subducting seamounts at the convergent margin of Costa Rica. J Geophys Res 103:7079–7093.

Bilek SL (2007) Influence of subducting topography on earthquake rupture, in The Seismogenic Zone of Subduction Thrust Faults, ed. T. Dixon and C. Moore, Columbia University Press, New York, 123–146.

Bilek SL, Engdahl, R (2007) Rupture characterization and aftershock relocations for the 1994 and 2006 tsunami earthquakes in the Java subduction zone. Geophys Res Lett 34 L20311 DOI:10.1029/2007GL031357.

Bilek SL, Schwartz SY, deShon HR (2003) Control of seafloor roughness on earthquake rupture behavior. Geol Soc Am 31:455–458.

Boyd TM, Nabelek JL (1988) Rupture process of the Andreanof Islands earthquake of May 7, 1986. Bull Seismol Soc Am 78:1653–1673.

Cande SC, LaBrecque JL, Larson RL, Pitman WC, Golovchenko X, Haxby WF (1989) Magnetic lineations of the World's ocean basins, scale 1:27,400,000. Am. Assoc. Pet. Geol (AAPG map), Tulsa, Oklahoma.

Cifuentes IL (1989) The 1960 Chilean earthquakes. J Geophys Res 94:665–680.

Cloos M (1992) Thrust-type subduction-zone earthquakes and seamount asperities: A physical model for seismic rupture. Geology 20:601–604.

Collot JY, Lewis K, Lamarche G, Lallemand S (2001) The giant Ruatoria debris avalanche on the northern Hikurangi margin, New Zealand: Result of oblique seamount subduction. J Geophys Res 106:19,271–19,297.

Collot JY, Marcaillou M, Sage F, Michaud F, Agudelo W, Charvis P, Graindorge D, Gutscher MA, Spence G (2004) Are rupture zone limits of great subduction earthquakes controlled by upper plate structures? Evidence from multichannel seismic reflection data acquired across the northern Ecuador-southwest Colombia margin. J Geophys Res 109 DOI:10.1029/2004JB003060.

Comte D, Pardo M (1991) Reappraisal of great historical earthquakes in the northern Chile and southern Peru seismic gaps. Nat. Hazards 4:23–44.

Curray JR, Emmel FJ, Moore DG (2003) The Bengal Fan: morphology, geometry, stratigraphy, history and processes, Marine and Petroleum. Geology 19:1191–1223.

Das S, Henry C (2003) Spatial relation between main earthquake slip and its aftershock distribution. Rev Geophys 41:1013, 3.1–3.23, doi:10.1029/2002RG000119.

Das S, Kostrov BV (1990) Inversion for seismic slip rate and distribution with stabilising constraints: Application to the 1986 Andreanof Islands earthquake. J Geophys Res 95:6899–6913.

Das S, Kostrov BV (1994) Diversity of solutions of the problem of earthquake faulting inversion. Application to SH waves for the great 1989 Macquarie Ridge earthquake. Phys Earth Planet Int 85:293–318.

Delisle G, Zeibig M (2007) Marine heat flow measurements in hard ground offshore Sumatra. Eos Trans Am Geophys Un 88:38–39.

DeMets C, Gordon RG, Argus DF, Stein S (1994) Effect of recent revisions to geomagnetic reversal time scale on estimates of current plate motions. Geophys Res Lett 21:2191–2194.

Deplus C et al. (1998) Direct evidence of active deformation in the eastern Indian oceanic plate. Geology 26:131–134.

Dominguez S, Malavieille J, Lallemand SE (2000) Deformation of accretionary wedges in response to seamount subduction: Insights from sandbox experiments. Tectonics 19:182–196.

Ekstrom GA, Engdahl ER (1989) Earthquake source parameters and stress distribution in the Adak Island region of the central Aleutian Islands, Alaska. J Geophys Res 94: 15,499–15,519.

GEBCO (2003, 2006) The GEBCO Digital Atlas, Centenary Edition (Cd-ROM). British Oceanographic Data Centre, Liverpool, UK.

Hegarty KA, Weissel JK (1988) Complexities in the development of the Caroline plate region, western equatorial Pacific. In Nairn AEM, Stehli FG, Uyeda S (Eds.) The Ocean Basins and their Margins. Vol 7B. Plenum, New York, 277–301.

Henry C, Das S (2002) The M_w 8.2 February 17, 1996 Biak, Indonesia earthquake: Rupture history, aftershocks and fault plane properties. J Geophys Res 107(B11)2312 DOI:10.1029/2001JB000796.

Henstock TJ, McNeill LC, Tappin DR (2006) Seafloor morphology of the Sumatran subduction zone: Surface rupture during megathrust earthquakes? Geology 34:485–488.

Hillier JK, Watts AB (2007) Global distribution of seamounts from ship-track bathymetry data. Geophys Res Lett 34, doi:10.1029/2007GL029874.

Houston H., Engdahl ER (1989) A comparison of the spatio-temporal distribution of moment release for the 1986 Andreanof Islands earthquake with relocated seismicity. Geophys Res Lett 16:1421–1424.

Hwang LJ, Kanamori H (1986) Source parameters of the May 7, 1986 Andreanof Islands earthquake. Geophys Res Lett 13:1426–1429.

Kanamori H (1970) Synthesis of long-period surface waves and its application to earthquake source studies–Kurile Islands earthquake of October 13, 1963. J Geophys Res 75:5011–5027.

Kelleher J, McCann W (1976) Buoyant zones, great earthquakes, and unstable boundaries of subduction. J Geophys Res 81:4885–4896.

Kisslinger C (1988) An experiment in earthquake prediction and the May 7, 1986 Andreanof Islands earthquake. Bull Seismol Soc Am 78:218–229.

Kodaira S, Takahashi N, Nakanishi A, Miura A, Kaneda Y (2000) Subducted seamount imaged in the rupture zone of the 1946 Nankaido earthquake. Science 289:104–106.

Ladge S, Gaedicke C, Djajadihardja Y (2006) Great Sumatran Earthquakes: MCS Images and bathymetry offshore Sumatra – first results of SEACAUSE Leg1 cruise SO186, proc. EGU meeting, Vienna, April 2006. Geophys Res Abstr 8:06774.

Larson RL, Carpenter GB, and Diebold JB (1978) A geophysical study of the Wharton basin near the Investigator Fracture Zone. J Geophys Res 83:773–782.

Lay et al. (2005) The great Sumatra–Andaman earthquake of 26 December 2004. Science 308:1127–1133.

Llenos AL, McGuire JJ (2007) Influence of forearc structure on the extent of great subduction zone earthquakes. J Geophys Res 112:B09301, DOI:10.1029/2007JB004944.

Macpherson GJ (1981) The Snow Mountain volcanic complex: an on-land seamount in the Franciscan terrain. California J Geol 91:73–92.

Maruyama S, Yamasaki M (1978) Paleozoic submarine volcanoes in the High-P/T metamorphosed Chichibu system of eastern Shikoku. Jpn J Volcanol Geotherm Res 4:199–216.

Moore GF, JR Curray (1980) Structure of the Sunda trench lower slope off Sumatra from multichannel seismic reflection data. Mar Geophys Res 4:319–340.

Mortera-Gutierrez CA, Scholl DW, Carlson RL (2003) Fault trends on the seaward slope of the Aleutian trench: Implications for a laterally changing stress field tied to a westward increase in oblique convergence. J Geophys Res 108 doi:10.1029/2001JB001433.

Robinson DP, Das S, Watts AB (2006) Earthquake rupture stalled by subducting fracture zone. Science 312:1203–1205.

Ruff LJ (1989) Do trench sediments affect great earthquake occurrence in subduction zones? Pageoph 129:263–282.

Sandwell DT, Smith WHF (1997) Marine gravity anomaly from Geosat and ERS-1 satellite altimetry. J Geophys Res 102:10,039–10,054.

Scholz CH, Small C (1997) The effect of seamount subduction on seismic coupling. Geology 25:487–490.

Spence W (1977) The Aleutian arc: tectonic blocks, episodic subduction, strain diffusion, and magma generation. J Geophys Res 82:213–230.

Spence W, Mendoza C, Engdahl ER, Choy GL, Norabuena E (1999) Seismic subduction of the Nazca Ridge as shown by the 1966–97 Peru earthquakes. Pure Appl Geophys 154:753–776.

Sykes LR (1971) Aftershock zones of great earthquakes, seismicity gaps, and earthquake prediction for Alaska and the Aleutians. J Geophys Res 76:8021–8041.

von Huene R et al (1995) Morphotectonics of the Pacific convergent margin of Costa Rica. Geol Soc Am (Spl Pap) 295:291–307.

von Huene R, Ranero CR, Weinrebe W (2000) Quaternary convergent margin tectonics of Costa Rica, segmentation of the Cocos Plate, and Central American volcanism. Tectonics 19:314–334.

Watts AB, Bodine JH, Ribe NM (1980) Observations of flexure and the geological evolution of the Pacific Ocean basin. Nature 283:532–537.

Weissel JK, Anderson RN (1978) Is there a Caroline plate? Earth Planet Sci Lett 41:143–158.

Great Earthquakes in Slow-Subduction, Low-Taper Margins

Marc-André Gutscher and Graham K. Westbrook

Abstract The seismic hazard presented by slow subduction zones is not well known. While there is a widely accepted apparent relation between "fast-young plate" subduction and great earthquake generation (e.g., Chile, 1960), the seismic record indicates that slow subduction zones are also capable of generating mega-thrust earthquakes ($M > 8.2$). Available data on the recurrence interval for slow subduction margins, suggests that repeat times are longer than for more rapid convergence margins (on the order of several hundred to a few thousand years). For several of these margins, however, no shallow dipping thrust earthquake focal mechanisms are observed and no mega-thrust earthquakes known either.

Slow subduction zones ($v \leq 4$ cm/year) are typically characterized by thick sedimentary sections on the incoming plate (2–6 km) and by a broad accretionary wedge (100–250 km). The "taper" of these accretionary wedges is mechanically related to the basal and internal friction and ranges from about 2° to 12°. Some wedges have extremely shallow mean surface and basal slopes (about 1–2° each, taper <4°) indicating a very weak decollement layer. These include: Barbados Ridge, Makran, Hikurangi, Mediterranean Ridge, Calabria, Gibraltar/Cadiz, and Cascadia/Washington. Nankai, Sumatra and E. Alaska have slightly higher tapers of about 5–7°. Most of these low-taper wedges have very slow to slow convergence rates (0.5–4 cm/year).

The presence of these wide accretionary wedges strongly affects the type and amount of deformation above the "up-dip limit" of the seismogenic zone. The thermally insulating effect of a wide and thick wedge of sediment produces a wide, shallow transition zone (between the 100°C and 150°C isotherms) as well as a substantial (up to 80 km wide) region between this and the front of the wedge, where the amount and timing of deformation is poorly understood. Indeed, recent seismological data from Nankai indicate "very-low-frequency" shallow-thrust earthquakes beneath the accretionary wedge, long considered to be "aseismic," underscoring the unusual mechanical behavior in the transition zone. As the rigidity of the high-porosity wedge sediments is low, for an earthquake of a given seismic moment, more co-seismic slip will occur and for a longer duration, than for a deeper earthquake in more consolidated units. Thus shallow earthquakes in the wedge are more efficient at generating a strong tsunami. Many of the margins with very broad accretionary wedges have produced extremely strong earthquakes (M9) in the past, as well as giant tsunamis.

M.-A. Gutscher
Université Européenne de Bretagne, Brest, Université de Brest; IUEM, Domaines Océaniques, UMR6538 CNRS, Plouzané, France; G.K. Westbrook, Dept. of Earth and Environmental Sciences, Univ. Birmingham, United Kingdom.

S. Lallemand and F. Funiciello (eds.), *Subduction Zone Geodynamics*,
DOI 10.1007/978-3-540-87974-9, © Springer-Verlag Berlin Heidelberg 2009

1 Introduction

The vast majority of seismicity occurs in subduction zones (Byrne et al., 1988), as well as nearly all very great ($M \geq 8.5$) earthquakes (Fig. 1). During the past several decades much effort has been devoted to understanding the factors controlling the frequency, location, and magnitude of these earthquakes (Kanamori, 1972; Ando, 1975; Ruff and Kanamori, 1980; Byrne et al., 1988; Tichelaar and Ruff, 1993; Hyndman and Wang, 1995; Goes, 1996; McCaffrey, 1997). Among the numerous factors investigated were: lateral segment boundaries (Kanamori, 1972; Ruff and Tichelaar, 1996; McCaffrey, 1997), thermal control of updip and downdip limits of the fault plane (Hyndman and Wang, 1995; Hyndman et al., 1995, Peacock and Wang, 1999, Oleskevich et al., 1999), asperity distribution (Ruff and Kanamori, 1983; Cloos, 1992; Bilek et al., 2003), sediment thickness on the down-going plate (Ruff and Kanamori, 1980; Ruff and Tichelaar, 1996; Cloos and Shreve, 1996; McCaffrey, 1997), dip-angle of the subducting plate (Uyeda and Kanamori, 1979; Ruff and Kanamori, 1980), unusual (flat slab) geometry of the subducting plate (Gutscher and Peacock, 2003), and segmentation related to upper plate structures (Collot et al., 2004).

A widely accepted hypothesis proposes that the strongest earthquakes occur in subduction zones where plate convergence velocity is high and the age of the subducting oceanic lithosphere is young, as these factors were thought to increase interplate coupling and thus increase the stress drop and maximum magnitude of a potential earthquake (Ruff and Kanamori, 1980, 1983). This model was partly inspired by and seemed to offer a plausible explanation for the strongest known earthquake, the 1960, M9.5 Chile-Valdivia earthquake. This tendency also appeared to be supported by the known great subduction earthquakes of the twentieth century, as recorded by global seismological networks (Fig. 2a). However, in recent years, the recognition and detailed study of great earthquakes from historical and paleoseismic records (Satake et al., 1996; Clague, 1997), as well as the great 2004 Sumatra-Andaman earthquake (Lay et al., 2005) have begun to cast doubt on this simple age-velocity relation.

While rapid subduction zones (with $v \geq 6$ cm/year, e.g., Kamchatka/Kuriles, Aleutians, Solomon Isl., Andes) in general produce most of the great earthquakes (Table 1), the period covering instrumental recording of earthquakes (beginning around 1900), is likely too short to adequately represent the long-term activity of all subduction zones worldwide. Numerous subduction zone segments exhibit recurrence of great earthquakes at periods exceeding 100 years (Goes, 1996) and thus may not have had any great thrust events in the early-mid twentieth century. For several subduction zones worldwide, (e.g., Antilles, Hellenic/Mediterranean, Hikurangi subduction zones) no great (M8) thrust type earthquakes are known and it is thus very difficult to properly assess the long-term seismic hazard presented by these margins. For these (and several other subduction margins), plate convergence velocities are low ($v \leq 4$ cm/year) and data on recurrence intervals are sparse. The purpose of this study is

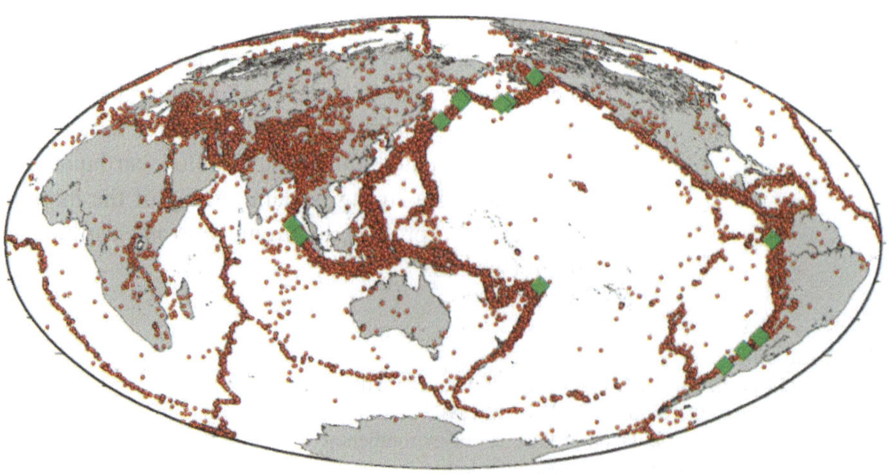

Fig. 1 Global seismicity ($M \geq 4$, depth <70 km, 1964–1998) shown as *red circles* (Engdahl et al., 1998) and very great M8.5 subduction earthquakes, shown as *green diamonds* (Engdahl and Villasenor, 2002; Lay et al., 2005)

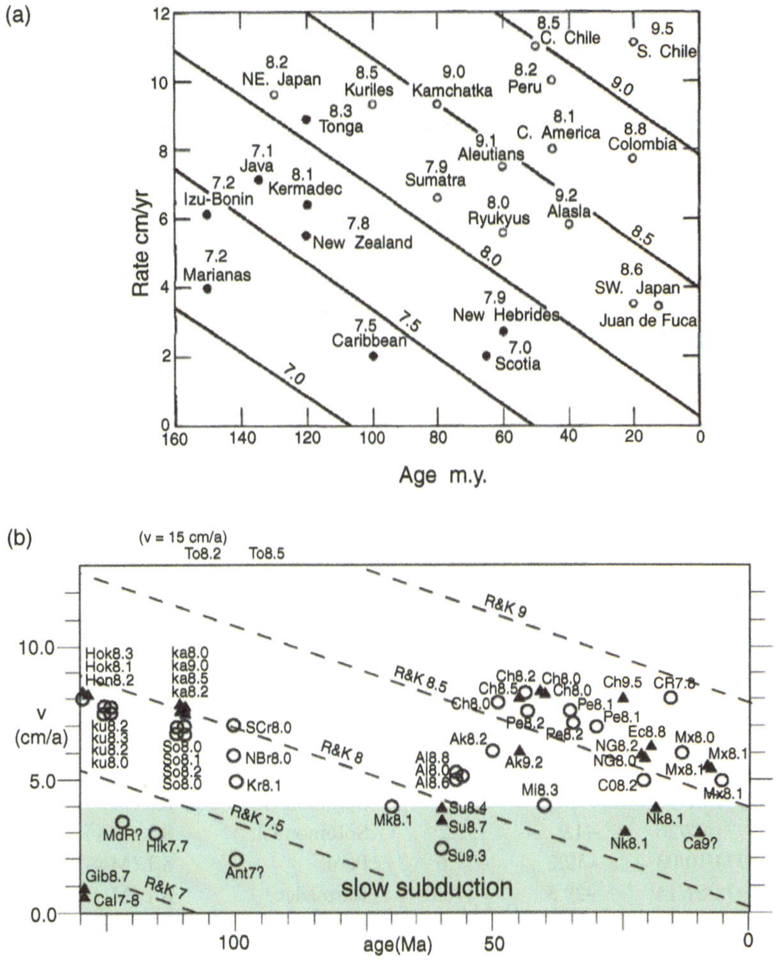

Fig. 2 (**a**) Age–velocity plot for great subduction zone earthquakes (Ruff and Kanamori, 1980), (**b**) Age–velocity plot, revised using all available modern data on M8 subduction earthquakes of past 100 years, as well as some historical events (see Table 1). *Triangles* are data from margins where information on recurrence interval are available. *Circles* show data from margins where no such information is known. The field of slow subduction margins is shown (by the gray shading)

Table 1 Great Subduction Earthquakes ($M \geq 8$) twentieth century and historical earthquakes

Date year/ month/day	Lat	Long	Region	Magn.	Recurr. time (years)	v_{orthog} (cm/year)	ocean lith. (age Ma)
1693/01/11	36.5	15	Sicily	8.0?	500	0.5?	130?
1700/01/26	47	−125	Cascadia	9.0?	300–1,000	4.0	10
1755/11/01	36	−8	SW Iberia	8.7?	1,500–2,000	0.5?	130?
1904/06/25	52.0	159.0	Kamchatka	8.0 (Ms)	48 (?)	7.5	115
1906/01/31	1.0	−81.3	Ecuador	8.8 (Mw)	40–70	6	20
1906/08/17	−7.0	149.0	New Britain	8.0 (Mw)		6?	100?
1906/08/17	−33.0	−72.0	Chile	8.5 (Mw)	79 (?)	8	40
1907/04/15	16.7	−99.2	Mexico	8.0 (Ms)		6	13
1914/05/26	−2.0	137.0	NW N.Guinea	8.0 (Ms)	82 (?)	6	20
1915/05/01	47.0	155.0	Kuriles	8.0 (Ms)		7.5	125
1917/06/26	−15.5	−173.0	Tonga	8.5 (Ms)		15*	100
1918/09/07	45.5	151.5	Kuriles	8.2 (Ms)		7.5	125
1919/04/30	−19.8	−172.2	Tonga	8.2 (Ms)		15*	100
1922/11/11	−28.5	−70.0	Chile	8.5 (Mw)		8	45
1923/02/03	54.0	161.0	Kamchatka	8.5 (Mw)	36 (?)	7.5	115
1924/04/14	6.5	126.5	Mindanao	8.3 (Ms)		4	40
1932/06/03	19.5	−104.3	Mexico	8.1 (Ms)	63 (?)	5.5	8
1934/07/18	−11.8	166.5	St. Cruz Isl.	8.1 (Ms)		7	100

(continued)

Table 1 (continued)

Date year/ month/day	Lat	Long	Region	Magn.	Recurr. time (years)	v_{orthog} (cm/year)	ocean lith. (age Ma)
1938/11/10	55.5	−158.0	Alaska	8.2 (Mw)		6	50
1939/01/30	-6.5	155.5	Solomon Isl.	8.0 (Ms)		7	110
1939/04/30	−10.5	158.5	Solomon Isl.	8.2 (Ms)		7	110
1940/05/24	−11.2	−77.8	Peru	8.2 (Mw)		7	35
1942/08/24	−14.5	−74.8	Peru	8.2 (Mw)		7.5	43
1943/04/06	−31.0	−71.3	Chile	8.2 (Mw)		8	43
1944/07/12	33.8	136.0	Nankai	8.1 (Mw)	90–150	3	25
1945/11/27	25.2	63.5	Makran	8.1 (Mw)		4	70
1946/12/20	32.5	134.5	Nankai	8.1 (Mw)	90–150	4	18
1950/10/05	11.0	−85.0	Costa Rica	7.8 (Mw)	50 (?)	8	16
1952/03/04	42.5	143.0	Hokkaido	8.1 (Mw)	51 (?)	8	130
1952/11/04	52.8	160.0	Kamchatka	9.0 (Mw)	48 (?)	7.5	115
1957/03/09	51.6	−175.0	Aleutians	8.6 (Mw)		5	55
1958/11/06	44.4	148.6	Kuriles	8.3 (Mw)		7.5	125
1959/05/04	53.2	159.8	Kamchatka	8.2 (Mw)	36 (?)	7.5	115
1960/05/22	−38.2	−73.5	Chile	9.5 (Mw)	100–300	8	25
1963/10/13	44.9	149.6	Kuriles	8.5 (Mw)		7.5	125
1964/03/28	61.1	−148.0	Alaska	9.2 (Mw)	50–150 (?)	6	45
1965/02/04	51.3	−178.6	Aleutians	8.8 (Mw)		5	55
1966/10/17	−10.7	−78.6	Peru	8.1 (Mw)		7	30
1968/05/16	40.9	143.4	Honshu	8.2 (Mw)		8	130
1969/08/11	43.6	147.2	Kuriles	8.2 (Mw)		7.5	125
1971/07/14	−5.5	153.9	Solomon Isl.	8.0 (Mw)		7	110
1971/07/26	−4.9	152.2	Solomon Isl.	8.1 (Mw)		7	110
1974/10/03	−12.2	−77.6	Peru	8.1 (Mw)		7.5	35
1976/01/14	−29.5	−178.0	Kermadec	8.1 (Ms)		12	100
1979/12/12	1.6	−79.4	Colombia	8.2 (Mw)	40–70	6	20
1985/03/03	−33.1	−71.9	Chile	8.0 (Mw)	79 (?)	8	48
1985/09/19	18.1	−103.0	Mexico	8.1 (Ms)		5	6
1986/05/07	51.3	−175.0	Aleutians	8.0 (Mw)		5	55
1995/07/30	−23.3	−70.2	Chile	8.0 (Mw)		8	48
1995/10/09	19.1	−104.2	Mexico	8.0 (Mw)	63 (?)	5.5	8
1996/02/17	−0.9	137.0	NW N. Guinea	8.2 (Mw)	82 (?)	6	20
2000/11/16	−4.0	152.1	N. Ireland	8.1 (Mw)		7.5	35
2001/06/23	−16.3	−73.6	Peru	8.4 (Mw)		7.5	35
2003/09/25	41.8	143.9	Hokkaido	8.3 (Mw)	51 (?)	9.5	90
2004/12/26	6.0	94.0	Sumatr/Andam	9.3 (Mw)	>300 ??	2–3	60
2005/03/28	2.0	97.0	Sumatra	8.7 (Mw)	140 (?)	3.4	60
2007/09/12	−4.0	102.0	Sumatra	8.4 (Mw)	180 (?)	3.7	60

M8 subduction earthquakes of the past 100 years (Engdahl and Villasenor, 2000) and historical earthquakes (Satake et al., 1996; Gutscher, 2004; Gutscher et al., 2006). Data on plate age and orthogonal velocity are taken mostly from (McCaffrey, 1997) and the Antilles subduction velocity is from (DeMets et al., 2000). Data on recurrence intervals are primarily from (Goes, 1996).

to examine these slow subduction zones and reassess their potential for generating great subduction earthquakes. As slow-convergence margins are commonly characterized by very large accretionary wedges, the seismogenic potential of these wedges and possible contribution to tsunami generation will also be investigated.

2 Age-Velocity, Magnitude Relationship

Since the advent of plate tectonics and the discovery of subduction zones, numerous authors have attempted to classify subduction zones and their seismogenic behavior on the basis of a few objective parameters. One of the earliest such attempts was the Marianas versus Andean type classification, where the differences were thought to be governed primarily by the age of the subducting lithosphere (Uyeda and Kanamori, 1979). For the Marianas, the age is great (>100 Ma), the slab-dip is steep, there is an associated active back-arc basin and there are no observed great thrust-type earthquakes. For the Andean margin, the age of the oceanic plate is young (<50 Ma), the slab dip is generally shallow and great thrust earthquakes have been observed along most of the margin. Later work, expanded upon this and proposed a combination of two parameters, plate age and subduction velocity, resulting in the well-known young-fast model for great earthquake generation (Ruff and Kanamori, 1980, 1983). The corollary is that slow subduction zones (with old slabs) do not produce great earthquakes. On the basis of available data in 1980 this hypothesis seemed to explain the occurrence and magnitudes of most great earthquakes (Fig. 2a). Given, our twenty-first century database of past great earthquakes, however, this model probably needs to be revised.

Indeed, inclusion of slow-convergence margins like Sumatra or Cascadia, where great M9 earthquakes occur, as well as more precise data on the age and velocities of subducting slabs worldwide paints a different picture of great earthquake occurrence (Fig. 2b). The Sumatra and Cascadia subduction zones plot in the lower central and lower right portion, respectively, of the age-velocity field, where the Ruff and Kanamori regression lines would predict earthquakes with maximum magnitudes between 8 and 8.5. However, the 2004 earthquake with an estimated magnitude of 9.1–9.3 (Lay et al., 2005; Stein and Okal, 2005), and the 1700 Cascadia earthquake with an estimated magnitude of 9.0 (Satake et al., 1996; Clague, 1997), do not obey the proposed relation. As a matter of fact, none of the four M9 events known (Kamchatka, 1952; Chile, 1960; Alaska, 1964; Sumatra, 2004), respects the initially proposed linear regression (Fig. 2b).

3 Velocity, Earthquake Magnitude and Recurrence

If we examine the maximum earthquake magnitude as a function of convergence velocity (Fig. 3a), no relationship can be discerned. Indeed both fast and slow subduction zones are capable of generating great earthquakes. However, when considering the recurrence interval between great earthquakes and the subduction velocity, a clear trend can be discerned (Fig. 3b). Where reliable data are available regarding recurrence interval, one notices that subduction zones with the slowest convergence (≤4 cm/year) are those exhibiting the longest recurrence interval (typically ≥100 years). This empirical relation is consistent with theoretical calculations of fault slip rates (Molnar, 1979; McCaffrey, 1997). For example, radiometric dating of earthquake triggered turbidites off Cascadia indicate a repeat time from several hundred to one thousand years (Goldfinger et al., 2003). For Northern Sumatra the interval between great earthquakes is not well known, but is estimated to be at least several hundred years (Lay et al., 2005). For other subduction zones, with even slower convergence (e.g., Calabria/Sicily or Cadiz/Gibraltar) the repeat times appear to range from several hundred to up to 2000 years, respectively (Lebreiro et al., 1997; Gutscher, 2004; Gutscher et al., 2006). There are numerous slow subduction zones worldwide (Mediterranean Ridge, Antilles, Hikurangi, Cascadia, etc.) where great thrust earthquakes (M8) have never been observed (Fig. 4). In some cases, historical and paleoseismic records indicate that great events have occurred in the past (e.g., Nankai, Cascadia).

4 Slow Subduction and the Seismogenic Zone

The characteristic morphology of slow subduction margins can differ greatly from rapid subduction zones, and may play a crucial role in determining the extent and behavior of the seismogenic fault zone. Several factors may contribute to controlling the geometry and overall morphology of subduction zones worldwide, the foremost being their plate-tectonic setting (intra-oceanic, or ocean-continent), available sediment supply, subducting plate-age and convergence velocity. Accordingly, numerous classification schemes have been proposed (Jarrard, 1986; von Huene and Scholl, 1991; Lallemand et al., 1994, 2005). Indeed, the nature of the upper plate seems to play an important role in controlling plate dip (for instance), but the age of the subducting slab plays a surprisingly minor role

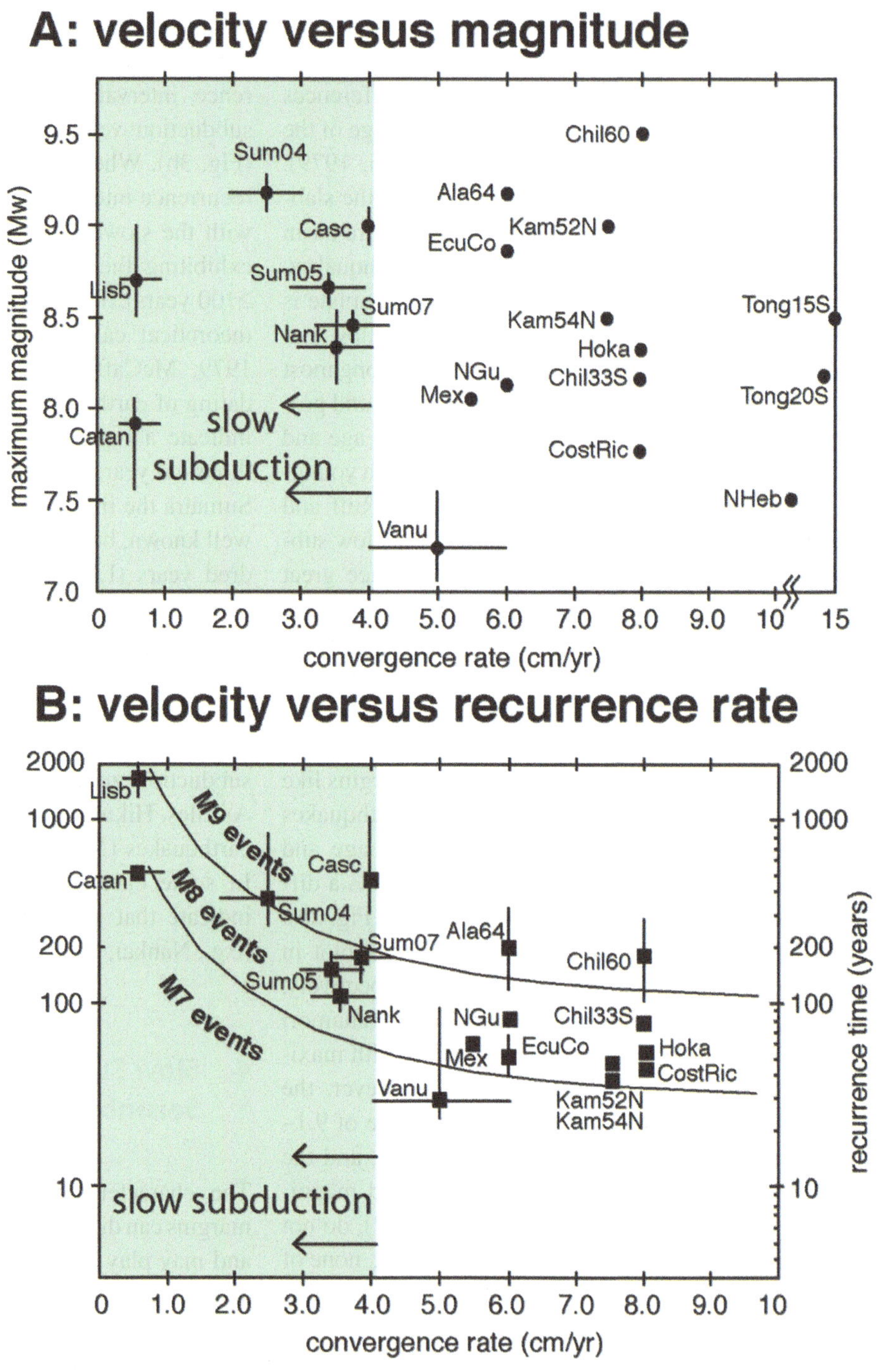

Fig. 3 (**a**) Velocity–magnitude plot for subduction zones and earthquakes in Table 1. Note that no distinct trend can be observed. (**b**) Velocity-recurrence interval plot for subduction zones where data is available on recurrence (see Table 1 and recurrence interval for Vanuatu is taken from Taylor et al., 1990). Generally, slow convergence corresponds to longer recurrence intervals

(Lallemand et al., 2005). The impact of plate-velocity in an absolute (hot-spot) reference frame was found to be significant in controlling the deep subduction geometry (slab dip) (Lallemand et al., 2005).

Typically, slow subduction margins are marked by very large accretionary wedges (100–250 km) and trench-arc distances of 350–500+ km (Fig. 5). In general, rapid subduction zones have a smaller trench-arc distance and smaller frontal wedges, which often exhibit erosional behavior (von Huene and Scholl, 1991; Lallemand et al., 1994) (Fig. 5). Of the thirteen slow subduction margins identified worldwide (Fig. 4), ten are characterized by very large accretionary wedges. The Southern Chile (Patagonia) margin

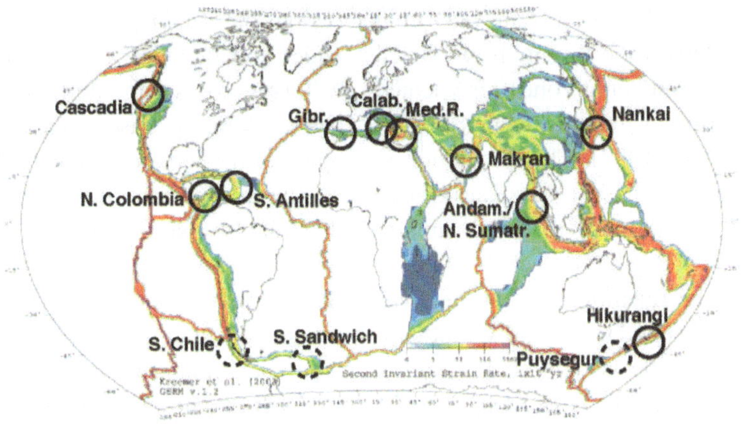

Fig. 4 Location map of slow subduction margins worldwide ($v \leq 4$ cm/a) indicated on the world strain map (Kreemer et al., 2003). *Dashed circle* shows slow subduction margins without a broad accretionary wedge; S. Chile (Polonia et al., 2007), South Sandwich and Puysegur. All others have broad (100–200 km wide) accretionary wedges. Accretionary wedge widths and references are: Hikurangi wedge – 150 km (Davey et al., 1986); Makran – 150 km (marine) to 300+ km (marine + land) (Kopp et al., 2000a; Ellouz-Zimmermann et al., 2007); S. Sumatra – 150 km (Kopp et al., 2000b), N. Sumatra – 200 km (Klingelhoefer et al., 2007); N. Colombia 150 km (Flinch et al., 2003); Cascadia 120 km (Flueh et al., 1998); Nankai 150 km (Moore et al., 1990; Park et al., 2002). References for the Gulf of Cadiz, Mediterranean Ridge, Calabria and Barbados Ridge are provided in caption to Fig. 8 and in the text)

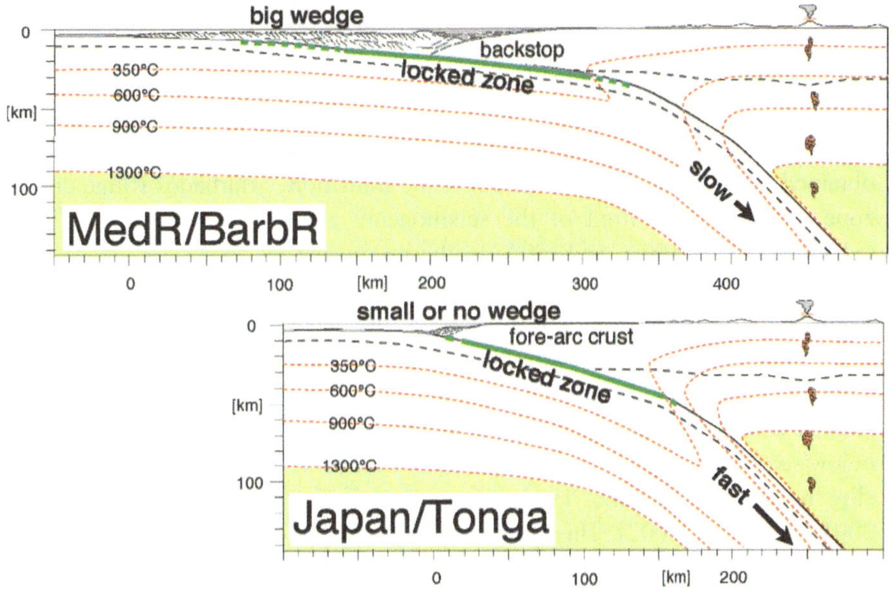

Fig. 5 Schematic representation showing the thermal impact of slow and fast subduction on the seismogenic zone. Schematic isotherms are shown. The principal differences are the presence and width of the accretionary wedge, the trench-arc distance, and the position and width of the seismogenic zone (roughly 10–30 km depth for slow subduction and 10–45 km depth for fast subduction). Note the broader transition zone (dashed) above the up-dip limit for typical slow subduction margins, defined by the 100°C and 150°C isotherms (see text)

has a smaller accretionary wedge (20–70 km wide), probably due to the recent subduction of the East Chile Rise and associated erosion of the margin (Polonia et al., 2007). South Sandwich and Puysegur (S. of New Zealand) are highly oblique and due to insufficient sediment supply have not developed significant accretionary wedges.

Furthermore, the location of the seismogenic zone can be expected to differ for slow and fast subduction (Fig. 5). For two given oceanic slabs of the same age, rapid subduction transfers more cold material into the upper mantle over a given time span, than slow subduction and is thus the thermal equivalent of "cold" subduction. Existing thermal modeling studies have shown that for rapid subduction of an old, cold slab, (the case of NE Japan) the thermally predicted downdip limit of the seismogenic zone (represented by the 350°C isotherm) can extend down to 60 km depth (Peacock and Wang, 1999). This prediction is confirmed by aftershock studies of the Sanriku-Oki M7.7 earthquake in 1994, showing that fault rupture extended to roughly 55 km depth (Hino et al., 2000).

The updip limit of the seismogenic zone is commonly considered to correspond to the 150°C isotherm (Hyndman et al., 1995; Oleskevich et al., 1999). As the onset of stick-slip behavior is thought to be controlled by several mineralogical/dehydration reactions (e.g., opal to quartz and smectite to illite) which occur over a temperature range of roughly 100–150°C (Moore and Saffer, 2001), these isotherms are commonly used to define a transition zone, where seismogenic behavior begins (Oleskevich et al., 1999) (Fig. 5). The position of the updip limit is more strongly dependent on the geothermal gradient in the subducting plate and less dependent on subduction rate, than the position of the downdip limit.

Several thermal models have been published for slow subduction zones with a wide accretionary wedge; the Barbados Ridge (Ferguson et al., 1993), the Gulf of Cadiz wedge (Thiebot and Gutscher, 2006) and the Calabrian arc (Gutscher et al., 2006). The results obtained were similar and indicate that the transition zone and the updip limit of the seismogenic zone (as defined by the 100°C and 150°C isotherms, respectively) are located at distances of about 80 and 140 km, from the deformation front. For these cases, the depth to the updip limit is about 10 km (below sea-level) and 6–7 km below the sea-floor. The updip limit for the Nankai trough has been calculated to be 8 and 6 km below sea-level for seismogenic slip and tsunamigenic slip, respectively (Satake, 1993; Moore and Saffer, 2001; Park et al., 2002). This is due to the young age and high temperatures of the subducting slab, causing the 100°C and 150°C isotherms to be situated at a shallower depth and within 30 km of the trench (Hyndman et al., 1995; Gutscher and Peacock, 2003). In general, we can conclude that the seismogenic zone usually starts at or near the contact between the forearc basement and the subducting plate (fairly close to the trench) and extends more deeply for fast subduction. For slow subduction margins, due to the thermal insulation effect of the very large accretionary wedge, combined with the shallow plate dip beneath this region, the seismogenic zone is shallower, but much wider and includes a much larger transition zone as well (Fig. 5). A large portion of the seismogenic zone passes along the base of the accretionary wedge, well updip of the forearc basement (backstop).

The main points implied by this simple model are illustrated by seismicity cross-sections from a representative sample of fast and slow-subduction margins (Fig. 6). Indeed both intra-oceanic and ocean-continent subduction margins are considered. The plate age of the subducting slab in all cases is nearly identical (≥100 Ma). The primary difference is the subduction velocity as well as the abundance of sediments at the trench. The seismic behavior of the Hellenic and Antilles subduction zones is very similar, both showing moderate to abundant intra-slab seismicity and a lack of thrust-type earthquakes along the plate interface (Fig. 6a and b). The nearly total absence of thrust earthquakes is confirmed by available focal mechanisms (CMT catalog). On the other hand, both the Japan trench and Tonga show highly abundant intra-slab seismicity, and a pronounced concentration of strong, shallow earthquakes along the plate interface, extending very close to the trench (Fig. 6c and d). In other words, the seismogenic zone is highly active over the past 35 years in Tonga and Japan, and essentially inactive beneath the Mediterranean Ridge and Barbados Ridge during this period of time.

5 Slow Subduction and Broad Accretionary Wedges

The presence (or absence) of an accretionary wedge is one of the defining characteristics of subduction zones. Worldwide, accretionary wedges are observed along 56% of convergent margins (Lallemand et al., 1994). The "taper" of these deforming wedges is mechanically related to the effective basal and internal friction (Davis et al., 1983) (where effective friction is defined as the product of effective normal stress and the coefficient of friction) and typically ranges from about 4° to 12° (Fig. 7). A class of accretionary wedges exists, with extremely shallow mean surface and basal slopes (about 1–2° each, taper <4°) indicating a very weak decollement layer. These include: Barbados Ridge, Makran, Hikurangi, Mediterranean Ridge, Calabria, Gibraltar/Cadiz, and Cascadia. Nankai, Sumatra and

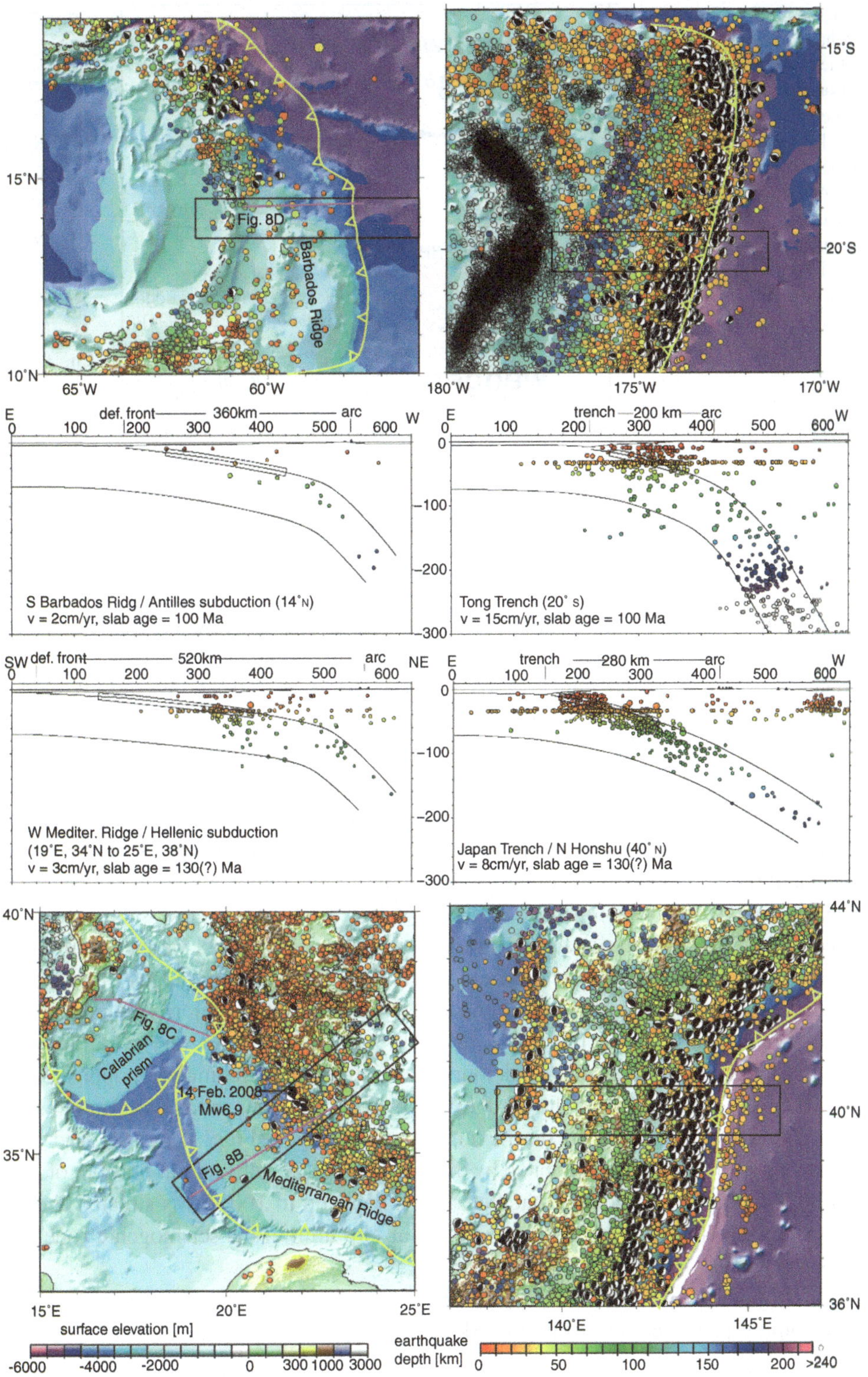

Fig. 6 Location maps and seismicity cross-sections from two slow and two fast subduction margins (USGS/NEIC PDE Catalog, January 1973–March 2008, $M \geq 4$) using a 1° (110 km) sampling width in all cases. All available shallow (≤50 km depth) thrust faulting focal mechanisms ($M \geq 5$ earthquakes, 1976–2008) are shown on maps (Global Centroid Moment Tensor Project – http://www.globalcmt.org). Relief is from the global 1 min Gebco grid (IOC et al., 2003). *Magenta lines* indicate the locations of three crustal scale cross sections shown in Fig. 8. Note the abundant shallow dipping thrust (subduction fault plane) earthquakes for Tonga and Japan, and the relative quiescence in the Antilles and Western Hellenic subduction zones. The recent Mw6.9 14 Feb. 2008 event is the stongest thrust earthquake in the CMT Catalog for the latter region. Upper left: Barbados Ridge/Antilles subduction; Lower left: Western Mediterranean Ridge/Hellenic subduction, w. Calabrian subduction also shown; Upper right: Tonga trench; Lower right Japan trench

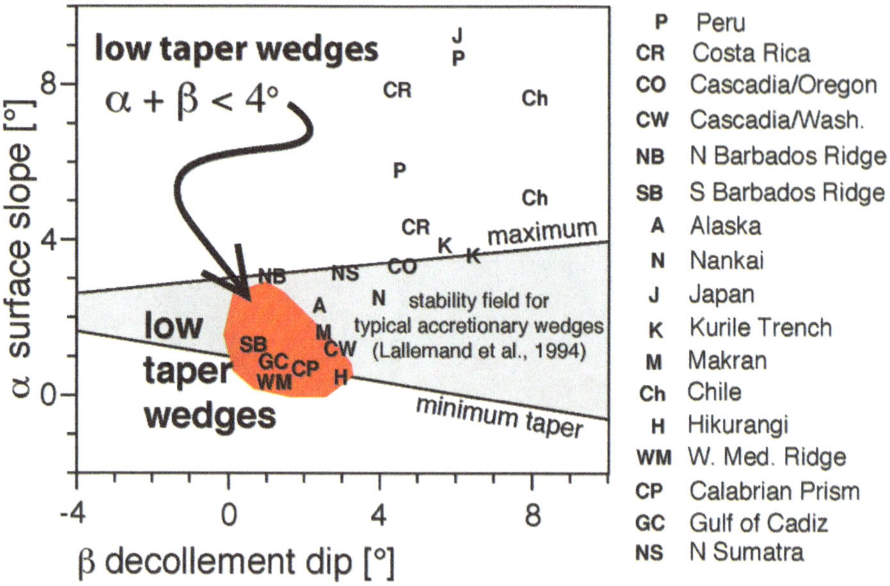

Fig. 7 Taper of accretionary and non-accretionary wedges worldwide and the stability field of typical accretionary wedges, (after Lallemand et al., 1994). Alpha is surface slope, Beta is plate dip and Taper is defined as Alpha + Beta. A low basal friction accretionary wedge undergoing regular accretion, should plot at the lower limit of the field (at the minimum taper)

SE Alaska have slightly higher tapers of about 5–7°. These subduction zones are typically characterized by thick sedimentary sections at the trench (2–6 km) and slow convergence (0.5–4 cm/year). South-eastern Alaska has a moderate to rapid convergence of 6 cm/year.

Available structural cross-sections from four of these margins confirm the shallow dip angle of the downgoing plate and the very wide, thick pile of sediments accumulated against the backstop (Fig. 8). The width of the Barbados Ridge accretionary wedge at 14°20′N (Ferguson et al., 1993) is 200 km and increases farther south where it reaches 270 km east of Grenada and Tobago. The accretionary wedge narrows to the north and is 100 km wide at 16°20′N, off Guadeloupe (Westbrook et al., 1988). The width of the wedge is so great, that its outer edge lies beyond the flexural bulge of the subducting lithosphere, as indicated by the free-air gravity high, with the consequence that there is no trench in front of it. All four margins show very broad accretionary wedges (roughly 200 km wide), constructed by accumulation of the thick sedimentary section of the incoming plate, ranging from 3 to 6 km total thickness.

The frontally accreted section ranges from about 2 to 3 km (above the decollement) and therefore, significant thickening must occur by basal accretion ("subcretion") as well (Westbrook et al., 1988; Westbrook and Reston, 2002; Reston et al., 2002). The surface slope and dip of the decollement and top basement surfaces are very low in all cases (close to 1°). This is mechanical evidence for a very low basal shear stress (Davis et al., 1983; Westbrook and Smith, 1983; Lallemand et al., 1994).

Interestingly, fluid flow modeling predicts that for sediments of equal permeability, rapid subduction should result in a poorly drained accretionary wedge, with high basal fluid pressure, low basal friction and thus a long, thin wedge (low taper) (Saffer and Bekins, 2002). Conversely, slow subduction is considered to produce well drained wedges, with high basal friction and shorter, steeper wedges (high taper). Empirically, however, for the majority of margins worldwide and in particular for the very wide accretionary margins examined here, a completely opposite trend is observed. In fact, the slow subduction margins have the lowest taper wedges (Figs. 7 and 8). And as a result of the low-taper, a substantial portion of the accretionary wedge (up to 100 km) is situated above the presumed up-dip limit of the seismogenic zone.

6 Discussion

The presence of a broad region of poorly consolidated sediments along the plate interface has important consequences for the generation of earthquakes and for potential tsunami generation as well. A direct impact of the shallow plate dip and slow subduction velocities

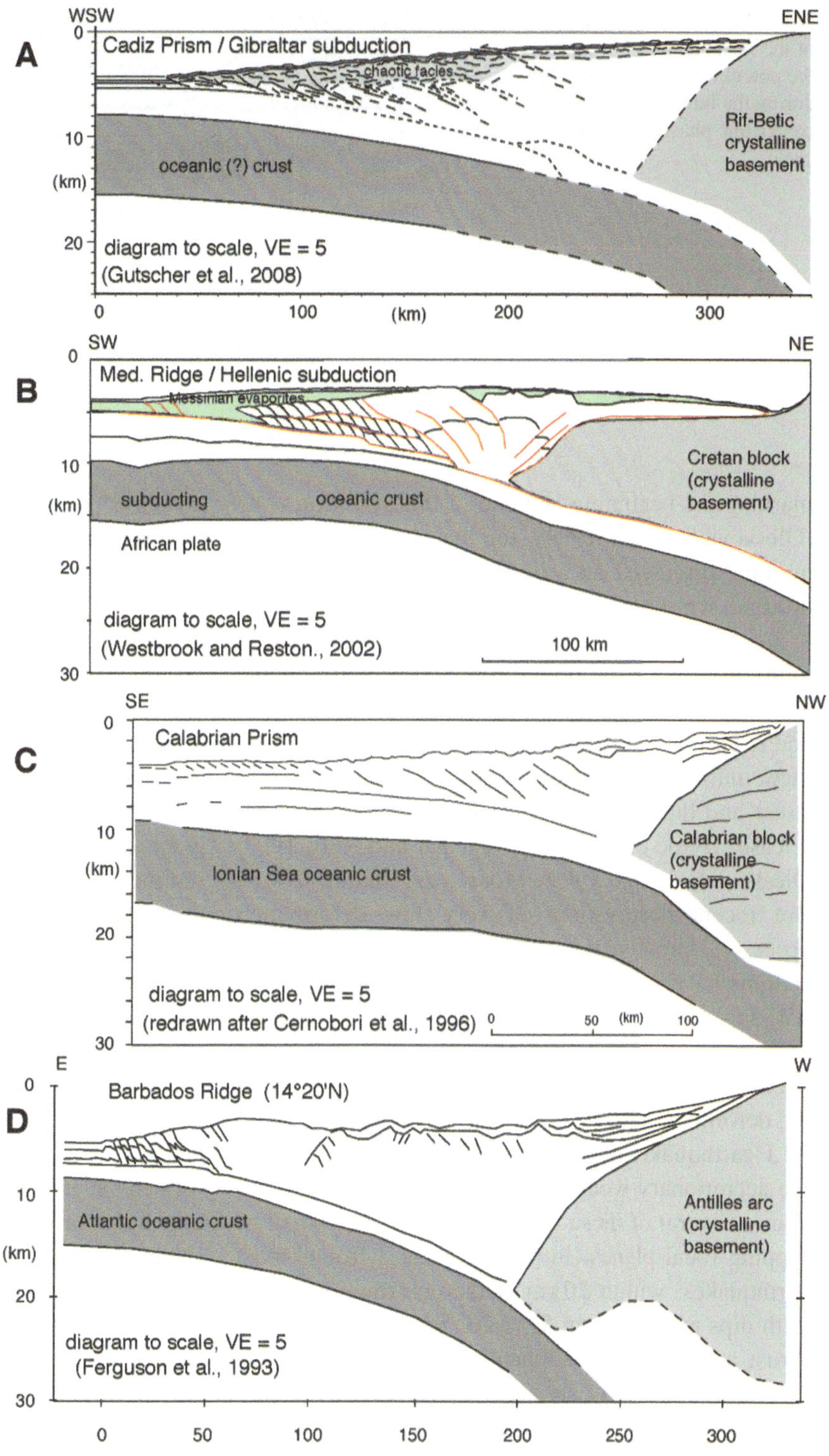

Fig. 8 Structural cross-sections of four slow subduction margins. Positions of the cross-sections are shown in Fig. 6, except Cadiz/Gibraltar (**a**) Gulf of Cadiz/Gibraltar (Gutscher et al., 2002, 2008); (**b**) Mediterranean Ridge/Hellenic subduction (after Westbrook and Reston, 2002); (**c**) Calabria (redrawn after Cernobori et al., 1996, using data from De Voogd et al., 1992); (**d**) Barbados Ridge/S. Antilles subduction 14°20'N (after Ferguson et al., 1993)

for the margins examined here, is that the mineralogical diagenetic and metamorphic processes believed to control the transition from aseismic to stick-slip behavior (Oleskevich et al., 1999; Moore and Saffer, 2001) will be spread out over a larger surface area and occur over a greater period of time (as the material is transported to appropriate temperature and pressure conditions for these reactions to take place). Thus, a much wider transition zone can be expected, for slow subduction (Fig. 5). If we take the 100°C and 150°C as

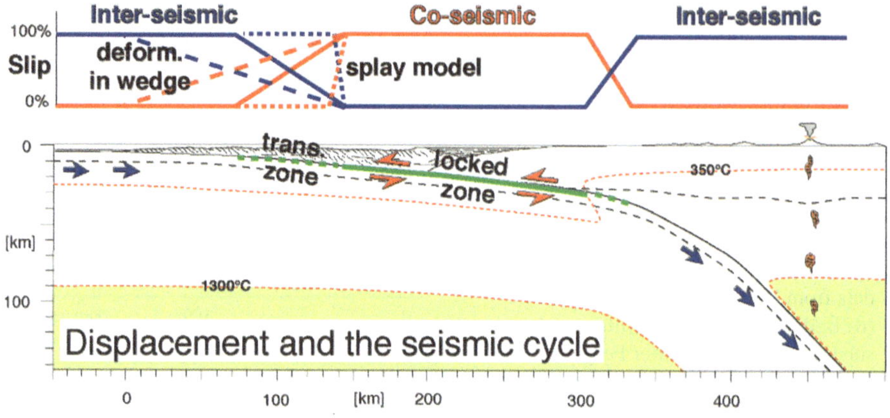

Fig. 9 Schematic cross-section of the seismogenic zone and the potential distribution of slip during the inter-seismic and co-seismic phases of the seismic cycle

marking the beginning and end of the transition zone (Oleskevich et al., 1999), for the slow-convergence margins discussed above, then we obtain a roughly 60 km down-dip width for the transition zone.

Furthermore, it is unclear whether the interplate fault zone above the up-dip limit of the seismogenic zone (including the transition zone) can accumulate and release elastic stress (Fig. 9). The commonly held view, until recently, was that this region could not store stress and thus did not contribute to seismic moment release (Byrne et al., 1988; Tichelaar and Ruff, 1993; Oleskevich et al., 1999; Moore and Saffer, 2001). But recent observations of very-slow deformation processes previously undetected by seismological stations have called this view into question. Offshore SW Japan, Very-Low Frequency (VLF) earthquakes with moment magnitudes M4 were detected in the Nankai Trough, located within the frontal 10–40 km of the deformation front (Ito and Obara, 2006a, b). These VLF earthquakes are located within or near the base of the accretionary wedge, near the plate boundary fault plane. Several of these earthquakes have very shallow dipping focal planes (in particular the 3 frontal-most earthquakes, within 20 km of the deformation front), with dips ranging from 5° to 20° and likely represent thrust motion along the plate boundary fault (Ito and Obara, 2006a). This atypical seismic activity underscores the unusual mechanical behavior of the "transition zone" as expressed by aseismic to seismologically detectable slow-slip events.

A thick sedimentary layer on the downgoing plate has long been proposed to influence plate coupling, although it has alternatively been argued to reduce coupling (Ruff and Kanamori, 1980) or increase it by allowing earthquake inducing asperities to subduct to greater depths and pressures (Cloos and Shreve, 1996). High sediment influx at the trench typically leads to construction of a large accretionary wedge (Lallemand et al., 1994). Thick sedimentary sections at the trench result from the interplay of subduction rate with the rate of sediment accumulation. The SE Alaska margin (site of the 1964 M9.2 Prince William Sound earthquake) is an example of moderately rapid subduction (6 cm/a) with a very high sedimentation rate at the trench (due to glaciation induced erosion of high topographic relief). Thus, while a range of subduction and sedimentation rates can lead to the accumulation of thick accretionary wedges, they are most common for slow subduction. A study of accretionary wedges worldwide found that large wedges (>40 km) are located along margins with an average convergence velocity of 3 cm/a (Lallemand et al., 1994).

Underconsolidated sediments along the interplate fault can also be expected to strongly influence tsunami generation in this zone. Source time studies performed on numerous earthquakes from six different trenches worldwide demonstrate that the shallowest sediments have the lowest rigidity (Bilek and Lay, 1999). While this result may seem unsurprising, the same study demonstrated that the source duration was longest for earthquakes occurring in this shallow region. Thus, earthquakes occurring in the region dominated by low-rigidity sediments have a longer duration and larger slip for an equivalent seismic moment (Bilek and Lay, 1999). They are particularly efficient at generating tsunamis as was shown by the slow-slip component in the northern portion of the fault plane of the Sumatra 2004 earthquake (Lay et al., 2005; Stein and Okal, 2005). In other words a magnitude 8 earthquake off the Southern Antilles or along the Mediterranean Ridge would be likely to generate a much more devastating tsunami than a magnitude 8 earthquake off Tonga, though the latter will occur much more often.

Finally, the fact that the Sumatra-Andman earthquake violated the proposed young-fast model (Ruff and Kanamori, 1980) was pointed out very recently, with the author drawing the completely opposite conclusion, that all subduction zones are potentially capable of generating M9 earthquakes (McCaffrey, 2008). On the basis of the subduction zones examined here, and the thermal structure characteristic of each type (fast-slow), we disagree. Tonga or NE Japan should never be capable of generating M9 earthquakes, for two main reasons. The first, is that frequent thrust earthquakes (M5-M7) occur here, releasing accumulated stress (Fig. 6). The maximum observed magnitudes here are M8–M8.5 (Table 1; McCaffrey, 2008). Given the fast subduction rates in both cases, our 100 year instrumental record covers 2–3 earthquake cycles. Taking into account the long historical records in Japan dating back 1,000 years, we can conclude that no M9 events have occurred during nearly 20 earthquake cycles off NE Japan. The second important reason is that thermally, rapid subduction (and an absence of an accretionary wedge) results in a narrower seismogenic zone, ~100 km for NE Japan, on the basis of a published thermal model (Peacock and Wang, 1999). For slow subduction zones and large accretionary wedges, the thermally predicted downdip width commonly exceeds 150 km (Gutscher et al., 2003, 2006; Thiebot and Gutscher, 2006). Increased downdip width implies a larger potential for the generation of great earthquakes.

7 Conclusions

The instrumental earthquake record of the past 100 years indicates that rapid subduction zones produce the vast majority of great earthquakes. However, the paucity of earthquakes observed in slow subduction zones is likely to be due to the longer recurrence intervals between great events, rather than a systematically different type of behavior (aseismic creep). The majority of slow subduction zones are marked by broad, thick accretionary wedges (100–250 km wide), sliding mechanically above a low friction decollement and consisting of low rigidity sediments. The transition zone from aseismic slip to stick-slip (seismogenic) behavior can be expected to be wider for these margins. During the inter-seismic phase of the seismic cycle, slow earthquakes and very-low frequency earthquakes can occur here. During great earthquakes, slip in the region of low rigidity sediments can be expected to contribute significantly to tsunami generation. It is thus crucial to improve our understanding of long-term kinematic motion and the typical recurrence times in such slow subduction zones in order to properly assess the natural hazard which they pose.

Acknowledgments We thank the organizers of the Montpellier Subduction Conference F. Funiciello and S. Lallemand for their efforts as well as the reviewers J.Y. Collot and Y. Aoki for helpful suggestions.

References

Ando, M., 1975. Source mechanisms and tectonic significance of historical earthquakes along the Nankai Trough, Japan. Tectonophysics, 27, 119–140.

Bilek, S.L., and Lay, T., 1999. Rigidity variations with depth along interplate megathrust faults in subduction zones. Nature, 400, 443–446.

Bilek, S.L., Schwartz, S.Y., and DeShon, H.R., 2003. Control of seafloor roughness on earthquake rupture behavior. Geology, 31, 455–458.

Byrne, D.E., Davis, D.M., and Sykes, L.R., 1988. Loci and maximum size of thrust earthquakes and the mechanics of the shallow region of subduction zones. Tectonics, 7, 833–857.

Cernobori, L., Hirn, A., McBride, J.H., Nicolich, R., Petronio, L., Romanelli, M., STREAMERS/PROFILES Working Groups, 1996. Crustal image of the Ionian basin and its Calabrian margins. Tectonophysics, 264, 175–189.

Clague, J.J., 1997. Evidence for large earthquakes at the Cascadia subduction zone. Rev. Geophys., 35, 439–460.

Cloos, M., 1992. Thrust-type subduction-zone earthquakes and seamount asperities: A physical model for seismic rupture. Geology, 20, 601–604.

Cloos, M., and Shreve, R.L., 1996. Shear-zone thickness and seismicity of Chilean- and Marianas-type subduction zones. Geology, 24, 107–110.

Collot, J.-Y., Marcaillou, B., Sage, F., Michaud, F., Agudelo, W., Charvis, P., Graindorge, D., Gutscher, M.-A., Spence, G., 2004. Are rupture zone limits of great subduction earthquakes controlled by upper plate structures? Evidence from multichannel seismic reflection data acquired across the northern Ecuador–southwest Colombia margin. J. Geophys. Res., 109, B11103, doi: 10.1029/2004JB003060.

Davey, F.J., Hampton, M., Childs, J., Fisher, M.A., Lewis, K., and Pettinga, J.R., 1986. Structure of a growing accretionary prism, Hikurangi margin, New Zealand. Geology, 14, 663–666.

Davis, D., Suppe, J., and Dahlen, F., 1983. Mechanics of fold and thrust belts and accretionary wedges. J. Geophys. Res., 88, 1153–1172.

DeMets, C., Jansma, P.E., Mattioli, G.S., Dixon, T.H., Farina, F., Bilham, R., Calais, E., and Mann, P., 2000. GPS geodetic constraints on Caribbean-North America plate motion. Geophys. Res. Lett., 27, 437–440.

De Voogd, B., Truffert, C., Chamot-Rooke, N., Lallemant, S., and Le Pichon, X., 1992. Two-ship deep seismic soundings in the basins of the Eastern Mediterranean Sea (Pasiphae cruise). Geophys. J. Int., 109, 536–552.

Ellouz-Zimmermann, N., Lallemant, S.J., Castilla, R., Mouchot, N., Leturmy, P., Battani, A., Buret, C., Cherel, L., Desaubliaux, G., Deville, E., Ferrand, J., Lugke, A., Mahieux, G., Mascle, G., Muhr, P., Pierson-Wickmann, A.-C., Robion, P., Schmitz, J., Danish, M., Hasany, S., Shazad, A., Tabreez, A., 2007. Offshore frontal part of the Makran accretionary prism (Pakistan) The CHAMAK survey. In Lacombe, O., Lavé, J., Roure, F., Verges, J. (Eds.), Thrust Belts and Foreland Basins. Frontiers in Earth Science, Springer, Berlin.

Engdahl, E.R., and Villasenor, A., 2002. Global Seismicity: 1900–1999. In Lee, W.H.K., Kanamori, H., Jennings, P.C., and Kisslinger, C. (Eds.), International Handbook of Earthquake and Engineering Seismology, Part A, Chapter 41, pp. 665–690, Academic Press.

Engdahl, E.R., van der Hilst, R.D., and Buland, R., 1998. Global teleseismic earthquake relocation with improved travel times and procedures for depth relocation. Bull. Seism. Soc. Am., 88, 722–743.

Ferguson, I.J., Westbrook, G.K., Langseth, M.G., and Thomas, G.P., 1993. Heat flow and thermal models of the Barbados ridge accretionary complex. J. Geophys. Res., 98, 4121–4142.

Flinch, J.F., Amaral, J., Doulcet, A., Mouly, B., Claudia Osorio, C., Pince, J.-M., 2003. Onnshore-offshore structure of the Northern Colombia accretionary complex. Proc. Amer. Assoc. Petr. Geol. Meeting, Barcelona, Spain Sept. 2003 (extended abstract, 5 p. 2 figs.).

Flueh, E.R., Fisher, M.A., Bialas, J., Childs, J.R., Klaeschen, D., Kukowski, N., Parsons, T., Scholl, D.W., ten Brink, U., Tréhu, A.M., and Vidal, N., 1998. New Seismic Images of the Cascadia Subduction Zone from Cruise SO108 – ORWELL: Tectonophysics, v. 293, p. 69–84.

Goes, S., 1996. Irregular recurrence of large earthquakes: An analysis of historic and paleoseismic catalogs. J. Geophys. Res., 101, 5739–5749.

Goldfinger, C., Nelson, C.H., Johnson, J.E., and Shipboard Scientific Party, 2003. Holocene earthquake records from the Cascadia subduction zone and northern San Andreas Fault based on precise dating of offshore turbidites. Ann. Rev. Earth Planet. Sci., 31, 555–577.

Gutscher, M.-A., 2004. What caused the Great Lisbon Earthquake? Science, 305, 1247–1248.

Gutscher, M.-A., and Peacock, S.M., 2003. Thermal models of flat subduction and the rupture zone of great subduction earthquakes. J. Geophys. Res., 108, B1, 2009, doi:10.1029/2001JB000787.

Gutscher, M.-A., Roger, J., Baptista, M.A., Miranda, J.M., Tinti, S., 2006. The source of the 1693 Catania earthquake and tsunami (Southern Italy): New evidence from tsunami modeling of a locked subduction fault plane. Geophys. Res. Lett., 33, L08309 10.1029/2005GL025442.

Gutscher, M.-A., Dominguez, S., Westbrook, G., Gente, P., Babonneau, N., Mulder, T., Gonthier, E., Bartolome, R., Luis, J., Rosas, F., Terrinha, P., and the Delila and DelSis Scientific Teams, 2008. Tectonic shortening and gravitational spreading in the Gulf of Cadiz accretionary wedge: Observations from multi-beam bathymetry and seismic profiling. Journal of Marine and Petroleum Geology, Sp. Vol. on Submarine instabilities (in press).

Hino, R., Ito, S., Shiobara, H., Shimamura, H., Sato, T., Kanazawa, T., Kasahara, J., and Hasegawa, A., 2000. Aftershock distribution of the 1994 Sanriku-oki earthquake (Mw7.7) revealed by ocean bottom seismographic observation. J. Geophys. Res., 105, 21697–21710.

Hyndman, R.D., and Wang, K., 1995. The rupture zone of Cascadia great earthquakes from current deformation and the thermal regime. J. Geophys. Res., 100, 22133–22154.

Hyndman, R.D., Yamano, M., and Wang, K., 1995. Thermal constraints on the seismogenic portion of the southwestern Japan subduction thrust. J. Geophys. Res., 100, 15373–15392.

IOC, IHO and BODC, 2003. "Centenary edition of the GEBCO Digital Atlas", published on CD-ROM on behalf of the Intergovernmental Oceanographic Commission and the International Hydrographic Organization as part of the General Bathymetric Chart of the Oceans; British Oceanographic Data Centre, Liverpool.

Ito, Y., and Obara, K., 2006a. Dynamic excitation of the accretionary prism excites very low frequency earthquakes. Geophys. Res. Lett., 33, L02311, doi:10.1029/2005GL025270.

Ito, Y., and Obara, K., 2006b. Very low frequency earthquakes within accretionary prisms are very low stress-drop earthquakes. Geophys. Res. Lett., 33, L09302, doi:10.1029/2005GL025883.

Jarrard, R.D., 1986. Relations among subduction parameters. Rev. Geophys., 24, 217–284.

Kanamori, H., 1972. Tectonic implications of the 1944 Tonankai and 1946 Nankaido earthquakes. Phys. Earth Planet. Inter., 5, 129–139.

Klingelhoefer, F., Dessa, J.-X., Graindorge, D., Gutscher, M.-A., Permana, H., Andre, C., Dean, S., Singh, S., and Chauhan, A., 2007. Crustal structure of the 2004 Great Sumatra Earthquake epicentral zone from wide-angle seismic data. AGU Fall Meeting, San Fransisco, Dec. 2007, EOS suppl., v. 87.

Kopp, C., Fruehn, J., Flueh, E.R., Reichert, C., Kukowski, N., Bialas, J., and Klaeschen, D., 2000a. Structure of the Makran subduction zone from wide-angle and reflection seismic data. Tectonophysics, 329, 171–191.

Kopp, H., Flueh, E.R., Klaeschen, D., Bialas, J., and Reichert, C., 2000b. Crustal structure of the central Sunda margin at the onset of oblique subduction. Geophys. J. Int., 147, 449–474.

Kreemer, C., Holt, W.E., and Haines, A.J., 2003. An integrated global model of present-day plate motions and plate boundary deformation, Geophys. J. Int., 154, 8–34.

Lallemand, S., Schnürle, P., and Malavieille, J., 1994. Coulomb theory applied to accretionary and nonaccretionary wedges: Possible causes for tectonic erosion and/or frontal accretion. J. Geophys. Res., 99, 12033–12055.

Lallemand, S., Heuret, A., and Boutelier, D., 2005. On the relationships between slab dip, back-arc stress, upper plate absolute motion and crustal nature in subduction zones.

Geochem. Geophys. Geosyst., 6, Q09006, doi:10.1029/2005GC000917.
Lay, T., Kanamori, H., Ammon, C.J., Nettles, M., Ward, S.N., Aster, R.C., Beck, S.L., Bilek, S.L., Brudzinski, M.R., Butler, R., DeShon, H.R., Ekstrom, G., Satake, J., and Sipkin, S., 2005. The Great Sumatra-Andaman earthquake of 26 December 2004. Science 308, 1127–1133.
Lebreiro, S.M., McCave, I.N., and Weaver, P., 1997. Late Quaternary turbidite emplacement on the Horseshoe abyssal plain (Iberian margin). J. Sediment. Res., 67, 856–870.
McCaffrey, R., 1997. Influences of recurrence times and fault zone temperatures on the age-rate dependence of subduction zone seismicity. J. Geophys. Res., 102, 22839–22854.
McCaffrey, R., 2008. Global frequency of magnitude 9 earthquakes. Geology, 36, 263–266.
Molnar, P., 1979. Earthquake recurrence intervals and plate tectonics. Bull. Seismol. Soc. Am., 69, 115–133.
Moore, J.C., and Saffer, D., 2001. Updip limit of the seismogenic zone beneath the accretionary prism of southwest Japan: An effect of diagenetic to low-grade metamorphic processes and increasing effective stress. Geology, 29, 183–186.
Moore, G.F., Shipley, T.H., Stoffa, P.L., Karig, D.E., Taira, A., Kuramoto, S., Tokuyama, H., and Suyehiro, K., 1990. Structure of the Nankai trough accretionary zone from multichannel seismic reflection data. J. Geophys. Res., 95, 8753–8765.
Oleskevich, D.A., Hyndman, R.D., and Wang, K., 1999. The updip and downdip limits to great subduction earthquakes: Thermal and structural models of Cascadia, south Alaska, SW Japan, and Chile. J. Geophys. Res., 104, 14965–14991.
Park, J.O., Tsuru, T., Kodaira, S., Cummins, P.R., and Kaneda, Y., 2002. Splay fault branching along the Nankai subduction zone. Science, 297, 1157–1160.
Peacock, S.M., and Wang, K., 1999. Seismic consequences of warm versus cool subduction zone metamorphism: Examples from northeast and southwest Japan. Science, 286, 937–939.
Polonia, A., Torelli, L., Brancolini, G., and Loreto, M.F., 2007. Tectonic erosion versus accretion along the Southern Chile Trench: Oblique subduction at a continental corner, Tectonics, v. 26, TC3005, doi:10.1029/2006TC001983.
Reston, T., Fruehn, J., von Huene, R., & the IMERSE Working Group, 2002. The structure and evolution of the western Mediterranean Ridge. Mar. Geol., 186, 83–111.
Ruff, L., 1996. Large earthquakes in subduction zones: Segment interaction and recurrence times. In Bebout, G.E., Scholl, D.W., Kirby, S.H., and Platt, J. (Eds.), Subduction: Top to Bottom, pp. 91–104, Geophysical Monograph 96, AGU.
Ruff, L., and Kanamori, H., 1980. Seismicity and the subduction process. Phys. Earth Planet. Inter. 23, 240–252.
Ruff, L., and Kanamori, H., 1983. Seismic coupling and uncoupling at subduction zones. Tectonophysics, 99, 99–117.
Ruff, L.J., and Tichelaar, B.W., 1996. What controls the seismogenic plate interface in subduction zones? in Subduction: Top to Bottom, ed. by G.E. Bebout, D.W. Scholl, S.H. Kirby and J. Platt, Geophysical Monograph 96, AGU, p. 105–111.
Saffer, D.M., and Bekins, B.A., 2002. Hydrological controls on the morphology and mechanics of accretionary wedges. Geology, 29, 183–186
Satake, K., 1993. Depth distribution of coseismic slip along the Nankai Trough, Japan, from joint inversion of geodetic and tsunami data. J. Geophys. Res. 98, 4553–4565.
Satake, K., Shimazaki, K., Tsuji, Y., and Ueda, K., 1996. Time and size of a giant earthquake in Cascadia inferred from Japanese tsunami records of January 1700. Nature, 379, 246–249.
Stein, S., and Okal, E.A., 2005. Speed and size of the Sumatra earthquake. Nature, 434, 581–582.
Taylor, F.W., Edwards, R.L., Wasserburg, G.J., and Frohlich, C., 1990. Seismic recurrence intervals and timing of aseismic subduction inferred from emerged corals and reefs of the Central Vanuatu (New Hebrides) frontal arc. J. Geophys. Res. 95, 393–408.
Thiebot, E., and Gutscher, M.-A., 2006. The Gibraltar Arc seismogenic zone (part1): Constraints on a shallow east dipping fault plane source for the 1755 Lisbon earthquake provided by seismic data, gravity and thermal modeling. Tectonophysics Sp. Vol. "Natural laboratories on seismogenic faults", 427, 135–152, doi:10.1016/j.tecto.2006.02.024.
Tichelaar, B.W., and Ruff, L.J., 1993. Depth of seismic coupling along subduction zones. J. Geophys. Res. 98, 2017–2037.
Uyeda, S., and Kanamori, H., 1979. Back-arc opening and the mode of subduction. J. Geophys. Res., 84, 1049–1061.
von Huene, R., and Scholl, D., 1991. Observations at convergent margins concerning sediment subduction, subduction erosion and the growth of continental crust. Rev. Geophys. 29, 279–316.
Westbrook, G.K., and Reston, T.J., 2002. The accretionary complex of the Mediterranean Ridge: Tectonics, fluid flow and the formation of brine lakes: An introduction. Mar. Geol., 186, 1–8.
Westbrook, G.K., and Smith, M.J., 1983. Long decollements and mud volcanoes: Evidence from the Barbados Ridge Complex for the role of high pore fluid pressure. Geology, 11, 279–283.
Westbrook, G.K., Ladd, J.W., Buhl, P., Bangs, N., and Tilley, G.J., 1988. Cross section of an accretionary wedge: Barbados Ridge complex. Geology, 16, 631–635.

Part IV

Seismogenic Zone Chaterization

Convergent Margin Structure in High-Quality Geophysical Images and Current Kinematic and Dynamic Models

Roland von Huene, Cesar R. Ranero and Dave W. Scholl

Abstract Understanding the mechanics of convergent margins is fundamental to assessing risks from earthquakes and trans-oceanic tsunamis. Marine observations of the past decade have advanced that understanding. A once commonly inferred accreted wedge extending from trench axes to shelves is now resolved into 3 domains of different mechanics in space, that vary during an earthquake cycle. The frontal prism increases weight on subducting materials elevating pore fluid pressure and reducing interplate friction. The middle prism is moderately stable and merges into the more stable margin framework of the inner prism beneath the upper slope and shelf. Significant accretion occurs as material from the frontal prism is added to the middle prism. Accretion is common along thickly (>1 km) sedimented trenches and slowly converging margins. Rapid convergence enhances the efficiency of sediment subduction and subduction erosion. The subduction channel on the lower plate accepts a finite amount of trench sediment and any excess is added to the frontal prism on the upper plate. Erosion beneath the middle slope contributes material to the subduction channel. Erosion and accretion can be coeval, for instance, subducted seamounts erode the upper plate as adjacent sediment accretes. The change in strain during interseismic locking that is released during coseismic slip, changes the dynamics of each segment in time. This helps explain extensional normal faults in a converging plate environment. Recent observations provide information for a unifying framework concept to aid interpretations of both accreting and eroding margins.

Keywords Convergent margins • Subduction erosion • Accretion

1 Introduction

Many concepts regarding subduction zones were derived from interpretations of seismic data with 1970–1980s spatial resolution and depth of imaging. Examining such interpretations with scientific drilling found that 70% of the tested geophysical interpretations were fundamentally wrong. These interpretations lacked adequate constraints and various model dependant interpretations were permissible. The high vertical exaggeration and travel-time rather than depth images in older seismic data distort structure and limit imaging of dipping reflectors. A lack of coherent structural images was commonly attributed to an inability of seismic techniques to image mélanges. Even with simple modern 3D images only 20% of a large test group of geoscientists arrived at correct tectonic interpretations in a community benchmarking exercise (Bond et al., 2007). Improved seismic imaging integrated with multibeam bathymetry reveals much more

R. von Huene
US Geological Survey and University of California, Davis, 4300 Carlson Way, Diamond Springs, CA 95619, USA,
rhuene@mindspring.com

S. Lallemand and F. Funiciello (eds.), *Subduction Zone Geodynamics*,
DOI 10.1007/978-3-540-87974-9, © Springer-Verlag Berlin Heidelberg 2009

structure allowing many earlier simplifying assumptions to be replaced with data and interpretations to be reconsidered. Yet tectonic models abound and no single one contains a unifying concept or model.

The original "plate tectonic" assumption that entire convergent margins form by steady state accretion producing a broad imbricate stack is replaced by observations showing segmentation of structure requiring areas with different materials and processes. Discrete margin segmentation was proposed 20 years ago (Cloos and Shreve, 1988; Taira, 1994). Recent data resolve an unstable frontal prism and a more stable middle slope, as for instance in the classical Sumatran accretionary wedge (Kopp and Kukowski, 2003) where earlier interpretations became a type example of a steadily accreting body (Hamilton, 1988). Segmentation is observed in improved seismic images at margins along the Americas and Japan (i.e., Ranero and von Huene, 2000; Christenson et al., 2003; von Huene and Ranero, 2003; Bangs et al., 2006; Ranero et al., 2006; Moore, 2007; Kimura et al., 2007). A two segment upper plate and subduction channel model introduced by Shreve and Cloos (1986) and Cloos and Shreve (1988) was based on fluid mechanics as proposed by Sorokhtin and Lobkovskiy (1976). In the subduction channel model a compressionally strained prism adjacent to the deformation front builds at the inlet where trench sediment subducts forming a subduction channel beneath the prism. Active contraction occurs mostly in this zone of compression rather than progressively across the entire margin. In the Cloos and Shreve model, the inlet controlled subduction channel thickness, and when inlet capacity was exceeded, the excess would accrete. The structural features of the model were not imaged in the seismic data of the 1970–1980 period but are now clearly observed.

The segments observed in seismic images are consistent with the Dynamic Coulomb wedge model (Wang and Hu, 2006). Conventional models are commonly without a reference to the geologic time associated with a margin's structural development. The dynamic wedge model shows the significance of a seismic cycle in understanding subduction zone mechanics. It explains how stress can affect segments differently during a cycle. During interseismic periods when the plate interface fault locks, the margin wedge landward of the frontal prism stores the elastic strain from plate convergence. The weak material and relatively small mass of the frontal prism cannot store much elastic strain so it is characterized by permanent deformation. During an earthquake the elastic strain released in the middle prism shoves against the frontal prism creating a coseismic pulse of deformation and elevated pore fluid pressure. This can explain features observed in multibeam seafloor morphology and depth processed seismic images. These two behaviors affect both accreting and eroding margins in similar manners.

In the literature, convergent margins are commonly divided into two pure end-members with accreting or eroding modes of mass transfer. Recent observations indicate a mix of accretion and erosion in space and time. An example is the Nankai margin where subducting seamounts erode the upper plate in an otherwise accretionary environment (Bangs et al., 2006). Despite a clear difference in mass transport, certain mechanical processes may share similarities.

1.1 Aims of this Chapter

In this review we expose to a broader readership the ideas advanced within a community of specialists. We develop a unifying fundamental framework in which to consider diverse observations. The major concepts we consider are the availability of trench sediment at the deformation front, the capacity of the subduction channel inlets to accept materials, the change in relative strength of the upper and lower plates, and an essential change of dynamics during coseismic and interseismic periods. We summarize kinematics in erosional and accretionary segments of the Chilean margin and relate these to convergent margin framework processes. Dynamics can be explained in accord with the dynamic Coulomb wedge concept (Wang and Hu, 2006). Differences in subduction zone behavior during interseismic locking and coseismic slip provide an explanation for puzzling observations of extensional faulting in a compressional convergent plate environment and helps to understand the paradox of a prism with contractile accretionary structure fronting eroding margins. Convergent margin structural complexity precludes a simple model that explains them all but some core concepts provide a framework that can be augmented and modified to fit the diversity of observations along various margins.

2 Upper Plate Rock Units

We apply non-generic terms to four upper plate rock units inferred from geology and structure in seismic images following Scholl and von Huene (2008). The generic terms "*accrete*" and "*erode*" imply material transfer between converging plates. We use them only if upper plate material provenance is known, whether sourced from the lower plate or repositioned upper plate material. Viewed from the trench axis landward (Fig. 1) the units are, (1) an actively deforming sediment body at the foot of the margin slope, the ***frontal prism*** which transitions into a (2) body of older and more consolidated material, the ***middle prism***, and (3) the ***inner prism*** consisting of the margin's rock framework. These units are covered by, (4) an overlying sedimentary ***slope apron***. On the subducting lower plate, the top layer of clastic material on basement comprises the ***subduction channel.*** Kimura et al. (2007) recognized a similar segmentation but termed it the outer wedge, transition zone, and inner wedge.

The term frontal prism came into wider use after drilling of the Costa Rica "accretionary prism" sampled only slope sediment and no lower plate sediment (Kimura et al., 1997). Lower plate sediment was not transferred to the upper plate and the pre-drilling interpretations that all landward dipping reflections represent accreted trench sediment was incorrect. In samples, accreted sediment lithology can be non-unique because trench axis sediment is compositionally the same as its slope sediment parent. Slope sediment trapped and kneaded into the imbricating frontal prism is repositioned and not accreted. Slope sediment crossing the frontal prism to reside briefly in the trench axis can be returned to the prism as the decollement propagates seaward of the deformation front. This accreted sediment has essentially the same lithology as slope sediment making a distinction from samples alone problematical. Where material transfer between plates is uncertain the non-generic terms "frontal and middle prisms" are convenient.

2.1 Frontal Prism

Convergent margin upper plates are bordered at the trench axis by a ***frontal prism,*** 5 to ~30 km wide (Fig. 1). In many seismic images, frontal prisms contain clear landward dipping reflections that indicate tectonic thickening by imbrication. Where the complexity of internal structure exceeds resolution of the applied seismic technique an imbricate structure is commonly assumed. Scientific drilling of accretionary frontal prisms along the Barbados and Nankai drill transects has sampled oceanic sediment transferred from the subducting plate, a transfer also seen in seismic images. Off Central America, Peru, and northern Chile, the term accretionary frontal prism is inappropriate because evidence that the sampled prism sediment was transferred from the lower to the upper plate is inconclusive. The frontal prism of accretionary margins either merges up slope into a older more consolidated *middle prism* through a 1–5 km wide contact zone or is separated from the middle prism by a fault zone. This boundary has been called a backstop but it differs from the one introduced by Byrne et al. (1988) which is stationary and separates igneous from sedimentary rocks.

2.2 Middle Prism

In accretionary margins middle prisms may be 40–100 km wide bodies presumed to contain trench sediment transferred from the frontal prism. Their internal structure is seldom imaged well by reflection seismic surveys. The overlying slope apron displays little active permanent contractile deformation in addition to what this unit inherited during residence in the frontal prism. These prisms are commonly bounded landward by a backstop of framework rock. At erosional margins the middle prism consists of decomposed and fractured framework rock with reduced seismic velocity (e.g., Chile, Central America). The fractured framework rock structure in seismic records is difficult to distinguish from accreted sediment as shown when they were drilled (i.e., Central America and Peru). Beneath middle slopes, material has greater rigidity and greater seismic velocity than frontal prisms. Large middle prisms commonly occur where margins have had thickly (>1 km) sedimented trenches for several Myr periods and where orthogonal convergence is less than ~40–50 km/Myr (e.g., Makran, and Nankai Trough) (Clift and Vannucchi, 2004).

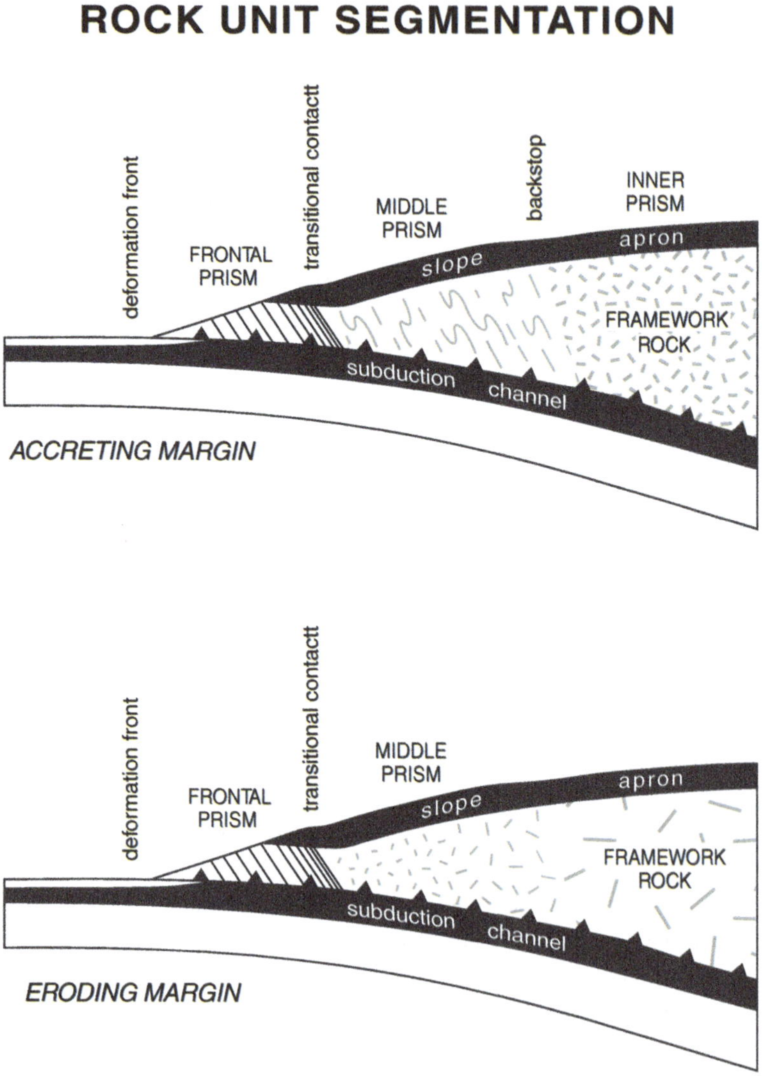

Fig. 1 Rock and sediment units of convergent margins. At accreting margins the rock framework (structurally the inner prism) backs a *middle prism* of accreted material. At eroding margins coherent framework rock becomes increasingly fractured down slope to form the *middle prism*. Frontal prisms can contain upper and/or lower plate materials

2.3 Inner Prism: Rock Framework

The margin's core, its ***rock framework***, consists of igneous or metamorphic basement and lithified sedimentary rock. These rocks are seen in coastal exposures and they extend seaward beneath the submerged margin as established with drill and surface samples. The framework rock is as diverse as Paleozoic and older igneous and metamorphic cratonic rock (i.e., Peru and Chile) and the magmatic massifs of Cenozoic arcs (i.e., Aleutian, Izu-Bonin, and Tonga). The rock framework locally contains metamorphosed fossil accretionary middle prisms. Structurally, the inner prism is strong and only broadly deformed, commonly at the edge of the shelf.

2.4 Slope Sediment Apron

The upper plate is covered by an apron of sediment that extends from the shelf to the lower trench slope. Scientific drilling at the erosional margins of northeast Japan, Tonga, Peru, Costa Rica, and Guatemala, revealed several km of long-term subsidence evidenced by shallow water indicators in the basal beds of the slope apron not far from the trench axis. The underlying basement surface is commonly an eroded wave-based unconformity and the overlying strata record a vertical displacement history. The thickness of aprons ranges from a few meters to 5 km (Clift and Vannucchi, 2004).

3. Geophysical Observations of Upper Plate Segmentation

A long standing assumption that margin wide wedges accreted linearly, consolidate, and increase in thickness up slope, fails to explain a variety of tectonic structures. In the past decade, upper plate deformational segments were resolved in multibeam seafloor morphology and in well processed seismic images (Fig. 2). In multibeam bathymetric surveys, an upper slope morphology sufficiently stable to develop many canyons changes to a moderately stable middle slope morphology and finally to the unstable frontal prism (Fig. 2a). Some associated seismic images illustrate a corresponding deterioration in reflector coherence down slope (Fig. 2c). In eroding margins, lowered seismic velocity and middle slope failure is explained as fracturing and progressive break-up of the framework rock. The presumed fractured material in an eroding margin has a seismic velocity distribution similar to the landward increased velocity resulting from sediment consolidation in an accreting one (Sallares and Ranero, 2005). Importantly, the seismic reflections in middle slope sediment aprons show little pervasive or strong permanent compressional deformation. In some drill cores the ages of little deformed strata are middle Miocene invalidating the argument that slope sedimentation is so rapid that it masks deformation (i.e., Ranero et al., 2007) Interplate shear as hundreds of km of lower plate subduct for <5 Ma should permanently deform the slope apron if the underlying middle prism were deforming significantly. The common loss of coherent slope apron reflections at the middle to frontal prism contact indicates deformation there (Fig. 2c).

Although the clearest change from frontal prism to middle slope has been seen in erosional margins it is also observed in improved seismic images and high resolution bathymetry of accreting margins. Across the Alaska accretionary margin the contact between the frontal and middle prisms is defined by a change in structural vergence from landward to seaward (von Huene and Klaeschen, 1999). Strain and fluid venting were quantified across this margin along a transect just north of the Kodiak group of islands. The rate of permanent deformation peaks in the frontal prism and drops four-fold to the middle prism along with expulsion of all but 20% of original pore fluid (von Huene and Klaeschen, 1999). Therefore middle prism strain must be largely elastic because it is associated with increased friction and microearthquake activity. The division between aseismic and seismogenic segments of convergent margin plate interfaces is consistent with segmentation. As mentioned previously, improved data across the "classical Sunda accretionary prism" (Hamilton, 1988) is separated into two prisms that back against a margin framework (Kopp and Kukowski, 2003). Wide-angle and reflection seismic images along with multibeam bathymetry show a little deformed slope apron covering a middle prism core containing Eocene rock. A splay fault system separates the middle slope from seafloor ridges of the frontal prism composed of material with low seismic velocity. This splay fault system produces a low continuous ridge along the 600 km lateral extent of the multibeam bathymetry acquired. Recent multibeam bathymetry and 3D seismic images across the Nankai margin show a splay fault system separating the active accreting frontal prism and a middle prism (Bangs et al., 2006; Moore, 2007). Forearc basin sediment on the middle prism is cut by numerous normal faults. The middle prism taper is shallower than the frontal prisms. Similar tectonic structure is observed along the accretionary Makran margin (Fruehn et al., 1997; Kopp et al.; 2000; Kukowski et al., 2001), the Cascadia off central Oregon (Gerdom et al., 2000), and south-central Chile (Ranero et al., 2006). A transition from the middle prism to the margin framework is observed in the canyon morphology that commonly diminishes sharply from the framework to the middle prism seafloor. Along the Chilean margin this corresponds to a change from landward dipping to seaward dipping faults (Ranero et al., 2006). Despite many second-order variations, a basic margin segmentation can be recognized in many recently acquired seismic images and multibeam bathymetry.

3.1 Width of Frontal Prisms

Most seismic images of frontal prisms show compressional structure and imbrication. Single seismic lines can be very misleading because damage to frontal prism structure from high relief plowing into the subduction zone can add more complexity than is resolved even with modern seismic acquisition. A multibeam

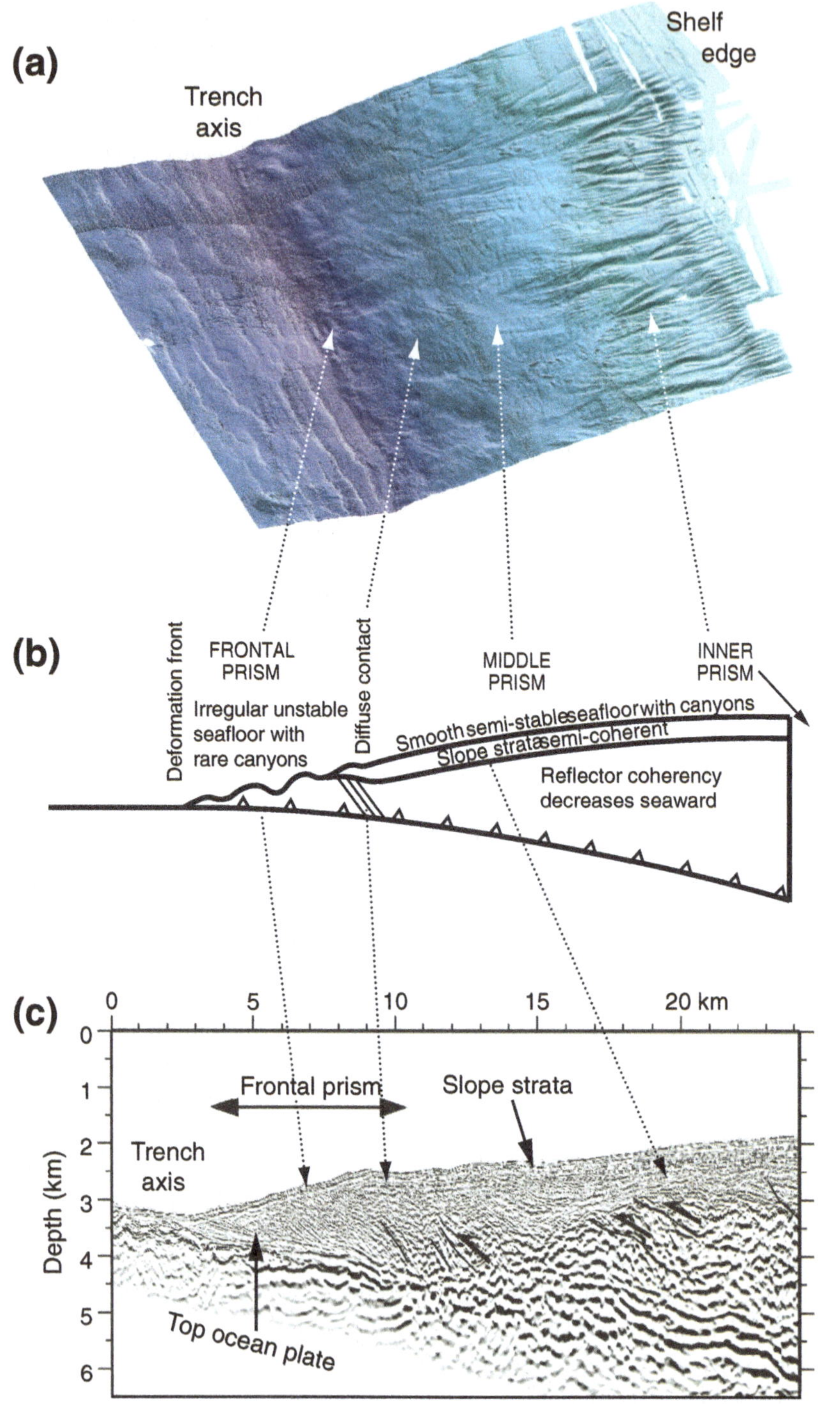

Fig. 2 The frontal prism, the middle (transition) prism, and the continental framework. (**a**) perspective view of multibeam bathymetry off northern Costa Rica from the trench to the shelf edge. The rough frontal prism seafloor without gullies is separated from a smoother middle prism with gullies and many small normal fault scarps. This morphology transitions to an upper slope with older and more sharply incised canyons. (**b**) Diagram of major features in the 2-prism model. (**c**) Pre-stack depth migrated seismic reflection image showing subsurface structure at a vertical exaggeration of 1.5×. Note abrupt change in prism taper. (After Ranero et al., 2007.)

morphology map view is helpful in understanding such complexity and how damaged zones heal. Frontal prisms are restored only to their original width and appear to be self limiting. Global observations indicate a restricted range in frontal prism width. Their taper and volumes measured in undistorted seismic depth images are not readily available so we approximated frontal prism size using prism width (Table 1, Fig. 3). A prism's landward contact was located using several of the following criteria.

Table 1 Compilation of geologic information on convergent margins used in Fig. 5

Global compilation of frontal prism width						
Margin sector	Length, (km)	Sed. Input (km)	Frontal prism	Middle prism	Subduct. Sed	Numb. lines multibeam
S. Chile	1,110	2.9	27	19		2 & mb
Chile 3-J to JFR	1,665	2	18	12	1.3	7 & mb
Chile JFR to Anto	1,445	0.4	10		1	6 mig + 5, mb
Peru N of Nazca	1,220	0.7	15		0.9	7 & mb
Middle America	1350	0.6	9		1	9 & mb
Mexico	1350	0.9	18		0.4	3 &mb
S Oregon	200	2	7	24	1.5	1 & mb
Oregon	770	4	14	9	2	2 & mb
NE Alaska	450	2	23	30	1.3	2 & mb
Kodiak-Shumigan	900	2	24	8	0.8	2 & mb
Aleutian to 180°	1,300	2	20	10		4
Kuril	1,320	0.7	14		0.8	4
N Japan	720	0.6	15		0.9	6 & mb
Nankai	700	1.3	24	52	1.1	6 & mb
Makran	900	7	30	250–350	3	1 & mb
Sunda Arc	2,500	1.5	22	75		4 & mb
IBM	3,500	0.5	10			2
Tonga	1,350	0.4	7			1 & mb

Column 2 – length km – is the extent of each margin. Column 3 – sed input km – is the trench sediment thickness at the deformation front. Column 4 – frontal prism – is the width of the frontal prism. Column 5 gives the width of the middle prism where present. Column 6 – subduct. sed – is the subducted sediment layer thickness in km where clearly defined. Column 7 – numb. line multibeam – is the number of transects available to establish thicknesses and extent of features – mb indicates where multibeam bathymetric information is available. The sum does not include IBM (Izu-Bonin-Mariana) which are italicized to indicate that information is not sufficient to meet standards used for margins listed above.

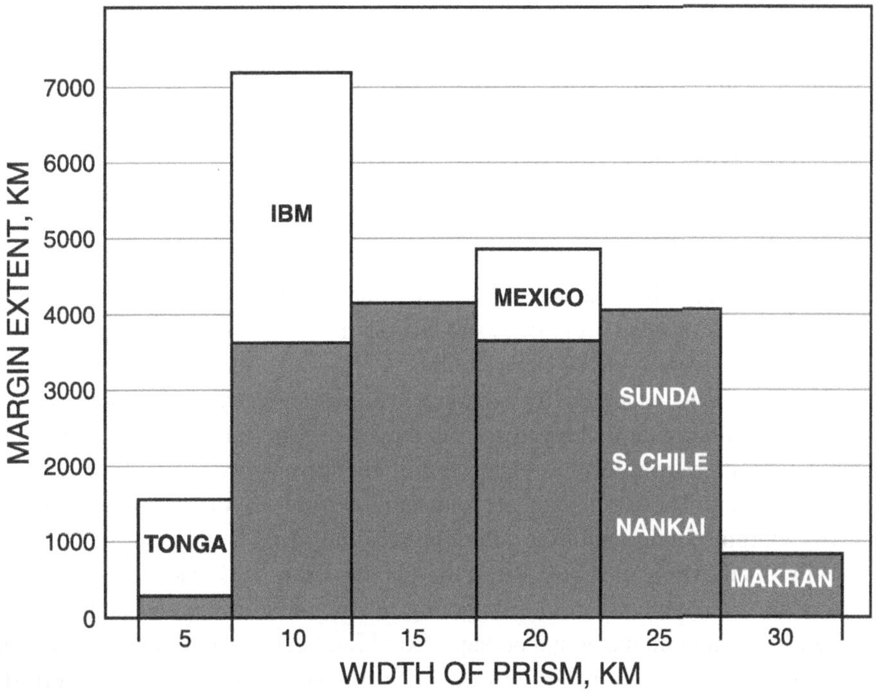

Fig. 3 Global compilation of frontal prism widths. Bars are widths from images that clearly constrain the frontal prism. The compilation includes seismic images across about 48% of margins globally excepting areas of collision and where convergence is more oblique than 30°. Unshaded areas show data with less certain widths because of inadequate seismic image quality

A change from relatively smooth to rough seafloor morphology commonly at a scarp.

A sharp change from little deformed to deformed and obscure slope apron strata.

A transition from semi-coherent to non-coherent basement reflectors of eroding margins.

Change of taper, physical properties, or an increased horizontal velocity gradient

An abrupt break in age or rock type indicating a tectonic boundary

Prism widths along convergent margins (Fig. 3) were estimated from more than 70 selected seismic records and from multibeam bathymetry. Their trench axis parallel extent was estimated from satellite derived bathymetry. Subducting ridges and areas with more than 30° oblique convergence were avoided. The compilation of clearly bounded prisms includes 48% of convergent margins globally. Insufficient information is available to assess the Tonga and Izu-Bonin-Mariana frontal prisms with the certainty desired. Nonetheless, because of their great lateral extent, the average of their sparse available data is shown (Fig. 3) to indicate how including them might affect the compiled data. The frontal prisms of margins currently eroding are on average about ~15 km wide whereas the currently accreting margins are about ~22 km wide. As noted by Clift and Vannucchi (2004) accreting margins commonly occur where convergence is less than 60 km/Myr and erosion occurs where convergence is more rapid. The Chile margin south of Valparaiso is an exception. Accretion south of Juan Fernandez Ridge became rapid as trench sediment fill thickness jumped from 1 km to more than 1.5 km (Onken et al., 2006). Accretion is a Pleistocene-Quaternary mode of tectonism following a Miocene period of dominantly erosional tectonism. Despite 70 km/Myr convergence, thick Pleistocene sediment is associated with current accretion showing that trench sediment thickness can overwhelm subduction channel capacity even where convergence is rapid.

Frontal prisms adapt readily to changes in the subducting seafloor morphology. Large embayments in the deformation front form as high relief on the lower plate subducts. These embayments return to their previous margin configuration as relief passes and are rapidly filled. Along the Costa Rican margin, the frontal prism embayments from subducting seamounts roughly 20 km in diameter and perhaps 2 km high heal in less than 0.2 Ma (von Huene et al., 2000). After accelerated erosion over the Nazca Ridge off Peru, the margin re-establishes its frontal prism in ~4 Myr (von Huene et al., 1996). The filling of embayments stops when pre-collisional conditions are re-established.

The restricted width of frontal prisms globally indicates that they are at an optimum size for the current convergence rate, mode of sedimentation, and taper. The self limiting indicators in nature were observed in sandbox modeling where different modes of material transfer became constant after adjusting to the given conditions (Kukowski and Onken, 2006).

Frontal prisms at erosional and sparsely sedimented margins probably contain abundant slope sediment as contrast to trench sediment building a dominantly accretionary prism. Thus the greater median width of accreted margin frontal prisms may reflect the influence of differing prism materials and manner of accumulation. If the frontal prism accretes sediment, its increased width is adjusted by shifting the frontal and middle prism contact seaward. In narrowing margins the frontal prism contact shifts landward. From our analysis we derive three principal observations: (1) Frontal prisms occur at both erosional and accretionary margins; (2) Frontal prism width is commonly between 5 to 30 km and rarely wider. (3) Frontal prisms appear to be self-limiting in accord with prevailing structure, material, and dynamics.

4 Observed Relation of Trench Sediment Thickness and Tectonic Processes

We now explore convergent margin kinematics, in particular the different environments associated with processes of accretion and erosion. Both accretion and erosion result from plate convergence so what differences in mass flux are associated with either tectonic type? Along the Chile subduction zone the effects of trench sediment supply and rate of plate convergence are well displayed. This subduction zone includes extremes from a sediment starved trench along its northern desert (<0.5 km thick trench sediment) to a sediment flooded (>1.5–3 km) trench settings along its glaciated southern coastal mountains. The effects of over- and under-abundance of sediment can be compared on either side of two ocean ridges that obstruct sediment transport down the northward deepening trench axis. Across the Juan Fernandez Ridge the change in trench sediment thickness of adjacent segments is 2 km. Across the South Chile Rise triple junction a four-times reduction in plate convergence allows a longer time for trench sediment deposition. South of the triple junction trench sediment 3.5 km thick has been imaged whereas just north the sediment is about 1 km thick. These areas were surveyed with extensive multibeam bathymetry and depth processed reflection and wide angle seismic data as summarized in Ranero et al. (2006). The empty trench off northern Chile illustrates a type example of an erosional margin.

4.1 The North Chile Example of a Erosional Margin

The northern Chile geophysical data includes precise aftershock locations of the 1995 Antofagasta earthquake ($M_w = 8.0$) recorded with a land-sea seismometer network (Fig. 4) (Husen et al., 2000; von Huene and Ranero, 2003; Sallares and Ranero, 2005). Multibeam bathymetry images canyons beginning on the shelf that deepen across the upper slope and are displaced by landward dipping normal faults (Fig. 3). Down slope this morphology merges into a middle slope smoothed by mass wasting debris cut by seaward dipping normal faults. The lower slope is characterized by broad ridges that are continuous with sharp well developed lower plate horsts extending into the subduction zone (Fig. 4).

The seaward dipping normal faults of the middle prism extend 4 km into basement where they merge along a detachment. Analysis of the detachment fault geometry with methods applied to slide bodies (Watts and Grilli, 2004) shows low frictional strength (von Huene and Ranero, 2003). Fractures permeate the entire upper plate basement. Mass-wasting debris migrates down slope and accumulates in a well developed frontal prism. The prism's 3.5 km/s P wave velocity indicates >15% fluid (Sallares and Ranero, 2005) and the geometry of conjugate faults (Davis and von Huene, 1987) indicates a strength similar to frontal prisms of accretionary margins composed of water rich trench sediment. The upper plate's crystalline inner prism has seismic velocities of 5 km/s along the coast and continental shelf, ~4.5 km/s in the mid-slope, and 3.5 km/s velocities beneath the lower slope without an apparent change in rock type.

Along northern Chile and other erosional margins, the contact between framework rock and the frontal prism is commonly a transition from coherent basement

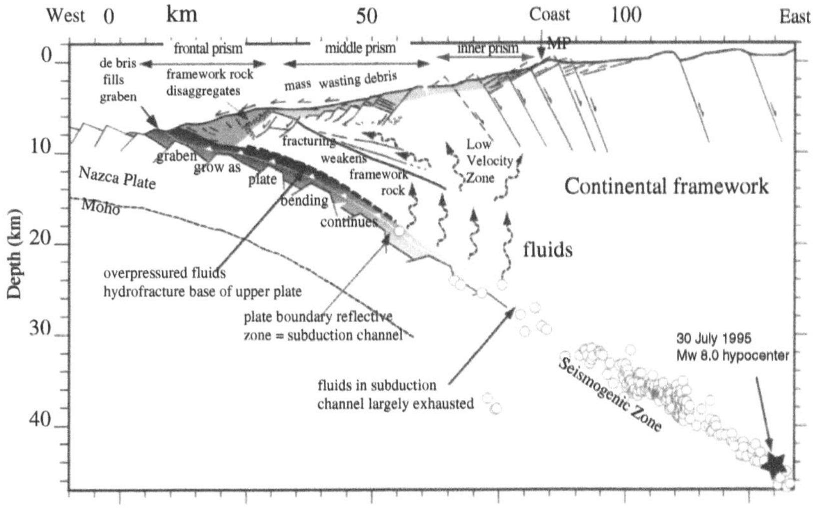

Fig. 4 Diagram of subduction erosion along the north Chile margin. In the upper plate, a decreasing rock strength is associated with 3 tectonic styles across the slope. Landward-dipping normal faults develop perhaps as material loss down slope progressively removes a constraining mass. In the middle slope seaward dipping normal faults develop from gravity tectonics in weakened framework rock. Mass wasting debris collects in a frontal prism and elevates pore fluid pressure allowing subduction of unconsolidated clastic debris. Clastic debris and basement fragments are inferred to fill a subduction channel. Normal faulting continues in the subducted ocean crust to accommodate continued plate bending in the subduction zone. (After von Huene and Ranero, 2003, Sallares and Ranero, 2005.)

reflections across the middle prism to a structureless seismic image (cf. Ranero and von Huene, 2000). The continuity of reflections weaken and fades away across the transition from a middle slope to the lower slope morphology. The middle slope morphology of fault scarps, rills, and ridges ends at the frontal prism where underlying slope reflectors decompose or break into short segments. Multibeam bathymetry commonly resolves frontal prism limits with better resolution than seismic images (cf. von Huene and Ranero, 2003).

Observations along the north Chile margin (von Huene and Ranero, 2003; Sallares and Ranero, 2005) reveal that: (1) erosional margins have frontal prisms that presumably elevate pore fluid pressure immediately above the underthrust lower plate and facilitate subduction of the slope materials spilling into the trench axis. (2) the upper plate framework rock decreasing P velocity seaward indicates material weakness to values found in middle prisms of accretionary margins; and (3) up-dip of the seismogenic zone, where slip is dominantly stable, the interplate fault is weakly coupled like those of accretionary margins. Slope material dominates the frontal prism and the subduction channel.

The crystaline basement ends at a transitional contact with a frontal prism, a contact that has been termed a "backstop." That term was formulated from concepts of accretionary wedges where the backstop is stationary (Byrne et al., 1988; Byrne et al., 1993). Along eroding margins this gradational contact migrates landward as material is removed. Despite its elegant simplicity and popular usage, the stable backstop concept does not apply here since over the long term erosional margins are extensionally faulted and material is removed rather than pushed or backed up against a rigid "bulldozer blade."

4.2 Trench Turbidite Thickness and Tectonics at the Juan Fernandez Ridge

The aseismic Juan Fernández Ridge (JFR), a seamount chain on the Nazca plate, subducts beneath the Central Chilean continental margin. The continental sediment source and plate convergence are the same on either side of the subducting ridge and only trench sediment volume changes. The crest of the subducting JFR uplifts the entire continental slope (Fig. 5) (Ranero et al., 2006; Laursen et al., 2002; von Huene et al., 1997). In the

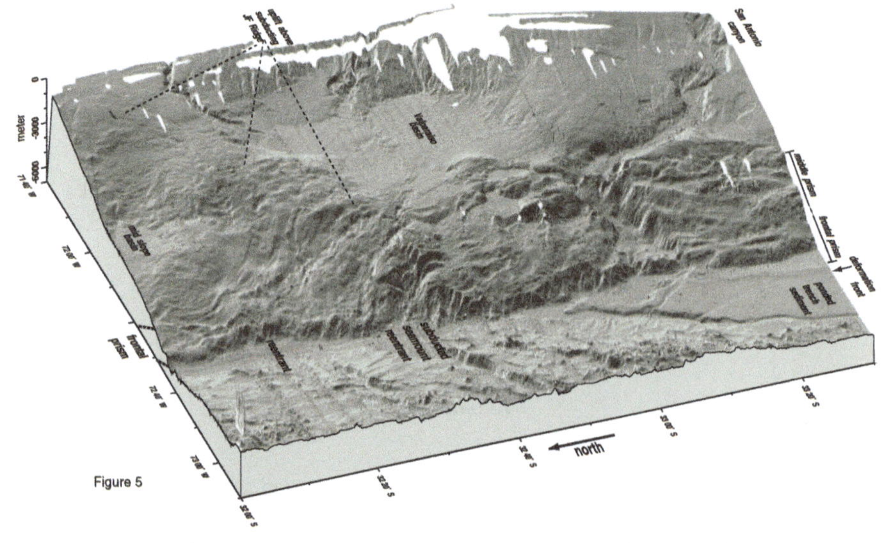

Fig. 5 Multibeam bathymetry of the central Chile margin where Juan Fernandez Ridge subducts. The ridge crest blocks axial sediment transport thereby separating a thickly sedimented from a modestly sedimented trench axis. After the ridge subducts its subducted continuation is marked by a low ridge diagonally crossing the slope. Frontal prism ridge and trough morphology south of Juan Fernandez ridge is replaced by a disorganized morphology north. The middle slope breaks-up over the trailing flank of the southward migrating ridge. Some sediment input to the subduction zone to the south accretes but north all sediment subducts despite the same rate of plate convergence. (After Ranero et al., 2006.)

trench the ridge relief impedes axial transport of sediment resulting in rapid sediment deposition upstream (south) and only modest sediment accumulation on the downstream (north) side. This results in accretion of trench sediment upstream, and essentially complete sediment subduction downstream (Ranero et al., 2006). The subducting ridge separates an erosional slope morphology to the north from an accretionary one to the south. The eroding continental slope frontal prism is 5–10 km wide and is presumed to reduce interplate friction so that all trench sediment subducts. Fluids from the subducted sediment escapes at the back of the frontal prism forming 100-m-high mud diapirs (Ranero et al., 2006). Once it has formed the frontal prism does not widen significantly as shown by more than 200 km of multibeam bathymetry north of the subducting ridge. A middle slope erosional margin morphology extends 1000 km north through the area off Antofagasta and beyond (von Huene and Ranero, 2003).

South of the JFR the trench axis contains >2 km of sediment, and a 10–25 km wide frontal prism. The ~20 km wide middle prism is covered by a little deformed seafloor. Sediment accretion is the dominant process along the lower slope and a 1.0–1.5 km thick layer of the lower trench sediment section subducts beneath the frontal prism (von Huene et al., 1997; Diaz-Naveas, 1999). Variants of this pattern extend southward for ~1500 km to where tectonics are modified by the subduction of the South Chile Rise (Bangs and Cande, 1997; Diaz-Naveas, 1999).

The thickness of sediment at the deformation front correlates with accretion and erosion. The upper sediment accretes and the bottom 1 km subducts as also noted by other investigators (Le Pichon et al., 1993; Clift and Vannucchi, 2004; Kukowski and Onken, 2006).

4.3 *Effect of Convergence Rates Observed Across the South Chile Rise*

Accretion and erosion on either side of the South Chile Rise illustrates the effects of convergence rate on tectonics (Fig. 6). South of the triple junction where the spreading South Chile Rise subducts, the Antarctic plate converges with South American at ~18 mm/year whereas north of the Rise convergence is 84 mm/year or roughly five times more rapid. The northward migrating Chile Rise collision left a truncated continental margin in the south. After the Chile Rise relief subducted, the continental slope subsided and sediment in the trench thickened and widened (Ranero et al., 2006). About 100 km south of the triple junction where the South Chile Rise subducted ~3–6 Myr ago, the frontal and middle prisms are ~60 km wide (Fig. 6). Slow convergence rates reduce sediment subduction and trench sediment has more time to accumulate. Where the Chile Rise currently collides with the continental margin, the frontal prism is only ~5 km wide (Bourgois et al., 2000; Behrmann et al., 1994). North of the Chile Rise where convergence is rapid, the trench deepens and has a relatively thin sediment fill (~1.5 km). The frontal prism here is ~10–15 km wide and sediment subduction is efficient (Bangs and Cande, 1997) (Fig. 6). Thick and thin trench sediment are associated with wide and narrow frontal prisms. The shallow dip of the subducting crust may also influence prism width. But a first order association is trench sediment thickness and prism width which here is a function of convergence rate.

5 Subduction Channel Character

The subduction channel, a layer directly below the plate interface, was originally conceived as filled with a slurry of sheared material (Shreve and Cloos, 1986). However, drilling through the frontal prism encountered narrow fault zone above little deformed sediment (cf. Kimura et al., 1997), consistent with recent seismic images (cf. Kinoshita et al., 2006). This called for modification of the original fluid mechanics analogy. At the Ecuadorian margin, good continuity and strong reflections allows exceptional velocity determinations from which physical properties estimates are derived (Calahorrano et al., 2008). Indicated is a division of the channel into 3 zones similar to upper plate segments. Such a segmentation is also proposed by Vannucchi et al. (2008) from study of a fossil subduction channel on land. The 3 zones of Calahorrano et al. (2008) indicate conditions in currently active subduction channel and the "Shallow, Intermediate and Deep" segments of Vannucchi et al. (2008) show a corresponding structure at outcrop scale. The channel character varies if sediment smoothes the subducting seafloor topography or if rough irregular seafloor has subducted (Sage et al., 2006).

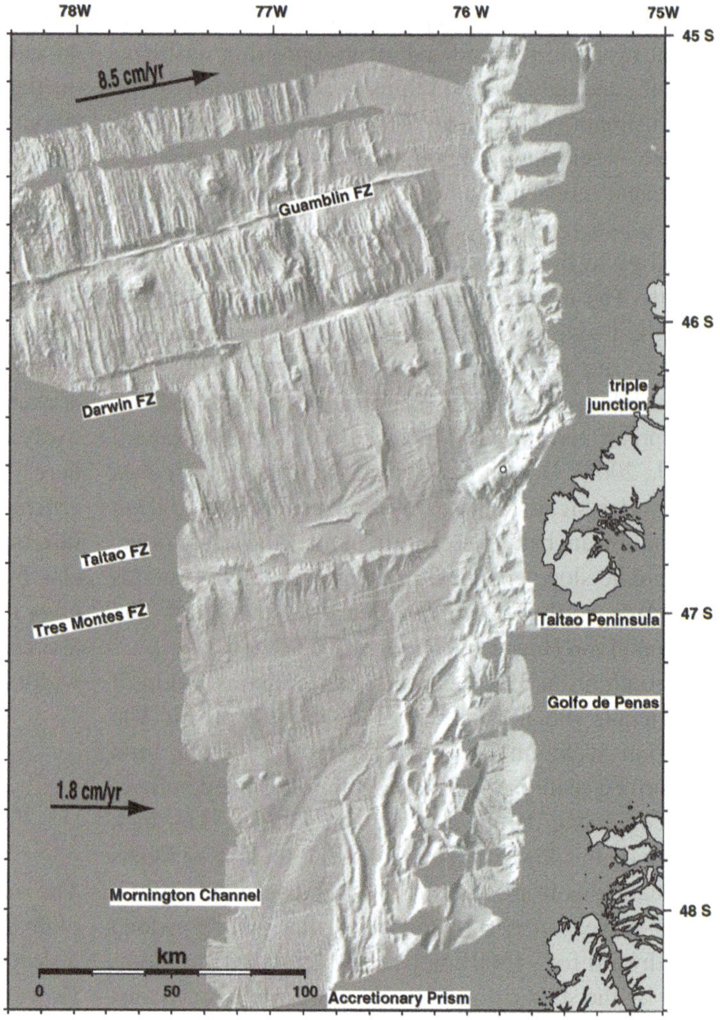

Fig. 6 Multibeam bathymetry at the triple junction of the south Chile Rise collision zone. Circles are ODP Leg 141 drill sites (Behrmann et al., 1994) and *numbered lines* are RC2902 seismic images (Ranero et al., 2006). The finger of land in the middle-right is the Taitao Peninsula opposite which the Chile Rise subducts. South of the Chile Rise convergence is slow allowing more time for trench sediment deposition. The resulting thick sediment enhances accretion as shown by the growth of ridges as the triple junction migrates north. Northward, where the margin is dominantly erosional, the trench axis is bordered by a narrow frontal prism backed against the margin framework. (From Ranero et al., 2006.)

In many seismic images, a minimally deformed stratified trench fill subducts beneath the frontal prism decollement. The decollement is locally a sharp reflection (cf. Bangs et al., 2006; Moore, 2007). Drill cores and logs off Costa Rica show a narrow interplate fault zone (20–30 m) at the top of a little deformed and under-consolidated sediment section (Kimura et al., 1997; Saffer, 2003). Compared with drainage paths from the frontal prism, fluid of the subduction channel must drain through a longer path and therefore dewatering is inferred to be slower.

Zone I of Calahorrano et al. (2008) and the shallow segment of Vannucchi et al. (2008) occur beneath the frontal prism. Seismic velocities increase rapidly in a landward direction from 1.8 to 2.6 km/s indicating relatively rapid fluid drainage but slower than from the overlying frontal prism. A weak interplate fault running through weak materials is consistent with little recorded microseismicity along the decollement. In rock outcrops, deformation is mainly extensional and the many small fractures probably served as conduits for fluid drainage. Quartz and calcite veins are absent. Vannucchi et al. (2008) propose accommodation of strain by pervasive failure along many small fractures during compaction.

Beneath the middle prism the channel image is commonly 3 or 4 thick (150–500 m) high-amplitude-low frequency reflections. The loss of resolution results from an increased sonic footprint with depth (Fresnell zone), increased attenuation and scatter of reflected high frequency energy. The change from the first to second zone or segment is transitional or abrupt. Off Ecuador the seismic velocity within this segment of the subduction channel stays relatively constant at 2.8 km/s and it is commonly a low velocity zone (Sage et al. 2006; Calahorrano et al., 2008). Sometimes

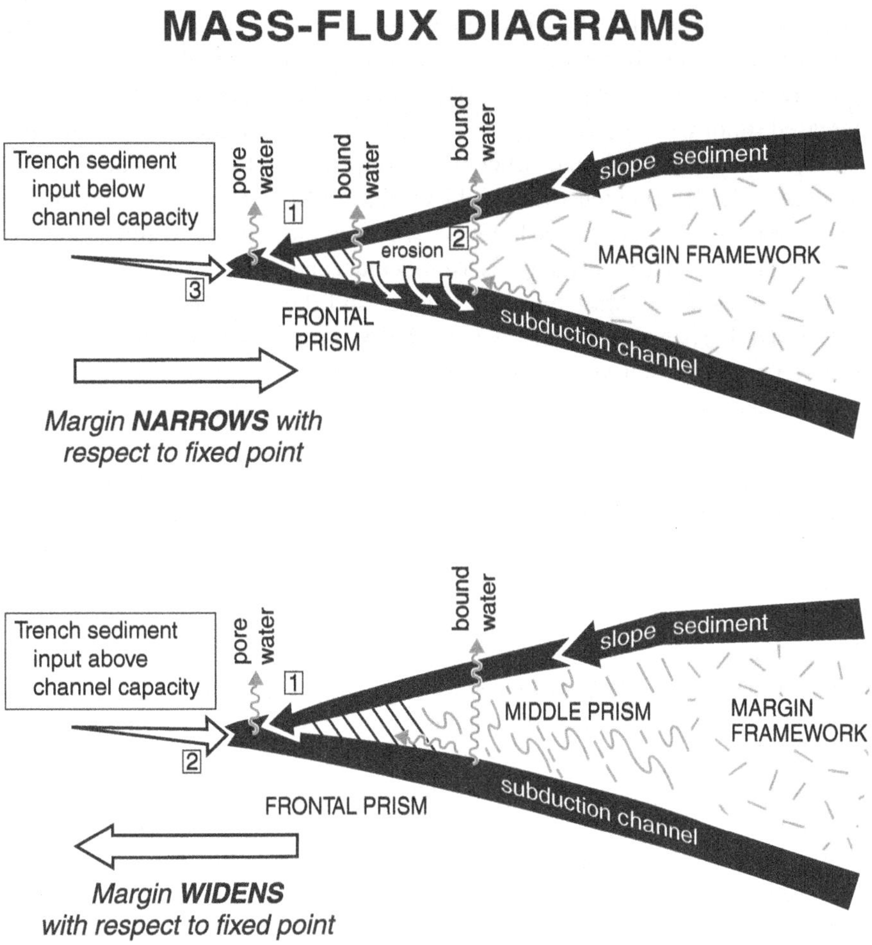

Fig. 7 Subduction channel mass transport. TOP Trench sediment thickness is less than subduction channel thickness and the contact between framework rock and the frontal prism migrates landward with respect to a fixed point. Material is input to the subduction channel from (1) the slope, (2) from basal erosion, and (3) at the deformation front. BOTTOM Trench sediment thickness exceeds subduction channel thickness and the excess accretes. Material input to the subduction channel is mainly at the deformation front. As the margin widens frontal prism material transfers to the middle prism and the prism contact migrates seaward

refraction data modeling yields thinner low velocity zones than the high amplitude-low frequency reflection images appear to indicate (Christeson et al., 1999; Takahashi et al., 2000). Perhaps the fault zone at the top of the channel is associated with concentrated fluids. In the absence of lower plate relief the reflective zone thickness is uniform indicating little internal slip. Differential subduction rates along thrust faulted segments within the channel would result in variable channel thicknesses. Seismic velocities in the second zone (Calahorrano et al., 2008) indicate a 28% porosity, more grain-to-grain contact, an increased rigidity, and a reduced rate of fluid drainage. Fluids that vent at the middle slope seafloor have the chemistry of chemically bound water released by clay dehydration reactions in underthrust sediment (Hensen et al., 2004). Inferred coseismic high fluid pressure pulses along the plate interface (Ranero et al., 2008) are interpreted to cause the hydrofracturing that facilitates basal erosion beneath the subsiding middle slope.

Beneath middle prisms the upper boundary images are commonly clear but not the lower boundary. Local sharp lower boundaries may indicate a floor thrust in addition to the roof thrust (cf. Sage et al., 2006; Lohrmann et al., 2006). A floor thrust was observed in outcrops of the subduction channel (Vannucchi et al. 2008).

Outcrops of the intermediate segment show a dense arrays of extensional calcite veins. The fault displacement occurs by many repeated small events and vein structure indicates cyclic fluid pressures (Vannucchi et al., 2008). The floor thrust begins to lock locally and produce contractile structure.

Zone II transitions into zone III where high amplitude plate interface reflectivity weakens and disappears (Ranero et al., 2008). It is also a region in which teleseismically recorded earthquakes nucleate. The loss of reflectivity and impedance contrasts is interpreted as decreased porosity and greater interplate friction. Calahorrano et al. (2008) point out that overburden loads are great enough to cause elastic deformation of clastic grains (70 to 110 MPa). The change from aseismic to seismogenic behavior appears to correspond with reduced fluid, increased compaction, and increased rigidity.

In outcrop, the extensional structure is overprinted by contractile deformation (Vannucchi et al., 2008). Temperature indicators approach 150°C which is the temperature inferred to mark the seismogenic zone. Locking of the floor thrust becomes evident and the roof thrust locks intermittently. Cyclic behavior is observed in the structure of veins and is inferred to show changes within earthquake cycles. Similar faults have been observed in rock outcrops of the Kodiak Island group of Alaska (Rowe, 2007).

Where abundant seafloor relief or ridges subduct, the thickness and physical properties of the subduction channel vary spatially. Over highs covered with little subducting sediment the porosity decreases more rapidly than across adjacent lows with thicker than normal sediment (Sage et al. 2006). Summit areas of subducting highs show erosion whereas surrounding lows have thick sediment. A probable differential friction in response to topography produces the implied patchyness off Ecuador (Sage et al., 2006) and the isolated centers of seismic activity of southeastern Costa Rica (Protti et al., 1995; Bilek and Lay, 2002; Bilek et al., 2003).

The original subduction channel concept analogues to fluid mechanics of a slurry now appears to include a multifaceted mechanical system. The zone of high amplitude reflections represents coherent stratification imaged with current techniques beneath the frontal prism and locally into the middle prism. This indicates a coherent mass at a stratal resolution of about a hundred meters. A small body of modern and fossil data show the expected increase in rigidity and friction with depth along the plate interface fault and reflectivity indicates that fluids drain until channel material has the seismic velocity of hard rock. This occurs roughly where aseismic behavior changes to seismogenic behavior (Ranero et al., 2008).

6 Updating the Subduction Channel Concept

The two prisms and a subduction channel of the original model are consistent with current observations. Since the concept was introduced, multiple inputs to the channel in addition to trench sediment, the structure of the major tectonic components, a lower material strength, and low interplate fault friction in the aseismic segment have been observed. Multiple inputs to the subduction channel up-dip of the seismogenic zone are: (1) the slope materials captured at the frontal prism seafloor and cycled through it, (2) chemically bound water released from subducting materials as temperatures and pressure increase, and (3) the clastic debris from basal and frontal erosion.

Observations of interplate and internal friction were measured along landslide slip planes that extend to the plate interface (von Huene et al., 2004) and derived from conjugate faults (Davis and von Huene, 1987). Critical taper, when constrained properly, indicates that interplate friction beneath the frontal prism is less than the static internal friction measured in samples (Wang and Hu, 2006; von Huene and Ranero, 2003; Lallemand et al., 1994). Strength of the material in a margin derived from taper have often been averaged beyond the area at critical taper. Disorganized and unstable seafloor of the frontal prism indicates an area at critical taper whereas the more stable middle slope commonly does not meet the critical taper criterion of a slope at or near failure. Interplate friction beneath the frontal prism derived from taper and fault geometry are as much as an order of magnitude less than assumed in the original model. Interplate friction sufficiently low for all trench sediment to subduct is observed in seismic images and was drilled at the Costa Rican margin (Kimura et al., 1997). Subduction of unconsolidated trench sediment beneath a poorly consolidated frontal prism requires the low basal friction consistent with "aseismicity" along the decollement of the frontal prism.

Channel thickness is a major uncertainty in quantifying the mass-flux of the subduction system. The low velocity plate interface reflective sequences associated with the subduction channel are considered indicators of hydration (Park et al., 2002). Hydration could extend above the channel into the fractured upper plate or below the channel into a porous upper ocean crust making a determination of thickness ambiguous. If interplate faults impede fluid migration at the top of the subduction channel, a zone of more concentrated fluid would occupy the thinner zone interpreted in wide angle seismic data (Christeson et al., 1999; Takahashi et al., 2000).

An indication of channel thickness is also derived from the mass required to balance material removed by basal subduction erosion to cause slope subsidence. A rate of material removal estimated from subsidence and convergence rates yields a subduction channel thickness (cf. Vannucchi et al., 2003, 2004; Scholl and von Huene, 2008). A nominal channel thickness estimate is roughly the 1 km lateral average of the high amplitude reflective layer.

7 Discussion

A unifying "framework" concept of convergent margin tectonics can be developed around the thickness of trench sediment, segmentation, a modified subduction channel concept, low interplate friction of the aseismic zone, and dynamic changes during an earthquake cycle.

7.1 *Tectonics and Trench Sediment Abundance*

Trench sediment abundance commonly correlates with accreting or eroding tectonics. In the subduction channel model a balance between channel capacity and the abundance of trench sediment is a first order control. Where trench sediment volume and plate convergence rate have been steady for million year periods, all trench sediment less than ~1 km thick subducts whereas a 2-km-thick trench sediment nearby looses the upper 0.5–1 km by accretion. This approximation appears to hold globally (Clift and Vannucchi, 2004). Short periods of disrupted axial sediment transport leave a less obvious record.

The mechanisms that control subduction zone channel thickness in nature are not fully understood. In studies of convergent margin tectonics with sandbox experiments, varying the height of a gate at the back of the box varies the subduction channel layer thickness (Lallemand et al., 1994; Kukowski and Onken, 2006; Lohrmann et al., 2006). When the gate opening equals the thickness of the sand layer representing trench sediment, all sand at the deformation front subducts whereas a greater or lesser gate opening allows erosion or accretion. Such a relationship between trench sediment and subduction channel thickness is seen in nature but an analogous controlling feature is not yet obvious. Mass flux control points along the subduction zone that have been identified in seismic images are few and imaging in this region requires the most powerful available techniques and a relatively simple geology. Some associations indicate possible controls. The middle prism where relatively rapid rates of subsidence are observed is probably a channel capacity control area. Another may result from character of the subducting plate. The subduction channel varies in thickness locally from basement relief and subducted horsts the height of the channel are overlain by a thin channel. Material strength above and below the plate interface fault zone and lithostatic load are probable controls. However until active subduction channels are sampled and physical properties are measured, understanding will probably advance with laboratory experiments and numerical modeling. Not withstanding uncertainties regarding controlling mechanisms, the balance between subduction channel thickness and trench axis sediment thickness is a first order control of erosion and accretion.

7.2 *Upper Plate Strain and Interplate Friction*

During earthquakes the dynamic friction of the seismogenic zone is presumably less than its friction during interseismic periods. Friction on faults during an earthquake are difficult to quantify and synthetic modeling or wave-form analysis commonly do not include fluid pressure pulses, opening and closing of fractures, and frictional heating including local flash-melting. The low fault and material strength in the frontal prism revealed by conjugate faults probably shows its dynamic strength during earthquakes. It is plausible that unstable morphology of the frontal prism forms principally during coseismic deformation. During interseismic periods deformation might occur from aseismic fault displacement and gravity tectonics. In the dynamic wedge theory, critical state is achieved during coseismic deformation in frontal prisms (Wang and Hu, 2006).

The paradox of a converging plate compressional environment and yet many small normal faults across the middle prism is most easily explained by transitory relaxation of the elastic strain during of shortly after an earthquake. When periodically relaxed sufficient for small extensional displacements to occur, normal faults will grow incrementally. The oscillation of compression and brief extension can occur during earthquake cycles.

7.3 The Limited Width of the Frontal Prism

A mechanism that limits frontal prism width is indicated by our compilation (Fig. 3). Self limiting is clearly observed where subducting seamounts form embayments. These breached areas fill very rapidly but only to their previous width. The narrow frontal prism above the 2 Km high subducting Nazca Ridge returns to its previous width in ~4 Myr. Rapid accretion that stops at a limit requires a change in controlling processes. A possible explanation involves the progressive change in physical properties up slope. In the trench axis, sediment has a normal vertical seismic velocity gradient that increases with depth. It parallels a vertical gradient of strength from weak at the top to relatively stronger at the bottom. This velocity/strength relation was measured in drill cores at DSDP site 499 in the Middle America Trench axis where the shear strength increased from near zero just below the seafloor to ~50 kPa at 200 m. In the same interval the seismic velocity increased ~300 m/s (Shipboard Party, 1982). At the deformation front of accreting margins the deep material that subducts is stronger than the shallower material that detaches and forms the overlying frontal prism. Even where all sediment subducts, the velocity gradient across the decollement steps from a lower one above it to a higher one below it (Saffer, 2003). After a clear decollement forms, the subducted section still retains its stratified structure and relatively greater strength whereas the weaker sediment of the frontal prism above it deforms. During some period of the earthquake cycle the decollement must be the weakest element in the system and yet one that at times is stronger than the prism in order to drive imbricate faulting.

Detailed velocity information from pre-stack depth migration processing (i.e., von Huene et al., 1999) shows that pore fluid in the frontal prism drains rapidly as the frontal prism thickens tectonically. On the other hand, below the plate interface the subducted sediment must drain through a longer path that retards fluid drainage and presumably development of greater sediment strength. As a margin wedge develops, the upper plate strength may increase faster than the subduction channel's and as this continues the step to a higher velocity across the decollement could disappear. Where the subduction channel is a clear low velocity the step has inverted. The interplate fault beneath the middle prism probably collects water released by clay dehydration from the subduction channel as described earlier. The relative plate velocity shift occurs at or near the contact between the frontal and middle prisms.

If velocity is a proxy for material strength then the middle prism may be as strong or stronger than the subduction channel when averaged over multiple earthquake cycles. If the subduction channel and the interplate fault become relatively weak, the shear strength from subduction becomes insufficient to produce much permanent compressional deformation in the upper plate. Fully understanding the interplay between upper and lower plate strength and that of the interplate shear during an earthquake cycle will require better resolution in spatial and temporal observations. Relative strength and dynamics in the frontal and middle prism transition is an interesting candidate explanation.

7.4 Dynamics

An interplay between state of stress and strength of the interplate fault zone are explained in the Dynamic Coulomb wedge theory (Wang and Hu, 2006). The Dynamic Coulomb wedge contains a frontal and middle prism with dissimilar bulk rheology. Therefore stress during a seismic cycle affects each segment differently. During interseismic periods the middle slope segment and the continental framework store elastic strain (Fig. 8b). Low relative rates of permanent deformation occur in the middle prism and framework because they can store more elastic strain than frontal prisms. During an earthquake the released elastic strain extends the mid slope segment and it shoves against the frontal prism creating a transient coseismic pulse of rapid deformation and elevated pore fluid pressure (Fig. 8a).

In many studies a time averaged steady state deformation is assumed. However, the variation in stress between interseismic and coseismic periods is significant as seen in the stress drop during an earthquake. Peak stress is the most likely time that major permanent deformation occurs. Physical parameters such as those from fluid migration will vary locally during an earthquake as waves of fluid pressure migrate along a rupture plane and change its strength profile. In fact it has been suggested that the plate interface becomes an aquatard during interseismic periods and with the release of strain it becomes an aquifer during and shortly after an event. Multibeam bathymetry with resolution greater than seismic reflection images shows structure from the most recent tectonics which best link it to current seismicity.

The mechanics of eroding and accreting margins are probably more similar than commonly envisioned (i.e., Stern, 2002) considering the following. (1) Both accreting and eroding margins have frontal prisms of similar width. (2) The taper angles alpha versus beta (Lallemand et al., 1994) if restricted to depth processed seismic records, scatter 4° or 5° along single margins whereas the maximum difference between accretionary and erosional margins is only 8° (von Huene and Ranero, 2003; Wang and Hu, 2006). Thus taper variability in a single margin can be half of the extreme difference between accretionary and erosional end-member tapers. (3) Frontal prism strength of each type is similar. Strength derived from angular relations of conjugate faults for the erosional northern Chile frontal prism lies between the values for the accretionary Aleutian and Oregon prisms and is the same as that of Nankai. (4) Locations of modeled 100–150°C isotherms are locally similar with respect to seismicity (Oleskevich et al., 1999).

Different seafloor morphologies apparent in multibeam bathymetric perspective diagrams probably reflect dominance of one middle prism tectonic process for longer periods or subduction of major seafloor relief as at Juan Fernandez Ridge. Although the rocks comprising each margin are petrologically different, a more dominant factor in end-member tectonics is the degree of framework rock fracturing in eroding margins and sediment consolidation in accreting ones. Locally, accretion and erosion interchange readily.

Material difference in a frontal prism appear to influence tectonics minimally. The few estimates from analysis of conjugate faults and landslide geometry show that a frontal prism of mass wasting rubble and a

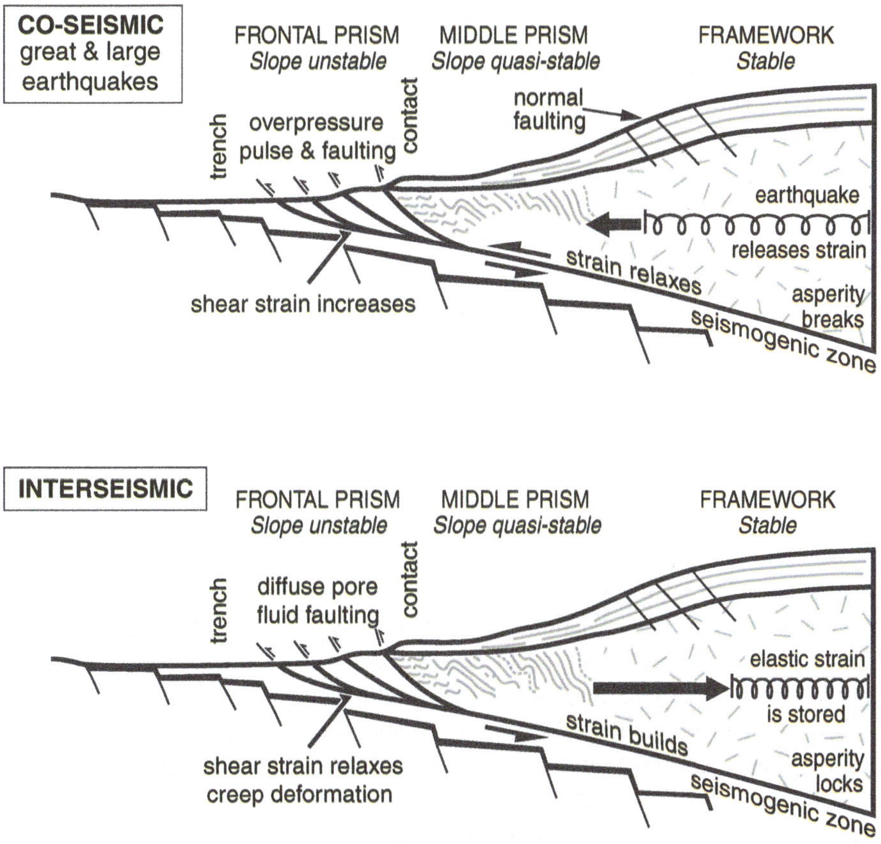

Fig. 8 Diagram illustrating the dynamic Coulomb Wedge theory (Wang and Hu, 2006). Elastic strain builds in the wedge beneath the middle and upper slope during the interseismic period (*lower*) and is released during the coseismic period (*upper*). Earthquake rupture propagating into the aseismic area will shove against the frontal prism and cause a pulse of deformation resulting in the seafloor instability revealed in high resolution multibeam bathymetry

prism of trench sediment have similar dynamic strengths (von Huene and Ranero, 2003; Wang and Hu, 2006).

8 Summary

Improved geophysical imaging has revealed structure that was obscure during early development of convergent margin models. With new information we draw ideas about accretion and erosion closer together in a more unified framework concept. The conventional "accretionary wedge" is resolved into three structural segments. A rapidly deforming frontal prism is separated by a splay fault or transitional contact from a moderately stable middle prism that in turn grades into a relatively stable inner prism of coherent margin framework structure. Eroding margins are also resolved into three segments beginning with a frontal prism structured like that of accreting margins and a similar inner prism. The moderately stable middle prism is commonly cut by normal faults, it has a reduced seismic velocity, and its morphology clearly differs from the parent margin framework of the inner prism. This tripartite framework characterizes most seismic images acquired with large modern systems and processed in depth, especially when integrated with multibeam bathymetry. Segmentation corresponds with an aseismic frontal region where few if any plate interface earthquakes are recorded that becomes seismogenic beneath the middle prism or inner prism.

The upper plate's basal boundary is the plate interface fault separating it from the underlying subduction channel. The channel is a layer of trench sediment and erosional debris on top of oceanic basement. The subduction channel is segmented as shown in a recent studies of seismic data and in studies of a fossil subduction channel in outcrops on land. These results indicate a division that roughly matches segments of the upper plate. The subducted material begins as a stratified sequence and its seismic velocity down dip is consistent with a progressive consolidation as would be expected. Much of the channel clastic material is stratified and not a slurry as originally conceived. A multifaceted mechanics can be anticipated once more detailed structural evidence and physical properties are resolved.

The mass flux (material movement and its source) and inferred mechanics are specific to each segment. Frontal prisms add relatively small volumes of material to a margin especially when composed of mostly repositioned slope materials. They widen or narrow readily in response to changes in character of the subducting plate: in other words they can accret or erode in response to a rough or smooth subducting seafloor. Their widths are commonly within a 5–30 km range globally and their constrained width, whether a trench axis is sediment flooded or starved, indicates a self limiting mechanism. Processes in the frontal prism reduce friction sufficient to allow subduction of soft trench sediment beneath poorly consolidated prism sediment. A fault within weak materials is weak, consistent with absence of plate interface seismicity.

Eroding and accreting middle prism are differentiated based on structure and mass flux. At accreting margins the frontal prism's highly deformed sediment is tectonically added to the middle prism which may in turn become part of the upper plate's older margin framework. At eroding margins, framework rock structure is commonly imaged down slope to the actively deforming frontal prism. However, seismic velocity much lower than its framework rock parent indicates pervasive fracturing across eroding margin middle prisms. Thus the greatest difference between eroding and accreting margins are observed in middle prisms whereas the frontal prism and inner prism seismic images are similar.

The subduction channel concept offers a simple and practical framework to visualize convergent margin mass flux. When trench axis sediment is thicker than subduction zones can accommodate the surplus material accretes, and when thinner the margin erodes. Observations indicted that trench sediment greater than ~1 ± 0.5 km thick is associated with accretion and less than ~1 ± 0.5 km thick with erosion, thereby generally bracketing channel thickness. This is consistent with thickness of high amplitude reflective sequence along plate boundary that is presumed to be the subduction channel (e.g., Nankai, Bangs et al., 2006; Ecuador, Sage et al., 2006). The relation between trench sediment thickness and subduction channel thickness is a first-order control on whether the margin's dominant mass flux is erosional or accretionary.

Eroding and accreting margins may be composed of different materials but their frontal prism width and critical tapers, plate interface friction, subduction channel thickness, porosity, and modeled temperature profiles are similar at current scales of investigation. Stress drops during earthquake rupture are similar so the plate interface dynamic strength and fluid pressures could reach similar peak values in both margin types. Physical properties derived from structure appear similar and it is peak dynamic forces that are most likely to shape major tectonic features.

The kinematic subduction channel concept is consistent with the Dynamic Coulomb Wedge theory (Wang and Hu, 2006) which specifies the time frame within which processes occur. The two prism time varying dynamic model can remove the paradox of an eroding margin with an "accretionary structured" imbricated frontal prism. It provides a better understanding of how plate convergence can be coeval with active normal faulting of the middle slope given the oscillatory relaxation of contractile strain during seismic cycles. Erosional and accretionary margins have more similar features than often inferred and it is the middle prism that displays basic rock and mechanical differences.

The tectonic effects of different inter-plate mass transfer schemes with regard to seismicity and tsunamis have been argued without an emerging consensus. Lack of consensus emphasizes the need for finer resolution in controlled source seismic data and a need for scientific drilling into plate interfaces to significantly advance understanding of convergent margin dynamics.

Acknowledgements We appreciate the discussions with Kelin Wang that greatly improved earlier manuscripts. R. Ranero is supported by Repsol. R von Huene received a USGS Bradley fellowship to help support this work.

References

Bangs, N.L., and S. Cande, (1997), The episodic development of a convergent margin inferred from structures and processes along the southern Chile margin. Tectonics, 16 (3), 489–505

Bangs, N.L.B., S.P.S. Gulick, and T.H. Shipley, (2006), Seamount subduction erosion in the Nankai Trough and its potential impact on the seismogenic zone. Geology, 34 (8), 701–704, doi; 10.1130/G22451.1

Behrmann, J.H., S.D. Lewis, S.C. Cande, and ODP Leg 141 Scientific Party, (1994), A synthesis of results from Leg 141 of the Ocean Drilling Program. Geol. Rundsch. 83, 832–852

Bilek, S.L., S.Y. Schwartz, and H.R. DeShon, (2003), Control of seafloor roughness on earthquake rupture behavior: Geology, 31 (5), 455–458

Bilek, S.L., and T. Lay, (2002), Tsunami earthquakes possibly widespread manifestations of frictional conditional stability. Geophys. Res. Lett. 29. p. art. No. 1673 Behrmann, J.H., S.D. Lewis, S.C. Cande, and ODP Leg 141 Scientific Party, (1994), A synthesis of results from Leg 141 of the Ocean Drilling Program. Geol. Rundsch. 83, 832–852

Bond, C.E., A.D. Gibbs, Z.K. Shipton, and S. Jones, (2007), What do you think this is? "Conceptual uncertainty" in geoscience interpretation. GSA Today 17, doi 10.1130/GSATO1711A

Bourgois, J., C. Guivel, Y. Lagabrielle, T. Calmus, J. Boulegue, and V. Daux, (2000), Glacial-interglacial trench supply variation, spreading-ridge subduction, and feedback controls on the Andean margin development at the Chile triple junction area (45–48°), J. Geophys. Res. 105, 8355–8386

Byrne, D.E., D.M. Davis, and L.R. Sykes, (1988), local and maximum size of thrust earthquakes and the mechanics of the shallow region of subduction zones, Tectonics 7, 833–857

Byrne, D.E., W.H. Wang, and D.M. Davis, (1993), Mechanical role of backstops in growth of forearcs, Tectonics, 12, 123–144

Calahorrano, B., V. Sallares, J.-Y. Collot, F. Sage, and C. Ranero, (2008), Nonlinear variations of the physical properties along the southern Ecuador subduction channel: Results from depth-migrated seismic data; Earth and Planet. Sci. Lett. 267; 453 doi:10.1016/j.epsl.2007.11.061

Christeson, G.L., K.D. McIntosh, T.H. Shipley, E.R. Flueh, and H. Goodde, (1999), Structure of the Costa Rica convergent margin, offshore Nicoya Peninsula, J. Geophys. Res., 104, 25, 443–25 468

Christeson, G.L., N.L. Bangs, and T.H. Shipley, (2003), Deep structure of an island arc backstop, Lesser Antilles subduction zone, J. Geophys. Res. 108 (B7), 2327, doi:10.1029/2002JB002243

Clift, P.D., and P. Vannucchi, (2004), Controls on tectonic accretion versus erosion in subduction zones: Implications for the Origin and recycling of the continental crust. Rev. Geophys.

Cloos, M., and R.L. Shreve, (1988), Subduction channel model of prism accretion, melange formation, sediment subduction, and subduction erosion at convergent plate margins: 2. Implications and discussion, Pageoph, 129, (¾), 501–545

Davis, D.M., and R. von Huene, (1987), Inferences on sediment strength and fault friction from structures of the Aleutian Trench, Geology 15, 517–522

Diaz-Naveas, J.L., (1999), Sediment subduction and accretion at the Chilean convergent margin between 35° and 40° S, PhD dissertation, Christian-Albrechts-Universität zu Kiel, 130 p,

Fruehn, J., R.S. White, and T.A. Minshull, (1997), Internal deformation and compaction of the Makran accretionary wedge, Terra Nova 9, 101–104

Gerdom, M., A.M. Trehu, E.R. Flueh, and D. Klaeschen, (2000), The continental margin off Oregon from seismic investigations, Tectonophysics 329, 79–97

Hamilton, W., (1988), Plate tectonics and island arcs, Geol. Soc. Am. Bull. 100, 1503–1527

Hensen, C., K. Wallmann, M. Schmidt, C.R. Ranero, and E. Suess, (2004), Fluid expulsion related to mud extrusion off Costa Rica –a window to the subducting slab, Geology 32, 201–204

Husen, S., E. Kissling, and E.R. Flueh, (2000), Local earthquake tomography of shallow subduction in north Chile:

A combined onshore and offshore study, J. Geophys. Res. 105, 28, 183–28,198
Kimura, G., E. Silver, et al., (1997), Proceedings Ocean Drilling Program Initial Reports, Volume 170, 458 p. (U.S. Government Printing Office, Washington, D.C.)
Kimura, G., Y. Kitamura, Y. Hashimoto, A. Yamaguchi, T. Shibata, K. Ujiie, and S. Okamoto, (2007), Transition of accretionary wedge structures around the up-dip limit of the seismogenic subduction zone, Earth Planet. Sci. Lett. 255, 471–484, doi10.101j.epsl.2007.01.005
Kinoshita, M., G. Moore, R. von Huene, H. Tobin, and C.R. Ranero, (2006), The seismogenic zone experiment, in Burger and Fujioka, eds, The impact of the Ocean Drilling Program, Oceanography, vol. 19, No. 4, p
Kopp, H., and N. Kukowski, (2003), Backstop geometry and accretionary mechanics of the Sunda margin, Tectonics, 22 (6), 1072, doi:10.1029/2002TC001420
Kopp, H., J. Fruehn, E.R. Flueh, C. Reichert, N. Kukowski, J. Bialas, and D. Klaeschen, (2000), Structure of the Makran subduction zone from wide-angle and reflection seismic data, Tectonophysics, 329, 171–191
Kukowski, N., T. Schillhorn, K. Huhn, U. von Rad, S. Husen, and E. Flueh, (2001), Morphotectonics and mechanics of the central Makran accretionary wedge off Pakistan, Marine Geology, 173, 1–19
Kukowski, N., and O. Onken, (2006), Subduction Erosion – the "Normal" Mode of Fore-Arc Material Transfer along the Chilean Margin?; The Andes, Springer, Berlin, Heidelberg New York; pp. 216–236
Lallemand, S.E., P. Schnurle, and J. Malavieille, (1994), Coulomb theory applied to accretionary and nonaccretionary wedges: Possible causes for tectonic erosion and/or frontal accretion: J. Geophys. Res., 99 (B6), 12, 033–12,055
Laursen, J., D.W. Scholl, and R. von Huene, (2002), Neotectonic deformation of the central Chile margin. Deepwater forearc basin formation in response to hot spot ridge and seamount subduction, Tectonics, 21 (5), 1038 doi;10.1029/2001 TC901023
Le Pichon, X., P. Henry, and S. Lallemant, (1993), Accretion and erosion in subduction zones: The role of fluids, Annu. Rev. Earth Planet. Sci. 21, 307–331
Lohrmann, J., N. Kukowski, C. Krawczyk, O. Onken, C. Sick, M. Sobiesiak, and A. Rietbrock, (2006), Subduction Channel Evolution in Brittle fore-Arc Wedges – a Combined Study with Scaled Sandbox Experiments, Seismological and Reflection Seismic Data and Geological Field Evidence: The Andes, Springer, Berlin, Heidelberg New York; pp. 237–262
Moore, G.F., (2007), Three-dimensional Splay Fault Geometry and Implications for Tsunami Generation, Science 318, 1128, DOI: 10.1126/Science.1147195
Oleskevich, D.A., R.D. Hyndman, and K. Wang, (1999), The updip and down-dip limits to great subduction earthquakes: Thermal and structural models of Cascadia, South Alaska, SW Japan, and Chile, J. Geophys. Res. 104 (14), 965–14,997
Oncken, O., D. Hindle, J. Kley, K. Elger, P. Victor, and K. Schemmann, (2006) Deformation of the Central Andean Upper Plate System – Facts, Fiction, and Constraints for Plateau Models; The Andes, Springer, Berlin, Heidelberg New York; pp. 3–27
Park, J.-O., T. Tsuru, N. Takahashi, T. Hori, S. Kodaira, A. Nakanishi, S. Miura, and Y. Kaneda, (2002), A deep strong reflector in the Nankai accretionary wedge from multichannel seismic data: Implications for underplating and interseismic shear stress release, J. Geophys. Res. 107 (B4), 10.1029/2001JB000262
Protti, M., F. Guendel, and K. McNally, (1995), Correlation between the age of the subducting Cocos plate and the geometry of the Wadati-Benioff zone under Nicaragua and Costa Rica, in Mann, P, ed., Geologic and Tectonic Development of the Caribbean Plate Boundary in Southern Central America: Geol. Soc. Am. Special Paper 295
Ranero, C.R., and R. von Huene, (2000), Subduction erosion along the Middle America convergent margin, Nature 404, 748–752
Ranero, C.R., R. von Huene, W. Weinrebe, and C. Reichert, (2006), Tectonic processes along the Chile convergent margin, in: The Andes, active subduction orogeny, Onken, O. and 7 others, eds, Springer verlag, Berlin-Heidelberg-New York, pp. 91–122
Ranero, C.R., R. von Huene, W. Weinrebe and U. Barckhausen, (2007), Convergent Margin Tectonics of Middle America: A Marine Perspective, in Central America: Geology, Resources and Hazards, (2007), Jochen Bundschuh and Guillermo E. Alvarado, eds, Taylor and Francis, London, Volume 1, pp. 239–265
Ranero, C.R., I. Grevemeyer, H. Sahling, U. Barckhausen, C. Hensen, K. Wallmann, W. Weinrebe, P. Vannucchi, R. von Huene, and K. McIntosh, (2008), The hydrogeological system of erosional convergent margins and its influence on tectonics and interplate Seismogenesis, Geochem. Geophys. Geosyst. 9 (1), 1–18, Q03S04, doi:10.1029/2007GC001679
Rowe, C., (2007), Comparison between Three Out-of-Sequence Thrusts from Japan and Alaska: Implications for Nankai Drilling Targest from the Rock Record. Scientific Drilling, special Issue No. 1, pp. 82–83
Saffer, D.M., (2003), Pore pressure development and progressive dewatering in underthrust sediments at the Costa Rican subduction margin: Comparison with northern Barbados and Nankai, J. Geophys. Res., 1089B50,2261, doi:10.1029? 2002JB001787
Sage, F., J.-Y. Collot, and C.R. Ranero, (2006), Interplate patchiness and subduction-erosion mechanisms: Evidence from depth-migrated seismic images at the central Ecuador convergent margin, Geology, 34 (12), 997–1000; doi. 10.1130/ G22790A.1
Sallares, V., and C.R. Ranero, (2005), Structure and tectonics of the erosional convergent margin off Antofagasta, north Chile (23°30'S), J. Geophys. Res. 110 (B06101), doi:10.1029//2004 JB003418
Scholl, D.W., and R. von Huene, (2008), Crustal recycling at modern subduction zones applied to the past – issues of growth and preservation of continental basement, mantle geochemistry, and supercontinent reconstruction, in Geol. Soc. Am. Memoir 200, 4-D Framework of Continental Crust (edited by Robert D. Hatcher, Jr., Marvin P. Carlson, John H. McBride, and José Ramón Martínez Catalán)
Shipboard Party, (1982), in Aubouin, J., and von Huene R., Init. Repts DSDP, 67, Washington DC (US Govt. Printing Office)
Shreve, R.L., and M. Cloos, (1986), Dynamics of sediment subduction, melange formation, and prism accretion, J. Geophys. Res. 91, 10,229–10,245
Sorokhtin, O.G., and L.I. Lobkovskiy, (1976), The mechanism of subduction of oceanic sediments into a zone of

underthrusting of lithospheric plates, Izv. Acad. Sci. USSR Phys. Solid Earth, Eng. Transl. 12, 289–293

Stern, R.J., (2002), Subduction zones, Rev. Geophys. 40.4, 3–1 – 3–48, doi10.1029/2001RG000108.

Taira, A., (1994), When Plates Collide, Convergent Margin Geology, Oceanus, Winter 1993/94, p 95–98

Takahashi, N., S. Kodaira, T. Tsuru, J.-O. Park, Y. Kaneda, H. Kinoshita, S. Abe, M. Nishino, and R. Hino, (2000), Detailed plate boundary structure off northeast Japan coast, Geophys. Res. Lett. 27 (13), 1977–1980

Vannucchi, P., C.R. Ranero, S. Galeotti, S.M. Straub, D.W. Scholl, and D. McDougall Ried, (2003), Fast rates of subduction erosion along the Costa Rica Pacific margin: Implications for non-steady state rates of crustal recycling at subduction zones, J. Geophys. Res. 108: doi:10.1029/2002JB002207

Vannucchi, P., R. Francesca, and B. Giuseppe, (2008), Geological record of fluid flow and seismogenesis along an erosive subducting plate boundary; Nature, v 451, 699–703, doi:10.1038/nautre06486

Vannucchi, P., S. Galeotti, P.D. Clift, V.R. Rantero, and R. von Huene, (2004), Long term subudction erosion along the Middle America Trench offshore Guatemala, Geology 32, 617–620

von Huene, R., and D. Klaeschen, (1999), Opposing gradients of permanent strain in the aseismic zone and elastic strain across the seismogenic zone of the Kodiak shelf and slope, Alaska, Tectonics, 18, 248–262, doi:10.1029/1998 TC900022

von Huene, R. and C.R. Ranero, (2003), Subduction erosion and basal friction along the sediment-starved convergent margin off Antofagasta, Chile, J. Geophys. Res. 108 (B2), 2079, doi:10,1029/2001JB001569, 2003

von Huene, R., I.A. Pecher, and M.A. Gutscher, (1996), Development of the accretionary prism along Peru and material flux after subduction of Nazca Ridge, Tectonics 15, 19–33

von Huene, R., J. Corvalan, E.R. Flueh, K. Hinz, J.A. Korstgard, C.R. Ranero, W. Weinrebe, and the CONDOR Scientists, (1997), Tectonic control of the subducting Juan Fernandez Ridge on the Andean margin near Valparaiso, Chile, Tectonics 16, 474–488

von Huene, R., C.R. Ranero, W. Weinrebe, and K. Hinz, (2000), Quaternary convergent margin tectonics of Costa Rica, segmentation of the Cocos Plate, and central American volcanism, Tectonics 19, m. 2, e14–334

von Huene, R., C.R. Ranero, and P. Watts, (2004) Tsunamigenic slope failure along the Middle America Trench in two tectonic settings, Mar. Geol. 203, 303–317

Watts, P., and S.R. Grilli, (2004), Tsunami generation by submarine mass failure I: Wavemaker modes, J. Wtrwy, Port, Coast, and Oc. Engrg., ASCE

Wang, K., and Y. Hu, (2006), Accretionary prisms in subduction earthquake cycles: The theory of dynamic Coulomb wedge, J. Geophys. Res. 111, B6, B06410, 16p, IOI 10,1029/2005 JB004094

Imaging Interseismic Locking at the Nankai Subduction Zone, Southwest Japan

Yosuke Aoki and Christopher H. Scholz

Abstract Recent geodetic and seismological observations have revealed that brittle-plastic transition zones at subduction zone interfaces are loci of slow slip episodes and nonvolcanic harmonic tremors. It is therefore important to estimate the depth range of brittle-plastic transition zones and how interplate locking changes with space and time within the brittle-plastic transition zone, not only for understanding the interseismic stress accumulation process in subduction zones but also to explore the causes of slow earthquakes and nonvolcanic harmonic tremors. With this point of view, the depth variations of interplate locking status in the Nankai trough, southwest Japan, were investigated by inverting vertical Global Positioning System (GPS) velocities. Although vertical component of GPS velocities have rarely been used as important information because of its higher noise level, we employed vertical, rather than, horizontal, velocities because the vertical deformation field will enable us to separate rigid plate motions from deformation due to interplate locking; horizontal displacements contain both rigid plate motion and deformation from interplate locking, whereas vertical displacements contain only the interplate locking effect. Interplate locking was estimated for three profiles: the source region of the 1946 Nankaido earthquake, that of the 1944 Tonankai earthquake, and the Tokai seismic gap. The results show a gradual decrease of interplate coupling between about 20–25 and 35–45 km depth with a plateau region between 25–35 km for all three profiles. These plateaus may be interpreted as stable–unstable boundary in rate- and state-dependent friction or represent the change in mechanism (say, frictionally to plastically) or material. The depth of the brittle-plastic transition zone in the Tokai profile seems to be shallower than other two probably because of the younger and hotter Philippine Sea plate in the Tokai area.

Keywords Brittle-plastic transition zone • Global Positioning System • Interplate locking • Nankai trough • Vertical deformation

1 Introduction

Subduction zones are quite variable ranging from extensional to compressional (Uyeda and Kanamori 1979). They also vary in seismic activity; compressional arcs such as the Chilean arc have a higher seismic coupling ratio than extensional arcs such as the Mariana arc, resulting in more great earthquakes in compressional arcs (e.g., Pacheco et al. 1993). Scholz and Campos (1995) found that seismic coupling in a subduction zone is correlated with the normal force acting on the plate interface, that is, higher seismic coupling is expected when the normal stress acting on the plate interface is higher, and lower seismic coupling is expected when lower normal stress is acting on the plate interface.

Y. Aoki
Earthquake Research Institute, University of Tokyo,
1-1 Yayoi 1, Bunkyo-ku, Tokyo 113-0032, Japan,
yaoki@eri.u-tokyo.ac.jp

S. Lallemand and F. Funiciello (eds.), *Subduction Zone Geodynamics*,
DOI 10.1007/978-3-540-87974-9, © Springer-Verlag Berlin Heidelberg 2009

Compressive subduction zones, where seismic coupling is high, are loci of great earthquakes. Over 90% of the total seismic moment in the world is released from subduction zones (Pacheco and Sykes 1992). Because a great interplate earthquake ruptures where plate interface is at least partially coupled, delineating the depth distribution of interplate locking of subduction zones is important in assessing seismic hazard.

A subduction zone is characterized by updip and downdip limits of the seismogenic zone (e.g., Byrne et al. 1988; Stern 2002). The updip limit of the seismogenic zone occurs because the shallower part is not capable of nucleating earthquakes due to unconsolidated sediments with high pore pressure in the accretionary prism (e.g., Byrne et al. 1988; Stern 2002). Vrolijk (1990) speculated that the updip limit of the seismogenic zone is controlled by a transformation from smectite, which deforms stably, to illite, which may exhibit stick–slip behavior. This transformation occurs at ~150°C, which is considered to mark the updip limit of the seismogenic zone.

The downdip limit of the seismogenic zone marks the transition from stick–slip to aseismic behavior. In continental faults, such as the San Andreas fault, the onset of quartz and feldspar plasticity are at 300°C and 450°C, respectively (e.g., Sibson 1984; Scholz 2002). These isotherms are thought to bound a brittle-plastic transition zone. Temperatures in the brittle-plastic transition zone of subduction zones are different in different subduction zones because the depth range of the brittle-plastic transition zone may be controlled by factors other than temperature, such as, the mineralogical composition of the plate interface. Comparison between geodetic data and heat flow modeling suggests that the brittle-plastic transition zone lies between 25 and 35 km depth for the Nankai trough (e.g., Hyndman et al. 1995).

Recent geodetic measurements have revealed that slow slip episodes occur in the deeper extension of the seismogenic zone, that is, brittle-plastic transition zone, of subduction zones all over the world such as in southwest Japan (e.g., Hirose et al. 1999; Ozawa et al. 2002; Miyazaki et al. 2003, 2006), northeast Japan (e.g., Heki et al. 1997; Kawasaki et al. 2001), Cascade (e.g., Dragert et al. 2001; Miller et al. 2002), Guerrero, Mexico (Kostoglodov et al. 2003; Larson et al. 2004; Yoshioka et al. 2004), Alaska (Ohta et al. 2006), and Hikurangi, New Zealand (Douglas et al. 2005). More detailed analyses revealed that these slow slips repeat more or less periodically in many subduction zones such as southwest Japan (e.g., Hirose and Obara 2005, 2006; Ozawa et al. 2007), Cascadia (Dragert et al. 2001; Miller et al. 2002), and New Zealand (Douglas et al. 2005).

Recent development in seismic network allowed us to discover deep low-frequency tremors in nonvolcanic regions. It was Obara (2002) that first discovered nonvolcanic tremors and subsequently these tremors are found in the Cascadia subduction zone (e.g., Rogers and Dragert 2003), as well as in the continental lower crust or upper mantle (e.g., Ohmi and Obara 2002; Nadeau and Dolenc 2005). Further studies unraveled some characteristics of the deep low-frequency tremors such as simultaneous occurrence of slow slips and tremors (e.g., Rogers and Dragert 2003; Obara et al. 2004; Hirose and Obara 2006), migration of epicenters (e.g., Obara et al. 2004; Kao et al. 2007), and the existence of similar tremors with lower frequency contents (Ito et al. 2007). A seismic reflection survey of Nedimović et al. (2003) speculated that the depth of the source of slow deformation inferred by Dragert et al. (2001) are in the brittle-plastic transition zone. More recently, Shelly et al. (2006) found a zone high V_p/V_s ratio in the brittle-plastic transition zone of the Nankai trough, suggesting that high pore pressure there may control the occurrence of the slow slips and tremors in brittle-plastic transition zones.

The mechanics of slow slips and deep low-frequency tremors seems still enigmatic except that dehydration from the subducting slab may be responsible for the generation of deep low-frequency tremors (e.g., Katsumata and Kamaya 2003); Ohmi and Obara (2002) suggested that focal mechanisms of deep low-frequency tremors have significant volume changes, while more recent studies of Ide et al. (2007a) and Shelly et al. (2007) demonstrated that these are purely shear slips. In addition, Ide et al. (2007b) constructed a scaling law for both slow slips and low-frequency tremors in which the seismic moment M_0 and the characteristic duration T are scaled as $M_0 \propto T$, in contrast to $M_0 \propto T^3$ for small "fast" earthquakes (e.g., Scholz 1994). They interpreted slow slips and deep low-frequency tremors as diffusion processes where the source length L and T are scaled as $L \propto T^{1/2}$, in contrast to $L \propto T$ for small "fast" earthquakes (e.g., Scholz 1994), but Watanabe et al. (2007) gave a different scaling law for deep low-frequency tremors where L is size-independent.

With this point of view, It is important to estimate the depth range of brittle-plastic transition zones and how interplate coupling changes within the brittle-plastic transition zones, not only for understanding the interseismic stress accumulation process in subduction zones but also for exploring the causes of slow slip and deep low-frequency tremors. Here we image depth variations of interplate locking status by inverting geodetic data to gain more insight into the rheology of the brittle-plastic transition zones in subduction zones.

We employ vertical, rather than horizontal, deformation field from Global Positioning System (GPS) data, although previous studies used horizontal deformation field to estimate interplate locking status (e.g., Miyazaki and Heki 2001; Wang et al. 2003). Employing only the horizontal velocities as data does not permit the separation of rigid plate motions and interplate coupling effect because horizontal displacements include both contributions. Thus rigid plate motions could thus be mapped into interplate locking, leading to a misunderstanding of its spatial distribution. To avoid this problem, Mazzotti et al. (2000) employed volumetric strain data obtained by spatial differentiation of horizontal velocities to invert for interplate locking. with a definition that rigid plate motion does not produce volumetric strains. However, the spatial resolution of interplate locking is then poor because the spatial differentiation of displacement data results in the introduction of additional noise to the data. These problems can be avoided by using vertical velocities to discriminate between the rigid plate motions and interplate locking effect. Here we neglected vertical deformation produced by inter-arc deformation but this assumption is justified because focal mechanisms of crustal earthquakes in southwest Japan arc suggest that the region is under a strike–slip regime which likely produces little vertical inter-arc deformation (Wang 2000; Townend and Zoback 2006).

The vertical component of GPS displacements has rarely been used for quantitative analyses because of its higher noise level mainly due to atmospheric disturbance of the GPS signal, but we apply Principal Component Analysis to (e.g., Menke 1989), which is commonly used in tide gage data analysis (e.g. Savage and Thatcher 1992; Savage 1995), to improve the signal-to-noise ratio of the time series. The velocities thus obtained from the "cleaned" time series are then combined with 25 year tide gage records to obtain absolute vertical velocities.

Aoki and Scholz (2003b) obtained depth variations of interplate locking at the source region of the 1946 Nankaido earthquake. They found a gradual decrease of interplate locking between 25 and 40 km depth with a plateau around a depth of 30 km. Here we also estimate interplate lockign for other two profiles, the source region of the 1944 Tonankai earthquake and the Tokai seismic gap, to check whether the characteristics of interplate locking shown by Aoki and Scholz (2003b) are a general property or not.

2 Kinematics, Dynamics, and Mechanics of Subduction Zones

We are going to discuss the mechanics of a subduction zone during an interseismic period in subsequent sections. Because how the kinematics of the plate interface inferred from geodetic data is translated into the dynamics, such frictional properties, and mechanics is not necessarily well explained in previous literatures (e.g., Wang et al. 2003; Wang and Dixon 2004), here we describe how we gain insights into the mechanics of the plate interface from geodetic data during an interseismic period.

Geodetic data gives us a snapshot of the kinematics of the plate interface during a seismic cycle. Geodetic data during an interseismic period thus provides us with a strain accumulation process during a seismic cycle. In subduction zones capable of generate great earthquakes, such as the Nankai trough, southwest Japan, the plate interface is locked at shallow depths (Fig. 1), although shallowest depths are not capable of nucleating earthquakes due to elevated pore pressure (e.g., Byrne et al. 1998; Stern 2002) but mechanically locked (McCaffrey 2002; Wang and Dixon 2004). At depth, the plate interface is unlocked because the elevated temperature promotes plastic deformation there.

The kinematic properties on the plate interface inferred from geodetic data is related to dynamic and mechanical properties via numerical simulations of seismic cycles. Tse and Rice (1986) simulated seismic cycles on a two-dimensional fault with depth-dependent fault properties, where rocks at shallower depths have velocity weakening properties and rocks at depth have velocity strengthening properties. Their simulation shows that during an interseismic period the plate interface is fully locked at shallower depths but not to a depth of the transition between velocity weakening to strengthening (Fig. 1). Also they show a broad transition from locked to unlocked at depths around the transition between velocity weakening to strengthening (Fig. 1). We define this transition zone appeared in the numerical simulation a brittle-plastic transition zone. Fig. 1 summarizes the relation among kinematic, dynamic, and mechanical representation of subduction zones.

Because a numerical simulation of Tse and Rice (1986) does not take slow slip transients into account, geodetic data in our analysis was picked to represent purely the strain accumulation process and not to include slow slip transients observed in a part of the Nankai trough in 2000 and later.

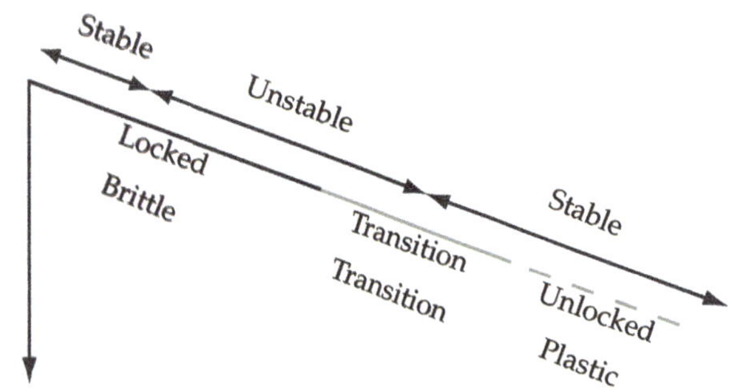

Fig. 1 A schematic representation of kinematics dynamics, and mechanics of a subduction zone during an interseismic period

3 The Nankai Trough

The Nankai trough (Fig. 2) is one of the most studied subduction zones. The Philippine Sea plate subducts beneath the Eurasian plate at the Nankai trough obliquely at ~50 mm per year (Seno et al. 1993; Zang et al. 2002). Stresses due to the interaction between the subducting Philippine Sea plate and overriding Eurasian plate are released by periodic great earthquakes with magnitude 8 or larger, most recent ones being the 1946 Nankaido and 1944 Tonankai earthquakes.

The Nankaido earthquake ruptured the western neighborhood of an area the Tonankai earthquake ruptured. Geodetic (e.g., Sagiya and Thatcher 1999) and tsunami data (e.g., Tanioka and Satake 2001; Baba and Cummins 2005) suggest that the rupture zone of the Nankaido earthquake extends more to the west of the rupture zone inferred from seismic waves (Kanamori 1972), implying the slow rupture, large afterslip, or both occurred to the west of the mainshock rupture.

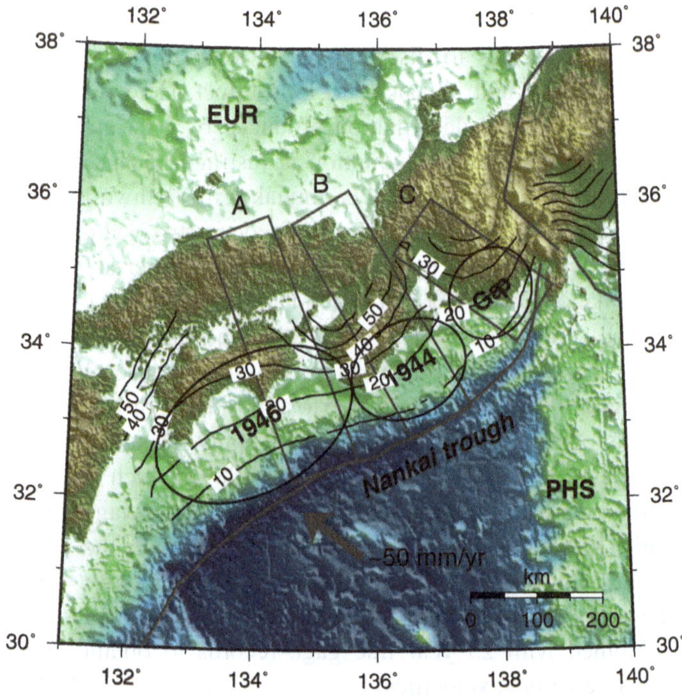

Fig. 2 Tectonic setting aroud the Nankai trough. EUR and PHS indicates the Eurasian and Philippine Sea plate, respectively. Profiles (A), (B), and (C) correspond to the Nankaido, Tonankai, and Tokai profiles in Figs. 7 and 8. Depth contours of subducting Philippine Sea plate and approximate source regions for the 1946 Nankaido, 1944 Tonankai, and Tokai seismic gap are also shown

Kodaira et al. (2000) found a subducting seamount near the western end of the rupture zone inferred from seismic waves (Kanamori 1972), suggesting that it may have controlled the seismic rupture during the Nankaido earthquake.

Ando (1975) compiled an earthquake history of this plate boundary going back to seventh century using felt reports and tsunamis and concluded that great earthquakes repeat there every 150 years or so. The eastern end of the Nankai trough, the Tokai seismic gap, has not ruptured for more than 150 years since 1854. a the next great earthquake is considered to be impending (Ishibashi 1981).

The Nankai trough is the ideal place to study the depth variation of interplate locking because there exists a dense continuous GPS network capable of imaging spatial variation of interplate locking. Aoki and Scholz (2003b) obtained the depth variation of interplate coupling at the source area of the 1946 Nankaido earthquake (line A; Fig. 2). They showed that a brittle-plastic transition zone extends between 25 to 40 km depth, consistent with heat flow modeling (e.g., Hyndman et al. 1995). The Nankaido earthquake ruptured down to 35 km depth (Sagiya and Thatcher 1999), to the middle of the brittle-plastic transition zone, consistent with a numerical simulations using fault friction laws (Tse and Rice 1986; Scholz 1998).

Aoki and Scholz (2003b) also showed that interplate locking does not decrease monotonically from 1 to 0 in the brittle-plastic transition zone but there is a small plateau region at 32–38 km depth range. Because a numerical simulation assuming frictional behavior from top to bottom predicts a monotonic reduction of interplate locking in brittle-plastic transition zone (Tse and Rice 1986), Aoki and Scholz (2003b) speculated that there is either a change in mechanism (say, from frictional to plastic) or phase, or both, at that depth range. Here we examine the depth variation of interplate locking in other profiles in the Nankai trough to check whether or not this is a general property.

4 Interseismic Vertical Velocities

We obtained an interseismic velocity field following a method of Aoki and Scholz (2003a) which is briefly described in this section; see Aoki and Scholz (2003a) for details. We used continuous GPS data operated by Geographical Survey Institute of Japan to obtain interseismic velocities. We selected GPS data between March 21, 1996, and December 31, 1999. Although the data in 2000 or later is available, we ignored these data to avoid the perturbation of the velocity field during an interseismic strain accumulation period from a volcanic activity in nothern Izu islands (e.g., Nishimura et al. 2001) and slow slip episodes in the Tokai seismic gap (Ozawa et al. 2002; Miyazaki et al. 2006).

The raw GPS data were first denoised by Principal Component Analysis, an effective method for extracting spatially correlated modes. It is also known as Empirical Orthogonal Function analysis, and has been commonly used to detect common modes signals in tidal data (e.g., Savage and Thatcher 1992; Savage 1995). It resolves simultaneously observed data into a superposition of several common modes, each of which are orthogonal. Figures 3 and 4 depict the eight lowest principal component modes and the ten highest

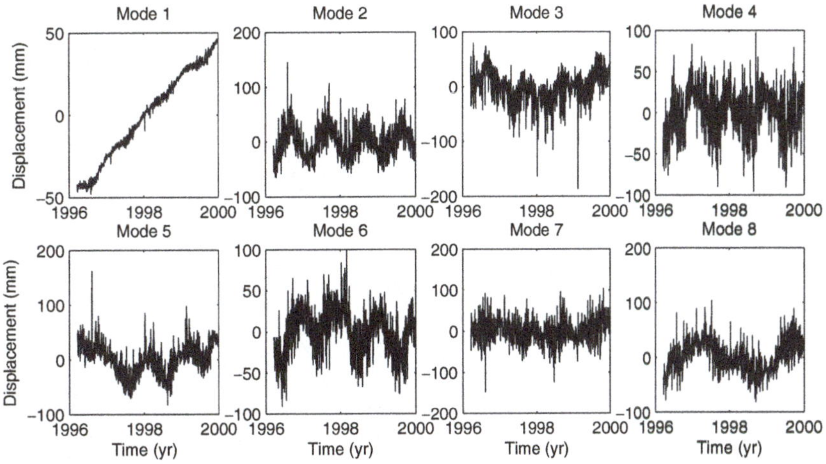

Fig. 3 Eight lowest modes of principal components of GPS time series

eigenvalues normalized by the highest one, respectively. Fig. 3 shows that Mode 1 and 2 correspond to linear interseismic deformation and deformation with an annual cycle. Here we extracted mode 1 only and neglected higher modes because eigenvalues of the higher modes are low (Fig. 4).

Although the procedure described above allows us to obtain a velocity field, it may not reflect the absolute vertical velocity field because the GPS data we used are processed to give an *a priori* velocity to a reference GPS site. The GPS vertical velocity field obtained here therefore needs to be combined with tide gage data to obtain absolute vertical velocities in this case because they represent absolute vertical velocities if corrected for global sea level change, here we took a value 1.8 ± 0.1 mm per year from Douglas (1991).

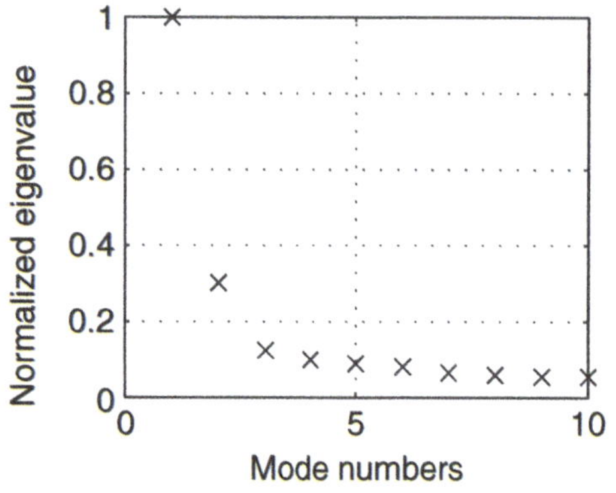

Fig. 4 Eigenvalues of ten lowest modes normalized by the highest eigenvalues of GPS time series

Figure 5 shows the distribution of vertical velocities in southwest Japan overlayed by detph contours of subducting Philippine Sea plate compiled from the distribution of microearthquakes (Hashimoto et al. 2004). The result shows a clear uplift pattern up to 8 mm per year parallel to the Nankai trough and subsidence pattern near the Suruga trough, indicating the overriding Eurasian plate is down-dragged by the subducting Philippine Sea plate. On the other hand, no clear deformation pattern is seen along the Pacific coast of Kyushu, probably indicating weak or no interaction between the subducting Philippine Sea plate and overriding Euraisn plate, as expected from lack of $M > 8$ earthquakes in the area (e.g., Shiono et al. 1980). The alternative interpretation of this is that the steep subduction the Philippne Sea plate there prohibits large deformation on land even if the plate interface is fully locked.

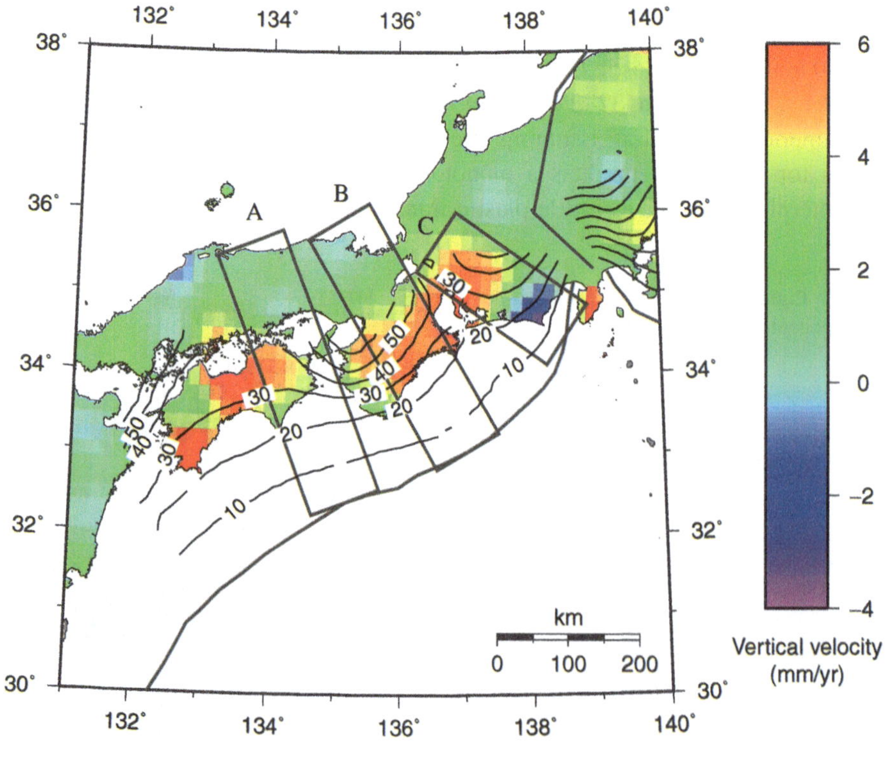

Fig. 5 Distribution of vertical velocities of southwest Japan with areas for the interplate locking analysis (Figs. 7, 8), in which (A), (B), and (C) correspond to Nankaido, Tonankai, and Tokai profiles, respectively. Depth contours of subducting Philippine Sea plate are also shown. Contour interval is 10 km

5 Method

We inverted GPS vertical velocities thus obtained for the depth variation of interplate locking for three profiles in the Nankai subduction zone, the source region of the Nankaido (line A; Fig. 2) and Tonankai (line B) earthquakes, and the Tokai seismic gap (line C). As noted above, vertical velocities are capable of separating rigid plate motion from interplate coupling because rigid plate motion produces no vertical motion (Aoki and Scholz 2003a, b) and vertical velocities are more sensitive than horizontal velocities to interplate locking on a shallow-dipping fault like the Nankai trough.

As in Aoki and Scholz (2003b), we employed a two-dimensional model for interplate coupling in each profile, in which the interplate locking coefficient is a function of depth only, to obtain better resolution of the depth variation by simplifying the problem. Vertical velocities of GPS sites within a 50 km band on each side around each profile, shown in Fig. 2, are used for the inversion. We employed the slip deficit model of Savage (1983) in which the overriding plate is down-dragged at the same velocity as the subducting plate when the interplate locking is full, while the overriding plate does not deform at all when the interplate locking is zero. The shape of the plate interface is taken from Hashimoto et al. (2004) who complied it from the distribution of microearthquakes. Figure 6 shows the geometries of the plate interface for the three profiles.

The observation equation is given by

$$d = Gm + \varepsilon, \tag{1}$$

where the data d and the model m are related by the data kernel G (Mansinha and Smylie 1971; Rani and Singh 1992), and ε is the error in the data which is

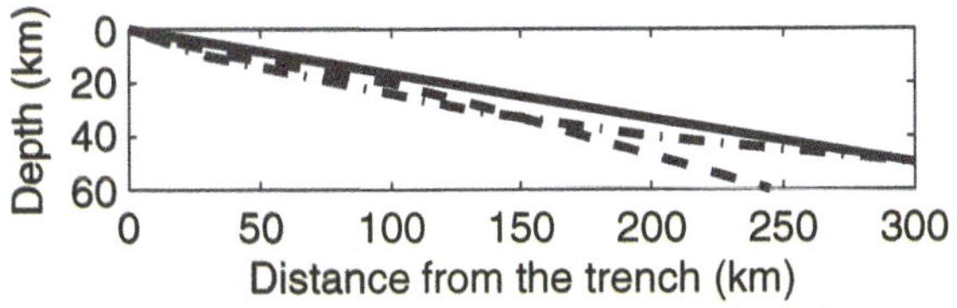

Fig. 6 Geometry of the plate interface for the Nankaido (*solid*), Tonankai (*dashed*), and Tokai (*dashdot*) profiles determined from the distribution of microearthquakes (Hashimoto et al. 2004)

assumed to have a Gaussian distribution with zero mean and covariance Σ. Our goal is to find m to minimize

$$E = (d - Gm)^T \Sigma_d^{-1} (d - Gm) + \lambda^2 (Dm)^T (Dm), \tag{2}$$

where Σ_d and D are the covariance matrix associated with d and the finite approximation of the second-order differentiation for the spatial smoothness of the solution, respectively. λ is a damping parameter, the larger value of which gives a smoother solution but less spatial resolution.

We determined the appropriate value of λ^2 by a mathematical method, not by a visual inspection, by applying the cross-validation method (Wahba 1990) to determine the value of λ^2. The cross-validation method is based on the idea that a good model well predicts unused from used data. The cross-validation sum of squares (CVSS) is given by

$$\mathrm{CVSS}(\lambda^2) = \frac{1}{N}\sum_{k=1}^{N}(d_k - d_k^p)^2 \tag{3}$$

where N is the number of data, d_k and d_k^p are the unused data and predicted data from the rest of $N - 1$ data. The optimal damping parameter is the one that minimizes CVSS.

We obtained two different depth variations of interplate locking. One is with a constraint that interplate locking must be between 0 (fully unlocked) and 1 (full locked, the upper plate is down-dragged by a velocity of relative plate convergence) with 1 at the surface and 0 at the bottom (60 km for the Tonankai profile and 50 km for the other two profiles). The other is with the same constraint as the first one plus the constraint that interplate locking must not increase with depth. Our boundary condition that interplate locking is 1 at the surface does not imply that we assume the shallowest part is capable of nucleating a great earthquake; a force balance calculation of McCaffrey (2002) demonstrated the breakdown of the slip deficit theory of Savage (1983) at the shallowest depth and the shallowest part of a subduction zone is down-dragged by the subducting plate if interplate locking is high at depth, even if the shallowest part is not capable of initiating a large earthquake. This indicates that geodetic data alone has not inherently a resolving power to explore interplate locking near the surface at a subduction zone.

6 Results and Discussion

Figures 7 and 8 show the inversion results. Figure 7 depicts the result with the boundary condition that interplate locking from 1 at the surface and 0 at the bottom (60 km for the Tonankai profiles and 50 km for the other profiles). Figure 8 depicts the result with the tighter constraints by adding a constraint that interplate coupling must decrease with depth from the Fig. 7 case. Again, note that our boundary condition does not

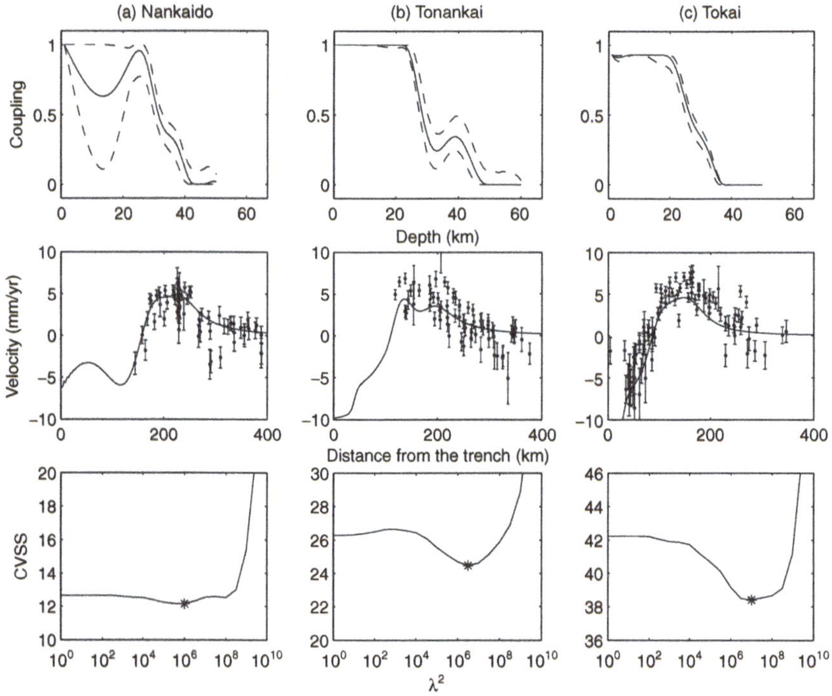

Fig. 7 Results of interplate locking analysis without for the (**a**) Nankaido, (**b**) Tonankai, and (**c**) Tokai profiles, respectively. Constraint that interplate coupling should decrease with depth is not applied in this case. These correspond to line A, B, and C in Fig. 2. The first row shows the depth variation of interplate coupling (*solid lines*) with 95% confidence bounds (*dashed lines*). The second row shows the comparison between observed data with 2σ uncertainties and calculated values. The third row shows the cross-validation sum of squares (CVSS) as a function of the smoothing parameter λ^2, where the larger value gives smoother solution. *Stars* indicate the optimum value of λ^2 with minimum CVSS

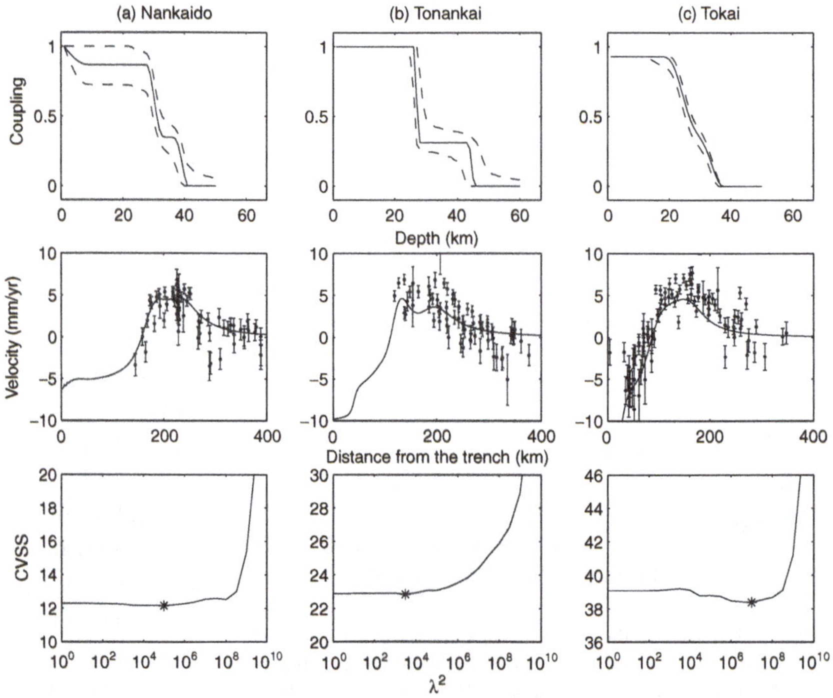

Fig. 8 Same as Fig. 7, except that the solution is constrained so that interplate locking must decrease with depth

imply that we assume the shallowest part is capable of nucleating a great earthquake, as described in the previous section.

Both inversion results (Figs. 7 and 8) show graduate reductions of interplate locking from 1 to 0 between about 20–25 and 35–45 km depth for all three profiles. As noted by Aoki and Scholz (2003b), this depth range for the brittle-plastic transition zone for the Nankaido profile (Figs. 7a, 8a) is comparable with a result from heat flow modeling (e.g., Hyndman et al. 1995) which suggests that the temperature of the plate interface is 350°C at 25 km and 450°C at 35 km.

What does our result imply for an interseismic strain accumulation model? There have been debates as to how the schizosphere and plastosphere behave to accumulate strain during the interseismic period. There are three basic models; the viscoelastic coupling model (e.g., Nur and Mavko 1974) in which a viscous plastosphere is overlain by a brittle schizosphere, the strong plastosphere model (e.g., Bourne et al.1998) in which deformation is governed by a strong plastosphere overlain by a weak schizosphere, and the deep slip model (Savage and Prescott 1978) in which the fault extends from a seismogenic schizosphere through a plastic plastosphere. Among the three, the deep slip model is supported by the geological interpretation that mylonite belts represent ductile shear zones in the plastosphere (Scholz 1988, 2002), results from seismic reflection studies that a fault extends into the lower crust (Henstock et al. 1997; Parsons 1998), and geodetic data (Gilbert et al. 1994). Our inversion result clearly endorse the likeliness of the deep slip model, as Savage (1995) pointed out that the viscoelastic coupling model is inconsistent with the tide gage data for the Nankai trough region. In the case of subduction, the strong plastosphere model can be ruled out because the loading is driven by the downgoing plate rather than the lower crust. Geodetic data cannot resolve the depth distribution of the seismic locking for a vertical strike–slip fault like the San Andreas fault (Mavko 1981) nor distinguish the three models shown above (Savage et al. 1999), but vertical deformation with GPS stations directly above a shallow dipping fault, as in the present case, are capable of resolving slip at depth sufficient to distinguish between the models.

If interplate locking is seen in terms of the relative velocity of the plate motion, we find that the brittle-plastic transition zone is acting like a deep-buried crack in which a crack tip is at the top of the brittle-plastic transition zone. The linear reduction of seismic coupling near the top of the brittle-plastic transition zone (Fig. 7) suggests that the brittle-plastic transition zone acts as a plastic rather than elastic crack. This is similar to the linear reduction of slip as a fault tip is approached (Dawers et al. 1993).

In all three profiles, interplate locking does not decrease monotonically with depth, but it decreases to a plateau region, after which it decreases rapidly to zero (Figs. 7, 8). Resolution tests for the Fig. 7 case (Fig. 9) show that these plateaus are likely to exist. The inversion results with tighter constraints also exhibits the plateau region (Fig. 8), confirming the existence of such plateau region. Although this feature has already seen by Aoki and Scholz (2003b), the fact that the plateaus are seen in each profile reinforces the the credibility of this feature. The plateaus are not expected from a seismic cycle simulation assuming frictional behavior from top to bottom, where interplate coupling is expected to decrease smoothly with depth (Tse and Rice 1986; Scholz 1998). If the brittle-plastic transition zone is flowing plastically, we would

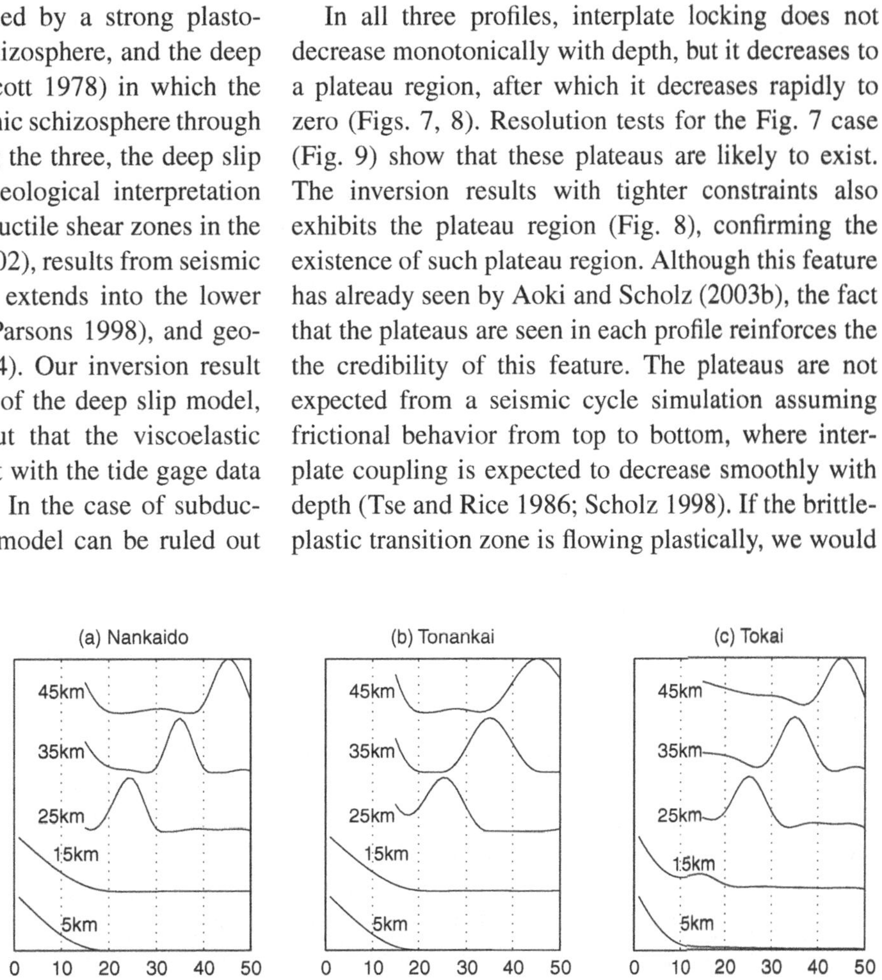

Fig. 9 Resolution kernels at various depths for (**a**) Nankaido, (**b**) Tonankai, and (**c**) Tokai profiles, respectively. If the resolution is perfect, the shape of the kernel will look like a delta function, that is, with a narrower peak, with better resolution of interplate locking. The results indicate interplate locking shallower than 20 km is not well resolved in all profiles

expect interplate locking to decrease exponentially with depth (e.g., Brace and Kohlstedt 1980). The existence of the plateaus suggests that there is either a change in mechanism or material, or both, at that depth. The depth range of the plateau for the Tokai profile (Fig. 7c) coincide well with that of the observed slow slip episodes in this region (e.g., Ozawa et al. 2002, Hirose and Obara 2005, 2006; Miyazaki et al. 2006). A recent seismic survey also suggests high pore pressure there, which may help to promote aseismic slip (Kodaira et al. 2004; Shelley et al. 2006). These may imply that ongoing slow slip in the Tokai seismic gap has something to do with high pore pressure and the plateau of interplate coupling detected in this study. One possibility for slow slip episodes is that they represent the slow oscillations that occur near the stable–unstable boundary in rate- and state-dependent friction (Scholz 2002, pp. 88–89). Provided that the slow slip episode is ongoing at a depth range where the plateau is seen, the plateau may mark the stable–unstable boundary in rate- and state-dependent friction. Recent receiver function analyses revelaed that the Moho depth of the overriding plate (Eurasian plate) along the Nankai trough is approximately 30–35 km (e.g., Shiomi et al. 2004), implying that a change in material properties of the upper plate may control the existence of the plateau and the occurrence of slow slip episodes.

Although the shapes of brittle-plastic transition zones are similar for all profiles, the depth ranges of them are slightly different; the brittle-plastic transition zone starts about 25 km depth for the Nankaido and Tonankai profiles (Figs. 7a, b, 8a, b), while it starts shallower, about 20 km depth, for the Tokai profile (Figs. 7c, 8c). Also, the brittle-plastic transition zone ends at about 36 km depth for the Tokai profile (Fig. 7c), while it extends below 40 km depth for both the Nankaido and Tonankai profiles (Figs. 7a, b, 8a, b). This difference is readily be explained by the difference in slab temperatures for each profile. Because the depth range of the brittle-plastic transition zone is controlled by temperature, cooler plates should have deeper brittle-plastic transition zones. We thus expect higher slab temperature for the Tokai profile than other two. In general, slab temperature is controlled by several factors including age, dip angle, and velocity of subducting slab (Molnar and England 1995); younger, steeper, and slower slab give higher temperature at a given depth. The age distribution of Philippine Sea plate is controversial; Taylor et al. (1991) and Okino et al. (1999) suggested that the Philippine Sea plate is younger to the east and older to the west, that is, it is younger than 10 Ma at the Tokai seismic gap and about 20 Ma at the source region of the Nankaido earthquake. On the other hand, recent heat flow measurements of Yamano et al. (2003) show the presence of an old ridge near the source region of the Nankaido earthquake, suggesting that the Philippine Sea plate is older to the east. Even if the age of the Philippine Sea plate is older for the Tokai profile than the Nankaido profile, it is probably reasonable to consider that the slab is hotter for the Tokai profile because the Tokai slab subducts more slowly and steeply.

Estimating the updip limit of the seismogenic zone is important for assessing seismic hazards caused by a great interplate earthquake. The ~ 10 km of the plate interface in the Nankai trough may not be locked because (1) pore pressure is generally high within the accretionary prism (Byrne et al. 1988) which seems indicated from a recent seismic reflection survey that shows a strong reflector in the Nankai accretionary prism above 9 km (Park et. al. 2002), and (2) geodetic data indicates that the rupture in the 1946 Nankaido earthquake did not likely reach to the surface (Sagiya and Thatcher 1999). Unfortunately, however, geodetic data cannot estimate the locking status at shallower depths because observed deformation is insensitive to the locking status at shallower depths. In addition, McCaffrey (2002) demonstrated from the force balance that the shallowest part of the plate interface is down-dragged regardless of the locking status at the shallow depths if interplate locking is strong at depth.

7 Conclusion

We derived the depth variation of interplate locking in three profiles along the Nankai trough, southwest Japan, from vertical GPS velocities. The coupling profiles for the Tonankai and Tokai regions have the same characteristics as found at the Nankaido profile (Aoki and Scholz 2003b). Thus these features, a broad gradual brittle-plastic transition zone with a plateau in the middle, appear to be general properties of the Nankai subduction zone. The depth range of the brittle-plastic transition zone are deeper for the eastern profile and shallower for the western profile, which implies hotter slab in the Tokai seismic gap than the source region of the Nankaido earthquake, which is consistent with slower and steeper subduction for the Tokai area.

Interplate locking does not decrease linearly with depth in the brittle-plastic transition zone, but there is a plateau in the middle of brittle-plastic transition zone, which is not expected from simple friction or plastic models of the transition. Thissuggests that there is a change in mechanism (from friction to plastic flow) or material, or both, in the middle of the brittle-plastic transition zone.

Although exploring updip limit of the seismogenic zone is important for assessing seismic hazard, geodetic data cannot estimate the coupling status at the shallowest depths because the slip deficit model employed in this study (Savage 1983) breaks down at the shallowest depths if interplate locking is strong at depth.

Acknowledgments We thank Serge Lallemand and Francesca Funiciello for organizing this volume. Comments by Wayne Thatcher, Stephane Mazzotti, and Serge Lallemand improved the manuscript. Some figures are created with the Generic Mapping Tools (Wessel and Smith 1998). Lamont-Doherty Earth Observatory contribution 7135.

References

Ando M (1975). Source mechanisms and tectonic significance of historical earthquakes along the Nankai trough, Japan. Tectonophysics 27:119–140

Aoki Y, Scholz CH (2003a). Vertical deformation of the Japanese islands, 1996–1999. J Geophys Res 108:2257, doi:10.1029/2002JB002129

Aoki Y, Scholz CH (2003b). Interseismic deformation at the Nankai subduction zone and the Median Tectonic Line, southwest Japan. J Geophys Res 108:2470, doi:10.1029/2003JB002441

Baba T, Cummins PR (2005). Contiguous rupture areas of two Nankai Trough earthquakes revealed by high-resolution tsunami waveform inversion. Geophys Res Lett 32:L08305, doi:10.1029/2004GL022320

Bourne SJ, England PC Parsons B (1998). The motion of crustal blocks driven by flow of the lower lithosphere and implications for slip rates of continental strike-slip faults. Nature 391:655–659

Brace WF, Kohlstedt D (1980). Limits on lithospheric stress imposed by laboratory experiments. J Geophys Res 85: 6248–6252

Byrne DE, Davis DM, Sykes LR (1988). Loci and maximumsize of thrust earthquakes and the mechanics of the shallow region of subduction zones. Tectonics 7:833–857

Dawers NH, Anders MH, Scholz CH (1993). Growth of normal faults – displacement-length scaling. Geology 21:1107–1110

Douglas BC (1991). Global sea level rise. J Geophys Res. 96: 6981–6992

Douglas A, Beavan J, Wallace L, Townend J (2005). Slow slip on the northern Hikurangi subduction interface, New Zealand. Geophys Res Lett 32:L16305, doi:10.1029/2005GL023607

Dragert H, Wang K, James TS (2001). A silent slip event on the deeper Cascadia subduction interface. Science 292: 1525–1528

Gilbert L, Scholz CH, Beavan J (1994). Strain localization along the San Andreas fault: consequences for loading mechanisms. J Geophys Res 99:975–984

Hashimoto C, Fukui K, Matsu'ura M (2004). 3-D modelling of plate interfaces and numerical simulation of long-term crustal deformation in and around Japan. Pure Appl Geophys 161:2053–2068

Heki K, Miyazaki S, Tsuji H (1997). Silent fault slip following an interplate thrust earthquake at the Japan trench. Nature 386:595–598

Henstock TJ, Levander A, Hole JA (1997). Deformation in the lower crust of the San Andreas fault system in northern California. Science 278:650–653

Hirose H, Obara K (2005). Repeating short- and long-term slow slip events with deep tremor activity around the Bungo channel region, southwest Japan. Earth Planet Space 57: 961–972

Hirose H, Obara K (2006). Short-term slow slip and correlated tremor episodes in the Tokai region, central Japan Geophys Res Lett. 33:L17311, doi:10.1029/2006GL026579

Hirose H, Hirahara K, Kimata F, Fujii N, Miyazaki S (1999). A slow thrust slip event following the two 1996 Hyuganada earthquakes beneath the Bungo Channel, southwest Japan. Geophys Res Lett 26:3237–3240

Hyndman RD, Wang K, Yamamo M (1995). Thermal constraints on the seismogenic portion of the southwestern Japan subduction thrust. J Geophys Res 100:15373–15392

Ide S, Shelly DR, Beroza GC (2007a). Mechanism of deep low frequency earthquakes: Further evidence that deep non-volcanic tremor is generated by shear slip on the plate interface. Geophys Res Lett 34:L03308, doi:10.1029/2006GL028890

Ide S, Beroza GC, Shelly DR, Uchide T (2007b). A scaling law for slow earthquakes. Nature 447:76–79

Ishibashi K (1981). Specification of soon-to-occur seismic faulting in the Tokai District, central Japan, based on seismotectonics. In: Simpson DW, Richards PG (eds) Earthquake prediction: an international review, American Geophysical Union, Washington DC, pp 297–332

Ito Y, Obara K, Shiomi K, Sekine S, Hirose H (2007). Slow earthquakes coincident with episodic tremors and slow slip events. Science 315:503–506

Kanamori H (1972). Tectonic implications of the 1944 Tonankai and 1946 Nankaido earthquakes. Phys Earth Planet Inter 5:129–139

Kao H, Shan SJ, Rogers G, Dragert H (2007). Migration characteristics of seismic tremors in the northern Cascadia margin. Geophys Res Lett 34:doi:10.1029/2006GL028430

Katsumata A, Kamaya N (2003). Low-frequency continuous tremor around the Moho discontinuity away from volcanoes in the southwest Japan. Geophys Res Lett 30:1020, doi:10.1029/2002GL015981

Kawasaki I, Asai Y, Tamura Y (2001). Space-time distribution of interplate moment release including slow earthquakes and the seismo-geodetic coupling in the Sanriku-oki region along the Japan trench. Tectonophysics 330:267–283
Kodaira S, Takahashi H, Nakanishi A, Miura S, Kaneda Y (2000). Subducted seamount imaged in the rupture zone of the 1946 Nankaido earthquake. Science 289:104–106
Kodaira S, Iidaka T, Kato A, Park JO, Iwasaki T, Kaneda Y (2004). High pore pressure may cause silent slip in the Nankai trough. Science 304:1295–1298
Kostoglodov V, Singh SK, Santiago JA, Franco SI, Larson KM, Lowry AR, Bilham R (2003). A large silent earthquake in the Guerrero seismic gap, Mexico. Geophys Res Lett. 30(15): 1807, doi:10.1029/2003GL017219
Larson KM, Lowry AR, Kostoglodov V, Hutton W, Sánchez O, Hudnut K, Suarez G (2004). Crustal deformation measurements in Guerrero, Mexico. J Geophys Res 109:B04409, doi:10.1029/2003JB002843
Mansinha L, Smylie DE (1971). The displacement fields of inclined faults. Bull Seismol Soc Amer 61:1433–1440
Mavco GM (1981). Mechanics of motion on major faults. Ann Rev Earth Planet Sci 9:81–111
Mazzotti S, Le Pichon X, Henry P, Miyazaki S (2000). Full interseismic locking of the Nankai and Japan-west Kurile subduction zones: An analysis of uniform elastic strain accumulation in Japan constrained by permanent GPS. J Geophys Res 105:13,159–13,177
McCaffrey R (2002). Crustal block rotations and plate coupling. In: Stein S, Freymueller JT (eds) Plate Boundary Zones, American Geophysical Union, Washington DC, pp 101–122
Menke W (1989). Geophysical Data Analysis: Discrete Inverse Theory, rev. ed. 289 pp, Academic Press, San Diego, CA
Miller MM, Melbourne T, Johnson DJ, Sumner WQ (2002). Periodic slow earthquakes from the Cascadia subduction zone. Science 295:2423
Miyazaki S, Heki K (2001). Crustal velocity field of southwest Japan: subduction and arc-arc collision. J Geophys Res 106:4305–4326
Miyazaki S, McGuire JJ, Segall P (2003). A transient subduction zone slip episode in southwest Japan observed by the nationwide GPS array. J Geophys Res 108:2087, doi:10.1029/2001JB000456
Miyazaki S, Segall P, McGuire JJ, Kato T, Hatanaka Y (2006). Spatial and temporal evolution of stress and slip rate during the 2000 Tokai slow earthquake. J Geophys Res 111:B03409, doi:10.1029/2004JB003426
Molnar P, England P (1995). Temperatures in zones of steady-state underthrusting of young oceanic lithosphere. Earth Planet Sci Lett 131:57–70
Nadeau RM, Dolenc D (2005). Nonvolcanic tremors deep beneath the San Andreas fault. Science 307:389
Nedimović, MR, Hyndman RD, Ramachandran K, and Spence GD (2003). Reflection signature of seismic and aseismic slip on the northern Cascadia subduction interface. Nature 424:416–420
Nur A, Mavko G (1974). Postseismic viscoelastic rebound. Science 183:204–206
Obara K (2002). Nonvolcanic deep tremor associated with subduction in southwest Japan. Science 296:1679–1681
Obara K, Hirose K, Yamamizu F, Kasahara K (2004). Episodic slow slip events accompanied by non-volcanic tremors in southwest Japan subduction zone. Geophys Res Lett 31:L23602, doi:10.1029/2004GL020848
Ohmi S, Obara K (2002). Deep low-frequency earthquakes beneath the focal region of the Mw 6.7 2000 Western Tottori earthquake. Geophys Res Lett. 29(16):1807, doi:10.1029/2001GL014469
Ohta Y, Freymueller JT, Hreinsdóttir, Suito H (2006). A large slow slip event and the depth of the seismogenic zone in the south central Alaska subduction zone. Earth Planet Sci Lett 247:108–116
Okino K, Ohara Y, Kasuga S, Kato Y (1999). The Philippine Sea: New survey results reveal the structure and the history of the marginal basins. Geophys Res Lett 26:2287–2290
Ozawa S, Murakami M, Kaidzu M, Tada T, Sagiya T, Hatanaka Y, Yarai H, Nishimura T (2002). Detection and monitoring of ongoing aseismic slip in the Tokai region, central Japan. Science 298:1009–1012
Ozawa S, Suito H, Imakiire T, Murakami M (2007). Spatiotemporal evolution of aseismic interplate slip between 1996 and 1998 and between 2002 and 2004, in Bungo channel, southwest Japan. J Geophys Res 112:B05409, doi:10.1029/2006JB004643
Parsons T (1998). Seismic-reflection evidence that the Hayward fault extends into the lower crust of the San Francisco Bay Area, California. Bull Seismol Soc Amer 88:1212–1223
Pacheco JF, Sykes LR (1992). Seismic moment catalog of large shallow earthquakes, 1900 to 1989. Bull Seismol Soc Amer 82:1306–1349
Pacheco JF, Sykes LR, Scholz CH (1993). Nature of seismic coupling along simple plate boundaries of the subduction type. J Geophys Res 98:14133–14159
Park JO, Tsuru T, Takahashi N, Hori T, Kodaira S, Nakanishi A, Miura S, Kaneda Y (2002). A deep strong reflector in the Nankai accretionary wedge from multichannel seismic data: Implications for underplating and interseismic shear stress release. J Geophys Res 107(B4):2061, doi:10.1029/2001JB000262
Rani S, Singh SJ (1992). Static deformation of a uniform half-space due to a long dip-slip fault. Geophys J Int 109: 469–476
Rogers G, Dragert H (2003). Episodic tremor and slip on the Cascadia subduction zone: The chatter of silent slip. Science 300:1942–1943
Sagiya T, Thatcher W (1999). Coseismic slip resolution along a plate boundary megathrust: The Nankai Trough, southwest Japan. J Geophys Res 104:1111–1129
Savage JC (1983). A dislocation model of strain accumulation and release at a subduction zone. J Geophys Res 88: 4984–4996
Savage JC (1995). Interseismic uplift at the Nankai subduction zone, southwest Japan, 1951–1990. J Geophys Res 100: 6339–6350
Savage JC, Prescott WH (1978). Asthenosphere readjustment and the earthquake cycle. J Geophys Res 83:3369–3376
Savage JC,Thatcher W (1992). Interseismic deformation at the Nankai Trough, Japan, subduction zone. J Geophys Res 97:11,117–11,135
Savage JC, Svarc JL, Prescott WH (1999). Geodetic estimates of fault slip rates in the San Francisco Bay area. J Geophys Res 104:4995–5002
Scholz CH (1988). The brittle-plastic transition and the depth of seismic faulting. Geol Rundsch 77:319–328

Scholz CH (1994). A reappraisal of large earthquake scaling. Bull Seismol Soc Amer 84:215–218

Scholz CH (1998). Earthquakes and friction laws. Nature 391:37–42

Scholz CH (2002). The mechanics of earthquakes and faulting, second ed, 471 pp, Cambridge University Press, Cambridge, UK

Scholz CH, Campos J (1995). On the mechanism of seismic decoupling and back arc spreading at subduction zones. J Geophys Res. 100:22103–22115

Seno T, Stein S, Gripp AE (1993). A model for the motion of the Philippine Sea Plate consistent with NUVEL-1 and geological data. J Geophys Res 98:17941–17948

Shelly DR, Beroza GC, Ide S, Nakamula S (2006). Low-frequency earthquakes in Shikoku, Japan, and the relationship to episodic tremor and slip. Nature 442:188–191

Shelly DR, Beroza GC, Ide S (2007). Non-volcanic tremor and low-frequency earthquake swarms. Nature 446: 305–307

Shiomi K., Sato H, Obara K, Ohtake M (2004). Configuration of subducting Philippine Sea plate beneath southwest Japan revealed from receiver function analysis based on the multivariate autoregressive model. J Geophys Res 109:B04308, doi:10.1029/2003JB002774

Shiono K, Mikumo T, Ishikawa Y (1980). Tectonics of the Kyushu-Ryukyu arc evidenced from seismicity and focal mechanisms of shallow to intermediate-depth earthquakes. J Phys Earth 28:17–43

Sibson R (1984). Roughness at the base of the seismogenic zone: contributing factors. J Geophys Res 89:5791–5799

Stern RJ (2002). Subduction zones. Rev Geophys 40:1012, doi:10.1029/2001RG000108

Tanioka Y, Satake K (2001). Coseismic slip distribution of the 1946 Nankai earthquake and aseismic slips caused by the earthquake. Earth Planet Space 53:235–241

Taylor B, Klaus A, Brown GR, Moore GF, Okamura Y, Murakami F (1991). Structural development of sumisu rift, Izu-Bonin arc. J Geophys Res 96:16113–16129

Townend J, Zoback MD (2006). Stress, strain, and mountain building in central Japan. J Geophys Res 111: B03411, doi:10.1029/2005JB003759

Tse ST, Rice JR (1986). Curstal earthquake instability in relation to the depth variation of frictional slip properties. J Geophys Res 91:9452–9472

Uyeda S, Kanamori H (1979). Back arc opening and the mode of subduction. J Geophys Res 84:1049–1061

Vrolijk P (1990). On the mechanical role of smectite in subduction zones. Geology 18:703–707

Wang K (2000). Stress-strain "paradox," plate coupling, and forearc seismicity at the Cascadia and Nankai subduction zones. Tectonophysics 319:321–338

Wang K, Dixon TH (2004). "Coupling" semantics and science in earthquakes research. EOS Trans Am Geophys Union 85:180

Wang K, Wells R, Mazzotti S, Hyndman RD, Sagiya T (2003). A revised dislocation model of interseismic deformation of the Cascadia subduction zone. J Geophys Res 108:2026, doi:10.1029/2001JB001227

Wahba G (1990). Spline Models for Observational Data. 169 pp., SIAM, Philadelphia, PA

Wessel P, Smith WHF (1998). New, improved version of Generic Mapping Tools released. EOS 79:579

Yamano M, Kinoshita M, Goto S, Matsubayashi O (2003). Extremely high heat flow anomaly in the middle part of the Nankai trough. Phys Chem Earth 28:487–497

Yoshioka S, Mikumo T, Kostoglodov V, Larson KM, Lowry AR, Singh SK (2004). Interplate coupling and a recent aseismic slow slip event in the Guerrero seismic gap of the Mexican subduction zone, as deduced from GPS data inversion using a Bayesian information criterion. Phys Earth Planet Inter 146:513–530

Zang SX, Chen QY, Ning JY, Shen ZK, Liu YG (2002). Motion of the Philippine Sea plate consistent with the NUVEL-1A model. Geophys J Int 150:809–819

Part V

Continental and Ridge Subduction Processes

Exhumation Processes in Oceanic and Continental Subduction Contexts: A Review

Stéphane Guillot, Keiko Hattori, Philippe Agard, Stéphane Schwartz and Olivier Vidal

Abstract Although the exhumation of high pressure (HP) and ultrahigh pressure (UHP) rocks is an integral process in subduction, it is a transient process, likely taking place during the perturbation in subduction zones. Exhumation of HP to UHP rocks requires the weakening of a subduction channel and the decoupling of the exhumed slice from the rest of the slab. Considering more than 60 occurrences of HP to UHP units of Phanerozic ages, we propose three major types of subduction zones:

Accretionary-type subduction zones exhume HP metasedimentary rocks by underplating. The exhumation is slow and can be long-lasting.

The serpentinite-type subduction zones exhume HP to UHP in a 1 to 10 km thick serpentinite subduction channel. The serpentinite matrix originates from both subducted abyssal peridotites and hydated mantle wedge. Exhumation velocity is low to intermediate and the exhumation is driven by the buoyancy and the low-viscosity of the serpentinite.

The continental-type subductions exhume UHP rocks of continental origin. The UHP rocks together with garnet-bearing peridotites form units from km-scale unit. The exhumation is fast, short-lived and occurs at the transition from oceanic subduction to continental subduction. It is driven by buoyancy forces and asthenospheric return flow.

Keywords Oceanic subduction • Continental subduction • Exhumation • HP to UHP rocks • Subduction channel

1 Introduction

Eclogites, HP-LT metamorphic rocks, have been reported since the first petrological description by Haüy (1822) and recognized from many locations in the world with ages ranging from Proterozoic to Phanerozic times (e.g., Godard, 2001 for review). The occurrences of pelitic rocks metamorphosed under eclogite facies conditions suggest that these rocks were subducted to great depths before exhumed (Compagnoni and Maeffo, 1973; Carswell, 1990). The discovery of coesite in Alpine metasediments (Chopin, 1984) introduced the term of UHP metamorphism and demonstrated that continental crust can be subducted to a depth greater than 100–120 km. Most Alpine-type HP to UHP-LT metamorphic rocks occur in peri-Pacific and peri-Mediterranean fold belts of Paleozoic to Tertiary ages (Fig. 1) and are characterized by geothermal gradients ranging between 4 and 10°/km (e.g., Maruyama

S. Guillot
University of Grenoble, OSUG – CNRS,
1381 rue de la Piscine 38041 Grenoble Cedex 9, France,
sguillot@ujf-grenoble.fr

S. Lallemand and F. Funiciello (eds.), *Subduction Zone Geodynamics*,
DOI 10.1007/978-3-540-87974-9,

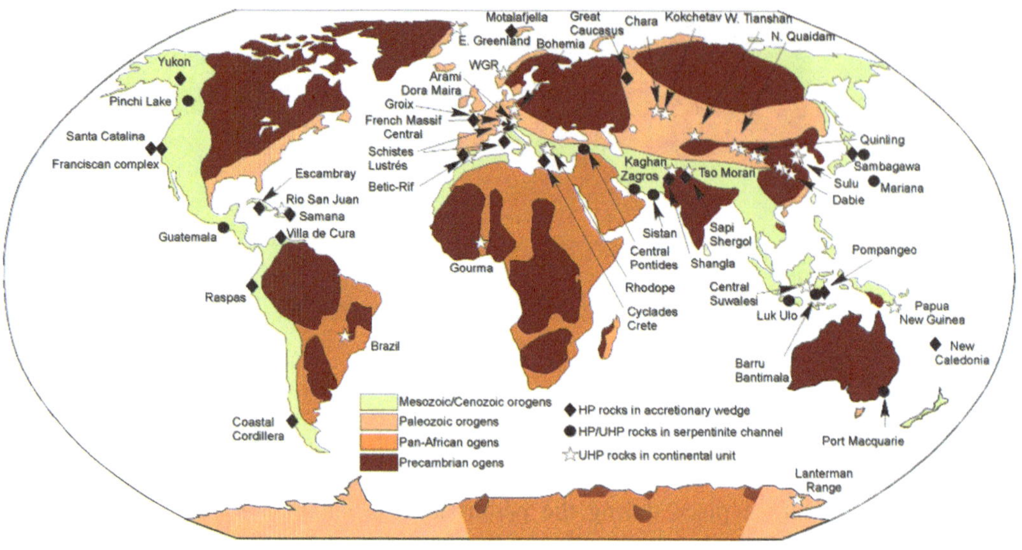

Fig. 1 Worldwide occurrences of HP and UHP massifs (modified after Liou et al., 2004; Tsujimori et al., 2006)

et al., 1996). The oldest UHP unit was described in the Gourma area (eastern Mali) in the Pan-African belt and dated at 620 Ma (Caby, 1994; Jahn et al., 2001). The oldest blueschist is also Late Proterozoic in age (Liou, 1990; Caby et al., 2008). The absence of older blueschists may be ascribed to overprinting by subsequent metamorphism or probably to hotter Earth during Archean and early Proterozoic time. The oldest lawsonite eclogite (e.g., Tsujimori et al., 2006) and oldest carpholite-bearing blueschist (Agard et al., 2005) are both early to middle Paleozoic in age, which is indeed compatible with decreased geothermal gradients to 6–7°/km in Phanerozoic subduction zones (Maruyama et al., 1996). Valli et al. (2004) estimated a geothermal gradient of 30°C/km for a Late Archean subduction zone and Moyen et al. (2006) reported a 3.2 Ga garnet-bearing amphibolite (1.2–1.5 GPa) from an arguably 15°/km subduction zone, similar to those observed during the Late Proterozoic (Maruyama et al., 1996). The oldest reported HP metamorphism (750°C, 1.8 GPa) was found in a 2.0 Ga eclogite terrane of Tanzania (Môller et al., 1995) likely formed as a result of continental subduction (Collins et al., 2004). The peak metamorphic condition indicates a moderate geothermal gradient of about 12°C/km, comparable to geotherms of modern subduction zones. In contrast, the youngest UHP rock on Earth was found in Papua New Guinea and is dated at 4 Ma (Baldwin et al., 2004). These data show that a study of HP to UHP rocks of various ages contributes to a better understanding of the evolution of thermal regimes of subduction zones since Precambrian time.

HP to UHP metamorphic rocks of continental or oceanic origins occur in convergent zones (Ernst, 2001).The pressure-temperature-time (P-T-t) paths suggest their subduction and subsequent return to the Earth's surface (e.g., Duchêne et al., 1997). Despite the growing amount of data on surface horizontal displacement, the vertical movements of the lithosphere and exhumation processes are still in debate. Proposed mechanisms for exhumation include channel flow (Cloos, 1982), corner flow (Platt, 1986), extensional collapse (Dewey et al., 1993), thrusting towards the foreland (Steck et al., 1998), buoyancy assisted by erosion and tectonic processes (Chemenda et al., 1995), compression of a soft zone between two rigid blocks (Thompson et al., 1997), serpentinite channel (Guillot et al., 2001), and coaxial extension associated with a decoupling fault (Jolivet et al., 2003).

Although these mechanisms may locally account for the exhumation of HP to UHP rocks in specific subduction zones, a general understanding of the major processes and associated settings that can explain the worldwide exhumation of HP to UHP rocks is still missing.

Detailed studies of HP and UHP rocks including kinematic analyses and dating of metamorphism provide invaluable information of thermomechanical processes in subduction zones (Coleman, 1971; Ernst, 1973). Such information can also constrain possible mechanisms of exhumation because different processes result in different styles of deformation and P-T-t paths of exhumed rocks. Furthermore, the data from HP and UHP rocks in conjunction with P-T-t paths predicted from numerical modelling provide key information related to the thermomechanical properties of subduction zones.

This paper reviews more than 60 occurrences of HP to UHP units of Phanerozic ages (Fig. 1), their protoliths, their P-T-t paths and their exhumation rates, and discusses the important factors controlling their exhumation and the possible significance of the so-called "subduction channel" in subduction zones.

1.1 Subduction Types

Bally (1981) defined two contrasting types of convergent zones: the Pacific- and Alpine-types. The Pacific-type subduction is characterized by long-lasting subduction of oceanic lithosphere. The Alpine-type first involves the consumption of an oceanic domain, similar to the Pacific-type subduction, followed by the subduction of continental margins. The continents involve in the Alpine-type could be large, such as those in the Alps, Variscides, Himalaya, Dabieshan or the Caledonides (Chopin, 1984; Lardeaux et al., 2001; Guillot et al., 2003, Yang et al., 2003; Hacker, 2007). Some continents are small, as those in Aegean (Jolivet et al., 2003) and Kazakstan (Hacker et al., 2003). Based on the the lithology, the peak P-T conditions, and the exhumation patterns of metamorphic rocks, we propose three types of dominant subduction regime to explain the different styles of exhumation observed: the accretionary-type, the serpentinite-type and the continental-type. The continental-type is similar to the Alpine-type defined by Bally (1981). A subduction zone may evolve from one type to others during its life and two different types may co-exist along one subduction zone.

1.2 Accretionary-Type Subduction

Forearc accretionary wedges (or prisms) develop in front of intra-oceanic arcs or continental arcs (Fig. 2). They are observed all along the Pacific subduction systems including the west coast of the North America (Alaskan–Cascades), the west coast of the South American (Ecuador-Chile), Japan, and Suwalesi. They also occur in the Barbados in the western Atlantic Ocean and Makran in the northern Indian Ocean (e.g., Lallemand, 1999). A major feature of accretionary wedges is the stacking of oceanic sediments by offscraping of the upper part of subducting plate or arc rocks eroded from the upper plate depending on the geometry of the buttress and the subduction angle (Cloos, 1982; Platt, 1986; Moore and Silver, 1987; Cloos and Shreve, 1988; Von Huene and Scholl, 1991)

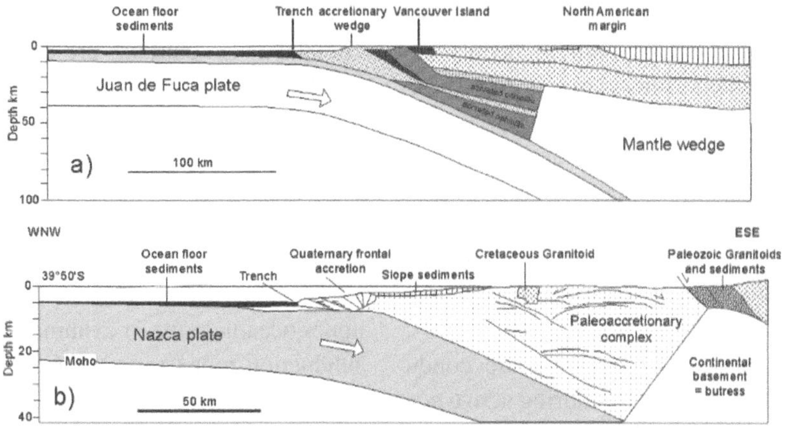

Fig. 2 (**a**) Schematic cross section through the Cascadia subduction zone beneath the Vancouver Island (modified after Hyndman, 1995). Note that the deep part of the accretionary wedge is comprised of imbricate slices of ophiolites and sediments. (**b**) Schematic cross section through the South-Central Chilean forearc based on reflection seismic data and offshore geology (modified after Glodny et al., 2005). Note that the paleoaccretionary wedge of Upper Paleozoic age show internal structures compatible with underthrusting at the base and extension at shallow level. Upper Paleozoic blueschists (~0.8 GPa) are being exhumed near the toe of the paleoaccretionary wedge

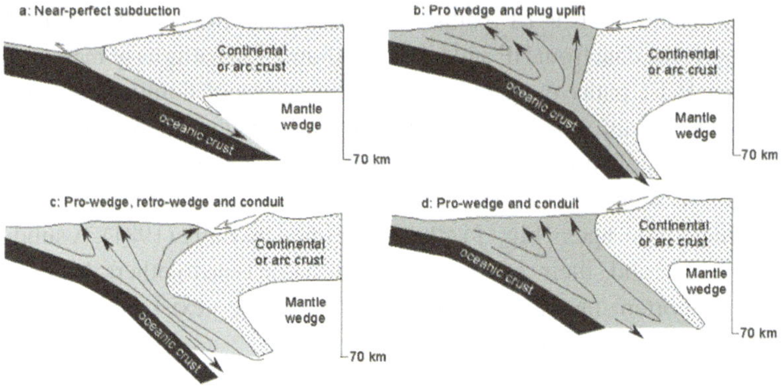

Fig. 3 Schematic geometry of accretionary prism (modified after Ernst, 2005). Note that the geometry controls the depth and origin of HP rocks either from upper or lower plates, and the exhumation trajectory. (**a**) In a narrow accretionary prism, a slab is parallel to the buttress, which prevents the exhumation of HP rocks. (**b**) In a wide accretionary wedge at shallow depth, the rocks are exhumed only from shallow depth. (**c**) In an intermediate model where an accretionary prism is wider than that in Fig. 3b, but narrower than that in Fig. 3d. (**d**) An accretionary wedge is wide at shallow and deep levels, which allow the exhumation of HP rocks from great depths at front, middle and rear of the wedge

(Fig. 3). The deepest part of an accretionary wedge is close to the buttress and about 20 km (ca. 0.6 GPa) in present-day subduction zones, but in exceptional cases about 40–60 km (1.1–1.6 GPa) in Chile (Glodny et al., 2005) and Cascadia beneath Vancouver Island (Hyndman, 1995) (Fig. 2).

Numerical simulation of an accretionary wedge (Beaumont et al., 1999; Allemand and Lardeaux, 1997; Yamato et al., 2007) shows that the initial geometry of a buttress or backstop (continental crust for active margin and arc crust for intra-oceanic subduction) affects the shape of an accretionary wedge and consequently the metamorphic pressures reached by the exhumed rocks. When a slab is parallel to a buttress, deeply subducted rocks are prevented from exhumation (Fig. 3a). On the other hand, a wide open wedge allows the exhumation of deeply subducted rocks that originated from the upper and lower plates (Fig. 3d). In intermediate geometries, deeply subducted rocks are exhumed close to the trench (pro-wedge exhumation), vertically (plug uplift) and also near the buttress (retro-wedge exhumation) (Ernst, 2005).

The geometries shown in Fig. 3b, c are conducive for the exhumation of HP rocks during active oceanic subduction. As already discussed, accretionary wedges are dominated by sediments derived from the upper and lower plates and contain exhumed HP-LT rocks. Two recent reviews by Tsujimori et al. (2006) and Agard et al. (submitted) list more than 20 HP and UHP massifs in the world belonging to this category (Table 1). All these units share many features, suggesting that similar processes were in operation during their metamorphism and exhumation of the HP rocks. These features are summarized below.

The HP-LT rocks in accretionary-type subduction zones are dominated by clastic sedimentary rocks with no mantle-derived material, suggesting that the protoliths of HP-LT rocks are sediments deposited on the sea floor and that the detritus of the sediments were supplied from the upper and lower plates (Plate 1a, Fig. 4a). For instance, upper crustal rocks of seamounts underwent HP-LT metamorphism in the Himalaya (e.g., Mahéo et al., 2006). The HP-LT unit in the Himalaya also contains arc-derived material, suggesting that the erosion of arc rocks was contemporaneous during the seamount subduction (e.g., Lallemand and Le Pichon, 1987; Von Huene and Cullota, 1989; Mahéo et al., 2006). The exhumation of only upper crustal rocks implies that lower crustal rocks of slabs are deeply subducted.

The relative abundance of oceanic sediments and upper oceanic crust in exhumed rocks varies from one subduction zone to another. Centimetric to hectometric blocks of mafic or ultramafic oceanic rocks is usually observed in a calcsilicate matrix. In the case of the Franciscan complex, this "melange" was previously interpreted as a tectonic melange developed along the subduction zone, leading to the concept of subduction channel where a soft matrix allows rigid blocks to be exhumed parallel to the subduction plane (Cloos, 1982;

Table 1 Compilation of available data on exhumed high pressure rocks in accretionary wedge

Unit	Samana Peninsula Dominican Republic	Escambray Cuba	Schistes Lustrés Corsica	Schistes Lustrés Western Alps	Franciscan C. California	Santa Catalina California	Coastal Cordillera Chile	Pam Peninsula New Caledonia	Motalafjella Spitsbergen	Shangla Pakistan	Groix France
Abbreviation	SA	ES	SLC	SLA	FC	SC	CC	PP	MS	SP	GX
Subduction context	Intra-oceanic	Intra-oceanic	Intra-oceanic	Intra-oceanic	Intra-oceanic	Intra-oceanic	Active margin	Intra-oceanic	Unknow	Intra-oceanic	Intra-oceanic
Tectonics of exhumed rocks	3 nappes	3 nappes	3 nappes	4 nappes	Several nappes	3 nappes	Sevreal nappes	>2 nappes	2 nappes	3 nappes	2 nappes
Lithology of exhumed rocks	Sedimentary rocks with serp and basic lenses	Sedimentary rocks with serp and basic lenses	Ophiolitic rocks with sedimentary cover	Sedimentary rocks with serpen-tinites and basic lenses	Ophiolitic rocks with sedimen-tary cover	Ophiolitic rocks with sedimen-tary cover	Sedimentary rocks with basic lenses	Ophiolitic rocks with sedimentary cover	Sedimentary rocks with basic lenses	Sedimentary and volcanic melange	Sedimentary rocks with basic lenses
Max P-T conditions	2.2–2.4 GPa	1.6–2.3 GPa	2.0 GPa	1.8–2.0 GPa	1.8–2.2 GPa	1.2 Gpa	1.1–1.6 GPa	2.0 GPa	1.8–2.4 GPa	0.7 Gpa	1.6–2.0 GPa
	610–625°C	530–620°C	390°C	500°C	360–445°C	600°C	600–760°C	460°C	580–640°C	400°C	500°C
Metamorphic age	Cretaceous-Eocene	Cretaceous-Eocene	Cretaceous-Eocene	Cretaceous-Eocene	Middle Jurassic	Cretaceous	Carboniferous	Eocene	Ordovician	Cretaceous	Devonian
Exhumation velocities	~1 mm/year	Unknown	<2 mm/year	<2 mm/year	5 mm/year	<2 mm/year	0.6 mm/yr	2–3 mm/year	Unknown	Unknown	2 mm/year
Exhumation timing	When the accretionary wedge collided	When the accretionary wedge collided	Intra-oceanic	Intra-oceanic	Intra-oceanic	Intra-oceanic	Active subduction	When the accretionary wedge collided	Unknown	Intra-oceanic	When the accretion-ary wedge collided
References	Goncalvez et al., 2000; Zack et al., 2004; Escuder-Viruete and Pérez-Estaùn, 2006	Schneider et al., 2004; Stanek et al., 2006	Caron and Péquignot, 1986	Agard et al., 2002; Tricart and Schwartz, 2007	Platt, 1986; Oh and Liou, 1990	Bebout and Barton, 1993; Sorensen, 1988; Anczkiewicz et al., 2004	Willner et al., 2004; Glodny et al., 2005	Cluzel et al., 2001; Fitzherbert et al., 2005	Hirajima et al., 1988; Agard et al., 2005	Jan, 1985; Anczkiewicz et al., 2000	Bosse et al., 2005

(continued)

Table 1 (continued)

Unit	Sapi-Shergol India	Sanbagawa Japan	Raspas Ecuador	Gourma Mali	Villa de Cura Venezuela	Pompangeo Suwalesi	Betic-Rif-Tell Spain-Morocco-Canada	Yukon	Great Caucasus Russia	Cyclades Greece	Crete
Abbreviation	SS	SJ	RE	GM	VC	PS	BRT	YC	GC	CG	C
Subduction context	Intra-oceanic	Intra-oceanic	Active margin	Active margin	Intra-oceanic	Intra-oceanic	Intra-oceanic	Active margin	Active margin	Intra-oceanic	Intra-oceanic
Tectonics of exhumed rocks	2 nappes	3 nappes	3 nappes	3 nappes	4 nappes	2 nappes	3 nappes	several nappes	3 nappes	4–6 nappes	
Unit	Samana Peninsula Dominican Republic	Escambray Cuba	Schistes Lustrés Corsica	Schistes Lustrés Western Alps	Franciscan C. California	Santa Catalina California	Coastal Cordillera Chile	Pam Peninsula New Caledonia	Motalafjella Spitsbergen	Shangla Pakistan	Groix France
Lithology of exhumed rocks	Sedimentary and volcanic melange	Sedimentary rocks with magmatic rocks	Ophiolitic rocks with sedimentary cover	Sedimentary rocks with volcanic rocks	Ophiolitic rocks	Sedimentary rocks with volcanic rocks	Sedimentary rocks	Ophiolitic rocks with sedimentary cover	Continental units	Sedimentary rocks	Sedimentary rocks
Max P-T conditions	0.9–1.0 GPa	2.0–2.1 GPa	1.9–2.0 GPa	1.6 GPa	0.8 GPa	1.2 GPa	1.8–2.0 GPa	1.5 GPa	1.6 GPa	2.0 GPa	1.6 GPa
	350–420°C	600°C	530–630°C	550–650°C	375°C	450°C	550–650°C	420–650°C	620–700°C	500°C	400°C
Metamorphic age	Cretaceous	Early Cretaceous	Cretaceous	Pan-African	Cretaceous	Cretaceous	Cretaceous-Paleocene	Carboniferous-Triassic	Devonian	Eocene	Miocene
Exhumation velocities	Unknow	<1 mm/year	2.5–5 mm/year	Unknown	Unknown	Unknown	2.8 mm/year	Unknown	~4 mm/year	2.6 mm/year	~5 mm/year
Exhumation timing	Intra-oceanic	Early stage of subduction	Active subduction	Unknown	Intra-oceanic	When the accre-tionary wedge collided	Intra-oceanic with slab retreat	When the accretionary wedge collided	When the accretionary wedge collided	Intra-oceanic with slab retreat	Intra-oceanic with slab retreat
References	Honegger et al., 1989 Mahéo et al., 2006	Wallis et al., 2004 Ko et al., 2005 Ota et al., 2004	Arculus et al., 1999 Gabriele et al., 2003	Caby et al., 2008	Avé Lallemant et al., 2005	Parkinson et al., 1998	Platt and Vissers, 1989 Augier et al., 2005	Erdmer et al., 1998 Philippot et al., 2001	Perchuk and Philippot, 1997	Jolivet et al., 2003 Forster and Lister, 2005	Jolivet et al., 2003

Plate 1 Field photographs of HP rocks in three subduction types. (**a**) Accretionary type subduction. Block of an hectometric blueschist (bs) corresponding to an oceanic olistolith embedded in a metasediemnatry matrix (sed) of Schistes Lustrés at Bric Bouchet, Queyras in western Alps. The contact between the blueschist and the metasediment is concordant and both recorded similar P-T conditions. (**b**) Serpentinite type. Metric blocks of blueschists (bs) of oceanic origin are embedded in a serpentinite matrix (ser). At the local and regional scale, each blueschist block recorded different peak metamorphic conditions in Northern Serpentinite mélange in Cuba. (**c**) Continental subduction context. Coesite-bearing eclogitic block (ecl) corresponding initially to a basaltic dyke emplaced in Permian sediment (sed) on the Indian continental margin. The intrusive contact between the dyke and sediments is preserved at the regional scale, and the Indian continental margin forms a coherent UHP unit of 100 *50 km at Tso Morari in western Himalaya

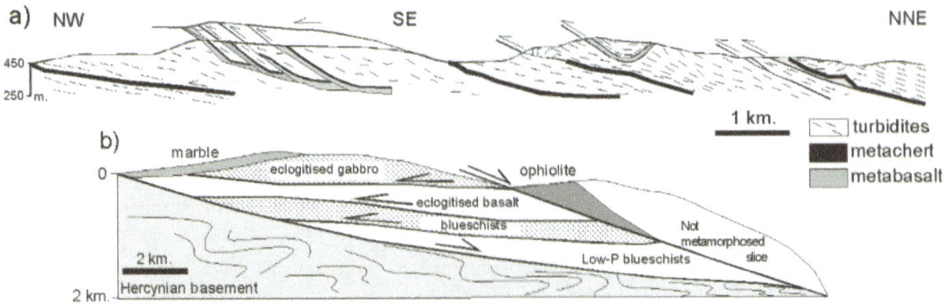

Fig. 4 (**a**) Nappes observed in the HP Franciscan complex. Note that the primary sedimentary contact between basalts and chert is preserved (modified after Kimura et al., 1996). (**b**) Schematic succession of HP nappes in the Cycladic blueschist belt in Greece (modified after Forster and Lister, 2005)

Shreve and Cloos, 1986). Ocean Drilling Project in the 80's documented that gabbros and dolerites are brecciated near ridges (Lagabrielle et al. 1981; Lagabrielle and Polino, 1985) (Plate 1a) and that breccias and olistoliths form by mixing of igneous rocks and sediments on the sea floor. Earlier, mineralogical studies suggested that lower metamorphic grade of the matrix than mafic lenses, but this interpretation has been questioned by recent studies showing that the matrix metasediments and lenses record similar P-T conditions (e.g., Kimura et al., 1996 for the Franciscan complex; Agard et al., 2002 for the Western Alps; Parra et al., 2002 for the Cyclades in Greece).

In term of geometry, several units are recognized in exhumed rocks with a thickness varying from the hectometre up to 5 km (e.g., two to four units in Kimura et al., 1996; Stanek et al., 2006). These units form nappes thrust towards the paleo-trench with lower metamorphic units overlain by higher metamorphic units (Fig. 4a). These nappes started to develop under HP-LT conditions, generally under blueschist conditions and ended under greenschist facies conditions, suggesting that the early exhumation is accommodated by thrusting (Fig. 4). Late extension that starts at the ductile-brittle transition commonly affects the nappes as documented in the Franciscan complex (Platt, 1986), the Samana complex in Dominican Republic (Goncalvez et al., 2000), the Cyclades in Greece (Jolivet et al., 2003) and the Piedmont complex in the western Alps (Tricart et al., 2004).

The maximum pressures recorded in exhumed rocks vary from 0.7 to 2.0 GPa and plot along geotherms ranging between 5 and 14°/km, which are similar to those of modern subduction zones (Fig. 5). Several eclogites show pressures equivalent to a depth of about 75 km. This is much deeper than the maximum depth, 20–40 km, observed in most active accretionary

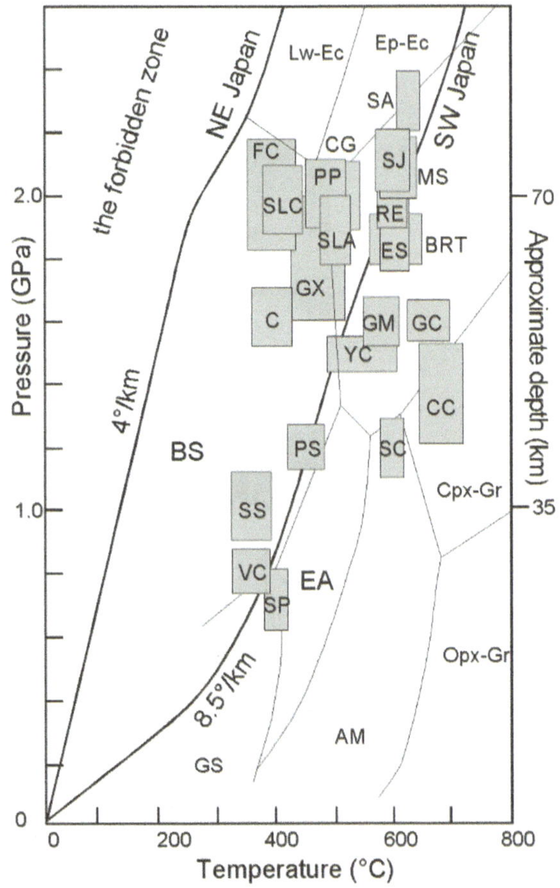

Fig. 5 Compilation of P-T data for accretionary wedge context (see Table 1 for abbreviations and references). Cold (4°/km, NE Japan) and hot (8.5°/km, SW Japan) geotherms from Peacock and Wang (1999). Lx-Ec: lawsonite eclogite; Ep-Ec: epidote eclogite; *Am-Ec* amphibole eclogite, *BS* blueschist, *Cpx-Gr* clinopyroxene granulite, *Opx-Gr* orthopyroxene granulite, *EA* epidote amphibolite, *AM* amphibolite, *GS* greenschist

wedges. Another common feature of HP-LT rocks is slow exhumation rates ranging between 1 and 5 mm/year (Table 1), which are independent of subduction velocities (e.g., Agard et al., submitted).

1.3 The Serpentinite-Subduction Channel

Recent petrological and geophysical evidence documented the presence of serpentinites in oceanic floor (e.g., Mével, 2003) and along active subduction zones (e.g., Furukawa, 1993; Maekawa et al., 1993; Bostock et al., 2002; Seno and Yamasaki, 2003). Serpentine minerals sensus lato display several distinctive characters: they contain up to 13 wt% of water and can be a major host of fluid-mobile elements in deep subduction zones (e.g., Schmidt and Poli, 1998; Hattori and Guillot, 2007) as they are stable under wide temperatures and pressures down to a depth of 150–170 km (Ulmer and Trommsdorff, 1995). They have a low density, about 2600 kg/m^3, a low viscosity of about 4.10^{19} Pas, a high poisson ratio (0.29) and low shear modulus (e.g., Moore and Lockner, 2007; Reynard et al., 2007; Hilairet et al., 2007). These physical properties allow serpentinites to be highly ductile to lubricate subduction planes (Guillot et al., 2001).

In paleo-subduction zones, serpentinites are commonly associated with HP-LT rocks and have been considered as fragments of oceanic lithosphere, and the contacts between eclogitic lenses and the matrix serpentinites are interpreted to be primary (e.g., Coleman, 1971). For instance, the high-pressure Monviso massif in the Western Alps has been considered as a continuous sequence of the Tethyan oceanic lithosphere (ophiolite). However, recent studies show that this massif represents a deep tectonic melange as individual eclogitic blocks record different P-T conditions (Fig. 6a) (Blake et al., 1995; Schwartz et al., 2000, 2001). Sixteen Phanerozic massifs are defined as serpentinite-type subduction complexes (Table 2). The Zermatt-Saas unit is included in this type because mafic bodies are intimately associated with serpentinites, although it is not a tectonic melange and has been interpreted as a complete ophiolite sequence (e.g., Li et al., 2004).

HP-LT units exhumed in serpentinite-type subduction zones are dominated by highly sheared serpentinites that contain blocks of metabasites (Plate 1b). The blocks are weakly deformed and range in size from metric to decametric. Some blocks are kilometric, as those in the Monviso massif, or centimetric, as those in the Voltri massif. The metasediments (metacherts, metagreywackes, metapelites, marbles) are highly deformed and minor in volume, less than 10% of the massifs (Fig. 6b).

The initial geometry is difficult to reconstruct because original contacts are no longer recognized in exhumed rocks. Nevertheless, it has been evaluated in two well studied locations; the Monviso and Voltri massifs in the Alps (Fig. 6). The Monviso massif is composed of six west-dipping tectono-metamorphic units of metabasalts and metagabbroic rocks, each of which is separated by west dipping normal shear zones containing serpentinites (Lombardo et al., 1978; Schwartz et al., 2000; Guillot et al., 2004; Fig. 6a). The basal unit is serpentinites with 400 m in thickness. The serpentinites that originated from lherzolite and minor harzburgite and dunite, are cut by sheared dykes of rodingitized gabbro and basalt. The serpentinite layer commonly contains metric to hectometric lenses of foliated eclogitic gabbro, ferrogabbro and metamorphosed plagiogranite. Considering the geometry, this basal serpentinite unit likely had an initial size of about 50 km × 10 km (Schwartz et al., 2001; Guillot et al., 2004). Five other units are composed of discontinuous layers of intensely deformed and recrystallized metagabbros. These metagabbros contain minor ultramafic cumulates and hydrated mantle peridotites (Messiga et al., 1999). Locally, greenschists and banded glaucophane-epidote metabasalts retain the pillow lava texture. The upper part of the massif exposes thin layers of carbonate-bearing micaschists (Schistes Lustrés) interbedded with the metabasites. The thickest section (~1.2 km) is composed of basalt breccia, pillow lavas, metagabbro and slices of serpentinites in upward direction. The serpentinites were metamorphosed under blueschist facies conditions. The Monviso massif is thus similar to a dismembered ophiolitic massif, yet each unit records different P-T conditions.

The Voltri massif in the western Alps is more akin to a mélange zone observed in British Columbia (Tsujimori et al., 2006) Cuba (Garcia-Casco et al., 2002), Dominican Republic (Krebs et al., 2008), and Turkey (Altherr et al., 2004). It is surrounded by highly sheared serpentinites and consists of chaotic mixture of meter-sized blocks of metagabbros, metabasites, métasediments and also serpentinites in the matrix of schistose chlorite-actinolite (Fig. 6b) (Vignaroli et al., 2005; Frederico et al., 2007). The serpentinte matrix both in the Monviso and Voltri massifs record HP conditions (Auzende et al., 2006).

Geochemical and petrological data suggest that mafic blocks in 18 serpentinite-type subduction complexes were derived from the subducted oceanic plate

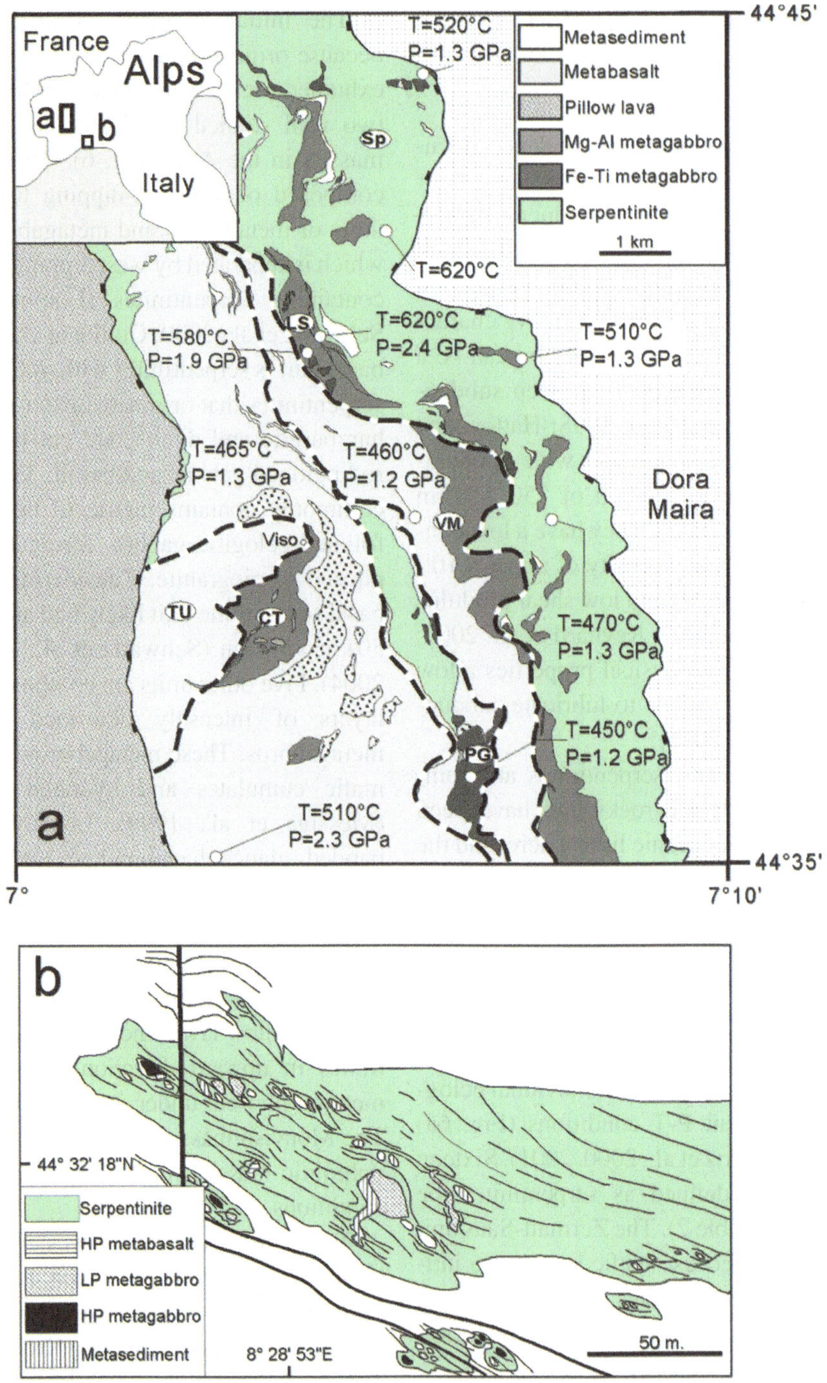

Fig. 6 (**a**) Lithological map of the central part of the Monviso in Western Alps (after Schwartz et al., 2001). At the regional scale, a network of normal shear zone underline by sheared serpentinites or sheared metasediments separated several metabasite blocks recording contrasted P-T conditions. The basal serpentinites unit (on the right side) forms a thick (400 m) serpentinite melange containing metric blocks of metabasalts, metagabbros of metasediments. *CT*: Costa Ticino; *PG*: Passo Galarino; *LS*: Lago Superiore; *TU*: Tour Real; *VM*: Viso Mozzo; *SP*: Basal Serpentinite; *TU*: (**b**) Geological and structural map of the Erro Tobio serpentine mélange (Southwestern Alps). Note the diversity of lithologies included in the serpentinite matrix (modified after Frederico et al., 2007)

Table 2 Compilation of available data on exhumed high pressure rocks in serpentinite subduction channel

Unit	South Motagua Guatemala	Central Pontides Turkey	Port Macquarie Australia	Pinchi Lake Canada	Barru Complex Suwalesi	Bantimala Complex Suwalesi	Luk Ulo Java	Zagros Iran	Sistan Iran
Abbreviation	SM	CP	PM	PL	BC	BAC	LU	ZA	SI
Subduction context	Intra-oceanic	Intra-oceanic	Intra-oceanic	Unknown	Active margin	Active margin	Active margin	Active margin	Active margin
Tectonics	Tectonic block within serpentinite melange	Tectonic block within serpentinite melange	Tectonic block within serpentinite melange	Tectonic block within serpentinite melange	Tectonic block within serpentinite melange	Tectonic block within serpentinite melange	Tectonic block within serpentinite melange	Tectonic block within serpentinite melange	Tectonic block within serpentinite melange
Lithology of exhumed rocks	Serpentinites, mafic rocks Metsediments	Serpentinites, mafic rocks	Serpentinites, mafic rocks	Serpentinites, mafic rocks	Serpentinite Metasediments	Serpentinites, mafic rocks Metsediments	Serpentinites, mafic rocks Metsediments	Serpentinites, mafic rocks Volcanoclastics	Serpentinites, mafic rocks Volcanoclastics
Protoliths of mafic rocks	MORB	MORB	MORB	MORB	MORB	MORB	MORB	MORB	MORB
Protoliths of serpentinites	Unknown	Probably mantle wedge	Mantle wedge	Unknown	Unknown	Unknown	Oceanic	Oceanic	Oceanic
Max P-T conditions	2.5 GPa 470°C	P > 1.4 GPa 400–430°C	2.0–2.4 GPa 420–570°C	2.2 GPa 450°C	2.1 GPa 520°C	2.4–2.7 GPa 580–620°C	2.2 GPa 365°C	1.8 GPa 500°C	1.9–2.2 GPa 600°C
Metamorphic age	Cretaceous	Cretaceous	Cretaceous	Triassic	Cretaceous	Cretaceous	Cretaceous	Cretaceous	Cretaceous
Exhumation velocities	~4 mm/yr	Unknown	Unknown	Unknown	Unknown	Unknown	Unknown	2–3 mm/year	Unknown
Exhumation timing	When the subduction melange collided with continent	Unknown	Intra-oceanic associated with change in stress field	Unknown	When the subduction melange collided with continent	When the subduction melange collided with continent	When the subduction melange collided with continent	Intra-oceanic change in subduction with rate	Intra-oceanic
References	Harlow et al., 2004 Tsujimori et al., 2006	Altherr et al., 2004	Aitchinson et al., 1994 Och et al., 2003	Ghent et al., 1993 Tsujimori et al., 2006	Parkinson et al., 1998	Parkinson et al., 1998	Kadarusman et al., 2007	Agard et al., 2006	Fotoohi Rad et al., 2005

(*continued*)

Table 2 (continued)

Rio San Juan Dominican Republic	Zaza Cuba	Monviso Italy	Zermatt Switzerland	Voltri Italy	Mariana	Omachi Izu Bonin	Borus-Chagan-Uzun Gorny Altai	Chara and Maksyutov Kazkhstan-Urals	Higashi-Akaishi Sanbagawa Japan
RSJ	ZZ	MV	ZS	VO	MA	OM	BCU	CM	SAJ
Intra-oceanic	Intra-oceanic	Intra-oceanic	Intra-oceanic	Intra-oceanic	Intra-oceanic	Intra-oceanic	Intra-oceanic	Intra-oceanic	Continental margin
Tectonic block within serpentinite melange	Tectonic block within serpentinite melange	Tectonic block within serpentinite melange	Coherent ophiolitic unit	Tectonic block within serpentinite melange	tectonic block within mud volcanoes	Tectonic block within serpentinite melange	Tectonic block within serpentinite melange	Tectonic block within serpentinite melange	Tectonic block within serpentinite melange
Peridotites serpentinites amphibolitic rocks	Serpentinites, mafic rocks	Serpentinites, mafic rocks metsediments	Serpentinites, mafic rocks metsediments	Serpentinites, mafic rocks metsediments	Serpentinites, mafic rocks	Serpentinites, mafic rocks	Serpentinites, mafic rocks metsediments	Serpentinites, mafic rocks metsediments	Peridotites serpentinites amphibolitic rocks
MORB	MORB	MORB	MORB	MORB	MORB	Andesite	MORB OIB	MORB OIB	MORB
Mantle wdege	Mantle wedge and oceanic	Oceanic	Oceanic	Oceanic	Mantle wedge	Mantle wedge	Oceanic & mantle wedge	Oceanic & mantle wedge	Mantle wedge
4.0 GPa	2.0 GPa	2.6 GPa	2.8 GPa	2.2 GPa	0.7 GPa	2.0 GPa	2.0 GPa	eclogites	3.8 GPa
1550°C	600°C	630°C	600°C	550°C	150–250°C	~650°C	660°C		810°C
Cretaceous	Cretaceous	Eocene	Eocene	Eocene	Present-day	Oligocene	Cambrian	Silurian	Early Cretaceous
~6 mm/year	Unknown	~10 mm/year	~10 mm/year	3–4 mm/year	Unknown	Unknown	Unknown	Unknown	Unknown
When the subduction melange collided with continent	When the subduction melange collided with continent	When the subduction melange collided with continent	When the subduction melange collided with continent	When the subduction melange collided with continent	Intra-oceanic	Intra-oceanic with back-arc extension	When the subduction melange collided with continent	When the subduction melange collided with continent	Syn-subduction
Abbott et al., 2006	Garcia-Casco et al., 2002	Blake et al., 1995	Reinecke, 1991	Hermann et al., 2000	Maekawa et al., 1993	Ueda et al., 2004	Dobretsov and Buslov, 2004	Dobretsov and Buslov, 2004	Enami et al., 2004
	Hattori and Guillot, 2007	Schwartz et al., 2000	Li et al., 2004	Frederico et al., 2007	Fryer et al., 1999				

(Table 2). The possible exception is the eclogitic blocks of andesite origin dredged near the Omachi forearc serpentinite diapir (Izu-Bonin arc) (Ueda et al., 2004).

Serpentinites in oceanic subduction zones mostly originated from abyssal peridotites and their hydration likely took place during the ridge hydrothermal activity, such as those in Java, Iran, and the Alps. Some serpentinites were derived from hydrated mantle wedges (Turkey, Australia, Mariana, Izu-Bonin) or both (Northern serpentinite mélange in Cuba and Dominican Republic) (Table 2).

Regarding the metamorphic conditions, most eclogitic blocks reached HP between 1.8 and 2.5 GPa and relatively low temperatures, which defines paleogeothermal gradients lower than 10°C/km (Fig. 7). Two localities provide evidence for deeper P-T conditions, at 3.2 and 4 GPa, respectively (SAJ, RSJ; Table 2; Fig. 7), as deduced from garnet peridotite blocks embedded in the serpentinite melange.

In the Western Alps and in the northern serpentinite mélange in the Dominican Republic, the maximum pressure of each block varies from 1.0 to 2.3 GPa (Schwartz et al., 2000; Frederico et al., 2007; Krebs et al., 2008), suggesting that their juxtaposition unlikely took place during exhumation. The metamorphic ages of different blocks show ranges in age; ±4 Ma in the Voltri massif in the western Alps (Frederico et al., 2007), ±15 Ma in the Monviso massif (e.g., Guillot et al., 2004) and 40 Ma in the Rio San Juan complex in Dominican Republic (Krebs et al., 2008). These variations likely reflect different depths and different times of metamorphism for blueschists or eclogitic blocks within the subduction channel. Finally exhumation velocities vary between 3 and 10 mm/year, which are faster than those recorded in accretionary wedge environment (Table 2). Again the exhumation velocity remains independent of the subduction velocity (e.g., Agard et al., submitted for publication).

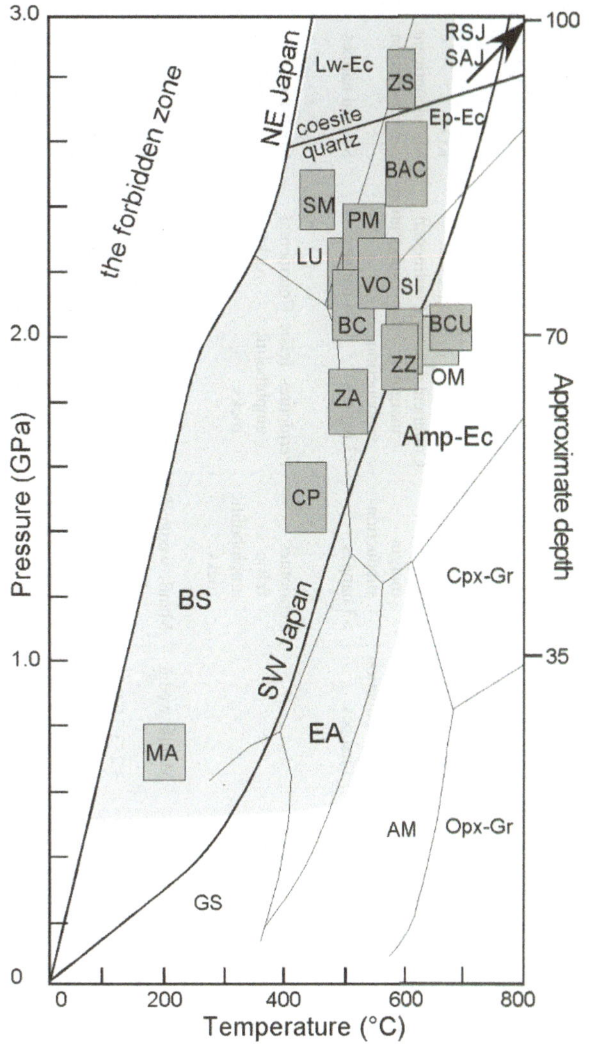

Fig. 7 Compilation of P-T data for serpentinite subduction channel context (see Table 2 and Fig. 5 for abbreviations and references). Grey area: stability field of antigorite after Ulmer and Trommsdorff (1995)

1.4 Continental-type Subduction

The discovery of coesite and microdiamond in subducted crustal rocks (Chopin, 1984; Smith, 1984; Sobolev and Shatsky, 1990) demonstrated that continental rocks can be subducted to depths of at least 100 km. Such UHP rocks have now been documented in most Phanerozoic mountain belts around the world (Fig. 1) but the mechanism by which these rocks were exhumed are still debated. This problem is not trivial because of the large sizes of some HP-UHP terranes (>50,000 km^2 in China and Norway), the large vertical displacement during their exhumation and the preservation of index minerals or assemblage (Grasemann et al., 1998; Hacker, 2007).

The protoliths of UHP rocks are predominantly upper continental crust (Table 3), such as granite gneisses, and metasedimentary rocks (quartzites, metapelites, and marbles). Mafic plutonic rocks are present in subduction zones, but they correspond to

Table 3 Compilation of available data on exhumed UHP rocks in continental and oceanic subduction complexes

Massif	Rhodope Greece	Ulten zone peridotites Italie, Alps	Dora Maira Italie	Alpes Arami Switzerland	d'Entrecasteaux Papua New Guinea	Lanterman Range Antarctica	North Quinlin-Dabie-Sulu China	South Quilin-Dabie China	Tianshan China	Kokchetav
Abbreviation	Rh	UZ	DM	AA	EN	LR	NQDS	SQD	TIA	KO
Subduction context	Microcontinent subduction	Continental subduction	Continental margin subduction	Continental subdcution	Continental margin subduction	Continental margin subduction	Continental margin subduction	Continental margin subduction	Continental subdcution	Microcontinent subduction
Tectonics	4 nappes	3 nappes	3 nappes	Several nappes	Metam. core complex	3 nappes	>3 nappes	<3 nappes	4 nappes	Several nappes
Lithology	Continental rocks with oceanic rocks	Garnet peridotites within gneiss and migmatites	Peridotites, felsic amphibolitic rocks	Peridotites, felsic amphibolitic rocks	Mafic lenses within gneisses		Peridotites, felsic amphibolitic rocks	Peridotites, felsic amphibolitic rocks	Continental rocks	Continental rocks
Protoliths of peridotites		Mantle wedge		Continental lithosphere		Mantle wedge	Mantle wedge & abyssal	Mantle wedge & abyssal		
P-T conditions	7 GPa 1100°C	2.7 GPa 850°C	3.4 GPa 675–775°C	3.2 GPa 840°C	2.0–2.6 GPa 870–930°C	3.2–3.3 GPa 764–820°C	4 GPa 700–800°C	>5.5 GPa 740–870°C	5.0 Gpa 560–600°C	6 Gpa 825–975°C
Metamorphic age	Cretaceous to Eocene	Early Paleozoic	Oligocene	Eocene	Miocene-Pliocene	Cambrian	Cambro-Ordovician	Triassic	Permian	Cambrian
Exhumation velocities	>8 mm/year	Unknown	>20 mm/year	Unknown	25 mm/year	>4 mm/year	6–8 mm/year	6 mm/year	Unknown	>18 mm/year
Exhumation timing	Microcontinent subduction	Collision	Syn-collision	Syn-collision	Back-arc spreading	Syn-subduction	Syn-collision	Syn-collision	Syn-collision	Syn-collision
Références	Liati, 2005 Perraki et al., 2006	Nimis and Morten, 2000	Rubatto and Hermann, 2001 Compagnoni and Rolfo, 2003	Nimis and Trommsdorf, 2001	Baldwin et al., 2004 Monteleone et al., 2007	Palmeri et al., 2007	Yang et al., 2003 Hacker, 2006	Yang et al., 2003 Hacker, 2006 Liu et al., 2006	Zhang et al., 2003	Hacker et al., 2003

Massif	Kaghan Pakistan	Tso Morari India	Sao Francisco Brazil	Gourma Mali	North Western gneiss region Norway	East Greenland	Bohemian massif Poland, Czech Republic	French Massif Central France	W. Gneiss Region Norway	Central Suwalesi
Abbreviation	KA	TSO	SAO	GOU	NWGR	EG	BO	FMC	WGR	CSU
Subduction context	Continental margin subduction	Continental margin subduction	Continental subdcution	Continental subdcution	Continental margin subduction	Continental margin subduction	Continental margin subduction	Continental margin subduction	Continental margin subduction	Oceanic subduction
Tectonics	3 nappes	3 nappes	2 nappes	3 nappes	2 nappes	1 nappe	5 nappes	3 nappes	2 nappes	>2 nappes
Lithology	Continental rocks	Upper continental rocks	Peridotites, felsic amphibolitic rocks	Continental rocks	Peridotites within gneisses	Continental rocks	Peridotites, felsic amphibolitic rocks	Peridotites, felsic amphibolitic rocks	Lower and upper continental rocks peridotites	Peridotites, serpentinites amphibolitic rocks
Protoliths of peridotites			Unknown		Mantle wedge		Mantle wedge & oceanic	Continental lithosphere		Oceanic
P-T conditions	3.0 GPa 720–820°C	3.9 Gpa 750–850°C	>2.8 GPA 750°C	>2.8 GPA 750°C	6.4–8 GPa 1200°C	3.6–4.2 GPa 928–972°C	up to 7.5 GPa up to 1335°C	3.0 GPa >750°C	3.5 GPa 700–800°C	>2.8 GPa 1100°C
Metamorphic age	Eocene	Paleocene	Pan-African	Pan-African	Silurian	Ordovician	Devonian	Devonian	Silurian	Cretaceous
Exhumation velocities	30–80 mm/year	17 mm/year	unknown	> > 4 mm/year	Unknown	Unknown	>6 mm/year	15 mm/year	8–12 mm/year	Unknow
Exhumation timing	Syn-collision	Syn-collision	Syn-collision	Syn-collision	Asthenospheric upwelling then syn-collision	Unknown	Syn-collision	Syn-collision	Syn-collision	Syn-subduction
Références	O'Brien et al., 2001	de Sigoyer et al., 2000	Vaugh and Parkinson, 2003	Caby, 1994	Van Roermund et al., 2001	Gilotti and Krogh Ravna, 2002	Massonne, 2003	Gardien et al., 1990	Terry et al., 2000	Parkinson et al., 1998
	Treloar et al., 2003	Guillot et al., 2003		Jahn et al., 2001	Drury et al., 2001		Medaris et al., 2004	Lardeaux et al., 2001	Labrousse et al., 2002	
	Parrish et al., 2006	Leech et al., 2005			Young et al., 2007				Root et al., 2005	

intrusions in shallow continental crust prior to the subduction as shown in Plate 3. The Caledonian UHP eclogites of Norway are considered to have originated from the lower crustal granulites of Precambrian age. However, the eclogitized granulites are associated with Precambrian gabbros, anorthosites and peridotites (Tucker et al., 1991). The occurrence of gabbro and peridotite with contemporanous gneiss suggests that these rocks were probably present in the continent-ocean transition where the lower crust is thin or totally absent. If this hypothesis is confirmed, it reinforces the idea that only the upper crust is exhumed.

Garnet peridotites have been described in UHP terranes in Phanerozoic continent-continent collision zones, including Dabie-Sulu terrane in China, Kokchetav massif in Kazakhstan, Western Gneiss Region in Norway, Alpe Arami in Switzerland and in the Palaeozoic belt of Europe (e.g., Medaris, 1999). They are classified into two types (e.g., Brueckner and Medaris, 2000; Zhang et al., 2000): (a) garnet peridotites originated from mantle wedges and tectonically incorporated within the subducting slab at great depth before its exhumation; (b) plagioclase-bearing cumulate ultramafic rocks emplaced at the base of the continental crust prior to the subduction and metamorphosed to garnet peridotites. In both cases, garnet peridotites are associated with continental rocks and their exhumation is explained by decoupling of the continental slice from the descending oceanic lithosphere due to the positive buoyancy of sialic continental rocks within the subduction channel (Van der Beuckel, 1992; Ernst, 1999, 2005).

The thickness of UHP domains varies widely. In the Western Alps, the Dora Maira UHP unit is 200 m in thickness and covers a surface area of about 25 km^2. In contrast, the Lower Paleozoic metamorphic domain in China forms an essentially continuous HP-UHP belt extending more than 4000 km from Quinlin to Dabie (Yang et al., 2003) with a thickness of 5 to 10 km (Hacker et al., 2000). Similarly, recent geochronological data confirm that the Western Gneiss Region in Norway forms a continuous HP-UHP unit of 200 × 400 × 5–10 km (Hacker, 2007; Young et al., 2007).

Primary magmatic texture in metamorphosed granite and volcanic rocks indicate locally low strain (Michard et al., 1993; de Sigoyer et al., 2004). High strain zones are found in most UHP units, such as the Dora Maira massif in the Alps (e.g., Michard et al., 1993), the Sulu and Dabie Shan massifs in China (e.g., Hacker et al., 2000; Zhao et al., 2005) and the Tso Morari and Kaghan units in Himalaya (e.g., Guillot et al., 2007). The evidence also indicates local high strain zone between these exhumed units from the dowgoing slabs. Based on the observation of the Himalayan UHP Tso Morari massif, Guillot et al. (2000) suggested, that the decoupling is controlled by the normal faults inherited from earlier rifts, making an upper crustal block easily dislodged from the rest of the subducting slab. This interpretation is analogous to that suggested by Jolivet et al. (2005) that the brittle to ductile transition of deformation plays an important role in decoupling the upper crust from the rest of the subducting lithosphere.

The progressive metamorphism of the Precambrian granulites in the Caledonian nappes of the Bergen Arc in Norway demonstrates the role of fluid circulation for eclogitization (Austrheim, 1994; Jolivet et al., 2005). Perhaps because of the presence of fluids, index minerals are poorly preserved in many UHP rocks of continental origin. This makes it difficult to define the boundary of an UHP domain. In fact, UHP rocks commonly record pressures ranging from about 2.5 GPa up to $P > 7$ GPa and temperatures varying between 500°C and 1335°C (Fig. 8). However, most UHP rocks of continental origin record pressures between 2.5 and 4.0 GPa (e.g., Hacker, 2006), and the UHP conditions are mostly recorded in garnet-bearing peridotites. The evidence suggests that mantle peridotites record an earlier event before their incorporation into the subduction channel due to an asthenospheric return flow (Fig. 8) (e.g., Spengler et al., 2006; Gorczyk et al., 2007).

UHP minerals are rarely preserved and mostly occur as relict in other minerals, such as coesite in garnet or omphacite or diamond in zircon. Nevertheless, their occurrence requires specific conditions, such as rapid cooling during decompression, rapid exhumation, fluid-absent condition during the exhumation and little deformation. These conditions are only locally attained so that the evidence for UHP conditions is only retained in lenses. UHP metamorphism records low geotherms ranging from 5 to 10°/km in most terranes, and down to 3.5°/km in the Forbidden Zone in China (Liou et al., 2000).

The P-T-t paths of UHP rocks are characterized by isothermal decompression until crustal depths (1.0 to 0.5 GPa). The absence of significant heat loss during the exhumation indicates their rapid exhumation, greater

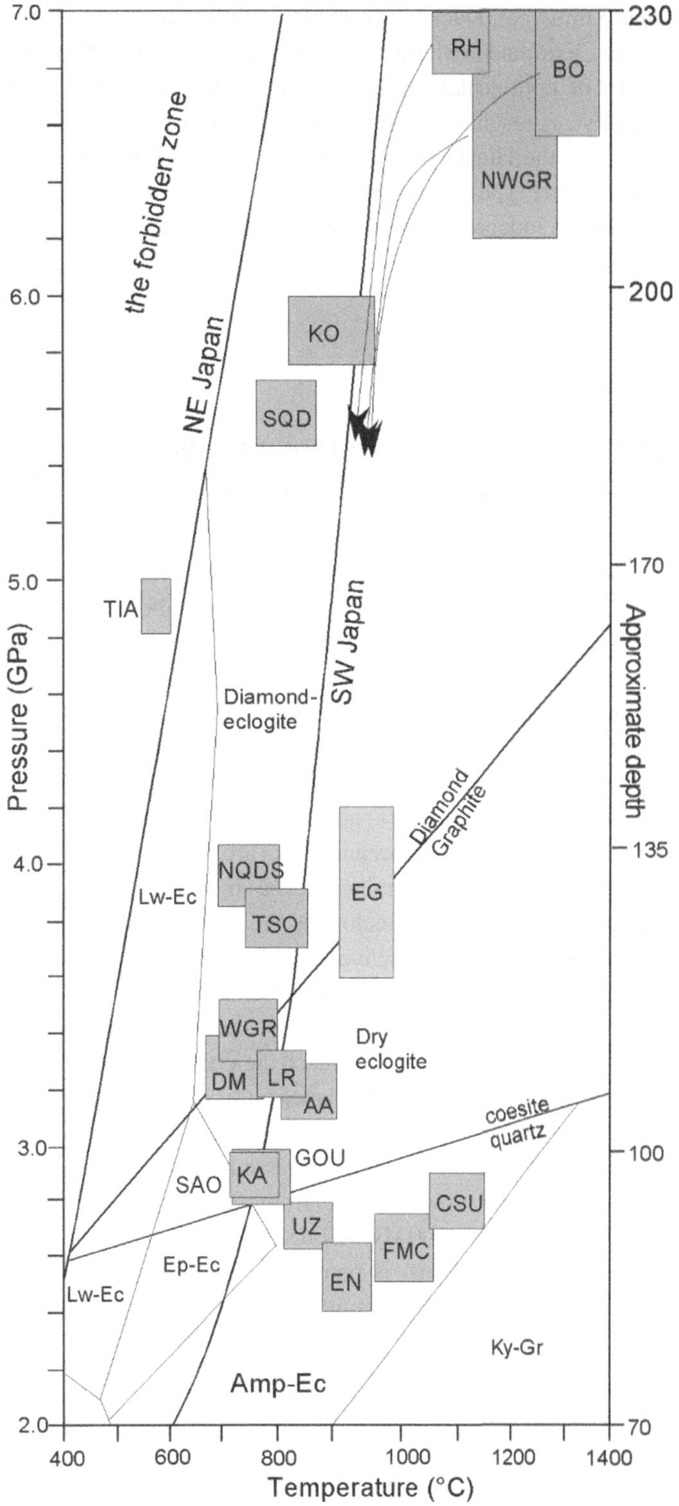

Fig. 8 Compiled P-T data for UHP rocks in continental and oceanic subduction complexes (see Table 3 for abbreviations of locations and references, and Fig. 5 for mineral abbreviations). The *arrows* show the asthenospheric upwelling of garnet peridotites before their integration in the subducting channels

than 3 mm/year (Duchêne et al., 1997; Grasemann et al., 1998). Estimated exhumation velocities in other UHP rocks of continental origin are also high, faster than 6 mm/year, reaching possibly up to 80 mm/year in the Alps and the Himalaya (Parrish et al., 2006) (Table 3). As for other types of subduction zones, the exhumation velocity is independent of the subduction velocity.

2 Discussion

2.1 *Subduction Environments and the Timing of Exhumation*

The common feature of the exhumation in the accretionary wedge environment and in the serpentinite-subduction channel environment is that both involve the subduction of oceanic lithosphere. The development of an accretionary wedge additionnally requires the offscraping of sediments derived from the lower plate or erosion of the upper plate. In the Western Alps, large proportions of these sediments (up to 50%) are exhumed, whereas only small fractions (<1%) of oceanic rocks are (Guillot et al., 2004; Agard et al., submitted). The metamorphosed oceanic rocks, blueschists-eclogites, are slowly exhumed (~ few mm/year) during active oceanic subduction. The peak pressures of those exhumed rocks are generally lower than 2.2 GPa, whereas peak pressures in serpentinite-subduction channel may reach the coesite stability field (2.8 Gpa, ZS; Fig. 7). The latter may contain garnet peridotites that were equilibrated at even higher pressures (~4 Gpa, RSJ; Fig. 7). Exhumation of sedimentary rocks lasts for a long time ranging from 25 Myr (Alpine Schistes Lustrés) to 100 Myr (Chile), whereas the exhumation of oceanic crust is commonly brief, less than 15–20 Myr (Agard et al., submitted). These authors have shown that the exhumation of oceanic lithosphere may occur shortly after the inception of subduction (Chile, Franciscan, Makran), in the midst of convergence (SE Zagros, NW Himalaya), or during the late stages of subduction (Western Alps, New Caledonia). Exhumed oceanic rocks are commonly associated with serpentinites. Exhumation velocities are also low, ranging between 1 and 5 mm/year. Exceptionally fast exhumation (~10 mm/year) in the western Alps is associated with later continental subduction (Agard et al., 2002; Guillot et al., 2004). The exhumation rates are independent of the subduction rates, confirming a decoupling between the subducting plate and the zone of exhumation.

Accretionary wedge and serpentinite subduction channel environments show two other major differences: the lithology and types of HP rocks. In accretionary wedge environment, HP rocks mostly originate from metasediments and form kilometric slices with continuous P-T conditions, frequently with higher pressure slices thrust over the lower pressure ones. On the other hand, serpentinite subduction channel is dominated by metabasites embedded in a sheared serpentinite matrix. These metabasite blocks record different P-T conditions, common in a tectonic mélange.

Note that an accretionary wedge and a serpentinite subduction channel may coexist in a single subduction zone at a given time, and exhumed rocks in these two settings may occur in close proximity in a subduction complex as shown in the Western Alps and the Franciscan (Fig. 9).

The continental-type subduction is accompanied by the exhumation of UHP rocks that were buried down to a depth between 100 and 200 km along cold geotherms. UHP rocks are exhumed rapidly (>6 mm/year) under isothermal conditions at the transition from oceanic subduction to continental collision. HP-UHP domains from 1 km to maximum 10 km thick nappe stacks over large surface areas (>50,000 km^2). Continuous UHP rocks of greater than 50 km in length are exposed in the Tso Morari area in the Himalaya, the Western Gneiss Region in Norway and Quinlin-Dabie in China. Due to its thickness and is positive buoyancy, the entire continental lithosphere cannot enter the subduction zone and stops within a couple of millions years after the initial contact of the continent with the trench. In the Himalaya, the thick buoyant upper Indian crust was blocked after 10 million years of continental subduction. It separated from the rest of the lithosphere and started to be stacked as nappes, which resulted in the high topographic relief in the area (Guillot et al., 2003). Similarly in the Caledonides, Hacker (2007) estimates that the UHP slab exhumed from mantle to crustal depth between 400 and 390 Ma.

As already discussed, accretionary wedge and serpentinite subduction channel can coexist in one subduction zone as observed in the Franciscan complex, the northern subduction complex in Dominican Republic, and the Western Alps. In other cases, serpentinite subduction channel may coexist with continental subduction, such as in Indonesia. The Alpine massif is

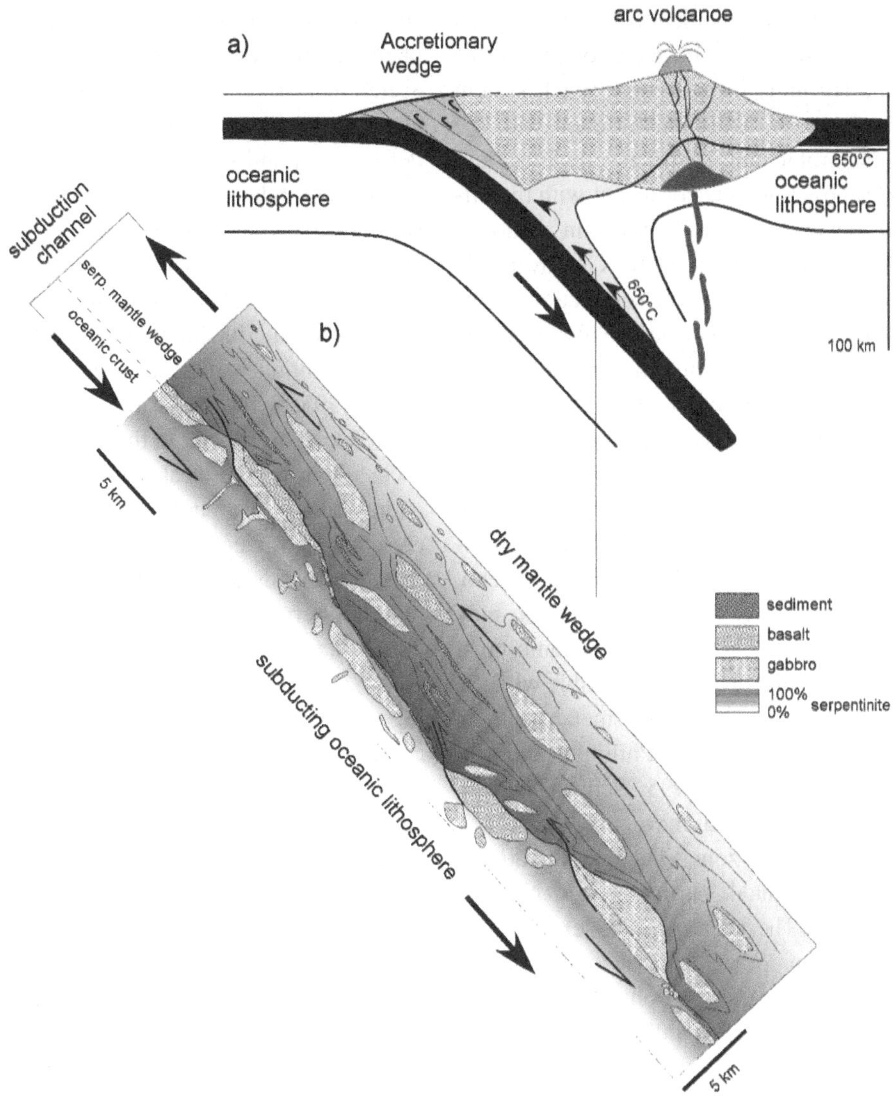

Fig. 9 (**a**) Schematic relationship between accretionary wedge and serpentinite subduction channel. The boundary between the serpentinite subduction channel is defined by the 650°C isotherm. (**b**) Detail of the serpentinite subduction channel. It forms a ~60 km long (from 40 to 100 km depth) soft channel between the dry (rigid) subducted oceanic lithosphere and the dry (rigid) mantle wedge. It is made of a melange of serpentinites deriving the hydrated oceanic lithosphere and from the hydration of the mantle wedge and contains exotic blocks of metabasalts, metasediments and metaggabros, mainly derived from the subducting oceanic lithosphere but also from the above arc system (e.g., Hattori and Guillot, 2007). Due to the low viscosity and low density of serpentinite mineral and the triangle shape of the serpentinite channel, the dowgoing material is progressively entrained upward (Guillot et al., 2001; Schwartz et al., 2001; Gerya et al., 2002)

probably the best exemple to depict the relationships between the three subduction types. Along a west-east traverse an accretionary wedge (the Schistes Lustés unit), a serpentinite channel (the Monviso unit) and a continental subduction unit (the Dora Maira unit) coexist. Several studies suggested that they correspond to the continuous evolution of a paleo-subduction zone from the Mid-Cretaceous to the Eocene (Agard et al., 2002; Schwartz et al., 2007; Yamato et al., 2007).

2.2 Essential Role of a Decoupling Zone

The opposite trajectories of exhumation and subduction require a decoupling zone within the subducting slabs. In fact, most of the subducted lithosphere is going down while only slices of their upper decoupled part make their way back to the surface, at certain time periods at the most (Agard et al., submitted). In oceanic subduction zones, the existence of a décollement layer is suggested at the ductile-britle transition. In accretionary wedge environment, the HP units form a nappe stack system thrusting in the direction of the paleo-trench with the lower grade metamorphic units at structurally lower levels (Fig. 4b). This observation is consistent with the off-scraping of the upper sedimentary layer in analogue models (Von Huene and Scholl, 1991; Glodny et al., 2005). In the case of serpentinite type subduction zones, several possible mechanisms contribute to the formation of the decoupling zone. Fluid migration in deep fractures is suggested to contribute to the formation of decoupling zones based on the occurrences of veins filled with eclogitic minerals (Philippot and Kienast, 1989). This field observation is compatible with geophysical data, suggesting dehydration-induced embrittlement of the subducting slab (Yamasaki and Seno, 2003). A serpentinized layer prior to subduction may become a decoupling zone between the oceanic crust and underlying lithospheric. This is compatible with the occurrence of boudinaged eclogitized metabasalts in serpentinites (e.g., Coleman, 1971; Philippot and Van Roermund, 1992; Blake et al., 1995; Schwartz et al., 2001; Vignaroli et al., 2005; Tsujimori et al., 2006).

In continental subduction zones, flat eclogitic ductile shear zones are documented, particularly in Norway (Jolivet et al., 2005). The shearing may exceeds the strength of the binding force of rocks and results in separating the buoyant upper crust from the denser lithosphere along a decollement. It is probable that quartzo-feldspathic upper crustal rocks at depths greater than 100km are hot enough to be separated from the rest of sialic lithosphere (Stöckert and Renner, 1998). The proposed interpretation is significantly different from the model suggesting that the whole crust is decoupled from the subducting slab and exhumed by buoyancy forces (e.g., Chemenda et al., 1995). In contrast, our proposed model follows the concepts of Cloos (1982) and Platt (1986) suggesting the presence of a decoupling zone within the upper part of the subducting slab.

2.3 A Weak Subduction Channel Required for the Exhumation of HP to UHP Rocks

Fluids released from subducting slabs during progressive metamorphism facilitate the lubrication of the subduction plane but also assist the formation of low strength metamorphic minerals, such as lawsonite and phengite (Stöckert and Renner, 1998). In the case of accretionary-type subduction zones, these weak minerals can form high-strain shear zones, which localize the deformation and separates an exhumed block from the rest of the subducting slab. Moreover, lenses and blocks surrounded by shear zones remain relatively free from deformation and preserve HP mineral assemblages during exhumation. This concept is particularly pertinent to rigid metabasite blocks in a soft matrix of serpentinites in a serpentinite subduction channel (Fig. 10) (Blake et al., 1995; Schwartz et al., 2001; Gerya et al., 2002; Frederico et al., 2007). In the exhumation of continental rocks, the role of a weak zone has not been adequately addressed because buoyant continental rocks is considered to be sufficient for their exhumation of UHP rocks (e.g., Chemenda et al., 1995; Ernst, 2006), but the occurrence of serpentinites along the interface between the Tso Morari UHP unit and the overlying rigid mantle wedge (e.g., Guillot et al., 2001) suggest that a lubricating weak zone may be important in the exhumation of continental rocks.

The occurrence of garnet-bearing peridotites further supports the presence of a weak subduction channel. They are exhumed in an oceanic subduction zones in two locations: Higashi-Akaishi peridotite body in the Sanbagawa metamorphic belt in Japan and the Cuaba peridotites in the Rio San Juan complex in Dominican Republic. They are probably extreme cases where buoyant partly hydrated mantle facilitated the exhumation of deep rocks. Numerical models of Gorczyk et al. (2007) show that the steady state subduction does not result in the exhumation of garnet-bearing peridotites, but that hydration of deep mantle wedge modifies its rheology to allow the upwelling of the asthenospheric mantle wedge and subsequent retreat of the subducting slab. Slab retreat would induce exhumation of a deep-seated melange from a depth of 100–150km that consists of UHP mafic rocks (subducted oceanic lithosphere), anhydrous peridotites, hydrated and partially molten peridotites of the mantle wedge.

2.4 Major Driving Forces for the Exhumation

It is easy to understand the cause of burial metamorphism, but the causes of exhumation are less easily understood. Exhumation of HP to UHP metamorphic

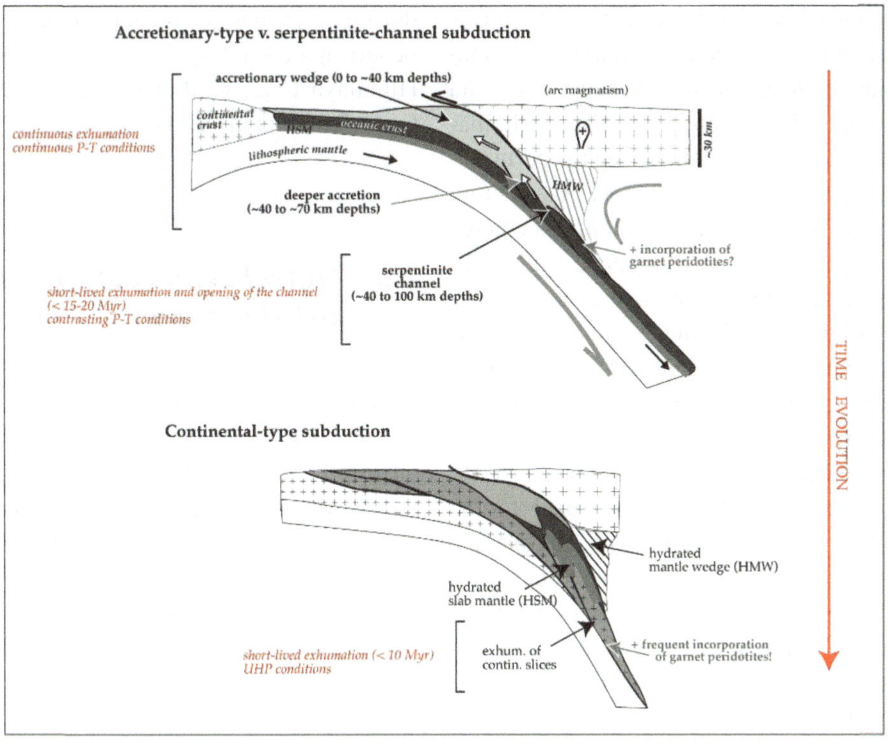

Fig. 10 Schematic model for the exhumation of HP-UHP rocks inferred from the tectonic evolution of the Piedmont zone (Western Alps) during Late-Cretaceous-Paleogene. During the oceanic subduction period: developped an accretionary wedge and a serpentinite channel. During short-lived period, HP rocks and locally UHP rocks (garnet bearing peridotites) exhumed. The implication of the continental lithosphere within the subduction increase the buoyancy, allowing the final exhumation of the accretionary wedge and the serpentinite channel. Slices of upper continental rocks incorporate mantle peridotites and exhumed rapidly within the suture zone

rocks requires a combination of several factors; buoyancy of rocks, a mechanism to reduce the boundary forces between these rocks, the descending lithosphere (Jolivet et al., 2005), erosion of overlying rocks, thrusting associated with normal faulting, and return flow inside the subduction channel. Metamorphism and fluids play an important role in changing the density of rocks, the balance of forces (Hacker, 1996; Le Pichon et al., 1997) and softening of eclogitized rocks (Hacker, 1996; Jolivet et al., 2005).

In the following paragraphs we will discuss these physical parameters for the exhumation of HP to UHP rocks in oceanic or continental context.

In a subduction zone, two types of forces contribute to the exhumation of HP to UHP rocks: the boundary forces related to the subduction zone itself and the internal forces induced by the density difference between the subducting slab and the surrounding rocks. The eclogitization of oceanic crust makes it denser than the surrounding mantle peridotites, and further promotes the subduction process. This implies that the exhumation of metamorphosed oceanic crust requires other factors than buoyancy. In accretionary wedges, metasediments with low densities (ca. 3,000 kg/m^3) are not easily subducted with dense metabasic rocks (>3,200 kg/m^3), which leads to their decoupling from the subducting salb and underplating at the base of the accretionary wedge. Platt (1986, 1987) proposed that this underplating process coupled with shallow extension would allow HP rocks to be exhumed to upper crustal levels. We argue against this proposal. First, most modern sedimentary accretionary wedges are not deep enough to produce HP rocks. The formation of these HP rocks requires unusually deep accretionary prisms. Secondly, exhumation frequently occurs during a short period in a given subduction zone (~15 Myr; Agard et al., 2006; Agard et al., submitted), suggesting that such a steady-state regime suggested by Platt would not lead to the exhumation of HP rocks. Fig. 3d shows the schematic section of the present-day

Cascadia accretionary wedge beneath the Vancouver Island (Fig. 2a), but it must be noted that no exhumation is happening at present within the active Cascadia accretionary wedge. This confirms that the exhumation of HP rocks likely takes place during the perturbation of subduction zones as suggested by Agard et al. (2006) and others. The perturbations include a change in subduction velocity or subduction angle, and docking of a seamount, an arc or a continental block.

Within a serpentinite subduction channel, the exhumation of dense mafic rocks is facilitated by low density (2,600 kg/m^3) and low viscosity of serpentine minerals. Low density of serpentinites results in diapiric ascent of serpentinites in the Mariana forearc (Fryer et al., 1999). Buoyancy of serpentinites likely contributed to the exhumation of Monviso massif in the Alps. The average density of the entire Monviso massif including eclogites is about 2,850 kg/m^3, which is lower than anhydrous mantle peridotites. Moreover, the low viscosity of serpentinite induces a dynamic flow inside the subduction channel (Schwartz et al., 2001; Gerya et al., 2002; Hilairet et al., 2007). A return flow within the serpentinite subduction channel can exhume dense eclogitic blocks (Cloos, 1982). This return flow is enhanced by the progressive dehydration of serpentinites at the depth where the subducting slab reaches the temperature of 650–700°C (Fig. 11). The temperature of 650–700°C is reached at a depth of about 100 km, ~2.8 GPa, which coincides with the maximum pressures recorded in eclogites in serpentinite melanges (Fig. 7).

In the continental-type subduction environment, the upper crust is firmly attached to the sinking lithosphere. Furthermore, sialic crustal rocks remain buoyant during subduction because their density is not significantly modified during subduction. Main hydrous phases, phengite and mica, are stable even during UHP metamorphism. Decoupling of a crustal slice from the descending slab requires the buoyancy forces exceeding the strength of the upper crust, which may occur at a depth of 90 to 140 km (Fig. 12b). Buoyancy difference between the continental rocks and oceanic lithosphere likely results in the separation of the two by thrusting along the subduction plane and normal faulting at shallower depth (Figs. 12a, b). Finally, the detachment of oceanic lithosphere further enhances the buoyant exhumation of the continental crust and sinking of the oceanic lithosphere (Fig. 12d) (Van der Beuckel, 1992; Davies and von Blanckenburg, 1995). However, this last model implies large uplift during exhumation incompatible with exhumation occurring beneath sea level as observed for exemple in Himalaya (e.g., Guillot et al., 2003) or in the Alps (e.g., Tricart et al., 2004).

2.5 Other Factors Contributing to Exhumation

Several other factors contribute to the exhumation of HP to UHP rocks. Slab retreat has been invoked as an important cause for the exhumation of HP rocks in the Mediterranean domain as it creates an extensional regime for the exhumation (e.g., Gautier et al., 1999; Jolivet et al., 2003). As previously discussed, slab retreat and associated back-arc extension are suggested to explain the exhumation of deep seated garnet-peridotites in the Dominican Republic (Gorczyk et al., 2007) and also the world youngest eclogite in Papua New Guinea (Monteleone et al., 2007). In the numerical model developed by Gorczyk et al. (2007), the role of asthenospheric upwelling is essential in rapid exhumation of deep-seated garnet-bearing peridotites. Similarly, the occurrence of majoritic garnet in the Western Gneiss Region of Norway (Van Roermund et al., 2001) and the recent discovery of coesite possibly replacing stishovite in non-metamorphic chromitite in southern Tibet (Yang et al., 2007) suggest that rocks originated from the deep upper mantle (>300 Km) are exhumed within suture zones. Exhumation of such deep rocks near the the mantle transition zone suggest a large-scale convection in the upper mantle.

Subduction angle is also important in controlling the production and exhumation of HP and UHP rocks. Guillot et al. (2007) estimated that the initial angle of continental subduction was greater than 40° in the western Himalayan syntaxis. Such a steep subduction is displayed in tomographic images to a depth of 200–300 km beneath the Hindu Kush and supported by seismic studies (Negredo et al., 2007). The evidence suggests that the UHP metamorphic rocks are being formed in the slab beneath the Hindu Kush at present (Searle et al., 2001). In contrast, tomographic images and seismic data show that the Indian continent subducted at a gentle angle of 9° beneath southern Tibet, reaching a depth of less than 80 km (<2.0–2.5 GPa), which precludes the formation of UHP rocks in the area (Guillot et al., 2008).

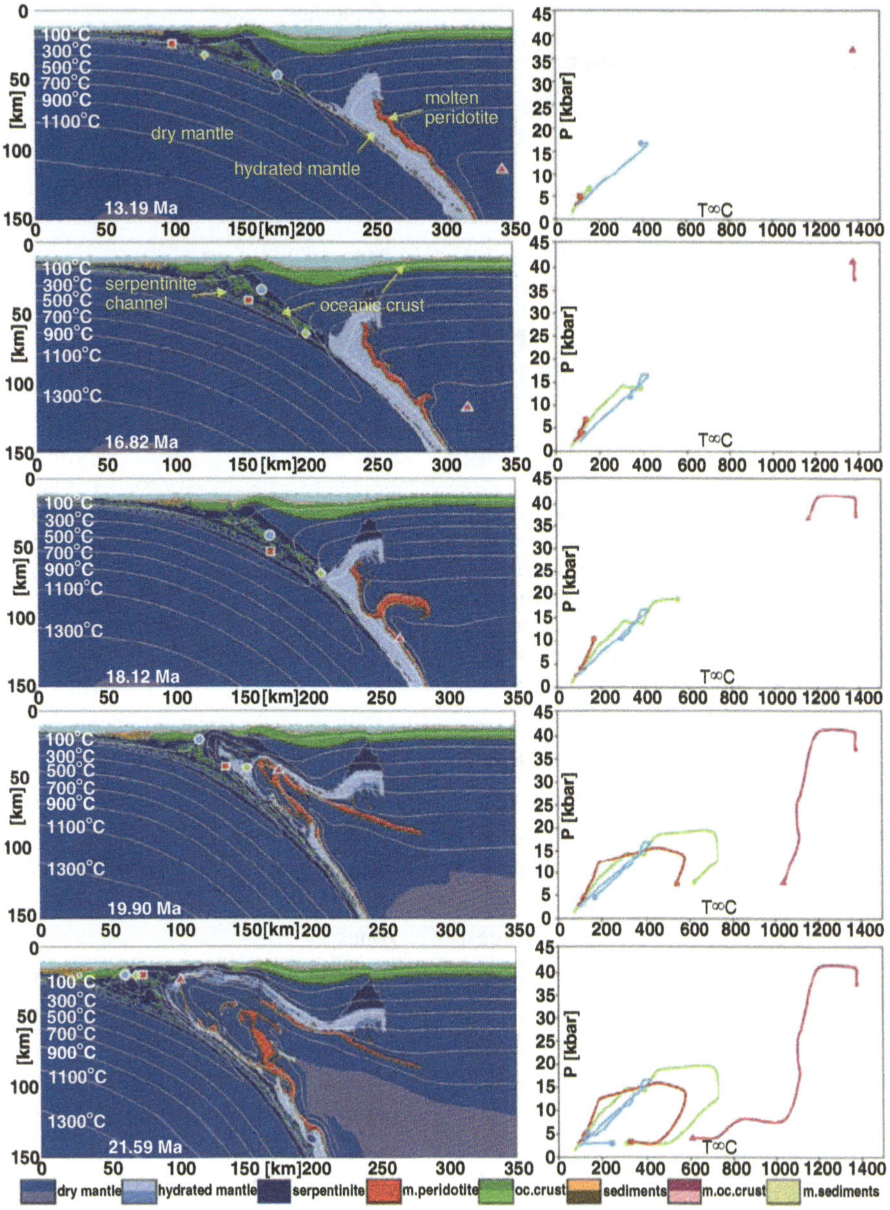

Fig. 11 Evolution of a subduction of oceanic lithosphere that formed at slow spreading ridge (after Gorczyk et al., 2007). The numerical model shows that the subduction produces a wide subduction channel composed of serpentinites because oceanic lithosphere formed at a slow-spreading ridge contain abundant serpentinites. The numerical model predicts two kinds of serpentinites within the subduction channel: (a) incoming hydrated abyssal peridotites and (b) hydrated, forearc mantle peridotites. The maximum depth of circulating material in the serpentinite subduction channel reaches a depth of 60 km (~2 GPa) and a temperature of 740°C, which corresponds to the upper stability limit of serpentine minerals (Ulmer and Trommsdorff, 1995). Progressive hydration of the mantle wedge modifies its rheology allowing the upwelling of the asthenospheric mantle wedge and subsequent retreat of the subducting slab. Slab retreat would induce exhumation of deep-seated melange from a depth of 100–150 km that consists of UHP mafic rocks (subducted oceanic lithosphere), anhydrous peridotites, hydrated and partially molten peridotites of the mantle wedge. This deep-seated melange would not reach the surface and stops at about 20 km depth beneath the already exhumed serpentinite melange

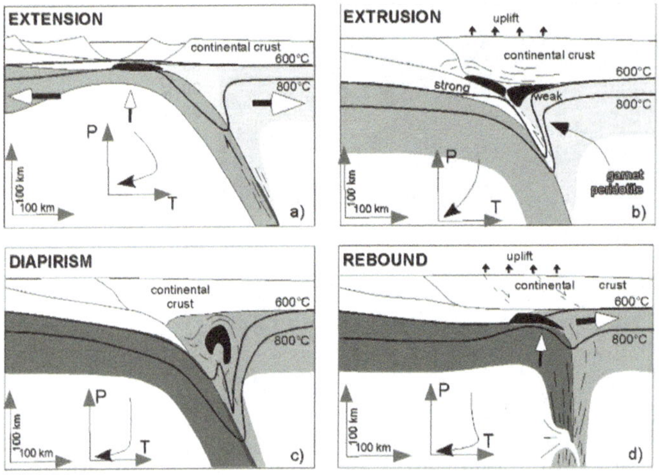

Fig. 12 Summary of UHP models (modified after Young et al., 2007). (**a**) Exhumation of subducted oceanic rocks by local extension. This model cannot exhume rocks from a depth greater than 100 km because a horizontal extension of greater than 200 km at crustal level is requires. Extrusion of delaminated dowgoing slab due to buoyancy forces with thrusting at the base and normal faulting at the top (Ernst, 2001, 2005). This model explains incorporation of deep mantle garnet peridotites. (**b**) Diapiric ascent of delaminated UHP rocks combined with horizontal compression in overlying plate (e.g., de Sigoyer et al., 2004). (**c**) Separation of continental rocks from descending oceanic lithosphere followed by flexural rebound of the UHP unit (Young et al., 2007)

The important role of steep subduction in the formation and subsequent exhumation of UHP rocks is further illustrated in the Alpine and Himalayan systems. The subducted European continental margin shows up to 1.2 GPa and 450°C in the central part and up to 3.5 GPa and 750°C in the southern part, suggesting that the dip steepened from 40° to 70° southward (e.g., Carry, 2007). As for the Himalayan system, the dips of subducting plate change along a transform fault and the exhumation of UHP rocks occur near the transform fault.

3 Conclusions

The P-T-t paths and protoliths of HP and UHP metamorphic rocks provide information relevant to a better understanding of subduction zones. The combination of data from metamorphic rocks with numerical models draws the following salient results:

- The exhumation of HP to UHP rocks including those originated from continental rocks is an integral part of subduction processes.
- Exhumation of rocks requires mechanically weak subduction channels that are comprised of sediments, hydrated peridotites or partial melt.
- The driving forces for exhumation are a combination of buoyancy and channel flow coupled with underplating of slabs. The former is the dominant force for the exhumation of continental rocks, whereas the latter prevails for the exhumation in oceanic subduction zones.
- Exhumation velocities are independent of plate velocities: (1) slow (<5 mm/year) exhumation of HP-LT metasediments ($P < 2.5$ GPa, $T < 600$°C) is a long-lasting process, in an accretionary prism; (2) slow to intermediate velocity ($1 < v < 10$ mm/year) exhumation of HP to UHP (< 3 GPa < 650°C) oceanic rocks is a discontinuous, transient process within a serpentinite subduction channel; (3) fast exhumation (up to 40 mm/year) of UHP (up to 6 GPa, 900°C) continental units is extremely short-lived (<10 My) and occurs in the mantle wedge combined with both asthenospheric return flow and buoyancy forces.
- Other parameters that affect the exhumation of HP to UHP rocks include slab retreat, and subduction dip angle. UHP rocks are not produced in subduction zones with gentle subduction angles.

Acknowledgments We thank W.G. Ernst and F. Rossetti for their helpful suggestions, which improved the quality of the manuscript. The work was supported by grants from "Dyeti" CNRS program and NSERC of Canada.

References

Abbott RN, Draper G and Broman BN (2006) P-T path for ultra-high-pressure garnet ultramafic rocks of the Cuaba gneiss, Rio San Juan complex, Dominican Republic. Int Geol Rev 28: 778–790.

Agard P, Monié P, Jolivet L and Goffé B (2002) Exhumation of the Schistes Lustrés complex: in situ laser probe Ar/Ar constraints and implications for the Western Alps. J Metam Geol 20: 599–618.

Agard P, Labrousse L, Elvevold S and Lepvrier C (2005) Discovery of Paleozoic Fe-Mg carpholite in Motalfjella, Svalbard Caledonides: a milestone for subduction-zone gradients. Geology 33: 761–764.

Agard P, Monié P, Gerber W, Omrani J, Molinaro M, Meyer B, Labrousse L, Vrielynck B, Jolivet L and Yamato P (2006) Transient, synobduction exhumation of Zagros blueschits inferred from P-T, deformation, time and kinematic constraints: implications for Neotethyan wedge dynamics. J Geophys Res 111 doi:10.1029/2005JB004103.

Agard P, Yamamoto P, Jolivet L and Burov E (2008) Discontinuous exhumation of oceanic crust: insights from blueschists and eclogites into the subduction channel: Earth Science Reviews, in press.

Aitchinson JC, Blake C, Flood PG and Jayko AS (1994) Paleozoic ophiolitic assemblages within the southern New England Orogen of eastern Australia: implications for growth of the Gondwana margin. Tectonics 13: 1135–1149.

Allemand P and Lardeaux JM (1997) Strain partioning and metamorphism in a deformable orogenic wedge: application to the Alpine belt. Tectonophysics 280: 157–169.

Altherr R, Marschall H and Ludwig T (2004). Evolution of a tourmaline-bearing lawsonite eclogite from the Elekdag area (Central Pontides, N Turkey): evidence for infiltration of slab-derived B-rich fluids during exhumation. Contrib Mineral Petrol 148: 409–425.

Anczkiewicz R, Burg JP, Villa IM and Meier M (2000) Late Cretaceous blueschist metamorphism in the Indus suture zone, Shangla region, Pakistan Himalaya. Tectonophysics 324: 111–134.

Anczkiewicz R, Platt JP, MF T and Wakabayashi J (2004) Franciscan subduction off to a slow start: evidence from high-precision Lu–Hf garnet ages on high grade-blocks. Earth Planet Sci Lett 225: 147–161.

Arculus RJ, Lapierre H and Jaillard E (1999) Geochemical window into subduction and accretion processes; Raspas metamorphic complex, Ecuador. Geology 27: 547–550.

Augier R, Agard P, Monié P, Jolivet L, Robin C and Booth-Rea G (2005) Exhumation, doming and slab retreat in the Betic Cordillera (SE Spain): in-situ $^{40}Ar/^{39}Ar$ ages and P-T-d-t paths for the Nevado-Filabride complex. J Metam Geol 23: 357–381.

Austrheim H (1994) Eclogitization of the deep crust in continent collision zones. C R Acad Sci 319: 761–774.

Auzende AL, Guillot S, Devouard B and Baronnet A (2006) Serpentinites in Alpine convergent setting: effects of metamorphic grade and deformation on microstructures. Eur J Mineral 18: 21–33.

Avé Lallemant HG and Sisson VB (2005) Exhumation of eclogites and blueschists in northern Venezuela: Constraints from kinematic analysis of deformation structures: in Avé Lallemant HG and Sisson VB eds., Caribbean/South American plate interactions, Venezuela: Geological Society of America Special Paper 394: 193-206.

Baldwin SL, Monteleone B, Webb LE, Fitzerald PG, Grove M and Hill EJ (2004) Pliocene eclogite exhumation at plate tectonic rates in eastern Papua New Guinea. Nature 431: 263–267.

Bally AW (1981) Thoughts on the tectonics of folded belts. In: Price NJ and McClay (eds) Thrust and Nappe Tectonics. Spec Pub Geol Soc London 9: 13–32.

Beaumont C, Ellis D and Pfiffner A (1999) Dynamics of sediment subduction-accretion at convergent margins: short-term modes, long-term deformation, and tectonic implications. J Geophys Res 104: 573–602.

Blake C, Moore DG and Jayko AS (1995) The role of the serpentinite melange in the unroofing of the UHP rocks: an example from the western Alps in Italy. In: from Coleman RG and Wang X (eds) Ultrahigh pressure metamorphism Cambridge University Press: 182–205.

Bosse V, Féraud G, Ballèvre M, Peucat JJ and Corsini M (2005) Rb-Sr and Ar-Ar ages in blueschists from the Ile de Groix (Armorican Massif, France): implications for closure mechanisms in isotopic systems. Chem Geol 220: 21–45.

Bostock MG, Hyndman RD, Rondenay S and Peacock SM (2002) An inverted continental Moho and serpentinization of the forearc mantle. Nature 417: 536–538.

Brueckner HK and Medaris LG (2000) A general model for the intrusion and evolution of 'mantle' garnet peridotites in high-pressure and ultrahigh pressure metamorphic terranes. J Metam Geol 18: 123–133.

Caby R (1994) Precambrian coesite from northern Mali: first record and implications for plate tectonics in the Trans-Saharan segment of the Pan-African belt. Eur J Mineral 6: 235–244.

Caby R, Buscail F, Dembélé D, Diakité S, Sacko S and Bal M (2008) Neoproterozoic garnet-glaucophanites and eclogites: new insights for subduction metamorphism of the Gourma fold- and- thrust belt (eastern Mali): Geological Society of London Special Publication 297: 203-216.

Caron JM and Péquignot G (1986) The transition between blueschists and lawsonite-bearing eclogites based on observations from Corsican metabasites. Lithos 19: 205–218.

Carswell DA (1990) Eclogite facies rocks. Blackie, Glasgow and London: 396p.

Carry N (2007) De la subduction continentale à l?exhumation dans les Alpes Penniques – Modélisation thermo-mécanique et paléogéographique.: Ph-D University of Rennes, 311 p.

Chemenda AI, Mattauer M, Malavieille J and Bokun AN (1995) A mechanism for syn-collisional rock exhumation and associated normal faulting: results from physical modelling. Earth Planet Sci Lett 132: 225–232.

Chopin C (1984) Coesite and pure pyrope in high-grade blueschists of the Western Alps. Contrib Minreal Petrol 86: 107–118.

Cloos M (1982) Flow melanges: numerical modelling and geological constraints on their origin in the Franciscan subduction complex. Geol Soc Am Bull 93: 330–345.

Cloos M and Shreve RL (1988) Subduction channel model of prism accretion, melange formation, sediment subduction, and subducting erosion at convergent plate margins: 1.

Background and description, in subduction zone Part 1. Pageophysics 128: 501–545.

Cluzel D, Aitchinson JC and Picard C (2001) Tectonic accretion and underplating of mafic terranes in the Late Eocene intra-oceanic fore-arc of New Caledonia (Southwest Pacific): geodynamic implications. Tectonophysics 340: 23–60.

Coleman RG (1971) Plate tectonic emplacement of upper mantle peridotites along continental edges. J Geophys Res 76: 1212–1222.

Collins AS, Reddy S, Buchan C and Mruma A (2004) Temporal constraints on Paleoproterozic eclogite formation and exhumation (Usagaran Orogen, Tanzania). Earth Planet Sci Lett 224: 175–192.

Compagnoni R and Maffeo B (1973) Jadeite-bearing metagranite l.s. and related rocks in the Monte Mucrone area (Sesia Lanzo zone, Western Italian Alps). Schweiz Mineral Petrog Mitt 53: 355–377.

Compagnoni R and Rolfo F (2003) UHPM units in the western Alps. EMU notes in Mineralogy 5: 13–49.

Davies J and von Blanckenburg F (1995) Slab breakoff: a model of lithosphere detachement and its test in the magmatism and deformation of collisional orogens. Earth Planet Sci Lett 129: 85–102.

de Sigoyer J, Chavagnac V, Blichert-Toft J, Villa IM, Luais B, Guillot S, Cosca M and Mascle G (2000) Dating the Indian continental subduction and collisional thickening in the northwest Himalaya: multichronology of the Tso Morari eclogites. Geology 28: 487–490.

de Sigoyer J, Guillot S and Dick P (2004) Exhumation Processes of the high-pressure low-temperature Tso Morari dome in a convergent context (eastern-Ladakh, NW-Himalaya). Tectonics 23: TC3003 – 10.1029/2002TC001492.

Dewey JF, Ryan PD and Andersen TB (1993) Orogenic uplift and collapse, crustal thickness, fabrics and metamorphic phases changes: the role of eclogites Geol Soc Spe Pub. In: Prochard HM, Alabaster T, Harris NBW and Neary CR (eds) Magmatic Processes and Plate tectonics, 325–343.

Dobretsov NL and Buslov MM (2004) Serpentinite mélanges associated with HP and UHP rocks in central Asia. Int Geol Rev 46: 957–980.

Drury MR, Van Roermund HLM, Carswell DA, De Smet JH, Van der Berg AP and Vlaar NJ (2001) Emplacement of Deep Upper-mantle rocks into cratonic lithophere by convection and diapiric upwelling. J Petrol 42: 131–140.

Duchêne S, Lardeaux JM and Albarède F (1997) Exhumation of eclogites: insights from depth-time path analysis. Tectonophysics 280: 125–140.

Enami M, Mizukami T and Yokoyam K (2004) Metamorphic evolution of garnet-bearing ultramafic rocks from the Gongen area, Sanbagwa belt, Japan. J Metam Geol 22: 1–15.

Erdmer P, Ghent E, Archibald D and Stout M (1998) Paleozoic and Mesozoic high-pressure metamorphism at the margin of ancestral North America in central Yukon. Geol Soc Am Bull 110: 615–629.

Ernst WG (1973) Blueschist metamorphism and P-T regimes in active subduction zones. Tectonophysics 17: 255–272.

Ernst WG (1999) Metamorphism, partial preservation, and exhumation of ultrahigh-pressure belts. Isl Arcs 8: 125–153.

Ernst WG (2001) Subduction, ultrahigh-pressure metamorphism, and regurgitation of buoyant crustal slices – implications for arcs and continental growth. Phys Earth Planet Inter 127: 253–275.

Ernst WG (2005) Alpine and Pacific styles of Phanerozic mountain building: subduction-zone petrogenesis of continental crust. Terra Nova 17: 165–188.

Ernst WG (2006) Preservation/exhumation of ultrahigh-pressure subduction complexes. Lithos 92: 321–335.

Escuder-Viruete J and Pérez-Estaùn A (2006) Subduction-related P-T path for eclogites and garnet glaucophanites from the Samanà Peninsula basement complex, northern Hispaniola. Int J Earth Sci 95: 995–1017.

Fitzherbert JA, Clarke GL and Powell R (2005) Preferential retrogression of high-*P* metasediments and the preservation of blueschist to eclogite facies metabasite during exhumation, Diahot terrane, NE New Caledonia. Lithos 83: 67–96.

Forster MA and Lister G (2005) Several distinct tectono-metamorphic slices in the Cycladic eclogite-blueschist belt, Greece. Contrib Mineral Petrol 150: 523–545.

Fotoohi Rad GR, Droop GTR, Amini S and Moazzen M (2005) Eclogites and blueschists of the Sistan Suture Zone, eastern Iran: a comparison of P–T histories from a subduction mélange. Lithos 84: 1–24.

Frederico L, Crispini L, Scambelluri M and Capponi G (2007) Ophiolite mélange zone records exhumation in a fossil subduction channel. Geology 35: 499–502.

Fryer P, Wheat CG and Mottl MJ (1999) Mariana blueschist mud volcanism: implications for conditions within the subduction zone. Geology 13: 103–106.

Furukawa Y (1993) Depth of the decoupling plate interface and thermal structure under arcs. J Geophys Res 98: 20005–20013.

Gabriele P, Ballèvre M, Jaillard E and Hernandez J (2003) Garnet-chloritoid-kyanite metapelites from the Raspas Complex (SW Ecuador): a key a key eclogite-facies assemblage. Eur J Mineral 15: 977–989.

Garcia-Casco AG, Torres-Roldan RL, Millan Trujillo G, Monié P and Schneider J (2002) Oscillatory zoning in eclogitic garnet and amphibole northern serpentinite melange, Cuba: a record of tectonic instability. J Metam Geol 20: 581–598.

Gardien V, Tegyey M, Lardeaux JM, Misseri M and Dufour E (1990) Crust-mantle relationships in the French Variscan chain: the example of the southern Monts du Lyonnais (MCF). J Metam Geol 8: 477–492.

Gautier P, Brun JP, Moriceau R, Sokoutis D, Martinod J and Jolivet L (1999) Timing, kinematics and cause of Aegean extension: a scenario based on a comparison with simple analogue experiments. Tectonophysics 315: 31–72.

Gerya TV, Stöckert B and Perchuck AL (2002) Exhumation of high pressure metamorphic minerals in subduction channels: a numerical simulations. Tectonics 21: 1056, doi:10.1029/2002TC001406.

Ghent E, Stout M and Erdmer P (1993) Pressure–temperature evolution of lawsonite-bearing eclogites, Pinchi Lake, British Columbia. J Metam Geol 11: 279–290.

Gilotti JA and Krogh Ravna EJ (2002) First evidence for ultrahigh-pressure metamorphism in the north- east Greenland Caledonides. Geology 30: 551–554.

Glodny J, Lohrmann J, Echtler H, Gräfe K, Seifert W, Collao S and Figueroa O (2005) Internal dynamics of a paleoaccretionary wedge: insights from combined isotope tectonochronology and sandbox modelling of the South-Central Chilean forearc. Earth Planet Sci Lett 231: 23–39.

Godard G (2001) Eclogites and their geodynamic interpretation: a history. J Geodyn 32: 165–203.

Goncalvez P, Guillot S, Nicollet C and Lardeaux JM (2000) Thrusting and sinistral wrenching in a pre-Eocene carribean accretionary wedge (Samana pensinsula -Dominican Republic). Geodin Acta 13: 119–132.

Gorczyk W, Guillot S, Gerya TV and Hattori K (2007) Asthenospheric upwelling, oceanic slab retreat and exhumation of UHP mantle rocks: insights from Greater Antilles. Geophys Res Lett 34: L211309, doi:10.1029/2007GL031059.

Grasemann B, Ratsbacher L and Hacker BR (1998) Exhumation of ultrahigh-presssure rocks: thermal boundary conditions and cooling history. In "When continent collide: Geodynamics and Geochemistry of Ultrahigh-pressure rocks" edited by BR Hacker and JG Liou, Springer New York, 21:117–139.

Guillot S, Hattori K and Sigoyer de J (2000) Mantle wedge serpentinization and exhumation of eclogites: insights from eastern Ladakh, northwest Himalaya. Geology 28: 199–202.

Guillot S, Hattori K, Sigoyer de J, Nägler T and Auzende AL (2001) Evidence of hydration of the mantle wedge and its role in the exhumation of eclogites. Earth Planet Sci Lett 193: 115–127.

Guillot S, Garzanti E, Baratoux D, Marquer D, Mahéo G and de Sigoyer J (2003) Reconstructing the total shortening history of the NW Himalaya. Geochem Geophys Geosyst 4(1) doi:10.1029/2002GC000484.

Guillot S, Schwartz S, Hattori K, Auzende A and Lardeaux JM (2004) The Monviso ophiolitic Massif (Western Alps), a section through a serpentinite subduction channel. In: Beltrando M, Lister G, Ganne J and Boullier A (eds) Evolution of the western Alps: insights from metamorphism, structural geology, tectonics and geochronology. J Virtual Explorer 16: Paper 6.

Guillot S, Replumaz A, Hattori K and Strzerzynski P (2007) Initial Geometry of Western Himalaya and Ultra-High Pressure Metamorphic Evolution. J Asian Earth Sci 30: 557–564.

Guillot S, Mahéo G, de Sigoyer J, Hattori KH and Pêcher A (2008) Tethyan and Indian subduction viewed from the Himalayan high- to ultrahigh-pressure metamorphic rocks. Tectonophyics doi:10.1016/j.tecto.2007.11.059.

Hacker BR (1996) Eclogite formation and the rheology, buoyancy, seismicity, and H2O content of oceanic crust. In: Bebout GE, Scholl DW, Kirby SH and Platt JP (eds) Geophysical Monograph Series, Subduction top to Bottom, 96: 337–346.

Hacker BR (2006) Pressure and Temperature of Ultrahigh-Pressure metamorphism: implications for UHP Tectonics and H20 in Subducting slabs. Int Geol Rev 48: 1053–1066.

Hacker BR (2007) Ascent of the ultrahigh-pressure Western Gneiss Region, Norway. In: Cloos M, Carlson WD, Gilbert MC, Liou JG and Sorensen SS (eds) Convergent Margin Terranes and Associated Regions: A tribute to WG Ernst: Geological Society of America Special Paper 419: doi: 101130/2006.211909.

Hacker BR, Ratschbacher L, Webb L, MCWilliams MO, Ireland T, Calvert A, Dong S, Wenk, HR and Chateigner D (2000) Exhumation of ultrahigh-pressure continental crust in east central China: Late Jurassic-Early Jurassic tectonic unroofing. J Geophys Res 105: 13339–13364.

Hacker BR, Calvert A, Zhang RY, Ernst WG and Liou JG (2003) Ultrarapid exhumation of ultrahigh-pressure diamond-bearing metasedimentary rocks of the Kokchetav Massif, Kazakhstan. Lithos 70: 61–75.

Harlow GE, Hemming SR, Avé Lallemant HG, Sisson VB and Sorensen SS (2004) Two high-pressure-low-temperature serpentinite-matrix mélange belts, Motagua fault zone, Guatemala: a record of Aptian and Maastrichian collisions. Geology 32: 17–20.

Hattori KH and Guillot S (2007) Geochemical character of serpentinites associated with high- to ultrahigh-pressure metamorphic rocks in the Alps, Cuba, and the Himalayas: recycling of elements in subduction zones. Geochem Geophys Geosyst 8: Q09010, doi:10.1029/2007GC001594.

Haüy RJ (1822) Traité de Minéralogie. Seconde édition, Bachelier et Huzard, Paris.

Hermann J, Müntener O and Scambelluri M (2000) The importance of serpentinite mylonites for subduction and exhumation of oceanic crust. Tectonophysics 327: 225–238.

Hilairet N, Reynard B, Wang Y, Daniel I, Merkel S, Nishiyama N and Petitgirard S (2007) High-Pressure Creep of Serpentine, Interseismic Deformation, and Initiation of Subduction. Science 318 doi: 10.1126/1148494.

Hirajima T, Banno S, Hiroi Y and Ohta Y (1988) Phase petrology of eclogites and related rocks from the Motalafjella high-pressure metamorphic complex in Spitsbergen (Artic ocean) and its significance. Lithos 22: 75–97.

Honegger K, Le Fort P, Mascle G and Zimmermann JL (1989) The blueschists along the Indus Suture Zone in Ladakh, NW Himalaya. J Metamorph Geol 7: 57–72.

Hyndman RD (1995) The lithoprobe corridor across the Vancouver Island and tectonic consequences of subduction. Can J Earth Sci 32: 1777–1802.

Jahn B-M, Caby R and Monie P (2001) The oldest UHP eclogites of the world: age of UHP metamorphism, nature of protoliths and tectonic implications. Chem Geol 178: 143–158.

Jan MQ (1985) High-P rocks along the suture zone around Indo-Pakistan plate and phase chemistry of blueschists from eastern Ladakh. Geol Bull Univ Peshawar 18: 1–40.

Jolivet L, Faccenna C, Goffé B, Burov E and Agard P (2003) Subduction tectonics and exhumation of high-pressure metamorphic rocks in the Mediterranean orogens. Am J Sci 303: 353–409.

Jolivet L, Raimbourg H, Labrousse L, Avigad D, Leroy Y, Austrheim, H and Andersen TB (2005) Softening trigerred by eclogitization, the first step toward exhumation during continental subduction. Earth Planet Sci Lett 237: 532–547.

Kadarusman A, Massonne HJ, Van Roermund H, Permana H and Munasri A (2007) P-T evolution of eclogites and bleuschists from the Luk Ulo Complex of Central Java, Indonesia. Int Geol Rev 49: 329–356.

Kimura G, Mayurama S, Isozaki Y and Terabayashi Y (1996) Well-preserved underplating structure of the jadeitited Franciscan complex, Pacheco, California. Geology 24: 75–78.

Ko ZW, Enami M and Aoya M (2005) Chloritoid-bearing basic schists from the Sanbagawa metamorphic belt, central Shikoku: their petrologic significance and tectonic implications. J Mineral Petrol Sci 100: 43–54.

Krebs M, Maresh WM, Schertl HP, Baumann A, Draper G, Ildeman B, Münker C and Trapp E (2008) The dynamics of intra-oceanic subduction zones: a direct comparison between

fossil petrological evidence (Rio San Juan Complex, Dominican Republic) and numerical simulation. Lithos doi: 10.1016/J.lithos.20007.09.003.
Labrousse L, Jolivet L, Agard P, Hébert R and Andersen TB (2002) Crustal scale boudinage and migmatization of gneiss during their exhumation in the UHP province of Western Norway. Terra Nova 14: 263–270.
Lagabrielle Y and Polino R (1985) Origine volcano-détritique de certaines prasinites des schistes lustrés du Queyras (France): arguments texturaux et géochimiques. Bull Soc Geol France 4: 461–471.
Lagabrielle Y, Auzende JM, Cornen G, Juteau T, Lensch G, Mével C, Nicolas A, Prichard H, Ribeiro A and Vanney JR (1981) Observations par submersible de croûte océanique affleurant sur le banc de Corringe (SW Portugal): evidences de processus de démantellement des gabbros en milieu sous marin. C Acad Sci 293: 827–832.
Lallemand S (1999) La subduction océanique. Gordon and Bridge Science Publishers: 194p.
Lallemand S and Le Pichon X (1987) Coulomb wedge models applied to the subduction of seamounts in the Japan Trench. Geology 15: 1065–1069.
Lardeaux JM, Ledru P, Daniel I and Duchêne S (2001) The Varisan French Massif Central – a new addition to the Ultra-high pressure 'club': exhumation and geodynamic consequences. Tectonophysics 332: 143–167.
Le Pichon X, Henry P and Goffe B (1997) Uplift of Tibet: from eclogites to granulites: implication for the Andean Plateau and the Variscan Belt. Tectonophysics 273: 57–76.
Lee CT and Chen WP (2007) Possible density segregation of subducted oceanic lithosphere along a weak serpentinite layer and implications for compositional stratification of the Earth's mantle. Earth Planet Sci Lett 255: 357–366.
Leech ML, Singh S, Jain AK, Klemperer SL and Manickavasagam RM (2005) The onset of India-Asia continental collision: early, steep subduction required by timing of UHP metamorphism in the western Himalaya. Earth Planet Sci Lett 234: 83–97.
Liati A (2005) Identification of repeated Alpine UHP metamorphic events by U-Pb SHRIMP geochronology and REE geochemitry of zircon: the Rhodope zone of Northern Greece. Contrib Mineral Petrol 150: 608–630.
Li XP, Rahn M and Bucher K (2004) Serpentinites of the Zermatt-Saas ophiolite complex and their texture evolution. J Metam Geol 22: 159–177.
Liou JG (1990) HP minerals from deeply subducted metamorphic rocks. In Ultra-High Pressure Mineralogy, RJ Hemley ed, Rev Mineral 37: 33–96.
Liou JG, Hacker BR and Zhang RY (2000) Into the Forbidden Zone. Science 287: 1215–1216.
Liou JG, Tsujimori T, Zhang RY, Katayama I and Marayuma S (2004) Global UHP metamorphism and continental subduction/collision: the Himalayan model. Int Geol Rev 46: 1–27.
Liu F, Gerdes A, Liou JG, Xue HM and Liang FH (2006) SHRIMP U-Pb zircon dating from Sulu-Dabie dolomitic marble, eastern China: constraints on prograde, ultrahigh-pressure and retrograde metamorphic ages. J Metam Geol 24: 569–589.
Liu J, Bohlen SR and Ernst WG (1996) Stability of hydrous phases in subducting oceanic crust. Earth Planet Sci Lett 143: 161–171.
Lombardo B, Nervo R, Compagnoni R, Messiga B, Kienast JR, Mevel C, Fiora L, Piccardo G and Lanza R (1978) Osservazioni preliminari sulle ofioliti metamorfiche del monviso (Alpi occidentali). R Soc Ital Mineral Petrol 34: 253–305.
Maekawa H, Shozui M, Ishii T, Fryer P and Pearce JA (1993) Blueschist metamorphism in an active subduction zone. Nature 364: 520–523.
Mahéo G, Fayoux C, Guillot S, Garzanti E, Capiez P and Mascle G (2006) Geochemistry of ophiolitic rocks and blueschists from the Sapi-Shergol mélange (Ladakh, NW Himalaya, India): implication for the timing of the closure of the Neo-Tethys ocean. J Asia Earth Sci 26: 695–707.
Maruyama S, Liou JG and Terabayashi M (1996) Blueschists and eclogites of the world and their exhumation. Int Geol Rev 38: 485–594.
Massonne HJ (2003) A comparison of the evolution of diamondiferous quartz-rich rocks from the Saxonian Erzgebirge and the Kokchetav Massif: are so-called diamondiferous gneisses magmatic rocks? Earth Planet Sci Lett 216: 347–364.
Medaris G, Wang H, Jelínek E, Mihaljevic M and Jakeš P (2004) Characteristics and origins of diverse Variscan peridotites in the Gföhl Nappe, Bohemian Massif, Czech Republic. Lithos 82: 1–23.
Medaris LG (1999) Garnet peridotites in Eurasian high-pressure and ultrahigh pressure terranes: a diversity of origins and thermal histories. Inter Geol Rev 41: 799–815.
Messiga B, Kienast J-R, Rebay G, Riccardi MP and Tribuzio R (1999) Cr-rich magnesio-chloritoid eclogites from the Monviso ophiolites (Western Alps, Italy). J Metam Geol 17: 287–299.
Mével C (2003) Serpentinization of abyssal peridotites at mid-ocean ridges: Comptes Rendus Géoscience, 335: 825–852.
Michard A, Chopin C and Henry C (1993) Compression versus extension in the exhumation of the Dora-Maira coesite-bearing unit, Western Alps, Italy. Tectonophysics 221: 173–193.
Môller A, Appel P, Mezger K and Schenk V (1995) Evidence for a 2 Ga subduction zone; eclogites in the Usagaran Belt of Tanzania Andreas Moeller, Peter Appel, Klaus Mezger, and Volker Schenk: Geology 23: 1067–1070.
Monteleone BD, Baldwin SL, Webb LE, Fitzerald PG, Grov M and Schmitt AK (2007) Late Miocene-Pliocene eclogite facies metamorphism, D'Entrecasteaux islands, SE Papua New Guinea. J Metam Geol 25: 245–265.
Moore DE and Lockner D (2007) Comparative deformation bahavior of minerals in serpentinized ultramafic rocks: application to the slab-mantle interface in subduction zones. Inter Geol Rev 49: 401–415.
Moore JC and Silver EA (1987) Continental margin tectonics: submarine accretionary prisms. Rev Geophys 25: 1305–1312.
Moyen JF, Stevens G and Kisters A (2006) Record of mid-Archaean subduction from metamorphism in the Barberton terrain, South Africa. Nature 442: doi:10.1038/nature04972.
Negredo AM, Replumaz A, Villasenor A and Guillot S (2007) Modelling the evolution of continental subduction in the Pamir-indu Kush region. Earth Planet Sci Lett 259: 212–225.
Nimis P and Morten L (2000) P-T evolution of crustal garnet peridotites and included pyroxenites from Nonsberg area

(upper Austroalpine), NE Italy: from the wedge to the slab. J Geodyn 30: 93–115.

Nimis P and Trommsdorf V (2001) Revised Thermobarometry of Alpe Arami and other garnet peridotites from the Central Alps. J Petrol 42: 103–115.

O'Brien P, Zotov N, Law R, Khan AM and Jan MQ (2001) Coesite in Himalayan eclogite and implication for models of India-Asia collision. Geology 29: 435–438.

Och DJ, Leitc EC, Caprarelli G and Watanabe T (2003) Blueschist and eclogite in tectonic melange, Port Macquarie, New South Wales, Australia. Mineral Mag 67: 609–624.

Oh CW and Liou JG (1990) Metamorphic evolution of two different eclogites in the Franciscan Complex, California, USA. Lithos 24: 41–53.

Ota T, Terabayashi M and Katayama I (2004) Thermobaric structure and metamorphic evolution of the Iratsu eclogite body in the Sanbagawa belt, central Shikoku, Japan. Lithos 73: 95–126.

Palmeri R, Ghiribelli B, Ranalli G, Talrico F and Ricci A (2007) UHP metamorphism and exhumation of garnet-bearing ultramafic rocks from the Lanterman range (northern Victoria Land, Antarctica). J Metam Geol 25: 225–243.

Parkinson CD, Miyazali K, Wakita K, Barber AJ and Carswell DA (1998) An overview and tectonic synthesis of the pre-Tertiray very high-pressure metamorphic and associated rocks of Java, Suwalesi and Kalimantan, Indonesia. Isl Arc 7: 184–200.

Parra T, Vidal O and Jolivet L (2002) Relation between the intensity of deformation and retrogression in blueschist metapelites of Tinos Island (Greece) evidenced by chlorite-mica local equilibria. Lithos 63: 41–66.

Parrish R, Gough SJ, Searle M and Dave W (2006) Plate velocity exhumation of ultrahigh-pressure eclogites in the Pakistan Himalaya. Geology 34: 989–992.

Pawley AR (1994) The pressure and stability limits of lawsonite: implication for H_20 recycling in subduction zones. Contrib Mineral Petrol 118: 99–108.

Peacock SM and Wang K (1999) Seismic consequences of warm versus cool subduction zone metamorphism: examples from northeast and southwest Japan. Science 286: 937–939.

Perchuk A and Philippot P (1997) Rapid cooling and exhumation of eclogitic rocks from the Great Caucassus, Russia. J Metam Geol 15: 299–310.

Perraki M, Proyer A, Mposkos E, Kaindl R and Hoinkes G (2006) Raman micro-spectroscopy on diamond, graphite and other carbon polymorphs from the UHP metamorphic Kimi complex of the Rhodope Metamorphic province, NE Greece. Earth Planet Sci Lett 241: 672–685.

Philippot P and Kienast JR (1989) Chemical-microstructural changes in eclogite-facies shear zones (Monviso, Western Alps, north Italy) as indicators of strain history and the mechanism and scale of mass transfer. Lithos 23: 179–200.

Philippot P and Van Roermund HLM (1992) Deformation processes in eclogitic rocks: evidence for the rheological delamination of the oceanic crust in deeper levels of subduction zones. J Struct Geol 14: 1059–1077.

Philippot P, Blichert-Toft J, Perchuk A, Costa S and Gerasimov VY (2001) Lu-Hf and Ar-Ar chronometry supports extreme rate of subduction zone metamorphism deduced from geospeedometry. Tectonophysics 342: 23–38.

Platt J and Vissers RLM (1989) Extensional collapse of thickened continental lithosphere: a working hypothesis for the Alboran sea and Gibraltar arc. Geology 17: 540–543.

Platt JP (1986) Dynamic of orogenic wedges and the uplift of high-pressure metamorphic rocks. Geol Soc Am Bull 97: 1037–1053.

Platt JP (1987) The uplift of high-pressure low temperature metamorphic rocks. Phil Trans Roy Soc London A321: 87–103.

Ranero CR, Morgan JP, McIntsoh K and Reichert C (2003) Bending-related faulting and mantle serpentinization at the Middle America trench. Nature 425: 367–373.

Reinecke T (1991) Very-high pressure metamorphism and uplift of coesite-bearing metasediments from the Zermatt-Saas zone, Western Alps. Eur J Mineral 3: 7–17.

Reynard B, Hilairet N, Balan E and Lazzeri M (2007) Elasticity of serpentines and extensive serpentinization in subduction zones. Geophys Res Lett 34: doi:10.1029/2007GL030176.

Root DB, Hacker BR, Gans PB, Ducea MN, Eide EA and Mosenfelder JL (2005) Discrete ultrahigh-pressure domains in the western Gneiss Region, Norway: implications for formation and exhumation. J Metam Geol 23: 45–61.

Rubatto D and Hermann J (2001) Exhumation as fast as subduction? Geology 29: 3–6.

Saumur BM, Hattori KH and Guillot S (2007) Protrusion of fore-arc mantle serpentinites together with HP and UHP rocks along major strike-slip faults in the northern subduction complex of Dominican Republic. Subduction zone geodynamics Conference, Montpellier.

Schmidt MW and Poli S (1998) Experimentally based water budgets for dehydrating slabs and consequences for arc magma generation: Earth and Planetary Science Letters 163: 361–379.

Schneider J, Bosch D, Monié P, Guillot S, Garcia-Casco AG, Lardeaux JM, Torres-Roldan RL and Millan T (2004) Origin and evolution of the Escambray massif (Central Cuba): an example of HP/LT rocks exhumed during intraoceanic subduction. J Metamorph Geol 22: 227–247.

Schwartz S, Lardeaux JM, Guillot S and Tricart P (2000) Diversité du métamorphisme éclogitique dans le massif ophiolitique du Monviso (Alpes Occidentales, Italie). Geodinam Acta 13: 169–188.

Schwartz S, Allemand P and Guillot S (2001) Numerical model of the effect of serpentinites on the exhumation of eclogitic rocks: insights from the Monviso ophiolitic massif (western Alps). Tectonophysics 42: 193–206.

Schwartz S, Lardeaux JM, Tricart P, Guillot S and Labrin E (2007) Diachronous exhumation of subducted HP metamorphic rocks from southwestern Alps: evidences from fission-track analysis. Terra Nova 19: 133–140.

Searle M, Hacker BR and Bilham R (2001) The Hindu Kush seismic zone as a paradigm for the creation of ultrahigh pressure diamond- and coesite- bearing continental rocks. J Geol 109: 143–53.

Seno T and Yamasaki T (2003) Low frequency tremors, intraslab and interplate earthquakes in southwest Japan, from a viewpoint of slab dehydration. Geophys Res Lett 30: 22 2171- doi 10.1029/2003GLO18349.

Shreve RL and Cloos M (1986) Dynamics of sediment subduction, melange formation and prism accretion. J Geophys Res 91: 10 229–10 245.

Smith DC (1984) Coesite in clinopyroxene in the Caledonides and its implication for geodynamics. Nature 310: 641–644.
Sobolev SV and Shatsky VS (1990) Diamond inclusions in garnets from metamorphic rocks: a new environment for diamond formation. Nature 343: 742–746.
Sorensen SS (1988) Petrology of amphibolite-facies mafic and ultramafic rocks from the Catalina Schist, southern California: metasomatism and migmatization in a subduction zone metamorphicsetting.JMetamGeol,6,doi:10.1111/j.1525-1314.1988.tb00431.x: 405–435.
Spengler D, Van Roermund H, Drury MR, Ottolini L, Mason PRD and Davies GF (2006) Derep origin and hot melting of an Archean orogenic peridotite massif in Norway. Nature 440: 913–917.
Stanek KP, Maresch WV, Grafe F, Grevel CH and Baumann A (2006) Structure, tectonics and metamorphic development of the Sancti Spiritus Dome (eastern Escambray massif, Central Cuba). Geol Acta 4: 151–170.
Steck A, Epard JL, Vannay JC, Hunziker J, Girard M, Moraro A and Robyr M (1998) Geological transect across the Tso Moarari and Spiti areas: the nappe structures of the Tethys Himalaya. Eclo Geol Helv 91: 103–121.
Stöcker B and Gerya TV (2005) Pre-collisional HP metamorphism and nappe tectonics at active continental margins: a numerical simulation. Terra Nova doi10.1111/j.1365-3112.2004.00589.
Stöckert B and Renner J (1998) Rheology of crustal rocks at ultrahigh pressure. In: Hacker BR and Liou JG (eds) When continent collide: Geodynamics and Geochemistry of Ultrahigh-pressure rocks, Kuwer Academic Publishers, Dordrecht: 57–95.
Terry MP Robinson P and Ravna EJK (2000) Kyanite eclogite thermobarometry and evidence for thrusting of UHP over HP metamorphic rocks, Nordoyane, Western Gneiss Region, Norway: American Mineralogist 85: 1637–1650.
Thompson A, Schulmann K and Jezek J (1997) Extrusion tectonics and elevation of lower crustal metamorphic rocks on convergent orogens. Geology 25: 491–494.
Treloar PJ, O'Brian PJ, Parrish RR and Khan AM (2003) Exhumation of early Tertiary, coesite-bearing eclogites from the Pakistan Himalaya. J Geol Soc London 160: 367–376.
Tricart P and Schwartz S (2007) A north-south section across the Queyras Schistes Lustrés (Piedmont zone, Western Alps): Syn-collision refolding of a subduction wedge. Eclogae Geol Helv 99 Doi 10.1007/s00015-006-1197-6.
Tricart P, Schwartz S, Sue C and Lardeaux JM (2004) Evidence for synextensional tilting and doming during during final exhumation from analysis multistage faults (Queyras, Schsites Lustrés, Western Alps). J Struct Geol 26: 1633–1645.
Tsujimori T, Sisson VB, Liou JG, Harlow GE and Sorensen SS (2006) Very-low-temperature record of the subduction processes: a review of worldwide lawsonite eclogites. Lithos 92: 609–624.
Tucker RD, Krogh TE and Rahei A (1991) Proterozic evolution and age province boundaries in the central part of the Western Gneiss Region: results of U-Pb dating of accessory minerals from Trondheimsford to Geiranger. In: Gower CF, Rivers R and Ryan B (eds) Mid-Proterozic laurentia Baltica. Geol Soc Spe Paper 38: 149–173.
Ueda H, Usuki T and Kuramoto Y (2004) Intraoceanic unroofing of eclogite facies rocks in the Omachi Seamount, Izu-Bonin frontal arc. Geology 32: 849–852.
Ulmer P and Trommsdorff V (1995) Serpentinite stability to mantle depths and subduction related magmatism. Science 268: 858–861.
Valli F, Guillot S and Hatori K (2004) Source and tectono-metamorphic evolution of mafic and pelitic metasedimentary rocks from the central Quetico metasedimentary belt, Archean Superior, Province of Canada. Precam Res 132: 155–177.
Van der Beuckel J (1992) Some thermo-mechanical aspects of the subduction of continental lithosphere. Tectonics 11: 316–329.
Van Roermund HLM, Drury MR, Barnhoorn A and de Ronde A (2001) Relict majoritic garnet microstructures from ultra-deep orogenic peridotites in Western Norway. J Petrol 42: 117–130.
Vaugh R and Parkinson D (2003) Comparison and tectonic significance of Pan-African HP/UHP eclogites from Pernambuco and Minas Gerais, Brazil. Geol Soc Am Abst Prog 35: 632.
Vignaroli G, Rossetti F, Bouybaouene M, Massonne HF, Theye T and Faccenna C (2005) A counter-clockwise P-T path for the Voltri massif eclogites (Ligurian Alps, Italy). J Metamorph Geol 23: doi:10/1111/j.152561314.2005.00592.
Von Huene R and Cullota R (1989) Tectonic erosion at the front of the Japan trench convergent margin. Tectonophysics 160: 75–90.
Von Huene R and Scholl D (1991) Observations at convergent margins concerning sediment subduction, subduction erosion, and the growth of continental crust. Rev Geophys 29: 279–316.
Wallis S, Moriyama Y and Tagam T (2004) Exhumation rates and age of metamorphism in the Sanbagawa belt: new constraints from zircon fission track analysis. J Metam Geol 22: 17–24.
Willner AP, Glodny J, Gerya TV, Godoy E and Massonne A (2004) A counterclockwise PTt-path of high pressure-low temperature rocks from the coastal Cordillera accretionary complex of South Central Chile: constraints for the earliest stage of subduction mass flow. Lithos 75: 283–310.
Yamasaki T and Seno T (2003) Double seismic zone and dehydration embrittlement of the subducting slab J Geophys Res 108: B4, 2212, doi: 101029/2002JB001918.
Yang JS, Xu Z, Dobrzhinetskaya LF, Green HW, Pei X, Shi R, Wu C, Wooden JL, Zhang JS, Wan Y and Li H (2003) Discovery of metamorphic diamonds in central China: an indication of a >4000-km-long zone of deep subduction resulting from multiple continental collisions. Terra Nova 15: 370–379.
Yang JS, Dobrzhinetskaya L, Bai WJ, Fanq QS, Robinson PT, Zhang, J and Green II HW (2007) Diamond- and coesite-bearing chromitites from the Luobusa ophiolite, Tibet Geology, 35, doi: 10.1130/G23766A.1.
Young DY, Hacker BR, Andersen TB and Corfu F (2007) Prograde amphibolite facies to ultrahigh-pressure transition from Nordfjord, western Norway: implications for exhumation tectonics. Tectonics 26: doi:10.1029/2004TC001781.
Yamato P, Agard P, Burov E, Le Pourhiet L, Jolivet L and Tiberi C (2007) Burial and exhumation in a subduction wedge: mutual constraints from thermomechanical and natural P-T-t data (Schistes Lustrés, western Alps). J Geophys Res 112: B07410, doi: 10.1029/2006JB004441.

Zack T, Rivers T, Brumm R and Kronz A (2004) Cold subduction of oceanic crust: implications from a lawsonite eclogite from the Dominican Republic. Eur J Mineral 16: 909–916.

Zhang L, Ellis DJ, Arculus RJ, Jiang W and Wei CJ (2003) Forbidden zone subduction of sediments to 150 km depth-the reaction of dolomiye + aragonite in the UHPM metapelites from western Tiansjhan, China. J Metam Geol 21: 523–529.

Zhang RY, Liou JG, Yang JS and Yui T-F (2000) Petrochemical constraints for dual origin of garnet peridotites from the Dabie-Sulu UHP terrane, eastern-central China. J Metam Geol 18: 149–166.

Zhao ZY, Wei CJ and Fang AM (2005) Plastic flow of coesite in a deep continental subduction regime: microstructures, deformation mechanisms and rheological implications. Earth Planet Sci Lett 237: 209–222.

Evolution of Subductions Indicated by Mélanges in Taiwan

Chung-Pai Chang, Jacques Angelier and Chi-Yue Huang

Abstract Two mélange zones occur in Taiwan: the Lichi Mélange in the Coastal Range of eastern Taiwan and the Kenting Mélange in the Hengchun Peninsula of southern Taiwan. Because of the southward propagation of the Taiwan orogen, these two mélanges now crop out at the western front of the Coastal Range (Lichi Mélange) and the west side of the southern Central Range (Kenting Mélange). These two mélanges are similar in appearance and controversial in origin, being interpreted either as subduction complexes, or as olistostromes. Remnants of neighbouring stratigraphic units, present as slices or clasts in the mélange zones, shed light on the nature and origin of these two mélanges and help deciphering the tectonic evolution of Taiwan orogen and its leading subductions.

Based on multiple lines of evidence, including fossil age, clay mineral composition and nature of exotic ophiolitic blocks, we conclude that the Kenting Mélange zone, which was probably active over a long period of Late Cenozoic time, was a direct result of subduction process, as a relict of subduction fault zone at the front of the Taiwan accretionary prism. The Lichi Mélange, for which the fossil age indicates a shorter Pliocene time span, was part of the Northern Luzon Trough forearc; it is regarded as a collision mélange formed during forearc closure in probable relation with arcward backthrusting, when the Luzon arc started to collide with the Eurasian continent margin following subduction of the oceanic crust of the forearc area. After comparing the geological situation of these two mélanges, we obtain a new perspective for the evolution of a trench-prism system, from subduction to arc-continent collision.

Keywords Subduction • Collision • Mélange • Taiwan

1 Introduction

From the geodynamic point of view, the Taiwan orogen is produced by the arc-continent collision, which leads by the Manila subduction system and occurred since about 5 Ma (Chi et al., 1981; Chang and Chi, 1983; Dorsey, 1985, 1988; Ho, 1986). As a newly emerged orogen, the Taiwan Island exhibits a young and complete accretionary prism sequence. This island is therefore an ideal area to understand the evolution of a subduction system from its active subduction stage to its collision stage (Suppe, 1984).

Mélange zones are generally found in mountains paralleling convergent plate margins. Investigating the processes responsible for fragmentation and mixing in mélanges is crucial. Onshore Taiwan, two mélange

C.-P. Chang
Center for Space and Remote Sensing Research, National Central University, Chungli, Taiwan, cpchang@csrsr.ncu.edu.tw

S. Lallemand and F. Funiciello (eds.), *Subduction Zone Geodynamics*,
DOI 10.1007/978-3-540-87974-9,

zones are present: the Lichi Mélange in the Coastal Range and the Kenting Mélange in the Hengchun Peninsula (Hsu, 1976; Tsan, 1974a, b). Because of their contrasting positions, respectively east and west of the main Central Range of Taiwan, these two mélange zones may provide important geological information about the history of convergence.

The aim of this paper is to discuss the evolution from the Luzon-Manila subduction systems to the Taiwan collision, taking advantage of the geological information about these two mélange zones. We thus study the evolution of the NNE-SSW convergent boundary of southern Taiwan, along the Luzon Arc system. The major convergent boundary of northeast Taiwan, related to the E-W trending Ryukyu subduction system, does not deserves consideration in this paper.

2 Geological Setting of Taiwan

The Taiwan island is surrounded by two subduction systems. To the northeast, the Philippine Sea plate is being consumed beneath the Eurasian continent along the Ryukyu arc and trench system. To the south, the South China Sea is subducting beneath the Philippine Sea plate along the Manila Trench and Luzon Arc system. Thus, in northeastern Taiwan the Eurasian plate overrides the Philippine Sea plate whereas in southeastern Taiwan the Philippine Sea plate overrides the Eurasian plate that includes the South China Sea with its 32–15 Ma old oceanic lithosphere (Fig. 1). Because of the opposite-vergent subductions and rapid plate convergence, the southeastern margin of the Chinese continent is now strongly deformed and highly uplifted in the Central Range of Taiwan.

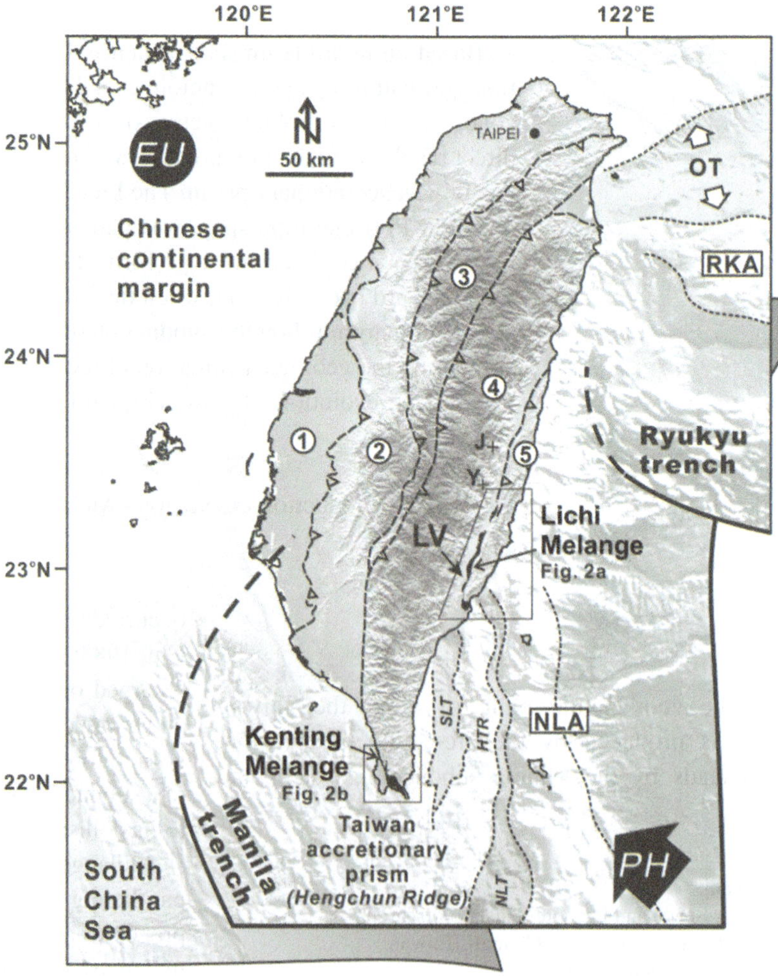

Fig. 1 Tectonic framework of the Taiwan orogenic belt between the Eurasian continent and the Philippine Sea plate. *Large black arrow* shows convergent direction between the volcanic arc (Northern Luzon Arc, NLA) and the Chinese continent margin. *Numbers* indicate main tectonostratigraphic units of Taiwan. *1*: Western Coastal Plain; *2*: Western Foothills; *3*: Hsüeshan Range; *4*: Central Range; 5: Coastal Range. *EU* Eurasian plate, *HTR* Huatung Ridge, *LV* Longitudinal Valley, *NLT* Northern Luzon Trough, *OT* Okinawa Through, *PH* Philippine Sea plate, *RKA* Ryukyu Arc, *SLT* Southern Longitudinal Trough. *J* Juishui Ophiolite; *Y* Yuli ophiolite. Locations of two mélanges in Taiwan are shown by areas in black (Figs. 2a, b)

The Taiwan island consists of five morphotectonic units, which are from west to east: the Coastal Plain, the Western Foothills, the Hsüehshan Range, the Central Range and the Coastal Range. The age and lithology of the first four units suggest that their geological formations belong to the upper crust and sedimentary basins of the Eurasian continent undergoing underthrusting along the Manila Trench system (Ho, 1988). In contrast, the Coastal Range in eastern Taiwan represents a northern segment of the northern Luzon arc that has been accreted onto the uplifted Eurasian continent (Chai, 1972; Biq, 1972), as shown in Fig. 1.

Between the Coastal Range and the Central Range, the Longitudinal Valley marks the "suture" between the Eurasian plate and the Philippine Sea plate. A large thrust fault, the Longitudinal Valley fault, is located on the eastern border of this valley. The Longitudinal Valley fault, as a reverse fault with a minor component of left-lateral motion, is one of the most active faults in Taiwan. Across this fault a steady creep of about 3 cm year^{-1} has been detected by geodetic and field means (Yu and Liu, 1989; Lee and Angelier, 1993; Angelier et al., 1997; Yu et al., 1997). Most authors consider the Longitudinal Valley as the major plate suture zone and a relic of the plate boundary between the Eurasian plate and Philippine Sea plate, which represented the main active structure before and during the arc-continent collision (Biq, 1971, 1973; Karig, 1973; Teng, 1981; Hsu, 1988; Chen, 1997). There is also evidence indicating that the Longitudinal Valley is the remnant, western part of the forearc basin which was a relatively quiet region before arc-continent collision and has been severely shortened (Hsu, 1956; Wang, 1976; Ernst, 1977; Ho, 1977; Liou et al., 1977; Page and Suppe, 1981; Barrier and Muller, 1984; Chang et al., 2000, 2001).

A major mélange zone, the Lichi Mélange, lying between the Coastal Range and the Longitudinal Valley, has long been considered as the key to the interpretation of the Longitudinal Valley origin (Hsu, 1956; Biq, 1971, 1973; Wang, 1976; Page and Suppe, 1981). On the opposite, western side of the Central Range at the southern tip of the Taiwan orogen, another mélange zone crops out: the Kenting Mélange. These two mélange zones, which show large deformation and major thrusting at their boundaries, have very similar features at first sight (Fig. 1). Some interesting geological issues are thus raised: did these two mélanges originate from separate basins? Did they result from the same process at the same time? If not, what was the specific origin and evolution of each of these mélanges? Which one formed first? The interpretation of these two major mélanges has strong inferences for any reconstruction of the geological history of Taiwan, especially for the evolution of the Longitudinal Valley, the so-called "suture" between the Eurasian continental margin and the Luzon arc, but also for the subduction history of Manila system and the structural development of the Hengchun Peninsula, which marks in space and time the transition from subduction to Taiwan collision.

To help deciphering this complex convergence history, we take advantage, in this paper, of previous geological findings and of our own observations to present a comprehensive comparison between these two mélanges and draw inferences in terms of convergence history and subduction-collision processes. We consequently focus on the age, composition and structure of the Lichi and Kenting mélange zones to clarify the role and formation process of typical mélanges in the growing Taiwan orogen.

3 Major Characteristics of Mélanges in Taiwan

3.1 *The Lichi Mélange*

The Lichi Mélange is mainly distributed along the southwestern flank of the Coastal Range with a length of about 65 km (Fig. 2a). It is composed of chaotic mudstones intermixed with "exotic" blocks of various size and lithology (Figs. 3a and 3a'). The exposures of the Lichi Mélange show a typical badland aspect and topography (Fig. 3a'). The most characteristic lithological feature of the Lichi Mélange lies in the presence of intensely sheared mudstones without distinctive stratification, and the most common mesoscopic structure is the scaly foliation (Figs. 4a, b). The curviplanar surfaces of this penetrative scaly foliation are generally polished and bear aligned pyrite and clay minerals and slickenside lineations (Hsu, 1976; Teng, 1981; Chen, 1991, 1997; Chang et al., 2000, 2001). Densely distributed shear surfaces are oblique to, and mechanically consistent with, the scaly foliation, thus

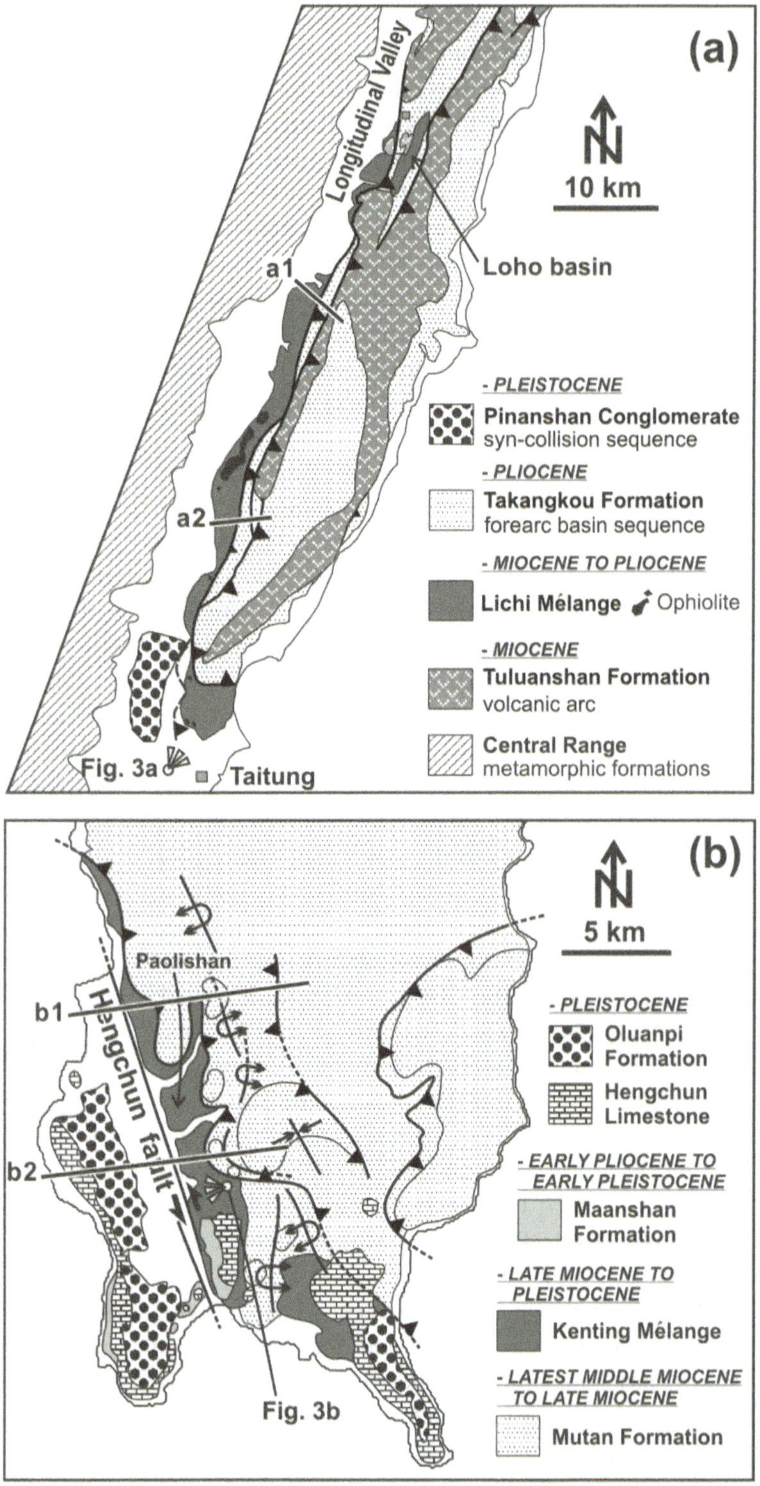

Fig. 2 (**a**) Geological map of the Lichi Mélange area. (**b**) Geological map of the Kenting Mélange area

producing typical lozengic structures in cross-section. Analysing the geometrical relationships between shear surfaces and scaly foliation enabled us to determine the direction and sense of shear deformation (Chang et al., 2000, 2001; Figs. 4a, b).

Although the Lichi Mélange is generally strongly sheared, many layers of pebbly mudstones and numerous coherent stratification evidences have been described (Figs. 4c, d; Chang, 1969; Liou et al., 1977; Page, 1978; Page and Suppe, 1981; Barrier and Muller,

Fig. 3 (**a**) Typical outcrops of the Lichi (a and a′) and Kenting Mélanges (**b**) Location in Fig. 2

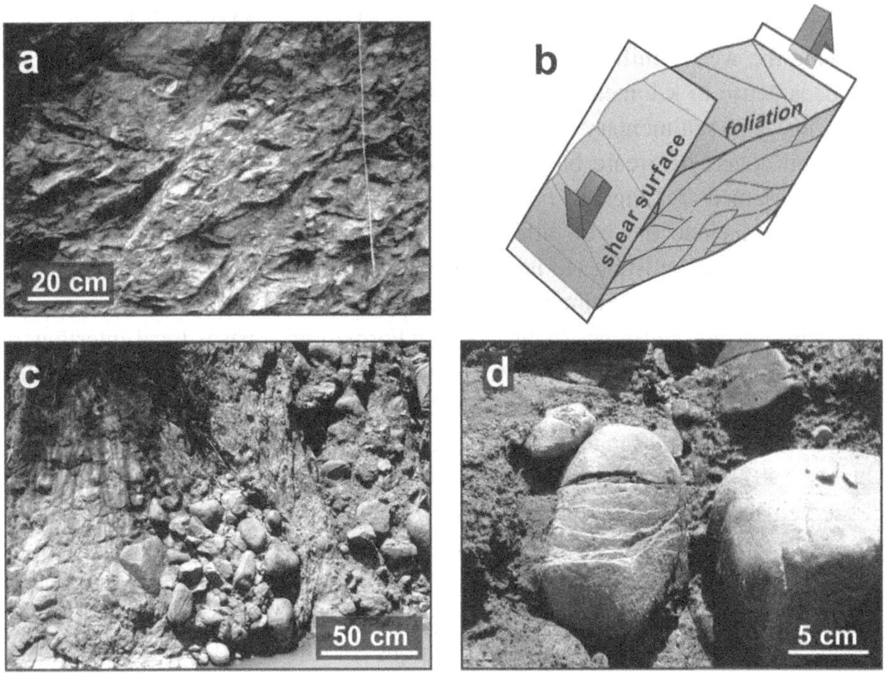

Fig. 4 (**a**) Scaly foliation associated with shear deformation in the argillaceous matrix of mélange. Note the presence of two main structural orientations that correspond to shear surfaces and foliation. (**b**) The sigmoid shape of scaly foliation indicates the sense of motion on the shear surfaces. (**c**) and (**d**) Pebbly layer in the Lichi Mélange. Location is indicated in geological profile a2 of Fig. 5

1984). In the Lichi Melange, the fragmentation and dismembering processes were widespread but could not erase some evidences of the initial stratigraphic relationships between the Lichi Mélange and the forearc basin strata, namely the lower Takangkou Formation. Particular attention was paid by many authors to the nature of the blocks inside the Lichi Mélange. These blocks, especially the large ones, are generally angular in shape. Most of them are metric or smaller in size, but some blocks reach hectometric and even kilometric size. Most small blocks (decametric or smaller) are heavily sheared and polished, but many large blocks appear almost internally undeformed (Teng et al., 1988).

In terms of lithology, the blocks considered as "exotic" include three major components (Biq, 1971; Wang, 1976; Hsu, 1976; Liou et al., 1977; Page and Suppe, 1981): andesite suite (andesite, volcanic breccias, tuffs, and volcaniclastic turbidites), ophiolitic suite (Figs. 3a and 3a'; serpentine, gabbro and pillow basalt) and sedimentary suite (sandstone, sandstone/shale interbeds, shale, and limestone). The andesite suite is derived from the Luzon volcanic arc, Miocene in age, as can be observed in the Tuluanshan Formation of the Coastal Range (Hsu, 1956). The origin of the dismembered ophiolite blocks in the Lichi Mélange may represent either the oceanic crust of normal South China Sea (Chung and Sun, 1992) or that of the Philippine Sea beneath the Luzon forearc/arc (Juan et al., 1980; Malavieille et al., 2002). Based on the analysis of the associated sediments, Suppe et al. (1977, 1981) pointed out that this ophiolite could not originate from a normal oceanic crust, but more likely represented an "irregular" oceanic crust.

The sedimentary blocks inside the Lichi Mélange include the weak-lithified Pliocene turbidites (blocks of decametric to kilometric size) with similar lithology, age and sedimentary turbidite structures as the remnant coherent forearc basin strata of the Coastal Range (named Takangkou Formation; Hsu, 1956). They also include the well lithified, whitish quartz-rich, feldspathic sandstones of late Miocene age (angular blocks of metric to kilometric size), very similar to the deep-sea fan sandstones (named Mutan Formation; Sung, 1991) that belong to the upper part of the accretionary prism in the Hengchun Peninsula (a southern extension of Central Range; Cheng et al., 1984; Huang et al., 1997). Remarkably, these sandstones are not observed in the coherent forearc basin turbidites of the Coastal Range, which are younger in age (Plio-Pleistocene).

The muddy matrix of the Lichi Mélange has also been analysed. Clay mineral compositions of the sheared mudstone are characterised by bearing illite (48.8%), chlorite (8.4%), kaolinite (13.9%), smectite (1.9%) and mixed-layer clay minerals (MLC, 27.0%) (Lin and Chen, 1986). Biostratigraphic studies revealed that the protolith of the Lichi Mélange was deposited in a narrow age range within Zone N19/20 of planktic foraminifers or Zone NN15 (3.5–3.7 Ma) of calcareous nannoplanktons (Chi et al., 1981; Chi, 1982; Barrier and Muller, 1984). This age is coeval to the lower remnant forearc basin sequence but is older than that of the upper forearc sequence of the Coastal Range (Zones N21–22 of planktic foraminifers, Chang, 1967; Zones NN16~19 of calcareous nannoplanktons, <3.5~1.15 Ma, Horng and Shea, 1996). This indicates that (1) the turbidites in the proto-Lichi Mélange were deposited in a short time span (~0.2 m.y); (2) the deformation (at ~3.5 Ma) of the proto-Lichi Mélange occurred very quickly after these turbidites were deposited (3.5~3.7 Ma); and (3) after deformation of the proto-Lichi Mélange, the young turbidites (3~1 Ma) were continuously deposited in the upper part of the remnant forearc basin (Huang et al., 2007).

The estimation of the thickness of the Lichi Mélange is known from a Chinese Petroleum Corporation well near Houtzeshan in the southernmost Coastal Range (Meng and Chiang, 1965). The well penetrated 1,056 m without reaching the base, which shows that the thickness of the Lichi Mélange is larger than 1 km. Note that because of the high deformation inside the Lichi Mélange this thickness should not be regarded as depositional. Besides, the well penetrated a number of slabs of basalt and other basic rocks embedded in the muddy matrix, demonstrating that the composition of the Lichi Mélange is chaotic at depth as well as at the surface (Hsu, 1976). The contact between the Lichi Mélange and other rock units of the Coastal Range is in most cases a fault, commonly a thrust (Fig. 2a; Hsu, 1976). It is very significant, however, that in some cases local interfingering occurs with the strata of the Takangkou Formation (Page and Suppe, 1981; Barrier and Muller, 1984).

The outcrops of the Lichi Mélange show large variation in intensity of scaly foliation, with in some cases a transition towards moderately deformed sediments that suggests a common origin with the adjacent flysch-type sediments of the lower Takangkou formation, Pliocene in age (Figs. 5a1, 2). Chang et al. (2000, 2001) thus distinguished four different domains within the Lichi Mélange and the adjacent formations. They include: the coherent unit (unit α), in which the internal stratal continuity is fully preserved; the broken unit (unit β), in which the rock is locally disrupted with most of the stratal continuity preserved; the dismembered unit (unit γ), in which disruption is complete so that no internal stratal continuity remains, but exotic blocks are still absent; the actual mélange (unit δ), in which the material is extensively sheared and exotic blocks are present. Based on this classification, Chang

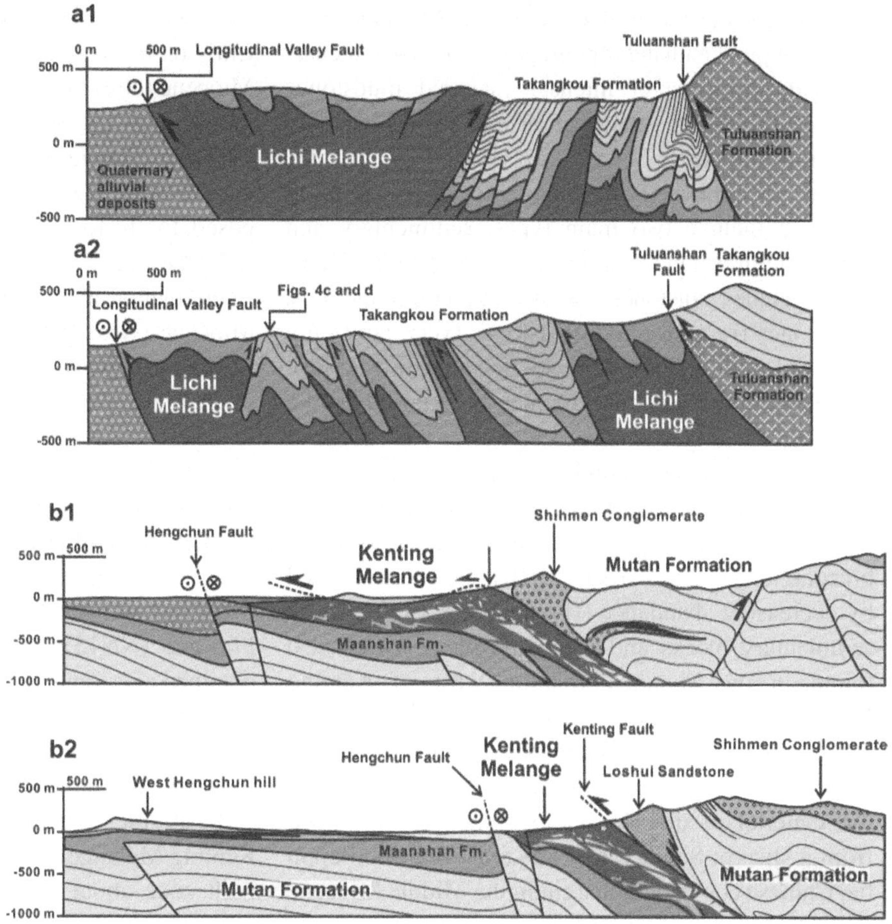

Fig. 5 Geological profiles of the Lichi (a1 and a2) and Kenting (b1 and b2) areas (modified after Chang et al., 2000, 2001, 2003). Locations of profiles are indicated in Fig. 2. The strike-slip component of strike-slip motion is indicated in the cross-sections by small circles with cross and dot

et al. (2000, 2001) carried out a detailed structural study in the southern Coastal Range and pointed out that the Lichi Mélange deserves interpretation as a sheared lower forearc basin sediment sequence. Because of the later development of the structure (thrusting, backthrusting and even diapiric extrusion), it is presently exposed along the western thrust boundary and some main fault zones (for example, that of the Loho basin, see location in Fig. 2a) in the Coastal Range (Figs. 5a1, 2).

Some surprising structural relationships have been observed in the geological profiles of Fig. 5a1, 2. Whereas many contacts between the Lichi Mélange and the Takangkou Formation are parallel to the adjacent strata, which suggests normal stratigraphic relationships, the younger Takangkou Formation is thrusted over the older Lichi Mélange in some places. This is the case of the Tuluanshan Fault in Fig. 5a2: the upthrust block, which includes the upper Takangkou Formation on top of the Tuluanshan andesitic Formation, overrides both the whole Takangkou Formation and the Lichi Mélange. This field reality could be explained through the occurrence of pre-existing backthrust followed by newly developing thrust (Chang et al., 2000, 2001), knowing that further west typical backthrust structures occur in the Lichi Mélange and Takangkou Formation (Fig. 5a). This process is discussed in the later Section "Forearc subduction stage".

3.2 The Kenting Mélange

The Kenting Mélange was defined by Tsan (1974a, b). It generally crops out in a narrow area of low hills that bound to the west the mountainous Miocene turbidites of the Hengchun Peninsula (Figs. 2b and 3b). In the Kenting Mélange, sheared polygenic clasts of millimeter to hundred-meter size were embedded in a scaly argillaceous matrix. In the field exposures, the Kenting

Mélange looks very similar to the Lichi Mélange: both exhibit typical badland topography (Fig. 3b) and characteristic presence of intensely sheared mudstones without distinctive stratification and pervasive scaly foliation (Fig. 4a).

The rock blocks and fragments of the Kenting Mélange include two main types, sedimentary and magmatic in origin (Pelletier, 1985). The sedimentary type includes siltstones, sandstones, conglomerates, limestones (rare) and red mudstones (very rare: only observed west of Paolishan, see location in Fig. 2b). Conglomerate blocks are frequently exposed in the mélange; the largest one is located near the village of Kenting as a several hundreds metres high isolated piton dominating the surrounding landscape. The magmatic type component includes pillow lavas, volcanic breccias and ultrabasic rocks (such as peridotite), listed by decreasing abundance (Huang et al., 1985). These sedimentary blocks and magmatic blocks were also found in the Mutan Formation west of Kenting Mélange. Pelletier and Stephan (1986) considered this as an evidence for subaerial exposure of a complex source terrain in the Miocene and further proposed that the material of the Hengchun Peninsula originated from a slice of South China Sea crust obducted upon the Chinese continental margin in the Middle Miocene time. No other clear evidence could be found in the Taiwan orogen. This obduction might have occurred 7–8 M.y. before the beginning of the last Taiwan collision collision. It is not considered in our reconstruction because it predated the Late Miocene to Present time span that we consider.

The clay mineral compositions of the sheared mudstone in the Kenting Mélange are characterised by bearing illite (57.8%), chlorite (10.3%), dickite (11.5%), montmorillonite (2.0%) and MLC (18.6%) (Lin and Wang, 2001). Contrary to the Lichi Mélange, the Kenting Mélange contains no kaolinite whereas dickite and montmorillonite are abundant.

In contrast with the Lichi Mélange, the age of the Kenting Mélange is controversial, because various ages were obtained by several researchers. First, Chang (1965, 1966) revealed, based on determinations of foraminifera, a Middle to Late Miocene age. Later, Chi (1982) produced the first detailed biostratigraphic study in the Kenting Mélange, with a variety of ages ranging from the Eocene (rare) to Miocene (very abundant); its most recent age determination was Late Miocene (NN 11). Page and Lan (1983), based on Chi's dating results (1982), proposed an age of Late Miocene (NN11) to middle Pliocene (NN15), which is the age of the base of the Maanshan Formation. However, Huang et al. (1983) found nannofossils of late Pliocene age in the Kenting Mélange, and proposed, for the formation of this mélange, a late Pliocene (NNH16) or even more recent age. Huang (1984) found planktonic foraminifera of different ages: Early Miocene (N5), Middle Miocene (N7 with N9), Late Miocene (N14 with N17), Pliocene (N19), and even early Pleistocene (N22). The Kenting Mélange has thus been considered by this author as an olistostrome of age from Middle Miocene to early Pleistocene (N22 being equivalent to NN19). Pelletier et al. (1985) and Pelletier and Stephan (1986) proposed, after examining approximately 200 samples from the Kenting Mélange, an age of Late Miocene (NN11) to middle Pliocene (NN15). In summary, the most frequently obtained ages range from Middle to Late Miocene, but many Pliocene ages (most of them indicating lower-middle Pliocene) have also been obtained and the early Pleistocene is present as pointed out by Huang (1984).

From the structural geology point of view, the Kenting Mélange can be considered as a mega-sheared fault zone about 1 km wide and 20 km long (Chang et al., 2003) that cuts across the Miocene Mutan Formation (Figs. 2b, 5b1 and 5b2). The geometrical distribution of the shear features in the Kenting Mélange suggests that this zone dips to the east with an angle of about 30° or less (Figs. 5b1, 2). Chang et al. (2003) named this shear zone as the "Kenting Fault". To the east, in the hanging wall of the mega-sheared fault zone, the Mutan Formation, which forms the main mountain range of the Hengchun peninsula, has been thrusted westward along the Kenting Mélange. To the west, in contrast with the sinuous upper boundary of the mélange zone, the western boundary of the Kenting Mélange appears as a linear structure (Fig. 2b), because it has been cut by the steeply east-dipping Hengchun Fault (Pelletier, 1985; Sung, 1991) (Figs. 2b, 5b1, 2). As the outcrops of the Hengchun Fault are generally covered by the quaternary sediments, the dip of this fault is mainly documented by a recent seismic reflection profile (Li et al., 2001). Moreover, because of the presence of both the Hengchun Fault and the Quaternary formations, the base of the mélange zone has never been directly observed.

Table 1 Comparison table of the Lichi Mélange and the Kenting Mélange

		Lichi Mélange	Kenting Mélange
Similarities	Matrix	Intensely sheared mudstones without distinctive stratification. Some layers with coherent stratification have been locally reported	
	Texture	Scaly foliation, dense pattern of small faults with slickenside lineations (mainly thrust and strike-slip in type, also normal)	
	Exotic blocks	Ophiolitic blocks and sedimentary blocks, metric to kilometric in size	
	Boundary structure	Bounded by west-verging thrusts at both eastern and western boundaries of mélange zone	
	Footwall	Quaternary fluvial deposit (both Lichi and Kenting) and coral reef (Kenting only) in the west	
	Tectonic regime	Mainly NW-SE compression indicated by fault slip data and other structures inside the mélange	
Differences	Clay mineral	Illite (48.8%), chlorite (8.4%), kaolinite (13.9%), smectite (1.9%) and mixed-layer clay minerals (27.0%)	Illite (57.8%), chlorite (10.3%), dickite (11.5%), montmorillonite (2.0%) and mixed-layer clay minerals (18.6%)
	Age of matrix	Rather well constrained, 3.5~3.7 Ma	Large dispersion of ages, approximate range 1 to10 Ma
	Exotic blocks	– Include both Miocene and Pliocene sedimentary rocks – Include Miocene volcanic rocks (with volcanic breccias, tuffs, and volcaniclastic turbidites) issued from the Luzon Arc – Not include ultrabasic rocks	– Include Miocene turbiditic blocks, but does not include Pliocene sedimentary rocks – Not include volcanic rocks – Include ultrabasic rocks
	Pebbly layer	Locally reported (especially in the rivers of the central segment)	Not found (but many conglomerate blocks exposed)
	Limestone included	Pliocene Kangkou Limestone (around the summit of the volcanic arc east to the mélange)	Pleistocene Hengchun Limestone (apron coral reef and lagoon phase limestone exposed in west of the mélange)
	Hanging wall strata	Miocene volcanic basement overlaid by Pliocene turbidites to the east	Miocene turbidite formation to the east
	Geographic location	Along the southwestern flank of the Coastal Range, east of the Central Range and Longitudinal Valley	In a narrow area of low hills that bound the main range of Hengchun Peninsula (souther extension of Central Range) to the west
	Geological location	Between the Taiwan accretionary prism/Central Range and the Luzon volcanic arc	Between the foreland of the Taiwan belt and the western flank of the southern Taiwan accretionary prism
	Origin of ophiolite	Basement of forearc basin (probably the Philippine Sea oceanic crust)	South China Sea oceanic crust
	Tectonic origin	Forearc mélange: originates from the thrusted strata at the front of a forearc basin west of, and attached to, the Luzon Arc	Formed at the front of an accretionary prism above the subduction zone, representing the relic of a submarine trench

A detailed comparison between the Lichi Mélange and the Kenting Mélange is presented in Table 1. We took their similarities and differences into account based on previously published information and our observations. Among the multiple characteristics summarised in this Table, some are of particular importance and will be invoked in the following discussion. They include the clay mineral composition, the origin of ophiolitic blocks, the particular nature of some sedimentary blocks and the age of the mélange matrix.

4 Significance of Clay Mineral Composition and Sources of Sedimentary Rock in Mélanges

According to a systematic study by Lin and Chen (1986), the clay mineral compositions of the muddy matrix and sedimentary blocks in various facies and exposures of the Lichi Mélange are very similar and contain illite, chlorite, kaolinite and little smectite.

This composition resembles that of the clay mineral assemblages of the Takangkou flysch-like formation of the remnant forearc basin of the Coastal Range, except for kaolinite. Lin and Chen (1986) claimed that the Lichi Mélange could have two sediment sources. One of these sources involves continental materials, which were rich in illite and chlorite and underwent transport from land into the marine basin, coming from the eastern Asian continent and/or the accretionary prism of the subduction zone (such as the ancient Central Range and its southern extension). The other source involves the volcanic arc material, which is rich in kaolinite and smectite and was transported from the volcanic arc in a tropical area (such as the Philippine islands). These two contrasting groups of source materials were simultaneously deposited in a single major sedimentary basin, that of future Lichi Mélange.

As indicated by the presence of large amounts of kaolinite in the Mélange, part of the source area for the primary strata precursor of the Lichi Mélange had a humid, warm climate with much rainfall. This context fits well with the Miocene paleo-position of the Luzon arc, far to the southeast of the present location of the Lichi Mélange in Taiwan. In contrast with the Lichi Mélange, the kaolinite is rare or almost absent in the clay fractions of the adjacent Takangkou Formation (the remnant forearc deposition in the Coastal Range). This suggests that contrary to the Lichi Mélange basin, the typical forearc basin received more material from a continental source to the west (like the ancient Central Range) than from a volcanic arc source to the east (Lin and Chen, 1986; Chang, 1996). Such a contrast favours an eastern location for the depocentre of the future Lichi Mélange during the Pliocene, as compared with the main forearc basin.

In terms of matrix clay contents, the most significant difference between the Lichi Mélange and the Kenting Mélange is the absence of kaolinite in the latter formation. The Kenting Mélange contains illite, chlorite, dickite, montmorillonite and MLC. A comparison between the Kenting Mélange and the Miocene Mutan Formation (that forms the main body of the Hengchun Peninsula) reveals closely similar compositions and the same polytypes of clay minerals. This similarity suggests that these two formations had a common sedimentary source, a terrain of acidic igneous and/or metamorphic rock with grade probably up to greenschist facies. Based on such observations, it is reasonable to consider the Kenting Mélange as the dismantled facies of the Mutan Formation affected by strong tectonic disruption and pervasive shear near a the west-verging low-angle thrust (Lin and Wang, 2001).

As could be expected according to the above interpretation, the sedimentary blocks contained in the Kenting Mélange principally consist of turbidites derived from the body of the Hengchun peninsula – the Miocene Mutan formation. This is another significant difference between the two mélanges: the sedimentary blocks in the Lichi Mélange display variety. In addition to the largest component made of rocks coming from the remnant forearc basin of the Coastal Range, one observes whitish, well-lithified, quartz-rich, feldspathic sandstone blocks in the Lichi Mélange. These sandstones were determined to be Late Miocene in age (Chi, 1982; Barrier and Muller, 1984), consistent with those of the Mutan Formation of the Hengchun peninsula (Chang, 1966; Huang et al., 1997) and clearly older than the Plio-Pleistocene remnant forearc basin strata and the Pliocene protolith of the Lichi Mélange. Assuming that these Miocene sandstones were actually derived from the Hengchun peninsula, two possible routes exist to explain their presence in the Lichi Mélange: (a) erosion of the exposed accretionary prism (that is, the future Central Range) before exhumation of the underthrust Eurasian continent materials in the last 2.5 Ma (Huang et al., 2006) followed by transport towards the North Luzon Trough (the forearc basin) by turbidity flow or mass slumping processes (Page and Suppe, 1981); or (b) eastward thrusting pushing the non-exposed Late Miocene accretionary prism materials upward into the deformed forearc strata (proto-Lichi Mélange like the modern Huatung Ridge) in the western part of the North Luzon Trough (Fig. 1). Because these particular Miocene standstones have never been observed in the weakly-sheared broken formation facies of the Lichi Mélange or in the remnant forearc basin sequences (as should have been the case had they been deposited in the Lichi Mélange basin), the first route is unlikely. In contrast, the second route may well account for the presence of exotic sedimentary blocks fully independent with respect to the turbiditic sediments of the Lichi Mélange basin and remnant forearc basin. The observation of marine seismic-reflection profiles suggests that such sandstones from the non-exposed accretionary prism may have been thrusted towards the western North Luzon Trough forearc basin to be incorporated in the deformed proto-Lichi Mélange (Huatung Ridge), instead of being derived from the exposed Taiwan accretionary prism by turbidity flows or mass slumping downward to the entire North Luzon Trough (Chang et al., 2001).

5 Origin of Ophiolite and Tectonic Evolution of Mélanges

After considering the geological situation of these two mélanges, it is appropriate to interpret their origin and evolution within the framework of the Late Cenozoic convergence in Taiwan. To this respect, the origin and significance of the ophiolites present in both the Lichi Mélange and the Kenting Mélange are key issues.

There is some agreement to consider that the ophiolite blocks in the Kenting Mélange represent the oceanic crust of the South China Sea (SCS), which was scraped into the Taiwan accretionary prism by thrusts and possible ductile flow (Biq, 1977; Lu and Hsu, 1992; Hsu, 1988). In contrast, the origin and the emplacement mechanism of the ophiolite blocks of the Lichi Mélange (the East Taiwan Ophiolite, hereafter ETO) is still a subject of debate. Three possible sources of the ophiolite blocks in the Lichi Mélange have been considered: (1) the South China Sea oceanic crust (Chung and Sun, 1992), (2) the oceanic basement beneath the Luzon arc/forearc of the Philippine Sea plate (Juan et al., 1980; Malavieille et al., 2002) and (3) the "irregular" oceanic crust produced at ridge or close to a transform zone (Suppe et al., 1977, 1981).

Had the ETO been part of the SCS oceanic crust, its age would be younger than late Oligocene because the SCS opened in 32~17Ma (Taylor and Hayes, 1980). Microfossils recovered from the thin red shale intercalated in the ETO sequence showed an age of 15Ma (Zone NN5 of calcareous nannoplanktons; Huang et al., 1979), close to the last phase of the SCS opening. This may favour the SCS oceanic crust source. The ETO blocks are generally massive and large in size, some reaching hectometric and even kilometric size (such as the major block near Kuanshan, in the southern Coastal Range). The sedimentary source of ETO, if a SCS source is assumed, should be very close to its present location; however, no ETO-like rock occurs in the present Central Range except some large masses near the Yuli and Juishui areas (see location in Fig. 1). These masses display relatively high metamorphic grade and have been incorporated in the syn-metamorphic deformation of the eastern Central Range, so that they probably have a different origin as compared with the ETO. For these reasons, the ETO blocks are unlikely to have been derived from the exposed accretionary prism (Central Range) as a result of sedimentary slumping processes. Interestingly, in the Lichi area the ETO blocks often coexist with typical Miocene whitish sandstone blocks. As discussed in the previous section, these particular sandstones were probably emplaced in the Lichi Mélange by eastward backthrusting processes, rather than simple mass slumping.

This observation may suggest a similar emplacement for the ETO, which could have also been thrusted eastward from the accretionary prism of the Hengchun Peninsula into the Lichi Mélange basin, during the early stage of the arc-continent collision. The hypothesis of a SCS origin for the ETO, however, encounters noticeable difficulties. Note first that unlike the widespread deep-sea fan sandstones of the Hengchun Peninsula (Huang et al., 1997), the ophiolite blocks in the Kenting Mélange occur in a very limited area (Fig. 2b), whereas the ETO is common along the southern Longitudinal Valley. More important, these ophiolite blocks in the Kenting Mélange markedly differ in composition from the ETO. As an example, the chromitite that was commonly found in the Kenting Mélange of the Hengchun Peninsula (Chu et al., 1988) was never observed in the ETO. We conclude that although a plausible emplacement mechanism can be invoked (the eastward backthrusting, such as for the sandstones probably issued from the Hengchun Peninsula domain) it is unlikely for the ETO to have originated from the West (the accretionary prism of the Hengchun Peninsula). A simpler hypothesis involves an eastern origin. The ETO would represent the oceanic crust underlying the forearc basin, an origin already considered by several authors (Juan et al., 1980; Malavieille et al., 2002). This oceanic crust would have been emplaced by thrusting processes that affected the Northern Luzon Trough or the Luzon arc basement during the final westward accretion of Luzon arc-forearc to eastern Taiwan in the last few million years. It is also possible that the ETO could be developed by suprasubduction (Pearce et al., 1984) by which the lower curst materials like serpentinite and peridotite can be found in the forearc basin setting like the modern Marianas (Fryer et al., 1985).

Suppe et al. (1977, 1981) observed that the plutonic parts of the ETO are generally breccias and have pelagic red clays intermixed. Because the red clays were deposited below the calcite compensation depth, which is too deep for most of the proposed origins of standard oceanic crust, they consequently interpreted these plutonic breccias as deep submarine scree deposits derived from a fault scarp in oceanic crust, probably an active ridge-ridge transform fault (Suppe et al., 1977, 1981). We accept their interpretation concerning the ETO is part of the forearc basin sequence, but our opinion differs concerning the site of the "fault scarp". According to Suppe et al. (1981), this site is located on the subducted South China Sea; however, this hypothesis meets the same problem as that of the SCS origin mentioned before. Another conceivable location for this "fault scarp" is the Ryukyu-Luzon Transform Fault (RLTF in Fig. 6b); a left-lateral transform fault that connected the Ryukyu trench and the Manila trench before the Taiwan orogeny (Angelier, 1990). Because according to our interpretation the Lichi Mélange originated from the lower forearc basin and formed by the follow-up tectonic processes, the presence of ETO, even originating from larger depths than the sediment, is easily explained. For the sake of simplificity, we consider the ETO as the forearc basement hereafter.

In summary, whereas the origin and evolution of the Kenting Mélange is relatively simple, that of the Lichi Mélange deserves discussion. In southern Taiwan, during the Plio-Pleistocene time the Miocene sediments of the Hengchun Peninsula were incorporated into the Taiwan accretionary prism along the Manila trench. At present, the front of the Manila trench had propagated, or jumped westward to the southwestern offshore Taiwan so that the affected units, including the Kenting Mélange, are now at a large distance east of the subduction front. This evolving structural situation with respect to the trench axis and subduction front could well explain why the Kenting Mélange, albeit mainly formed with Late Miocene sediments, contains Plio-Pleistocene elements. We infer that these younger ages may belong to mélange sub-zones that have been tectonically reworked and rejuvenated in marine sediments, during the westward migration of the prism front, a phenomenon that has been illustrated at a wider scale by out-of-sequence thrust sheets near the fronts of mountain belts.

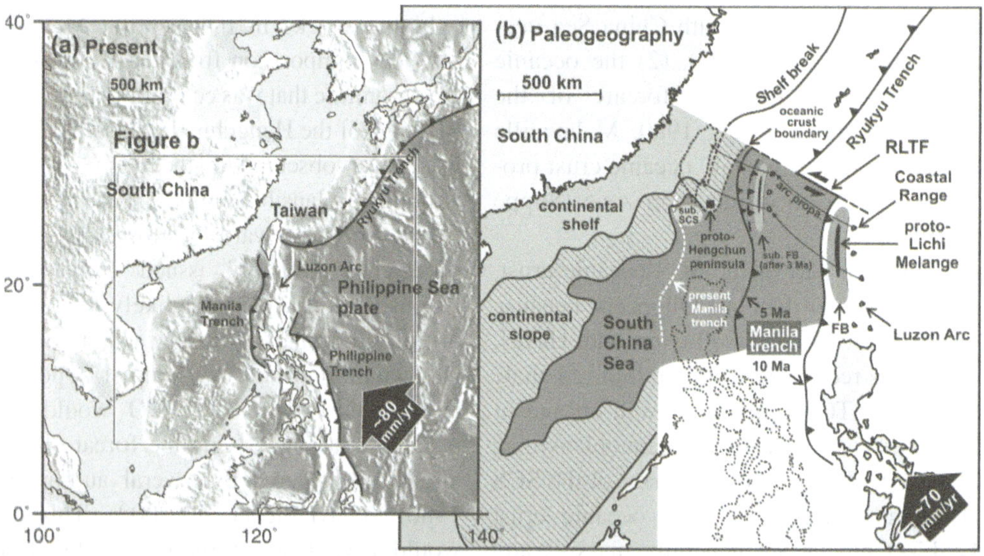

Fig. 6 (**a**) Topographic map of the Taiwan area. Principal subduction zones are shown, including: Ryukyu Trench, Manila Trench, and Philippine Trench. *Large black arrow* shows the movement of the Philippine Sea plate against the Eurasian plate. (**b**) Presumed paleogeographic configuration of Ryukyu-Luzon arcs and the trenches at around 10 Ma and 5 Ma. The present Luzon arc is shown by *blank islands* and *dashed lines*. The proto- Hengchun Peninsula (shown by *black rectangular*) was located on the base of continental slope southeastern to its present position. arc propa.: Moving path of the Luzon Arc on the Philippine Sea plate; *FB*, forearc basin, *RLTF* Ryukyu-Luzon Transform Fault, a left-lateral transform fault connected the Ryukyu trench and the Manila trench before the Taiwan orogeny (Angelier, 1990); sub. FB and sub. *SCS* subduction of forearc basin and South China Sea oceanic crust

From the tectonic point of view, the Kenting Mélange is thus considered herein as a mega-sheared fault zone located (when active) at the front of the ancient Taiwan accretionary prism (but now inside the prism, because of westward migration of most active thrusts). This fault zone cuts across the Miocene Mutan formation and may represent a relic of the Manila trench. Regarding the Lichi Mélange, if we accept the most reasonable hypothesis, that is the ETO originating from the forearc basement oceanic crust, the Lichi Mélange should be considered as part of the deformed North Luzon Trough forearc sequences. It would thus represent a collision mélange formed during the forearc closure, when the Luzon arc started to collide with the Eurasian continent. In this case, the simpler hypothesis would consist of explaining the genesis of the Lichi Mélange by the west-vergent thrusting process. Such an hypothesis cannot provide any good explanation for the presence of the blocks of white quartz-rich sandstone similar to the Mutan formation, which originate from the Hengchun Peninsula domain to the west. Moreover, considering that the Lichi Mélange uniquely results from the westward thrusting at the front of a subduction zone fails to explain why fossils constrain a short time span for the age of the mélange matrix during the Pliocene, especially why the Miocene is absent except in the blocks. For these reasons, we consider that the Lichi Mélange probably developed in connection with the along arcward (eastward) backthrusting that is well documented along the eastern flank of the Central Range of Taiwan (Chang et al., 2007). Invoking this backthrusting process as a major element responsible for the development of the Lichi Mélange accounts for this relatively recent age, as compared with the long-term convergence between the Luzon arc system and the Eurasian continental margin. It is also consistent with the short lifetime of the Lichi Mélange and the presence of sandstones coming from the eastern slope of the accretionary prism to the west.

We finally point out that a variety of independent observations concur to highlight the differences between the two mélanges of Taiwan. Not only does the composition of sedimentary and ophiolitic blocks suggest differences in origin, but also the age range difference between the sediment matrixes of the Lichi and the Kenting mélanges shed light on a very significant difference in terms of their basin location and origin.

6 From Subduction to Collision: Evolution of the Manila Subduction System

Figure 6 shows the present (Fig. 6a) and ancient (Fig. 6b) tectonic configuration around the Taiwan island. Before considering the subduction evolution of the Manila system, one needs to discuss the origin of the South China Sea. The rifting of the Paleocene-Eocene times led to the development of a series of NE-SW trending grabens along the SE Asian continental margin (Sun, 1982). From the Oligocene to the Middle Miocene, widespread rifting and sea-floor spreading resulted in the formation of the oceanic lithosphere of the South China Sea, which ceased about 16 Ma ago (Taylor and Hayes, 1983). Soon after it was generated in the Early Miocene, the South China Sea oceanic lithosphere was subducted beneath the west-moving Philippine Sea plate along the Manila trench (Fig. 6b). The present Luzon arc thus formed in the western edge zone of the Philippine Sea plate. Continued subduction finally resulted in the collision between the Luzon arc and the Eurasian continent in the Late Miocene (Suppe, 1984; Ho, 1986).

From the kinematic point of view, it is likely that, between the northern tip of the Luzon arc and the southwestern tip of the Ryukyu subduction zone, a large zone of left-lateral transform motion accommodated the opposite-vergent subductions, as the Taiwan-Luzon transform zone (Angelier, 1990) (Fig. 6b). The present Taiwan island is a product of arc-continent collision (Chai, 1972; Biq, 1973; Bowin et al., 1978) (Fig. 6a), and because this arc-continent collision is an oblique collision, the orogenic belt has been propagating southward (Suppe, 1984).

After considering the geometric situation of plates, the distribution of continental margin, the geological situation of the Kenting and Lichi Mélanges, and the kinematic state, the tectonic evolution of the Manila subduction system, from its subduction to collision, can be reconstructed as follows.

6.1 *Pure Subduction Stage (Fig. 7a)*

This stage corresponds to Miocene time (before 5 Ma) in Taiwan. The South China Sea oceanic crust was sub-

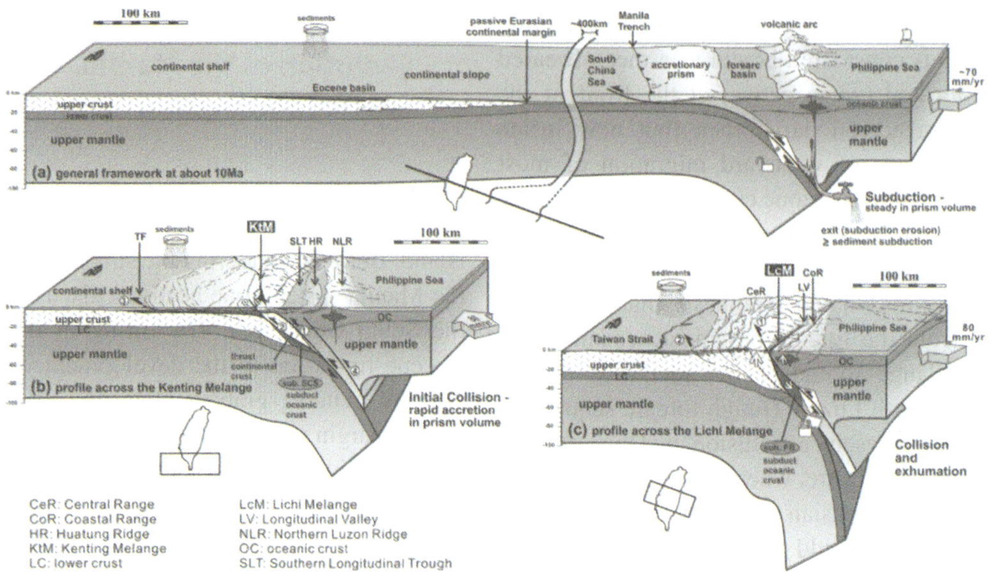

Fig. 7 Lithospheric profile of the Taiwan orogen from subduction to collision. (**a**) General framework at about 10 Ma. (**b**) Profile cuts across the present Kenting Mélange. *1*: older relic of submarine trench; *2*: younger relic of submarine trench; *3*: present thrust front; *4*: new generated lithospheric fault. (**c**) Profile cuts across the present Lichi Mélange. *1*: Longitudinal Valley fault; *2*: thrust front. sub. FB and sub. SCS: subduction of forearc basin and South China Sea oceanic crust (see paleogeographic location in Fig. 6b)

ducting beneath the Philippine Sea plate to generate an arc–trench system and a forearc basin. At this stage, the Manila trench was the active plate boundary between the Eurasian plate and the Philippine Sea plate. An accretionary prism, composed of deep-marine sediments originally deposited on the South China Sea oceanic crust, developed east of the Manila trench; this accretionary prism propagated westward gradually and smoothly. The growth rate of the accretionary prism increased when the subduction front was approaching the continental margin, because of the increasing abundance of sediments. However, the main Taiwan arc-continent collision had not yet occurred.

6.2 Trench Jump Stage (Fig. 7b)

This stage corresponds to Plio-Pleistocene time (from 5 Ma to around 2 Ma) in Taiwan. Considering the orogenic propagation from north to south along the main mountain trend of Taiwan, the general tectonic configuration of this stage can be observed today in the cross-section of southern Taiwan, cutting through the Hengchun peninsula and the Kenting Mélange (Fig. 7b). At this stage, the subduction front of the Manila trench (especially Fault 1 in Fig. 7b) reached the Chinese continental margin. Because of the presence of the continental crust to the west, the westward propagation of the submarine trench and accretionary prism met increasing mechanical resistance. As the most likely scenario, the trench developed itself first along the innately weak zone in the continental crust (Fault 2 in Fig. 7b), and then transformed into the thrust front (TF or Fault 3 in Fig. 7b) and propagated (or jumped) into the thick sediment layer on the continental shelf. A large slice of continental crust has thus been scraped off and incorporated into the accretionary prism by propagation of the Manila trench. This evolution well accounts for the relatively long history and present-day position of the Kenting Mélange as a major thrust zone that has occasionally been re-activated after the upper Miocene.

We consider the paleogeographic configuration as an important factor while generating the Kenting Mélange. According to the reconstruction of Fig. 6b, the original continental margin was not straight: before the subduction, part of the continental slope was protruding in the area SE to the present Hengchun peninsula (Fig. 6b, proto-Hengchun peninsula). When the submarine trench (Fault 1 in Fig. 7b) propagated and reached this area, the protruding continental crust of this proto-Hengchun peninsula and the oceanic crust of South China Sea west to it experienced thrusting, part of the oceanic crust (sub. SCS in Figs. 6b and 7b) being broken and subducted along a large thrust fault (Fault 2 in Fig. 7b, resembling a relay of jumping trench). This phenomenon probably commenced from the Plio-Pleistocene time (the initial stage of arc-continent collision, Chi et al., 1981; Chang and Chi, 1983; Dorsey, 1985, 1988) and left some ophiolite blocks in the Kenting Mélange (scraped off by thrusting). Then, as a result of the increasing resistance of the continental crust, the trench zone changed from an active subduction plate boundary into a separated thrust zone (Fault 3 in Fig. 7b) in the western foreland area.

Based on the stratigraphic analysis, Pelletier and Stephan (1986) and Stephan et al. (1986) considered the Hengchun stratigraphy to have formed on the Chinese continental margin, close to its present location. However, the Hengchun peninsula lies now 100 km east of the thrust front and in the uppermost position of the present accretionary prism (Fig. 7b). Suppe (1988) therefore considered that the Hengchun peninsula has been incorporated into the wedge relatively early in its history. He further proposed a collision with rifted fragments of China on the south side of the South China Sea in the Miocene to explain the stratigraphic and tectonic records of the Hengchun Peninsula. However, there is no clear evidence for the existence of a micro-continent in the present Taiwan orogen. The protruding continental margin and the trench jump process proposed in this paper can be regarded as a compromise but provides a reasonable solution to explain this dilemma.

Later during this stage, east of the accretionary prism of the southern Central Range and Hengchun Peninsula, the narrow domain left by continuing convergence between the Luzon arc and the accretionary prism experienced increasing shortening, which produced a new lithospheric fault (Fault 4 in Fig. 7b) below the forearc area as well as series of back-thrusts in the eastern accretionary prism and forearc basin, this increasing compression uplifted the precursor of the Lichi Mélange, that is the Huatung Ridge (HR in Fig. 7b. Chang et al., 2000, 2001; Huang et al., 2007).

6.3 *Forearc Subduction Stage (Fig. 7c)*

This stage started ~3.5 Ma ago in eastern Taiwan, when the trench jump began to be blocked. This fits well the independent observation concerning the very beginning of the paroxysmal collision in central-southern Taiwan (e.g., Chang and Chi, 1983). The mature tectonic configuration of this stage can be observed today in a cross-section cutting across the Coastal Range and the Lichi Mélange (Fig. 6c). At this stage, the Longitudinal Valley fault (LV; Fault 1 in Fig. 7c) replaced the submarine trench and the thrust front of the western foreland (Fault 2 in Fig. 7c), as the most active structure accommodating shortening in Taiwan. This LV fault occupies the site of arc-prism boundary, and represents the principal present-day zone of weakness and shortening in the Taiwan orogenic belt.

Continuous shortening caused the basement of the forearc basin (sub. FB in Figs. 6b and 7c; probably the oceanic crust of the Philippine Sea) to subduct beneath the volcanic arc along the LV fault. The material in the western forearc basin (including the Pliocene forearc sediments, the Miocene Mutan Formation and part of the forearc basement) was extruded and exposed at the western border of the Coastal Range to form the Lichi Mélange. In the initial period, the backthrusts formed mechanically consistent conjugate systems of reverse faults with the thrusts in front of the accretionary prism, and hence played an important role in uplifting the forearc material. However, the present-day attitude of faults in Lichi Mélange principally results from the latest evolution, which also involved major westward thrusting and eastward backthrusting as local shown by geological cross-sections (Fig. 5a). At the latest stage, strong shortening and west-verging thrusting resulted in a rearrangement of the regional structural pattern. This final evolution, with major west-verging thrusts cutting through a complex structure incorporating backthrusts, explains why the younger Takankou Formation could be locally thrusted westwards over the older Lichi Mélange along the Tuluanshan Fault (Fig. 5a2), a situation difficult to understand otherwise.

6.4 *Mélanges as Mega-Thrust Zones*

We finally consider that the Lichi and Kenting Mélanges are two distinct mega-fault zones that represent two separate sites of subduction. On the west side of the major accretionary prism of southern Taiwan (the southern Central Range and Hengchun Peninsula), the Kenting Mélange is one of the relics of the Manila trench, where the South China Sea oceanic crust subducts beneath

Luzon and South Taiwan. East of this main accretionary prism, the Lichi Mélange marked a newly developed subduction zone, where the oceanic crust of the forearc basement of the Luzon arc system was subducting. Note that according to this reconstruction this oceanic crust did not develop in the South China Sea but was probably part of the Philippine Sea Plate.

The large difference between the fossil ages in the matrix of Taiwan mélanges has been a controversial subject for a long time. For the Kenting Mélange, the fossil age ranges from Miocene (~10 Ma) to Pleistocene (~1 Ma) time. On the contrary, for the Lichi Mélange, the fossil age concentrates between 3.5~3.7 Ma. To us, this difference is quite significant. The most simple and direct explanation is that whereas the Kenting Mélange was generated during a longer time span the Lichi Mélange was produced relatively quickly. However, considering the tectonic origin and deformation of mélanges and their relationship with the adjacent strata, this inference deserves discussion. In the model presented before, the Kenting Mélange marks the relic of a subduction fault zone. When the newly formed shear zone cut across the Miocene Formation deposited prior to thrust development, fossils of Miocene time were incorporated; later, when the fault cut through more recent sediments on the sea floor, new sediments and fossils were incorporated in the mélange. We conclude that the variety of fossil ages in the Kenting Mélange results from its origin as a propagating subduction fault. In contrast, the Lichi Mélange was principally generated by compressional deformation process (thrusting and folding) occurring in the forearc basin during the stage of arc-continent collision. The material of the Lichi Mélange was mainly issued from the sediments of the lower forearc basin and adjacent strata. Because of this rather quick development there was no way for the Lichi mélange to gather a large range of fossil ages from the affected sediments. The fossil records are thus consistent with our interpretation of the Kenting Mélange as a subduction mélange and the Lichi Mélange as a collision mélange.

7 Conclusion

The Lichi and Kenting Mélanges of Taiwan play a key role in reconstruction of the subduction history of the Manila-Luzon trench-prism system. Examining the geological evidences collected in the mélanges and the results of field observation in light of the general evolution of the convergence from subduction to arc-continent collision enables us to present the reconstruction summarised in Fig. 7.

All evidences, including the structural configuration, clay mineral composition, age(s) of matrix, nature of exotic blocks, geological position and relationship with adjacent strata, suggest different origins for these two mélanges. West of the accretionary prism, the Kenting Mélange is a relic of the Manila trench, along which the South China Sea oceanic crust was subducted since the middle-late Miocene. East of the accretionary prism, the Lichi Mélange is located at the arc-prism boundary, west of the Luzon volcanic arc and far from the South China Sea. A considerable amount of shortening has occurred between the accretionary prism and the volcanic arc during the arc-continent collision (soon after 3.5 Ma) and induced a smaller-scale subduction which consumed the oceanic crust underlying the forearc basin area (probably belonging to the Philippine Sea plate). The present Lichi Mélange can be considered as the product of this shortening and subduction, but its main development occurred in conjunction with the incipient collision that produced widespread backthrusting on the eastern side of the accretionary prism.

Based on the mélanges origin and evolution, the tectonic history of the Manila subduction system can be reconstructed in more detail than before, including a pure subduction stage (before 5 Ma), a trench jump stage (from 5 to around 2 Ma) and a forearc subduction stage (since the last 3.5 Ma). The mélanges that developed on both sides of the major accretionary prism of southern Taiwan markedly differ. To the west, the Kenting mélange was active during a long period of subduction of the South China Sea floor. The presence of continental material in the Hengchun Peninsula and the exotic block composition in the Kenting Mélange indicate that the paleogeography of the continental margin, curved and protruded, was a major factor inducing trench jump. To the east, the Lichi Mélange provides an example of mélange active during a relatively short period, developing by tectonic thrusting during the closure of a forearc basin in an active arc-continent collision context.

Acknowledgements This work is supported by the National Science Council, under the Grant No. 94-2119-M-008-006, and also by the cooperation framework I.F.T.-N.C.S. (Institut

Français à Taipei and National Science Council of Taiwan) and the C.N.R.S.-N.S.C. International Laboratory ADEPT. We wish to thank Dr. John Suppe and Dr. Patrick de Luca for careful reviews, constructive criticism and wise suggestion on this paper. Helpful suggestions were made by T.-Y. Chang, B. Kou, W.-T. Liang, and J. Malavieille.

References

Angelier J (1990) Foreword, Special Issue "Geodynamic Evolution of the Eastern Eurasian Margin". Tectonophysics 183:II–X

Angelier J, Chu HT, Lee JC (1997) Shear concentration in a collision zone: kinematics of the Chihshang Fault as revealed by outcrop-scale quantification of active faulting, Longitudinal Valley, eastern Taiwan. Tectonophysics 274: 117–143

Barrier E, Muller C (1984) New observations and discussion on the origin and age of the Lichi Mélange. Mem Geol Soc China 6:303–326

Biq C (1971) Comparison of mélange tectonics in Taiwan and in some other mountain belts. Petrol Geol Taiwan 9:79–106

Biq C (1972) Dual-trench structure in the Taiwan-Luzon region. Proc Geol Soc China 15:65–75

Biq C (1973) Kinematic pattern of Taiwan as an example of actual continent-arc collision. Report of the Seminar on Seismology, US-ROC Cooperative Science Program 25:149–166

Biq C (1977) The Kenting Mélange and the Manila Trench. Proc Geol Soc China 20:119–122

Bowin C, Lu RS, Lee CS, Schouten H (1978) Plate convergence and accretion in Taiwan-Luzon region. Am Assoc Petr Geol Bull 62:1643–1672

Chai BHT (1972) Structure and tectonic evolution of Taiwan. Am J Sci 272:389–422

Chang CP (1996) Geological relationship between the Lichi Mélange in the Coastal Range and the Huatung Ridge in the active collision zone off southeastern Taiwan. MS Thesis, Inst Geol, Nat Taiwan Univ, pp 1–120

Chang CP, Angelier J, Huang CY (2000) Origin and evolution of a mélange: the active plate boundary and suture zone of the Longitudinal Valley, Taiwan. Tectonophysics 325: 43–62

Chang CP, Angelier J, Huang CY, Liu CS (2001) Structure evolution and significance of a mélange in a collision belt: the Lichi Mélange and the Taiwan arc-continent collision. Geol Mag 138:633–651

Chang CP, Angelier J, Lee TQ, Huang CY (2003) From continental margin extension to collision orogen: structural development of the Hengchun Peninsula and tectonic rotation, southern Taiwan. Tectonophysics 361:61–82

Chang CP, Angelier J, Lu CY (2007) Polyphase deformation in a newly emerged accretionary prism: folding, faulting and rotation in the southern Taiwan mountain range. Tectonophysics, 10.1016/j.tecto.2007.11.002.

Chang LS (1965) A Biostratigraphic Study of the Tertiary in the Hengchun Peninsula, Taiwan, Based on Smaller Foraminifera (II: Middle Part). Proc Geol Soc China 8:9–18

Chang LS (1966) A biostratigraphic study of the Tertiary in the Hengchun peninsula, Taiwan, based on smaller foraminifera (III: Southern Part). Proc Geol Soc China 9:53–63

Chang LS (1967) A biostratigraphic study of the Tertiary in the Coastal Range, eastern Taiwan, based on smaller foraminifera (I: Southern Part). Proc Geol Soc China 10:64–76

Chang LS (1969) A biostratigraphic study of the Tertiary in the Coastal Range, eastern Taiwan, based on smaller foraminifera (III: Middle Part). Proc Geol Soc China 12:89–101

Chang SSL, Chi WR (1983) Neogene nannoplankton biostratigraphy in Taiwan and the tectonic implications. Proc Geol Taiwan 19:93–147

Chen WS (1991) Origin of the Lichi Mélange in the Coastal Range, eastern Taiwan. Spec Publ Central Geol Surv 5: 257–266

Chen WS (1997) Mesoscopic structures developed in the Lichi Mélange during the arc-continent collision in the Taiwan region. J Geol Soc China 40:415–434

Cheng YM, Huang CY, Yeh JJ, Chen WS (1984) The Loshui Formation: deeper-water sandstone of the Hengchun peninsula, Southern Taiwan. Acta Geol Taiwan 22:100–117

Chi WR (1982) The calcareous nannofossils of the Lichi mélange and the Kenting mélange and their significance in the interpretation of plate tectonics of the Taiwan region. Ti-Chih (Geology) 4:99–112 (in Chinese with English abstract)

Chi WR, Namson J, Suppe J (1981) Stratigraphic record of plate interactions in the Coastal Range of eastern Taiwan. Mem Geol Soc China 4:155–194

Chu HT, Shen P, Jeng RC (1988) The origin of chromitite from the Kenting Mélange, southern Taiwan. Proc Geol Soc China 31:33–52

Chung SL, Sun SS (1992) A new genetic model for the East Taiwan Ophiolite and its implications for Dupal domains in the Northern Hemisphere. Earth Planet Sci Lett 109: 133–145

Dorsey RJ (1985) Petrography of Neogene sandstones from the Coastal Range of eastern Taiwan: response to arc-continent collision. Petrol Geol Taiwan 21:184–215

Dorsey RJ (1988) Provenance evolution and unroofing history of a modern arc-continent collision: evidence from petrography of Plio-Pleistocene sandstones, eastern Taiwan. J Sedi Petrol 58:208–218

Ernst WG (1977) Olistostromes and included ophiolite debris from the Coastal Range of eastern Taiwan. Mem Geol Soc China 2:97–114

Fryer P, Ambos EL, Hussong DM (1985) Origin and emplacement of Mariana forearc seamounts. Geology 13:774–777

Ho CS (1977) Mélanges in the Neogene sequence of Taiwan. Mem Geol Soc China 2:85–96

Ho CS (1986) A synthesis of the geologic evolution of Taiwan. Tectonophysics 125:1–16

Ho CS (1988) An introduction to the geology of Taiwan and explanatory text of the geologic map of Taiwan. Central Geol Surv Taipei, Taiwan, pp 1–192

Horng CS, Shea KS (1996) Dating of the Plio-Pleistocene rapidly deposited sequence based on integrated magneto-biostratigraphy: a case study of the Madagida-Chi section, Coastal Range, eastern Taiwan. J Geol Soc China 39:31–58

Hsu TL (1956) Geology of the Coastal Range, eastern Taiwan. Bull Geol Surv Taiwan 8:39–63

Hsu TL (1976) The Lichi Mélange in the Coastal Range framework. Bull Geol Surv Taiwan 25:87–95
Hsu KJ (1988) Mélanges and the mélange tectonics of Taiwan. Proc Geol Soc China 31:87–92
Huang CY (1984) Field Trip Guide to the Hengchun Peninsula, Southern Taiwan. Field guide book, Sino-French Colloquium on Geodynamics of the Eurasian Philippine Sea Plate Boundary, Taipei, pp 77–94
Huang CY, Cheng YM, Jeh CC (1985) Genesis of the Kenting Formation in the Hengchun Peninsula, southern Taiwan. Ti-Chih (Geology) 6:21–38 (in Chinese with English abstract)
Huang CY, Wu WY, Chang CP, Tsao S, Yuan PB, Lin CW, Xia KY (1997) Tectonic evolution of accretionary prism in the arc-continent collision terrane of Taiwan. Tectonophysics 281:31–51
Huang CY, Yuan PB, Tsao SJ (2006) Temporal and spatial records of active arc-continent collision in Taiwan: a synthesis. Geol Soc Am Bull 118:274–288
Huang CY, Chien CW, Yao B, Chang CP (2007) The Lichi Mélange: a collision mélange formation along early arcward backthrusts during forearc basin closure, Taiwan arc-continent collision. Geol Soc Am special paper (in press)
Huang TC, Chen MP, Chi WR (1979) Calcareous nannofossils from the red shale of the ophiolite-mélange complex, eastern Taiwan. Mem Geol Soc China 3:131–138
Huang TC, Ting JS, Muller C (1983) A Note on Pliocene Microfossils from the Kenting Mélange. Proc Geol Soc China 26:57–66
Juan VC, Lo HJ, Chen CC (1980) Genetic relationships and emplacement of the exotic basic rocks enclosed in the Lichi mélange, east Coastal Range, Taiwan. Proc Geol Soc China 23:56–68
Karig DE (1973) Plate convergence between the Philippines and the Ryukyu Islands. Mar Geol 14:153–168
Lee JC, Angelier J (1993) Localisation des déformations actives et traitements des données géodésiques: l'exemple de la faille de la Vallée Longitudinale, Taiwan. Bull Soc Géol France 164:533–570
Li YH, Guo TR, Hsu BH, Hwang YC (2001) Seismic survey in Hengchun peninsula. Proc Chinese Geophys Soc Symp, Taipei:347–356
Lin SB, Chen GT (1986) Clay minerals from the Lichi Mélange and its adjacent formations in the Coastal Range, eastern Taiwan. Acta Geologica Taiwanica 24:319–356
Lin SB, Wang YR (2001) Clay minerals in the rock formations on the Hengchun Peninsula, southern Taiwan, and their tectonic implications. Western Pacific Earth Sci 2: 157–174
Liou JG, Suppe J, Ernst WG (1977) Conglomerates and pebbly mudstones in the Lichi Mélange, eastern Taiwan. Mem Geol Soc China 2:115–128
Lu CY, Hsu KJ (1992) Tectonic Evolution of the Taiwan Mountain Belt. Petrol Geol Taiwan 27:15–35
Malavieille J, Lallemand SE, Dominguez S, Deschamps A, Lu CY, Liu CS, Schnurle P, ACT Scientific Crew (2002) Arc-continent collision in Taiwan: new marine observations and tectonic evolution. Geol Soc Am Special Paper 358: 187–211
Meng CY, Chiang SC (1965) Subsurface Data from Wildcat SS-1, Shihshan, Taitung. Petrol Geol Taiwan 4:283–286
Page BM (1978) Franciscan mélange compared with olistostromes of Taiwan and Italy. Tectonophysics 47:223–246
Page BM, Lan CY (1983) The Kenting Mélange and its record of tectonic events. Mem Geol Soc China 5:227–248
Page BM, Suppe J (1981) The Pliocene Lichi Mélange of Taiwan: its plate tectonic and olistostromal origin. Am J Sci 281:193–227
Pearce JA, Lippard SJ, Robert S (1984) Characteristics and tectonic significance of suprasubduction zone ophiolites. In: Kokelaar BP, Howell MF (eds) Marginal Basin Geology. Geol Soc [London] Special Publication 16, pp 77–94
Pelletier B (1985) De la fosse de Manille à la chaîne de Taiwan: édtude géologique aux confins d'une subduction et d'une collision active, modèle géodynamique. Ph. D thesis, Université de Brest, Brest, pp 1–268
Pelletier B, Stephan JF (1986) Middle Miocene obduction and Late Miocene beginning of collision registered in the Hengchun Peninsula: Geodynamic implications for the evolution of Taiwan. Mem Geol Soc China 7:301–324
Pelletier B, Stephan JF, Blanchet R, Muller C, Hu HN (1985). L'emergence d'une zone de collision active à la pointe sud de Taiwan (penisule de Hengchun). Tectonique superposées et mise en evidence d'une obduction miocène moyen. Bull Soc Géol France 8:161–171
Stephan JF, Blanchet R, Rangin C, Pelletier B, Letouzey J, Muller C (1986) Geodynamic evolution of the Taiwan-Luzon-Mindoro belt since late Eocene. Mem Geol Soc China 7:69–90
Sun SC (1982) The Tertiary basins of offshore Taiwan. Proc ASCOPE Conf and Exhib 2nd, Manila:125–135.
Sung Q (1991) Some characteristics of sedimentary blocks in the Lichi Mélange, Coastal Range, Taiwan. Spec Publ Cent Geol Surv 1:231–256
Suppe J (1984) Kinematics of arc-continent collision, flipping of subduction and back-arc spreading near Taiwan. Mem Geol Soc China 6: 21–34
Suppe J (1988) Tectonics of arc-continent collision on both sides of the South China Sea: Taiwan and Mindoro. Acta Geologica Taiwanica 26:1–18
Suppe J, Lan CY, Hendel EM, Liou JG (1977) Paleogeographic interpretation of red shales within the East Taiwan Ophiolite. Petrol Geol Taiwan 14:109–120
Suppe J, Liou JG, Ernst WG (1981) Paleogeographic origin of the Miocene East Taiwan Ophiolite. Am J Sci 281: 228–246
Taylor B, Hayes DE (1980) The tectonic evolution of the South China Sea Basin. In: Hayes DE (ed). The Tectonic and Geologic Evolution of the Southeast Asian Seas and Islands (part I). Amer Geophys Union Monogr 23, pp 89–104
Taylor B, Hayes DE (1983) Origin and history of the South China Sea Basin. In: Hayes DE (ed). The Tectonic and Geological Evolution of Southeast Asian Seas and Islands (Part II). Am Geophys Union Monogr 20, pp 129–155
Teng LS (1981) On the origin and tectonic significance of the Lichi Formation, Coastal Range, eastern Taiwan. Ti-Chih (Geology) 2:51–62 (in Chinese with English abstract)

Teng LS, Chen WS, Wang Y, Song SL, Lo HJ (1988) Toward a comprehensive stratigraphic system of the Coastal Range, eastern Taiwan. Acta Geologica Taiwanica 26:19–36

Tsan SF (1974a) The Kenting Formation: a note of Hengchun peninsula stratigraphy. Proc Geol Soc China 17:131–134

Tsan SF (1974b) Stratigraphy and structure of the Hengchun peninsula, with special reference to a miocene olistostrome. Bull Geol Surv Taiwan 24:99–108

Wang CS (1976) The Lichi Formation of the Coastal Range and arc-continent collision in eastern Taiwan. Bull Geol Surv Taiwan 25:73–86

Yu SB, Liu CC (1989) Fault creep on the central segment of the Longitudinal Valley Fault, Eastern Taiwan. Proc Geol Soc China 32:209–231

Yu SB, Chen HY, Kuo LC (1997) Velocity field of GPS stations in Taiwan area. Tectonophysics 274:41–59

Subduction of an Active Spreading Ridge Beneath Southern South America: A Review of the Cenozoic Geological Records from the Andean Foreland, Central Patagonia (46–47°S)

Bruno Scalabrino, Yves Lagabrielle, Aude de la Rupelle, Jacques Malavieille, Mireille Polvé, Felipe Espinoza, Diego Morata and Manuel Suarez

Abstract The Chile-Argentina Patagonian Cordillera is a natural laboratory to study the interactions between oceanic and continental lithosphere during the subduction of an active spreading ridge beneath a continent. Subduction of the South Chile spreading ridge, which separates the Nazca plate from the Antarctic plate, started around 15–14 Ma at the southern tip of Patagonia. Presently, the southernmost segment of the Chile Ridge enters the Peru-Chile trench at 46°S, at the site of the Chile Triple Junction (CTJ). We review the main events which occurred on land in the CTJ region (46–47°S), related with processes of ridge subduction. We summarize tectonic, sedimentary, and magmatic features in a 30 Ma-to Present chronological table. A pre-ridge subduction stage, from 30 to 15 Ma, is characterized by the onset and growth of Patagonian relief and by a shift from marine to continental detrital sedimentation in the foreland at 20–22 Ma. The change from pre-ridge subduction to ridge subduction is marked on land by a transition from calc-alkaline to alkaline volcanism, at 14–12 Ma, and by the onset of eruption of very large flood basalt provinces (future volcanic plateaus) following rapid erosion of the eastern foreland belt. Post-plateau basaltic volcanism (<4 Ma) is coeval with a period of tectonic and morphological rejuvenation during which the eastern foreland of the Cordillera has been affected by extensional/transtensional tectonics. We place these events in the framework of a tectonomagmatic model involving the opening of slab windows due to both slab tear and ridge axis subduction.

Keywords Central Patagonia • South Chile Ridge • Subduction • Cenozoic • Geological records • Compressional to extensional tectonics • Geodynamic model

---

B. Scalabrino
CNRS-UMR 5243 Géosciences Montpellier, Université Montpellier 2, CC 60, Place Eugène Bataillon, Montpellier Cedex 5 34095, France, scalabrino@gm.univ-montp2.fr

S. Lallemand and F. Funiciello (eds.), *Subduction Zone Geodynamics*,
DOI 10.1007/978-3-540-87974-9, © Springer-Verlag Berlin Heidelberg 2009

1 Introduction

During the Cenozoic, the Patagonian Cordillera, which forms the southern segment of the Andes, north of the Andes of Tierra del Fuego, was built coevally with the subduction of the southern part of the Farallon-Nazca plate followed by subduction of the southern part of the Chile Spreading Ridge, which separates the Nazca plate from the Antarctic plate. The South Chile Ridge entered the south Chile trench, bounding the South American plate, at around 15 Ma at the southern tip of Patagonia. The Chile Triple Junction (CTJ), now located at 46°12′S, is the point where the Nazca, Antarctic, and South American plates meet (Fig. 1). The subduction of successive spreading segments of the South Chile Ridge led to the opening of a slab

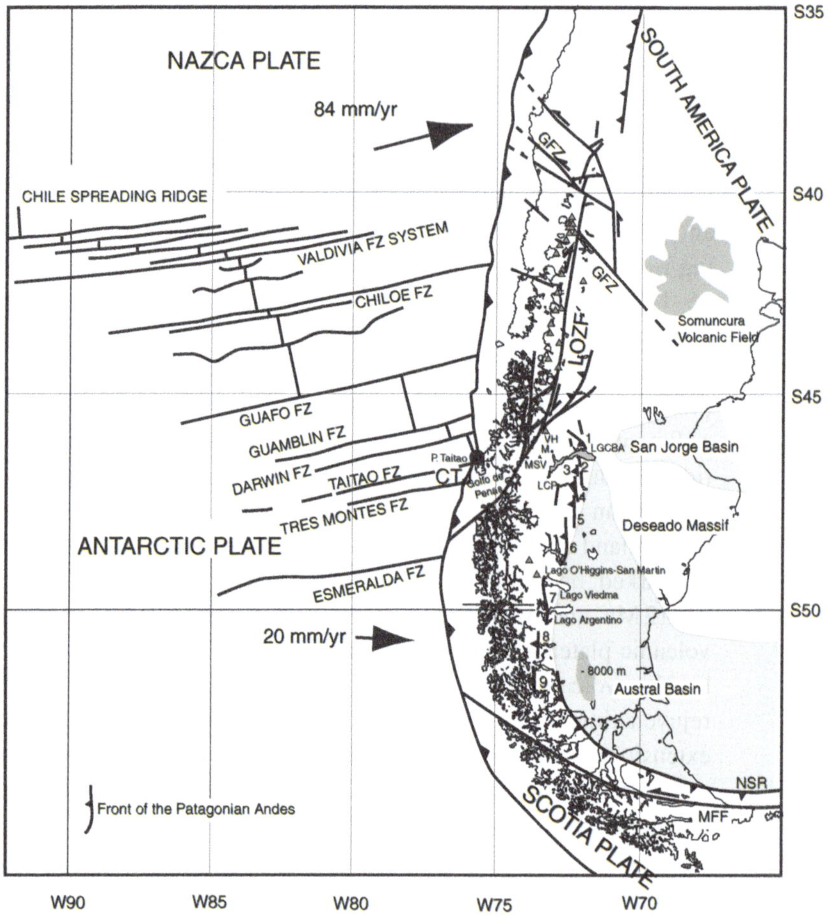

Fig. 1 Plate tectonic setting of southern South America between 35 and 55°S and main features of the subducting South Chile Ridge, including transform fault zones (*FZ*) and active spreading segments (*thick black lines*) (adapted from Tebbens et al., 1997; Ramos, 1989; DeMets et al., 1990; Cembrano et al., 1996; Guivel et al., 1999). Tectonic data are compiled after: Ramos, 1989; Tebbens et al., 1997; Cembrano et al., 1996, 2002; Coutand et al., 1999; Diraison et al., 2000; Melnick et al., 2002; Lagabrielle et al., 2004 and Ghiglione and Ramos, 2005). Location of Somuncura volcanic field is after De Ignacio et al. (2001). The segments of the tectonic front of the Cordillera south of the latitude of the Chile Triple Junction are labelled *1–9*: (*1*) Balmaceda-Portezuelo segment; (*2*) Chile Chico segment; (*3*) Las Horquetas segment; (*4*) Paso Roballos segment; (*5*) Lago Posadas segment; (*6*) Lago San Martin segment; (*7*) Lago Viedma segment; (*8*) Paine segment; and (*9*) Cordillera Riesco segment. *CTJ* Chile Triple Junction, *LGCBA* Lake General Carrera Buenos Aires, *LCP* Lake Cochrane-Pueyrredon, *MSV* Monte San Valentin, *VH* Volcan Hudson, *M* Murta basalts, *LOZF* Liquine Ofqui Fault Zone, *GTZ* Gastre Fault Zone, *MFF* Magallanes-Fagnano Fault, *NSR* western tip of the North Scotia Ridge, corresponding to the North Scotia orogenic front

window beneath the South American upper plate, triggering large scale changes in the rheology of the lithosphere due to dynamic upwelling of hot mantle beneath the entire region (Lagabrielle et al., 2004). A similar situation is reported for the Antarctic Peninsula and the western Scotia Sea, where active ridge segments have been subducting coevally (Garret and Storey, 1987).

The Patagonian Cordillera is almost topographically continuous from 54°S to 40°S. However, between 45°S and 47°S, at the latitude of the CTJ in Central Patagonia, it is characterized by several topographic anomalies. This unusual region includes both the highest mountains (Monte San Valentin, 4058 m a.s.l.) and a series of internal deep transverse incisions occupied by large post-glacial lakes (Fig. 1) (Lake General Carrera Buenos Aires: LGCBA; Lake Cochrane Pueyrredon: LCP; Lake Lapparent). The bottom of the largest lakes lie well below sea level, with lakeshores close to +200 m. Therefore, the axial region of the Central Patagonia Cordillera has an average low elevation, rather than showing the highest peaks, as along the Central and Northern Andes, and in most mountain belts worldwide. It has been suggested that these internal belt basins have tectonically controlled edges, although evidence of recent deformation is scarce due to the strong imprint of glacial processes during landscape evolution (Lagabrielle et al., 2004, 2007; Scalabrino et al., 2007). Such peculiar depressions are locally associated with very recent basaltic volcanism implying a MORB-like source from a buried spreading centre at depth (e.g., Guivel et al., 2006).

It has been suggested that the evolution of the Southern Andes contains a succession of events related to the various steps of subduction of the South Chile Ridge. The aim of this article is first to provide a synthesis of available plate kinematic models of the region (Sect. 2), and second to summarize the history of the Patagonian Cordillera at the latitude of the CTJ (46–47°S) since 30 Ma, based on regional geological records (Sect. 3). This history can be divided into two main periods: a pre-ridge subduction stage of "normal subduction" (30–15 Ma), and a ridge subduction stage involving opening of slab windows, due to both slab tear and ridge axis subduction (15 Ma-Present). This two-period history is in turn divided into a six-stage model, as proposed in the last section.

2 Subduction of Active Ridges and Plate Kinematics of the South American-Antarctica Junction Since 60 Ma: A Synthesis of Available Models

North of the CTJ, the present-day relative motion vector between the Nazca and the South American plates is oriented N80 (Gripp and Gordon, 1990; DeMets et al., 1990) with a magnitude of 84 mm/year (Fig. 1). Oblique subduction has led to tectonic partitioning in the Cordillera, and a part of the convergence is now accommodated along the Liquine-Ofqui fault zone north of the CTJ (Hervé, 1994; Cembrano et al., 1996, 2002; Roseneau et al., 2006) (Fig. 1). South of the CTJ, the current convergence rate between the Antarctic and the South American plates is 20 mm/year, in an E-W direction (Gripp and Gordon, 1990; DeMets et al., 1990) (Fig. 1).

The Patagonian and Antarctic Peninsulas have undergone a long history of ridge subduction throughout the Cenozoic. Figure 2 shows a series of cartoons based on a compilation of various reconstructions gathered from the literature, highlighting the changes in plate configuration of the Antarctica-Patagonia junction region since 60 Ma. This figure shows for the first time on a large scale all the active spreading centres that existed in the area including both the Pacific and the Atlantic sides of South America, and the successive key steps of the evolution this domain.

In the early Cenozoic, the South American and Antarctic plates were welded, constituting a single large continental plate. The Phoenix oceanic plate was subducting beneath this large plate at a rate of about 102 mm/a (McCarron and Larter, 1998) (Fig. 2). The Antarctica-Phoenix active spreading center was subducting under the western Antarctica Peninsula (Ramos, 2005; Eagles, 2003; Eagles et al., 2005).

Around 50 Ma, subduction of the Farallon-Phoenix spreading center under the South America plate began in the N20° direction (Pardo-Casas and Molnar, 1987). This ridge progressively migrated southward along the South American trench, leading to the opening of a first slab window below the continent, as recorded by the alkaline Eocene plateau basalts of Central Patagonia (Ramos and Kay, 1992; Espinoza et al., 2005). This stage was also characterized by the onset of the dislocation of the South American and Antarctic plates south of the future Tierra del Fuego. The relative motion between South America

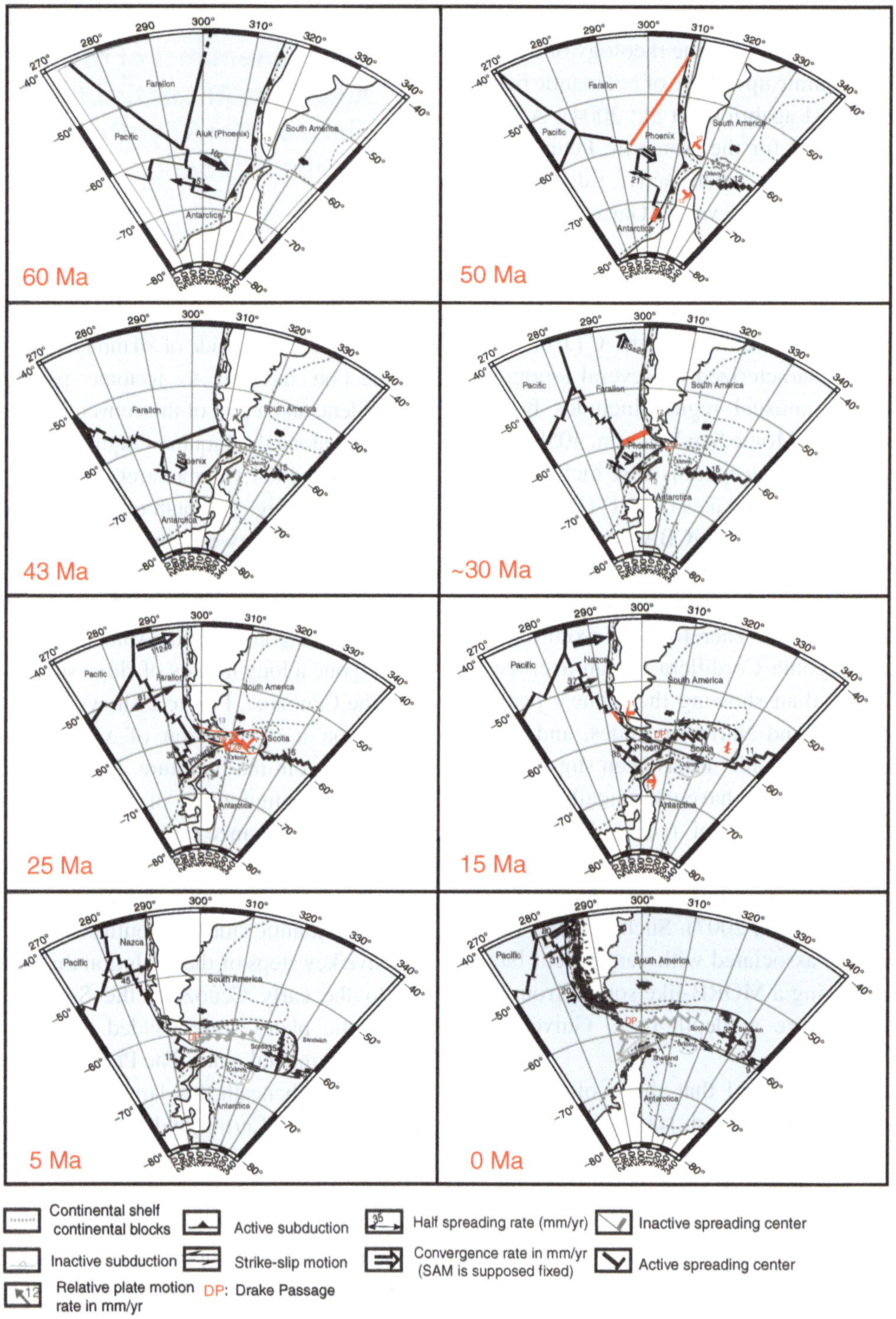

Fig. 2 Simplified kinematic reconstructions of the junction between the South American and Antarctic Peninsulas since 60 Ma, emphasizing the relationships between the subductions of numerous active spreading ridges, the migrations of several triple junctions and the onset of oceanic spreading within the Scotia Sea region. These reconstructions are based on a compilation of plate kinematic models available for the region. The South American plate is considered fixed

and Antarctica increased from 3 to 24 km/Ma, and was oriented WNW-ESE (Livermore et al., 2005), causing small continental blocks to detach from the continental shelf near the South American and Antarctic peninsula junction (Fig. 2). A decrease in the Phoenix plate convergence rate to 42 mm/a occurred as a segment of the Phoenix-Antarctic active spreading center approached the Antarctic trench (McCarron and Larter, 1998).

Around 43 Ma, dislocation between South America and Antarctica continued. The Farallon-Phoenix spreading center was still subducting beneath the South American plate and migrating southwards. The Phoenix plate was converging on the Antarctic Peninsula at a rate of 28 mm/a (McCarron and Larter, 1998) (Fig. 2).

The Oligocene was characterized by the onset of oceanic spreading in the Scotia Sea, along the West Scotia spreading center (oldest anomalies at 30 Ma; e.g., Eagles et al., 2003, 2005; Lawver and Gahagan, 2003; Livermore et al., 2005, 2007) (Fig. 2). A significant change in the convergence rate and direction of the Nazca (ex-Farallon plate) plate with respect to South America occurred in this period: the earlier rate of 35 ± 25 mm/a at 40 Ma increased to 112 ± 8 mm/a at 25 Ma (Pardo-Casas and Molnar, 1987), and the plate underwent clockwise rotation of about 70°, with a new convergence direction of N90° (Pardo-Casas and Molnar, 1987). These changes led the South Chile Ridge to lie near the Chile trench, in a slightly oblique position. The Farallon-Phoenix-South American triple junction reached Tierra del Fuego in the Early Oligocene; this event was followed by the cessation of Phoenix-Farallon spreading center activity (Fig. 2).

The subduction of southern segments of the South Chile Ridge at 15–14, 14–13, 12, 6, 3, and 0.3 Ma (Herron et al., 1981; Cande and Leslie, 1986; Cande et al., 1987; Nelson et al., 1994; Tebbens and Cande, 1997; Tebbens et al., 1997) progressively buried at depth, led to the opening of a large slab window beneath Patagonia (Fig. 4) (Ramos and Kay, 1992; Gorring et al., 1997; Gorring and Kay, 2001).

This complex evolution, which implies the subduction of various spreading ridges since 60 Ma, obviously had a significant effect on various characteristics of the mantle convecting beneath this region. Disruption of the Antarctic-Patagonian connection and rapid seafloor spreading in the Scotia region may be viewed as consequences of this evolution. The successive migration of two spreading centers beneath southern South America also left a clear volcanic mark along the Patagonian Cordillera, including a very large gap in Cenozoic typical calc-alkaline products, the development of basaltic plateau basalts (50 Ma to Present), and the occurrence of a volcanic arc of adakitic edifices outlining the Antarctic edge of the current slab window (Hole and Larter, 1993; Stern and Kilian, 1996; Stern, 2004; D'Orazio et al., 2005) (Fig. 4).

3 Subduction of the South Chile Ridge: Geological Records from the Patagonian Cordillera

Since the Late Paleozoic, the western margin of South America has been characterized by eastward subduction of oceanic crust (Bell and Suarez, 2000). As a consequence, Jurassic subduction-related acidic volcanics and volcanosedimentary rocks of the Ibañez Group unconformably overlie the Paleozoic basement rocks. The western part of the Patagonian belt is characterized by exposures of Patagonian Batholith calc-alkaline granitoids, which were emplaced over a continuum subduction environment from the Late Jurassic to the Cretaceous (Pankhurst et al., 2000; Suarez and de la Cruz, 2001) and were exhumed during the Cenozoic (Thomson et al., 2001).

In the following section, we detail the major sedimentary, magmatic and tectonic features of the Central Patagonian Cordillera, focusing particularly on the internal part and eastern side of the belt. These features represent geologic records related to pre-ridge subduction to ridge subduction stages of the Cordillera, over the period 30–0 Ma.

3.1 Sedimentary Records from the Central Patagonian Cordillera

South of 45°S, Cenozoic sedimentation is represented by foreland molasse deposits and internal inverted basins that are roughly the same age along strike (Fig. 3). All records from regions north of the Tierra del Fuego are consistent with a major constriction of the Cordillera before Chile Ridge subduction, that is 30–14 Ma (Marenssi et al., 2003; Lagabrielle et al., 2004). Three main steps in the development of the Cenozoic clastic sequences can be described.

In a first step, Patagonia experienced generalized regression that initiated during the Late Eocene-Early Oligocene, related to the onset of uplift due to first compressional events in the Cordillera. The maximum of this phase, characterized by the deposition of clastic continental sequences in the proto-Cordillera, occurred around 30 Ma. Such a clastic sequence is clearly observed in the internal part of the Central Patagonian belt, south of Lake General Carrera Buenos Aires, in the tectonically inverted Cosmelli basin

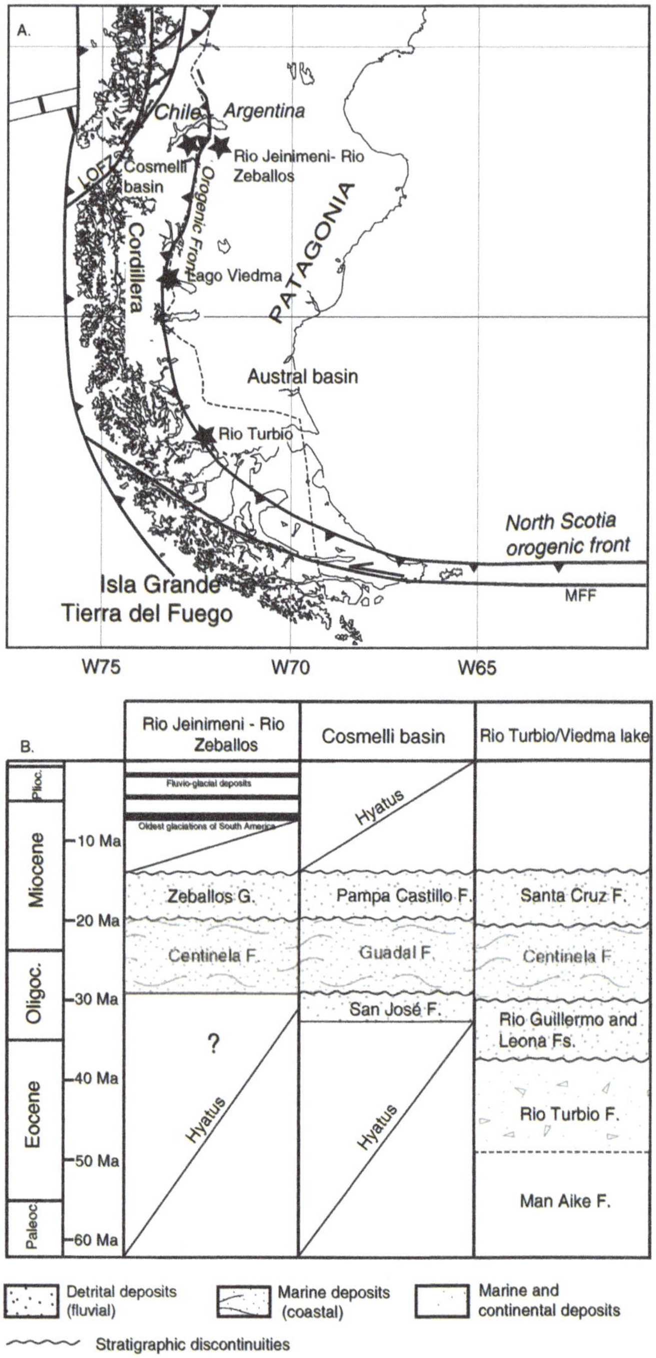

Fig. 3 **(a)** Simplified map of Central and Southern Patagonia showing key-localities (black stars) used to synthesize the main sedimentary records discussed in the text. MFF: Magallanes-Fagnano Fault. **(b)** Simplified stratigraphic sections of some key-localities located in Figure A. The orogenic front between central Patagonia and Tierra del Fuego is also shown

(Flint et al., 1994) where it corresponds to the San José Formation (Fig. 3).

In a second step, an important transgressive phase occurred during the Late Oligocene and Early Miocene (29–20 Ma), in which all of southern Patagonia was invaded by the *Patagoniano Sea*, including regions now located in the core of the belt, close to the Pacific coast. This stage of relative high sea-level is typically recorded by the Centinela Formation (Fig. 3), characterized by fossiliferous, shallow-water, coastal sediments, well exposed in the foothills region of Patagonia (Santa Cruz Province). It is best explained by foreland basin subsidence due to flexural response to crustal thickening in the western Cordillera. In the Cosmelli basin, this marine sequence corresponds to the Guadal Formation deposited from Late Oligocene to Early Miocene (Fig. 3), and overlies the San José Formation (Flint et al., 1994). The Guadal Formation often shows contractional syn-sedimentary deformation, such as internal thrusts and duplex systems.

In a third step, continuous shortening and uplift of the Cordillera led to a shift from shallow marine to renewed continental fluvial conditions, still recording an important outflow of detrital sediments. These voluminous erosional products now form the well-known Santa Cruz Formation of the Patagonian foothills of Argentina (Fig. 3). An equivalent of the Santa Cruz Formation, the Pampa Castillo Formation (Fig. 3), is exposed in the Cosmelli basin and includes terrestrial sandstones, claystones, and minor conglomerates. The transition from the Guadal to the Pampa Castillo Formation marks the withdrawal of the middle Cenozoic Patagonia seaway after it reached its maximum northward and westward extents (Flynn et al., 2002). At the latitude of the CTJ, the morphotectonic front of the belt exposes continental beds of the Rio Zeballos Group, which correspond to the Santa Cruz Formation (Fig. 3). In the Rio Jeinimeni-Rio Zeballos valleys, the Rio Zeballos Group stratigraphically overlies the Centinela Formation. This consists of a 1500 m thick sequence of fluvial sandstones, siltstones, and conglomerates that record rapid erosion of uplifted relief (Escosteguy et al., 2002). At Cerro Zeballos, at the head of Rio Zeballos, these sandstones are capped by lava flows dated to 16 and 14 Ma (Espinoza et al., 2006). South of the Lake General Carrera Buenos Aires, the sub-horizontal basal flows of the volcanic meseta Lago Buenos Aires, dated around 12 Ma (Guivel et al., 2006), rest unconformably over these deposits, confirming that erosion and compressional tectonics ceased completely between 14 and 12 Ma (Lagabrielle et al., 2004).

South of Lake General Carrera Buenos Aires, in the Lake Cochrane-Pueyrredon region, the continental deposits of the Santa Cruz Formation are represented by clastic fluvial deposits dominated by sand-, silt- and claystone beds locally with conglomerates, resting on the Centinela Formation (Fig. 3). The age of the Santa Cruz Formation in this region has been constrained to be between 22 and 14 Ma by tuffaceous deposits (Blisniuk et al., 2005).

The overall stratigraphy of the Cenozoic detrital formations exposed in Central Patagonia is very close to the successions exposed in more southern Patagonian regions. South of 50°S, in the southwestern corner of the Santa Cruz province, the Rio Turbio Formation of Middle Eocene age comprises marine and terrestrial sandstones and conglomerates. It is followed by the Rio Guillermo and Rio Leona fluvial deposits (Upper Eocene-Lower Oligocene), equivalent to the San Jose Formation (Fig. 3). These continental sequences are unconformably overlain by the marine Centinela Formation. Isotopic dating of oysters and Ar/Ar data yield ages between 27 and 20 Ma for the Centinela Formation (Parras et al., 2004; Guerstein et al., 2004), thus dating the major Cenozoic transgression in all Patagonia. A similar succession is found in the Viedma Lake region (Marenssi et al., 2003) (Fig. 3).

The Patagonian Cordillera is also characterized by a complex succession of Upper Miocene to Quaternary glacial deposits (Fig. 3). Some of these, dated between 7 and 3 Ma, are still preserved as tills interbedded within lava flow successions of Meseta Lago Buenos Aires or close to it (Mercer and Sutter, 1982; Lagabrielle et al., 2007). A complete review of such deposits is beyond the scope of this paper.

3.2 *Volcanic and Plutonic Records of the Central Patagonian Belt*

The volcanic-plutonic history of the Patagonian Cordillera at the latitude of the CTJ is characterized by successive magmatic events related to the successive subduction of two active spreading ridges, the Farallon-Phoenix (Aluk) ridge and the Chile Spreading Ridge.

South of the Lake General Carrera Buenos Aires, the Meseta Chile Chico (2160 m a.s.l, Fig. 4) exposes a Cenozoic succession unconformably overlying the Jurassic volcanic rocks of the Ibanez Group (Suarez and de la Cruz, 2000; Suarez et al., 2000a, b). This succession, 1000 m thick, consists of: a) Late Paleocene-Eocene basalts with alkaline and sub-alkaline tholeiite affinities (OIB signature, Espinoza, 2003) dated to 57–34 Ma (Charrier et al., 1979; Petford et al., 1996; Flynn et al., 2002); these overlie late Paleocene-Lower Eocene fluvial deposits and are related to a slab window that opened during the older Farallon-Aluk ridge subduction (Espinoza, 2003); and b) Late Miocene to Pliocene basalts overlying Late Oligocene-Early Miocene marine sediments of the Guadal Formation. These basaltic flows are 400 m thick and have alkaline and sub-alkaline affinities; they give K/Ar ages ranging between 4. ± 0.8 and 9.8 ± 0.1 Ma (Espinoza et al.,

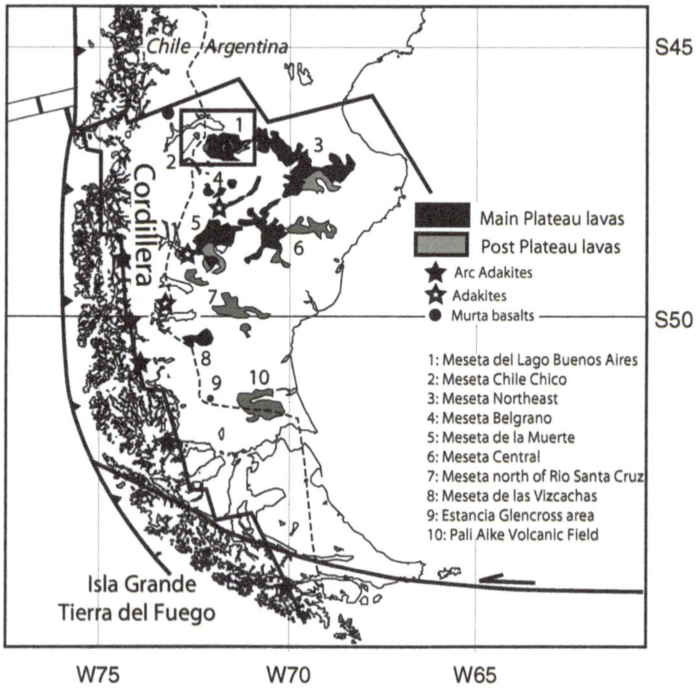

Fig. 4 Simplified map of Central and Southern Patagonia showing the distribution of Cenozoic Patagonian lavas: Main Plateau (black) and Post Plateau lavas (grey). The map also shows the location of the Patagonian adakites (*black and white stars*) and the Murta Quaternary basalts (*black circle*). The predicted extent of the Patagonian slab window is represented by black lines (after Breitsprecher and Thorkelson, submitted). *MFF* Magallanes-Fagnano Fault. The black frame delineates the area of Fig. 5

2003) (Figs. 5a, b). These are related to the subduction of the South Chile Ridge and were probably connected to lava flows of the Meseta Lago Buenos Aires, located some km to the southeast. The youngest flows of Meseta Chile Chico correlate with basaltic flows exposed on the poorly known Avellanos paleo-surface, which dominates the northern side of Lake General Carrera Buenos Aires (Lagabrielle et al., 2007).

At the morphotectonic front of the Central Patagonian belt, a basaltic flood volcanism event post-dates a period of intense peneplanation of the tectonic front (Lagabrielle et al., 2004). The remnants of these flows now form large basaltic plateaus or mesetas that morphologically dominate the eastern pampa region. The largest plateau in Central Patagonia is the flat-topped Meseta del Lago Buenos Aires (MLBA; Figs. 4–5), mainly formed by a first-stage OIB-like basaltic magmatism dated from 12.4 to 5 Ma (Main-Plateau basalts, Gorring et al., 1997; Guivel et al., 2006). Sporadic volcanic activity occurred later, as shown by numerous scoria cones, maars and lava flows younger that 5 Ma emplaced over the Miocene basalts (Post-Plateau basalts; Gorring et al., 2003; Brown et al., 2004; Guivel et al., 2006, Fig. 5b). Flows are interbedded with fluvial deposits and glacial sediments, confirming that basalts were emplaced along a main drainage area that is now topographically inverted. This later volcanism event is thought to be the expression of opening of a slab window due to the presence of subducted segments below them (Gorring et al., 2003). However, discrepancies between the ages of the oldest flows (around 12 Ma) and the date of the onset of migration of the subducted segment of the South Chile Ridge at the latitude of the Meseta Lago Buenos Aires led some authors to propose a model involving slab tear within the Nazca plate prior to ridge subduction (Guivel et al., 2006).

More recently, the northern side of Lake General Carrera Buenos Aires was marked by the emplacement of the Murta Quaternary E-MORB type basalts (Fig. 4). The source of these basalts has clear affinities with oceanic mantle (Demant et al., 1998; Guivel et al., 2006), confirming the presence of an opened slab window below the region. Only a few km to the NW of the Murta basalts, the Cerro Hudson active arc-volcano is the southernmost volcano of the South Volcanic Zone and represents the limit of the volcanic arc gap (Stern and Kilian, 1996; Stern, 2004). Due to the presence of relatively primitive basaltic magmas forming the lower part of this edifice, this volcano has also been interpreted as a consequence of opening of a slab window related to the subduction of a young segment of the South Chile Ridge (Gutiérrez et al., 2005).

Numerous examples of large volcanic fields related to slab window development are also observed south of the latitude of the CTJ. First, immediately south of

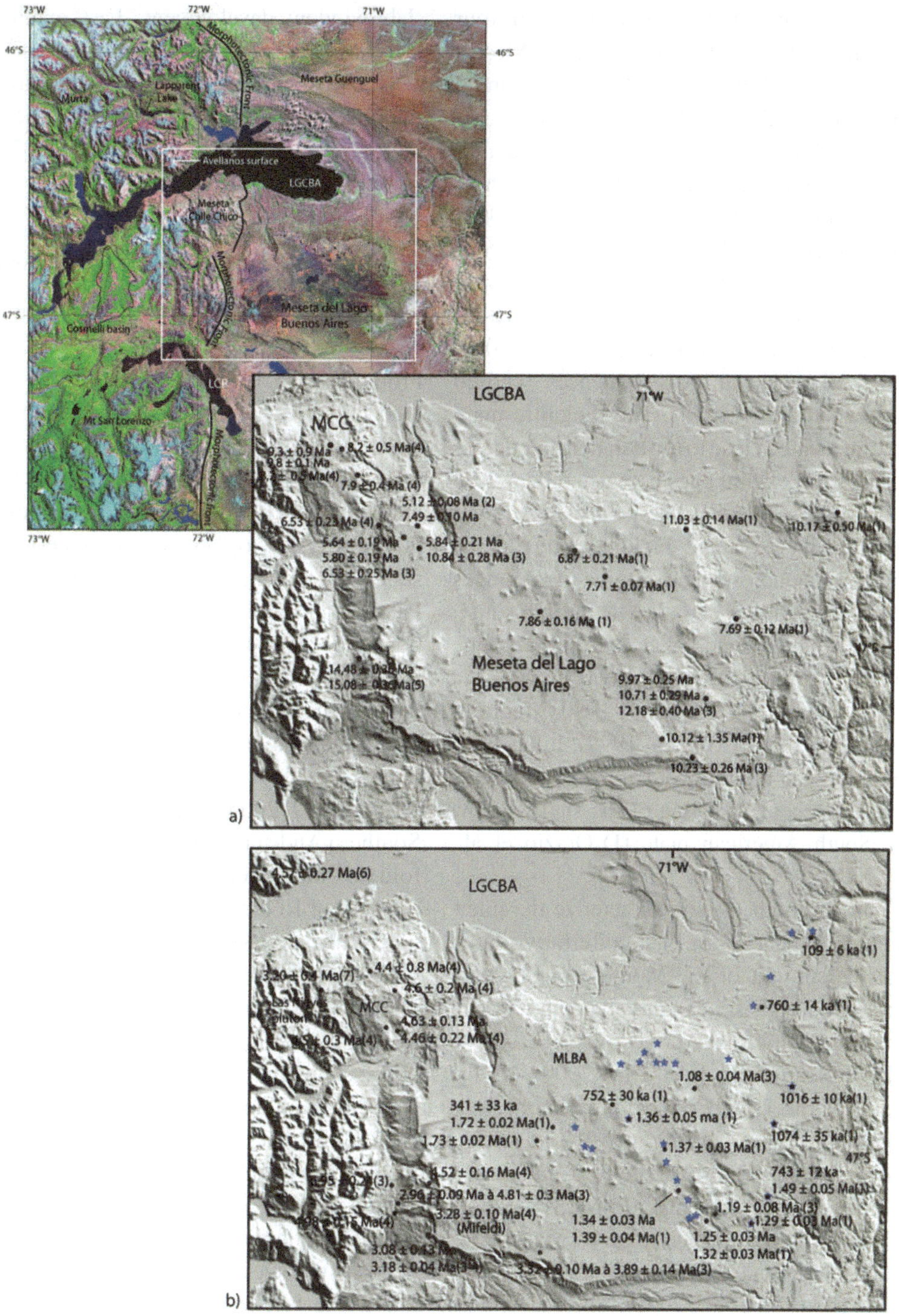

Fig. 5 Landsat 7 image of the internal part of Central Patagonia at the latitude of the Chile Triple Junction (see location in Fig. 4) and a compilation of published ages of volcanic rocks shown on a DEM of the Meseta del Lago Buenos Aires region (after Scalabrino et al., in preparation). Main Plateau **(a)** and Post Plateau **(b)** volcanics are shown separately. Black dots in Figure (a) correspond to dated samples. The origin of data is marked as numbers in brackets after each age, as follows: (1) Brown et al., 2004; (2) Mercer and Sutter, 1982; (3) Guivel et al., 2006; (4) Espinoza et al., 2005; (5) Espinoza, 2003; (6) Pelleter, 2003; (7) Suarez et al., 2000b. Grey stars in Figure (b) are the locations of the Pliocene-Pleistocene Post-Plateau lavas studied by Gorring et al., (1997). LGCBA Lake General Carrera Buenos Aires, MLBA Meseta del Lago Buenos Aires, MCC Meseta Chile Chico

the Meseta Lago Buenos Aires, around 48–49°S, four plateau lavas are present: Meseta Northeast, Meseta Central, Meseta de la Muerte, and Meseta Belgrano (Fig. 4). These all exhibit a typical bi-modal succession, including: a) Late Miocene to Early Pliocene tholeiitic voluminous main plateau sequences with whole rock Ar/Ar ages ranging from 13.88 ± 0.32 to 4.99 ± 0.11 Ma (Gorring et al., 1997), and b) Latest Miocene to Plio-Pleistocene alkaline minor post-plateau lava sequences dated from 6.22 ± 0.09 to 1.96 ± 0.16 Ma (Gorring et al., 1997). D'Orazio et al. (2000, 2001, 2005) found similar plateau lavas sequences exposed between 50°S and 52°S (Fig. 4). The tholeiitic main plateau lavas are located close to the Estancia Glencross (at around 52°S) and exhibit ages between 9.02 ± 1.03 and 7.34 ± 1.01 Ma, while alkaline post-plateau lavas of the Camusu Aike Volcanic Field located on the northern side of Glencross meseta give Ar/Ar ages between 2.95 ± 0.06 and 2.51 ± 0.09 Ma. Further south, in the Pali Aike volcanic field, alkaline to sub-alkaline basalts give Ar/Ar ages of 3.78 ± 0.17–0.17 Ma (Mercer, 1976; Linares and Gonzalez, 1990; Meglioli, 1992; Singer et al., 1997; Corbella, 1999). It has been suggested that these southernmost Patagonian mesetas are related to the recent widening of the slab window beneath the South American plate (D'Orazio et al., 2000, 2001, 2005).

Various plutonic bodies also characterize the internal Patagonian belt near 47°S. First, a Tertiary pluton crops out at the summit of the Monte San Lorenzo, forming the second highest peak in Patagonia (3706 m a.s.l), which yields biotite K/Ar ages of 6.6 ± 0.4 Ma (Welkner, 1999; Suarez and de la Cruz, 2000). On the southern coast of the LGCBA, a young pluton crops out in the Paso Las Llaves, which gives biotite Ar/Ar dates of 9.6 ± 0.5 and 9.6 ± 0.4 Ma (Petford and Turner, 1996), a Rb/Sr isochron of 10.3 ± 0.4 Ma (Pankhurst et al., 2000), a biotite K/Ar age of 10 ± 1.1 Ma (Suarez and de la Cruz, 2000) and a zircon fission track date of 9.7 ± 0.4 Ma (Thomson et al., 2001). The Las Nieves plutonic body crops out close to the Meseta Chile Chico (Fig. 5b), and gives a biotite K/Ar age of 3.2 ± 0.4 Ma (Suarez et al., 2000a, b) and an apatite fission track date of 4.3 ± 0.7 Ma (Morata et al., 2002). Recently, a small alkaline body exposed on the western side of the meseta del Lago Buenos Aires, the Mifeldi pluton, gave a whole rock K/Ar age of 3.28 ± 0.1 Ma (Espinoza et al., 2005) (Fig. 5b). These Pliocene plutons have been interpreted as the result of a thermal anomaly linked to slab window development. Fast exhumation of the Mifeldi pluton may be a consequence of tectonic activity and rapid erosion of an important lineament running along the Patagonia front, the Rio Zeballos fault zone (Lagabrielle et al., 2004, 2007).

3.3 Tectonic Records from the Central Patagonian Cordillera

From the CTJ to north of the Tierra del Fuego, the morphotectonic segments of the Patagonian Cordillera are 20–100 km long and more often show right lateral offset (Ramos, 1989; Coutand et al., 1999; Diraison et al., 2000; Thomson et al., 2001; Lagabrielle et al., 2004) (Figs. 1–6). As reported above, it has been suggested that the first stage of constriction of Central Patagonia started close to the Eocene-Oligocene boundary (32–24 Ma) (e.g., Marenssi et al., 2003). To the south, in Southern Patagonia or in the Fuegian Andes, the tectonic history of the belt is different because of the presence of an old back-arc basin with a mafic sea-floor (Rocas Verdes basin) (Dalziel, 1981; Ramos 1989, 2005; Ghiglione and Ramos, 2005). The Southern Andean foothills correspond to a thin-skinned fold and thrust belt built up from the Late Cretaceous (closure of Rocas Verdes basin) (Dalziel, 1981) to the Late Miocene (Kraemer et al., 1998; Coutand et al., 1999). Later, this portion of the Patagonian belt was affected by a major Neogene extensional phase affecting the Magallanes foreland, giving rise to graben systems (Diraison et al., 1997) (Fig. 6).

At the latitude of the CTJ, the eastern orogenic front is marked by a topographic jump of several hundred meters, due to an easterly vergent thrust system, juxtaposing internal Paleozoic and Mesozoic units of the main Cordillera over the Mesozoic-Cenozoic foreland deposits to the east (Ray, 1996; Suarez and de la Cruz, 2000; Lagabrielle et al., 2004). This compressive stage started at around 32 Ma, as shown by contractional syn-sedimentary deformation recorded by the fluvial deposits of the San Jose and Rio Leona Formations (Upper Eocene-Late Oligocene) and by marine sediments of the Guadal and Centinela Formations (Lower Oligocene-Early Miocene). As reported earlier, the change from continental to marine depositional environment is related to foreland basin subsidence due to flexural response to crustal thickening in the western

Cordillera. The ages of main uplift and exhumation of the central Patagonian Andes are confirmed by apatite and zircon fission track ages that show accelerated denudation between 30 and 12–8 Ma and up to 200 km eastward migration of the locus of maximum denudation during this period (Thomson et al., 2001). The age of the frontal thrust is now well constrained by calc-alkaline flows dated at 16–14 Ma that conformably overlie the eroded upper beds of the Zeballos Group (Espinoza et al., 2006). Therefore, it appears that compressional tectonics ceased in Central Patagonia at around 14 Ma, which is the time when the South Chile Ridge entered the Chile trench.

During the Latest Miocene-Pliocene, the eastern foreland of the Patagonian Cordillera experienced an episode of relief inversion, revealed by the presence of relict planar surfaces capping the flat-topped mesetas (Avellanos surface, meseta Chile Chico, meseta Lago Buenos Aires, meseta del Guenguel). Before this event, the overall topography was smoother and not so deeply incised as the present-day (Lagabrielle et al., 2004). The transverse incisions of the General Carrera Buenos Aires and Cochrane-Pueyrredon lakes, the N-S oriented valleys paralleling the front of the Cordillera (Rio Jeinimeni and Rio Ghio valleys), and the tectonically-controlled depressed area in the core of the belt did not exist before 5–4 Ma (Lagabrielle et al., 2004; Scalabrino et al., 2007) (Fig. 6). An extensional or transtensional tectonic event occurred after 5 Ma, reactivating ENE and NNW-trending lineaments (Lagabrielle et al., 2007; Scalabrino et al., 2007). Faults are well-observed west and south of meseta Lago Buenos Aires (Fig. 6) as well as on the edges of the Cenozoic inverted pop-down Cosmelli basin. These faults have a complex polyphase history and can be interpreted as back-arc extensional fault that controlled the Cretaceous basins, were slightly inverted during the Late Oligocene-Middle Miocene compressional stage, and have been partially reactivated with a normal to strike-slip component since 5 Ma (Scalabrino et al., 2007). However, since Patagonia was covered by thick ice sheets during the Pliocene-Quaternary glacial periods, the most obvious marks of recent tectonic activity have been erased by widespread glacial erosion. This explains why little evidence of recent active tectonics has been observed to date. The extensional regime that caused local to regional collapse of the belt and relief inversion in Central Patagonia was induced by possible weakening of the South American lithosphere in response to the presence of hot mantle at depth related to opening of a slab window (Lagabrielle et al., 2004, 2007). Such an extensional pattern related to a subducted ridge at depth was also shown by Garret and Storey (1987) in the Antarctic Peninsula.

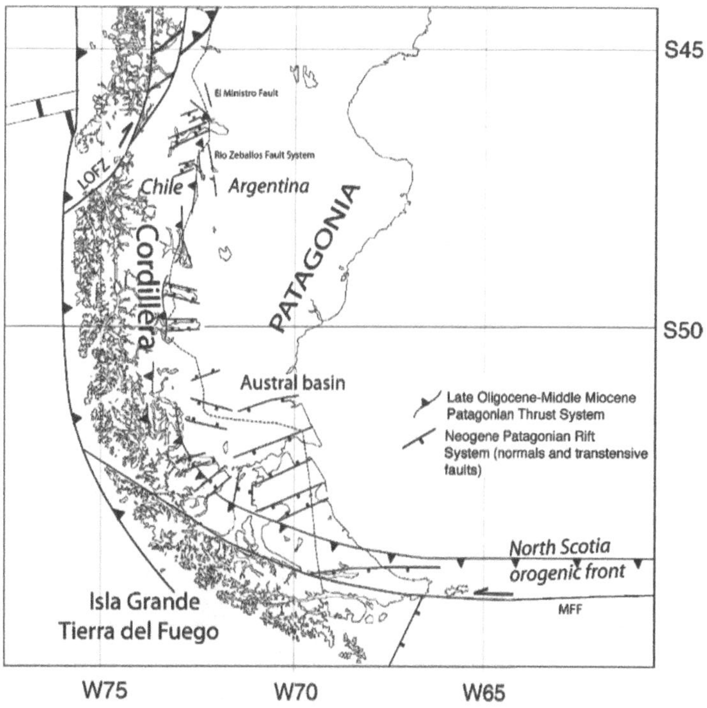

Fig. 6 Simplified structural map of Central and Southern Patagonia showing the main neotectonic features. Segments of the now inactive thrust system are represented by *grey lines*, while Neogene rift systems and extensional areas are marked by *black lines*. Data are compiled after: Ramos, 1989; Coutand et al., 1999; Diraison et al., 2000; Lagabrielle et al., 2004, 2007; Ghiglione and Ramos, 2005 and Scalabrino et al., in preparation *MFF* Magallanes-Fagnano Fault, *LOFZ* Liquine Ofqui Fault Zone

4 Subduction of the South Chile Ridge and Its Impact on the Patagonian Cordillera at the CTJ Latitude: A Six-stage Evolution

In this section, we summarize the onland effects of the subduction of successive segments of the South Chile Ridge in a chronological table for the period 30 Ma to Present (Fig. 7). This table includes tectonic, sedimentological, and magmatic features and is based on data collected over more than 10 years by our group, together with data from other workers.

- First stage (before 30 Ma): This represents the pre-ridge subduction stage. The convergence of the Nazca (ex-Farallon) plate is nearly frontal relative to the Chile trench. A shallow sea with cool water is locally present at the site of the current Patagonian Cordillera. During this period of "normal subduction," no significant magmatic and tectonic activity is observed, probably because of the remnant of a wide slab window due to southward migration of the Phoenix-Farallon spreading ridge before 30 Ma (Figs. 7–8).
- Second stage (around 25 Ma): The Patagonian Cordillera experiences onset of compressional deformation (syn-sedimentary deformation of upper marine sequences – Guadal formation, Fig. 7). This may be due to the increase of the Nazca convergence rate (35–112 mm/year; Pardo-Casas and Molnar, 1987), and the approach of the South Chile Ridge positive bathymetric high, causing increased tectonic coupling between the Nazca and South American plates.
- Third stage (20–15 Ma): A long segment of the South Chile Ridge is very close to the Chile trench (Figs. 7–8). An increase in tectonic coupling causes uplift and compression in the internal Cordillera. During this period, sedimentation shifts from marine to continental deposits (Upper Guadal and Galera formations, fluvial humid to arid, Fig. 7) and easterly vergent frontal thrust systems bring external units over molasse deposits.
- Fourth stage (16–14 Ma): The tip of a segment of the South Chile Ridge enters the Chile trench at around 55°S. In the Patagonian Cordillera, compression resumes and a major phase of erosion results in the rapid peneplanation of a wide frontal part belt on its eastern side (Fig. 7). Very scarce calc-alkaline volcanic lavas erupt (Cerro Zeballos) with a transition to the first alkaline flows that are emplaced on the eastern side of the Cordillera, over the eroded surface.

Fig. 7 Chronological table including the main sedimentary, tectonic and magmatic events of the Patagonian Cordillera at the latitude of the CTJ, from 30 to 0 Ma, in relation to the plate tectonic settings (left column)

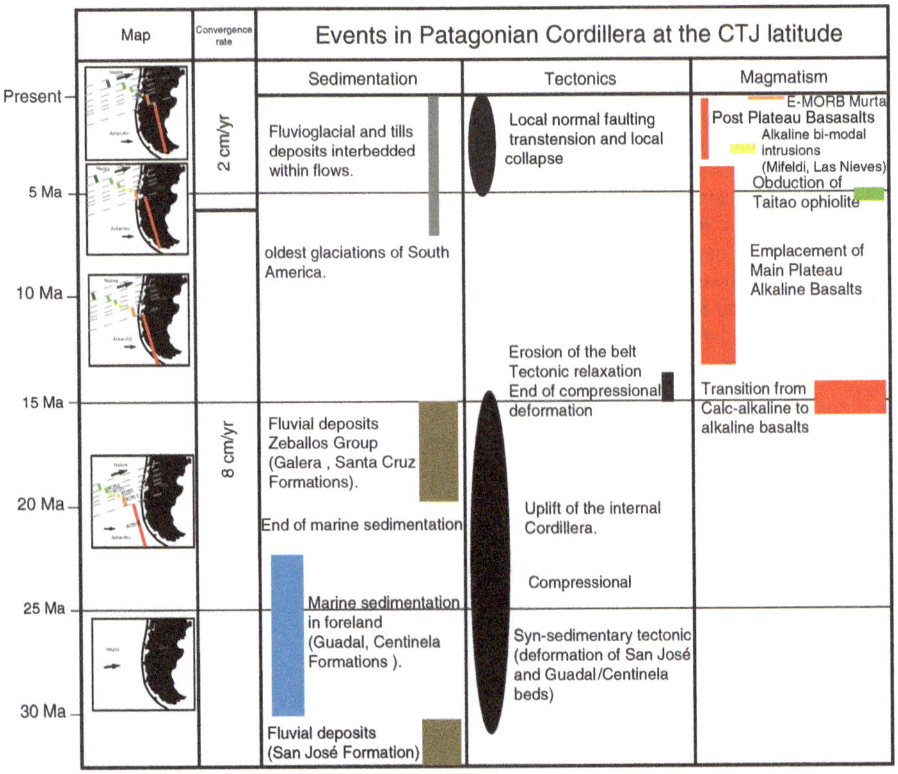

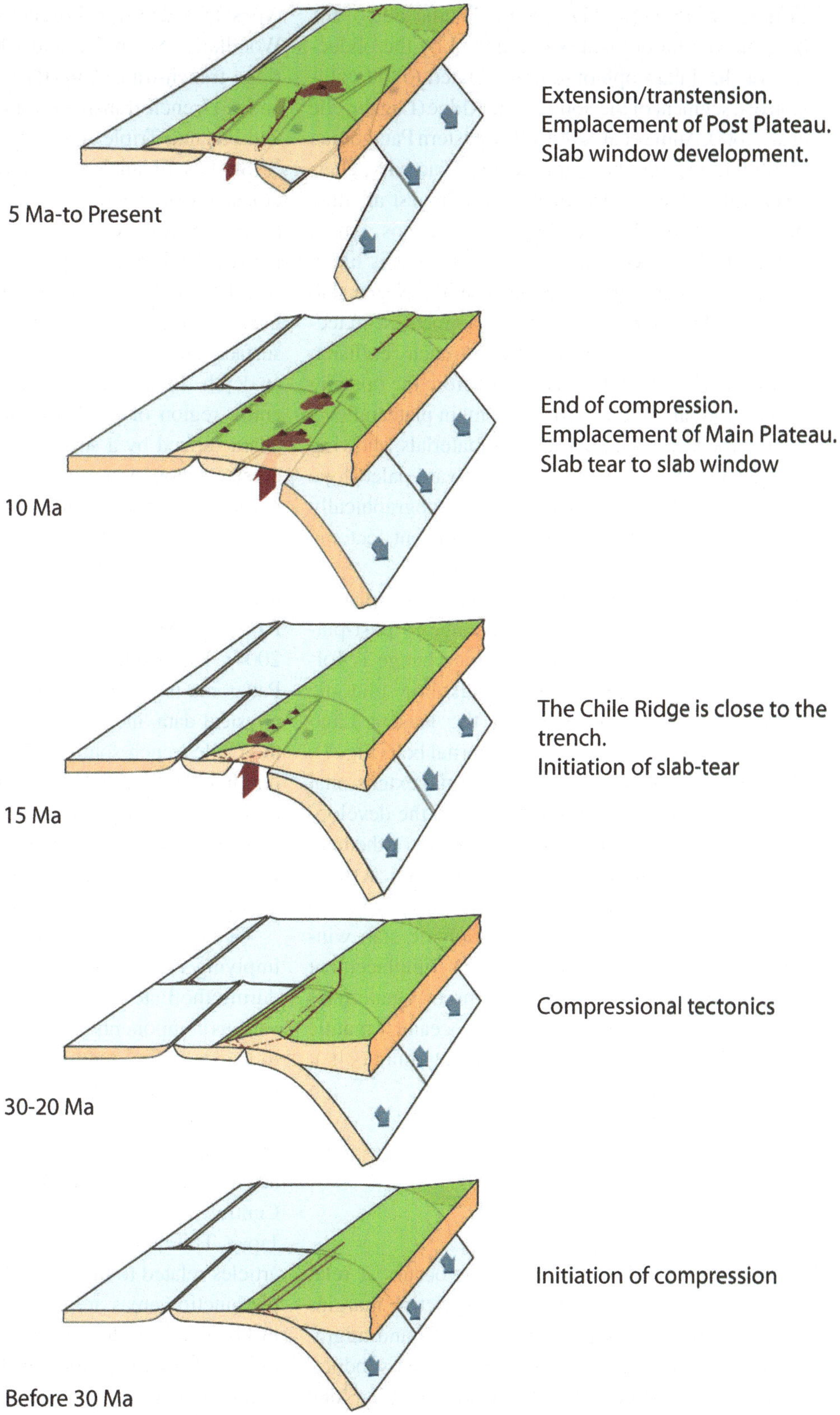

Fig. 8 Simplified geodynamical models showing the six main stages of the evolution of the Patagonian Cordillera, including opening of a slab-tear-related to true slab window during the subduction of the active Chile spreading ridge

- Fifth stage (15 to 5–4 Ma): On the Pacific side of the belt, the coastal domain is dominated by the obduction of the Taitao ophiolite units related to the collision of a segment of the South Chile Ridge (Lagabrielle et al., 2000, Anma et al., 2006). The eastern Patagonian Cordillera records the main magmatic activity, characterized by the emplacement of the largest alkaline plateau basalts (Meseta del Lago Buenos Aires; Meseta Chile Chico, Figs. 7–8). This event is likely linked to the development of a tear in the Nazca slab (Fig. 8), which is a consequence of an increase in tectonic coupling during the ridge-trench collision (Guivel et al., 2006) that occurred before the opening of the "normal" slab window. The main plateau flows are interbedded with fluvioglacial materials, showing that glaciers were flowing on an eastward paleoslope that is now highly dissected and topographically inverted. During this period, no significant tectonic activity is known in the Cordillera domain.
- Sixth stage (5–4 Ma to Present): The main plateau basalt volcanism ceases with a change to post-plateau volcanic activity. This volcanic change is followed by the emplacement of alkaline plutonic bodies in the western flank of the meseta Lago Buenos Aires, and locally in the internal belt (Fig. 7). The consequences of a transtensional/extensional phase are local collapse of the belt and the development of deep depressions in the axial part of the belt. We suggest that the tectono-magmatic activity recorded from 5 to 4 Ma to Present is closely linked to the development of an asthenospheric slab window beneath the study area (Fig. 8). Emplacement of the Murta basalts in the bottoms of Quaternary glacial valleys confirms that a hot "oceanic" mantle is present beneath the axis of the Patagonian belt at the latitude of the CTJ.

5 Discussion and Conclusions

Subduction of active spreading ridges occurs at relatively few locations along present-day active margins worldwide. However, due to plate motions and migrations of triple junctions, interaction between subducting ridges and continents has drastically modified long portions of the circum-Pacific margin during the Cretaceous and Tertiary. Modern triple junctions that include a ridge along the Pacific margin are of three types: (1) ridge-trench-trench (RTT): British Columbia, Woodlark, Antarctica and Chile Triple Junctions, (2) ridge-trench-transform (RTTr): Rivera Triple Junction, and (3) trench-transform-transform (TTT): Mendocino and Panama Triple Junctions.

Studies of ancient systems that may expose mid-crustal rocks provide important perspectives on the long-term temporal evolution of a ridge subduction event. However, it is necessary to investigate modern examples of ridge subduction because robust links can be established between magmato-tectonic manifestations at the surface and well-constrained locations of slab windows at depth. In the case of the Chile triple junction, the entire region of the Patagonia-Antarctic connection is characterized by a strong thermal anomaly due to successive subduction of various spreading ridges, as summarized by our reconstructions in Fig. 2. Global models show that a thin lithosphere is present beneath southern South America and Antarctica (Artemieva, 2006) and that a region of abnormal heat flow is present in the Patagonia-Antarctic connection (Shapiro and Ritzwoller, 2004). The development of a slab window beneath all of Patagonia is proven or suggested by a number of geophysical data, including a gap in seismic activity along the Chile trench south of 46°S, a region of abnormally hot mantle revealed by seismic tomography (Heintz et al., 2005) and abnormal terrestrial heat flow values along the Patagonian Andes (Hamza and Munoz, 1996). Figure 9 compiles data showing the extent of the thermal anomaly at the scale of the Antarctic-Patagonian region.

Geological records from ancient triple junctions implying ridge subduction, such as Alaska and Japan during the Late Cretaceous and Tertiary, show that the various components of subduction have changed through time. Therefore, long term variations in the configurations of subduction parameters are expected in these complex tectonic settings. Table 1 gives a summary of the major geological features linked to the subduction of active spreading ridges beneath Alaska, Northern and Central America, Chile, Antarctica, Woodlark and Japan. This table summarizes data collected from review articles related to these type-localities. It shows that triple junction areas are first characterized by rapid shifts in kinematics inducing temporal and spatial variations in the structural history of the upper plate. The most common magmatic impacts of ridge subduction include a gap in normal orogenic volcanism, and back-arc magmatism with an alkalic signature. Such features are well expressed in southern Chile, where correlations can be

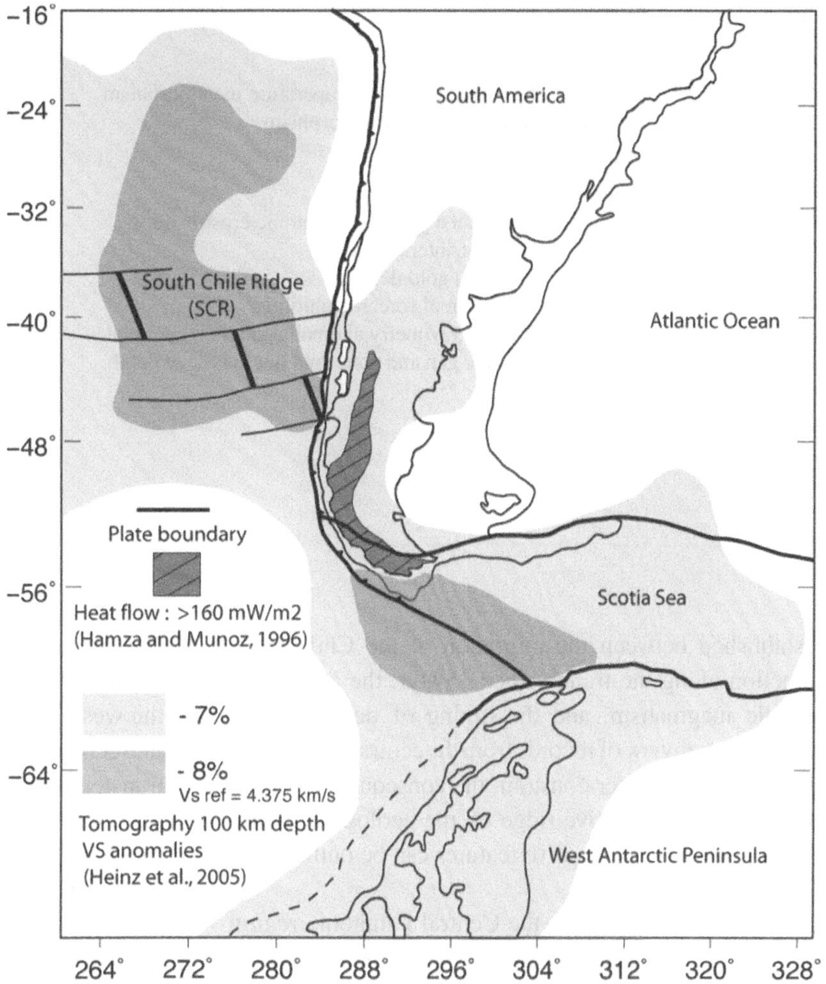

Fig. 9 Compilation map of SV wave heterogeneity at 100 km (red regions, after Heintz et al., 2005) and heat flow measurements in Central and Southern Patagonia (dashed region, after Hamza and Munoz, 1996)

Table 1 Summary of geological and geophysical impacts of Cretaceous to Present ridge subductions in the Circum-Pacific area

Geological processes	Characteristics	Some examples
Tectonics	Rapid shifts in kinematics in one region Temporal and spatial variations in structural history Extensional tectonism	Alaska and Northern America (Bradley et al. 2003). Antarctica Peninsula (Garret and Storey, 1987) South Chile (Lagabrielle et al. 2004, 2007; Scalabrino et al. this paper)
Magmatism	Abnormal near-trench volcanism Gap in normal arc magmatism Back-arc magmatism with alkalic signature localized N-MORB basaltic volcanism	Alaska and Northern America (Kusky et al. 2003; Sisson and Palvis, 1993; Sisson et al., 2003) California (Thorkelson et al., 1990; Cole and Basu, 1992, 1995) Woodlark (Johnson et al., 1987) Antarctica Peninsula (Hole and Larter, 1993) Japan (Hibbard and Karig, 1990, Maeda and Kagami, 1983) South Chile (Lagabrielle et al., 1994; Gorring et al., 1997; Guivel, 1999; Guivel et al., 2006, Espinoza et al., 2005, 2006)

(*continued*)

Table 1 (continued)

Geological processes	Characteristics	Some examples
Metamorphism	Low pressure/high temperature metamorphism Buchan series metamorphism	Japan (Osaozawa, 1992) Alaska and Northern Cordillera (Bowman et al., 2003; Zumsteg et al., 2003) South Chile (Bell and Suarez, 2000)
Others	Punctuated periods of tectonic erosion during ridge interactions Regional gold deposits Accretion of forearc ophiolites Strong gravimetry anomalies on the upper plate Seismic gap and abnormal hot mantle at depth	South Chile (Behrmann et al., 1994) Alaska (Haeussler et al., 1995, 2003) Chile (Lagabrielle et al., 2000; Anma et al., 2006) Alaska (Nelson and Forsythe, 1989; Kusky and Young, 1999) South Chile (Murdie et al., 2000) Japan (Sakagushi, 1996) South Chile (Hamza and Munoz, 1996; Murdie et al., 2000; Heintz et al., 2005; Artemieva, 2006) and Antarctica (Shapiro and Ritzwoller, 2004)

established between the migration of the Chile triple junction along the trench since 15 Myr, the age of the alkalic magmatism, and the timing of deformational events. Our review of records from the central Patagonian foreland helps better constrain the consequences of the subduction of an active ridge on the geological evolution of the upper plate. Five features can be outlined:

1. During the Cenozoic, the Central Patagonia region shows a simple correlation of sedimentary and tectonic events. Cenozoic sedimentation can be simply explained by foreland subsidence followed by contractional tectonics and uplift of the belt during the pre-ridge subduction stage.
2. In Southern and Central Patagonia, compressional tectonics ceased around 14 Ma, when the active Chile ridge entered the Chile trench. This tectonic relaxation may have been induced by a drastic decrease in the tectonic coupling at the trench, inducing weakening of the horizontal forces that maintained active shortening.
3. The spatial and temporal distribution of Cenozoic Patagonian basaltic plateaus (main-plateau basalts) cannot be simply explained by slab window development beneath the South American plate. A slab tear model ahead of subducted spreading ridge segments has been proposed by Guivel et al. (2006), but alternative models may imply return of asthenospheric mantle flow such as that described for the Somuncura volcanic field in Northern Patagonia (41°S) (De Ignacio et al., 2001).
4. The limits of the slab window below Patagonia can be followed by the presence of the adakitic belt on the western side (e.g., Stern, 2004) and the basaltic mesetas on the eastern side. Recent extensional and transtensional tectonics in Central and Southern Patagonia record important changes in lithosphere rheology that occurred after subduction of the Chile Ridge. The areas where this new tectonic regime develops are likely to delineate the central region of the slab-window opened at depth; the same is true for the Pali-Aike field in southernmost Patagonia (Stern et al., 1990).
5. The Central Patagonia Cordillera is characterized by anomalous topography and the occurrence of several tectonically-controlled transverse depressions bounded by regions of high elevation, which is opposite of what is expected for an orogenic belt in a classical subduction context. This facet warrants further development, in particular because it is not clear whether a phase of doming existed before local collapse of transverse depressions. This is important as doming may be the only topographic consequence of slab window opening. However, previous studies in the Antarctica Peninsula and Alaska have reported extensional tectonics in the upper plate following ridge subduction (Garret and Storey, 1987; Bradley et al., 2003). Therefore, Central Patagonia should remain a suitable location to study the balance between coupling at the trench and vertical forces due to abnormal mantle at depth in controlling the topography of subduction-related mountain belts.

Acknowledgements This research was funded by grants from the Chilean-French ECOS-CONICYT program (project C05U01) and CNRS-INSU DyETI (France) program. We thank R. Stern and C. Faccenna for constructive and helpful reviews.

References

Anma, R., Armstrong, R., Danhara, T., Orihashi, Y., Iwano, H., 2006. Zircon sensitive high mass-resolution U-Pb and fission-track ages for gabbros and sheeted dykes of the Taitao ophiolite, Southern Chile, and their tectonic implications. Island Arc 15, 130–142.

Artemieva, I.M., 2006. Global 1°×1° therma model TC1 for the continental lithosphere: Implications for the lithosphere secular evolution. Tectonophysics 416, 245–277.

Bangs, N.L., Cande, S.C., Lewis, S.D., Miller, J.J., 1992. Structural framework of the Chile margin at the Chile Ridge collision zone. Proceedings of Ocean Drilling Program Initial Report 141, 11–21.

Bell, M., Suarez, M., 2000. The Rio Lacteo Formation of southern Chile. Late Paleozoic orogeny in the Andes of the southernmost South America. Journal of South American Earth Sciences 13, 133–145.

Behrmann, J.H., Leslie, S.D., Cande, S.C., 1994. ODP Leg 141 scientific party. Tectonics and geology of spreading ridge subduction at the Chile Triple Junction: a synthesis of results from Leg 141 of the Ocean Drilling Program. Geol. Rundsch, 83, 832–853.

Blisniuk, P.M., Stern, L.A., Chamberlain, C.P., Idleman, B., and Zeitler, P.K., 2005. Climatic and ecologic changes during Miocene surface uplift in the Southern Patagonian Andes. Earth Planet Sciences Letters, 230, 125–142.

Bowman, J.R., Sisson, V.B., Valley, J.W., Pavlis, T.L., 2003. Oxygen isotope constraints on fluid infiltration associated with high-temperature-low-pressure metamorphism (Chugach metamorphic complex) within the Eocene southern Alaska forearc. In Sisson, V.B., Roeske, S.M., and Palvis, T.L., eds., Geology of a transpressional orogen developed during ridge-trench interaction along the North pacific margin: Boulder, Colorado, Geological Society of America Special paper 371, p. 237–253.

Bradley, D., Kusky, T., Haeussler, P., Goldfard, R., Miller, M., Dumoulin, J., Nelson, S.W., and Karl, S., 2003. Geologic signature of early Tertiary ridge subduction in Alaska. In Sisson, V.B., Roeske, S.M., and Palvis, T.L., eds., Geology of a transpressional orogen developed during ridge-trench interaction along the North pacific margin: Boulder, Colorado, Geological Society of America Special paper 371, p. 19–49.

Brown, L.L., Singer, B.S., Gorring, M.L., 2004. Paleomagnetism and $^{40}Ar/^{39}Ar$ chronology of lavas from Meseta del Lago Buenos Aires, Patagonia. Geochem. Geophys. Geosyst. 5 (1), Q01H04. Doi:10.1029/2003GC000526.

Cande, S.C., Leslie, R.B., 1986. Late Cenozoic tectonics of the Southern Chile Trench. Journal of Geophysical Reseach 91, 471–496.

Cande, S.C., Leslie, R.B., Parra, J.C., Hobart, M., 1987. Interaction between the Chile ridge and the Chile trench: geophysical and geothermal evidence. Journal of Geophysical Research 92, 495–520.

Cembrano, J., Hervé, F., Lavenu, A., 1996. The Liquine-Ofqui fault zone: A long-lived intra-arc fault system in Southern Chile. Tectonophysics 259, 207–224.

Cembrano, J., Lavenu, A., Reynolds, P., Arancibia, G., Lopez, G., Sanhueza, A., 2002. Late Cenozoic transpressional ductile deformation north of Nazca-South America-Antarctica triple junction. Tectonophysics 354, 289–314.

Charrier, R., Linares, E., Niemeyer, H., Skarmeta, J., 1979. K-Ar ages of basalt flows of the Meseta Buenos Aires in the southern Chile and their relation to the Southeast Pacific triple junction. Geology 7, 436–439.

Cole, R.B., and Basu, A.R., 1992. Middle Tertiary volcanism during ridge-trench interactions in Western California. Science, 258: 793–796.

Cole, R.B., and Basu, A.R., 1995. Nd-Sr geochemistry and tectonics of ridge subduction and middle Cenozoic volcanism in Western California. Geological Society of America Bulletin, 18, 167–179.

Corbella, H., 1999. Dataciones radiometricas en Pali Aike, patagonia Austral In: XIV Congreso Geologico Argentino, Actas, II, 265–268.

Coutand, I., Diraison, M., Cobbold, P.R., Gapais, D., Rossello, E., 1999. Structure and kinematic of a foothills transect, Lago Viedma, southern Andes (49°30'S). Journal of South American Earth Sciences (Oxford) 12, 1–15.

Dalziel, I.W., 1981. Back-arc extension in the southern Andes: areview and critical reappraisal. Phil. Trans. R. Soc. Lond. A300, 319–335.

Demant, A., Belmar, M., Hervé, F., Pankhurst, R.J., Suarez, M., 1998. Pétrologie et géochimie des basaltes de Murta: une éruption sous-glaciaire dans les Andes patagoniennes (46°S). Relation avec la subduction de la ride du Chili. C. R. Acad Sci., Paris 327, 795–801.

DeMets, C., Gordon, A.E., Aegus, D.F., Stein, S., 1990. Current plate motions. Geophysical Journal International 101, 425–478.

Diraison, M., Cobbold, P.R., Gapais, D., Rossello, E.A., 1997. Magellan strait: Part of a Neogene rift system. Geology 25, 703–706.

Diraison, M., Cobbold, P.R., Gapais, D., Rossello, E.A., Le Corre, C., 2000. Cenozoic crustal thickening, wrenching and rifting in the foothills of the southernmost Andes. Tectonophysics 316, issue 1-2, 91–119.

De Ignacio, C., Lopez, I., Oyarzun, R., Marquez, A., 2001. The Northern Patagonia Somuncura plateau basalts: a product of slab-induced, shallow asthenospheric upwelling? Terra Nova, 13, 117–121, 2001.

D'Orazio, M., Agostini, S., Mazzarini, F., Innocenti, F., Manetti, P., Haller, M., Lahsen, A., 2000. The Pali Aike Volcanic Field. Patagonia: slab-window magmatism near the tip of South America. Tectonophysics 321, 407–427.

D'Orazio, M., Agostini, S., Innocenti, F., Haller, M., Manetti, P., Mazzarini, F., 2001. Slab-window-related magmatism from southernmost South America: The Late Miocene mafic volcanics from the Estancia Glencross area (~52°S Argentina-Chile). Lithos 57, 67–89.

D'Orazio, M., Innocenti, F., Manetti, P., Haller, M.J., Di Vincenzo, G., Tonarini, S., 2005. The Late Pliocene mafic lavas from the Camusù Aike volcanic field (~50°S,

Argentina): Evidence for geochemical variability in slab window magmatism. Journal of South American Earth Sciences 18, 107–124.

Eagles, G., 2003. Plate tectonics of the Antarctic-Phoenix plate system since 15 Ma. Earth and Planetary Science Letters, 88, 289–307.

Eagles, G., Livermore, R.A., Fairhead, J.D., Morris, P., 2005. Tectonic evolution of the west Scotia Sea. Journal of Geophysical Research, 11, B02401.

Escosteguy, L., Dal Molin, C., Franchi, M., Geuna, S., Lapido, O., 2002. Estratigrafia de la cuenca de los rios el Zeballos y Jeinimeni, noroeste de la Provincia de Santa Cruz. Actas del XV Congresso Geologico Argentino. El Calafate.

Espinoza, F., 2003. Petrologia y geoquimica de los basaltos cenozoicos de la meseta de Chile Chico, 46°35'S-46°47'S- 71°46'-72°02'W XI region de Aysen, Chile. Unpublished MSc thesis, Universidad de Chile, Santiago, Chile. 220 pp.

Espinoza, F., Morata, D., Pelleter, E., Maury, R.C., Suarez, M., Lagabrielle, Y., Polvé, M., Bellon, H., Cotten, J., de la Cruz, R., Guivel, C., 2005. Petrogenesis of the Eocene and Mio-Pliocene alkaline basaltic magmatism in Meseta Chile Chico, Southern Patagonia, Chile: evidence for the participation of two slab windows. Lithos 82 (3–4), 315–343.

Espinoza, F., Morata, D., Polvé, M., Maury, R.C., Cotten, J., Bellon, H., Guivel, C., Lagabrielle, Y., Suárez, M., and Rosselló, E., 2006. Mio-Pliocene magmatic variability in the central Patagonia back-arc region (47.5ªS). Backbone of the Americas – Patagonia to Alaska. Geological Society of America, Mendoza (Argentina), 3–7 Abril, CD-rom.

Flint, F.F., Prior, D.J., Agar, S.M., Turner, P., 1994. Stratigraphic and structural evolution of the tertiary Cosmelli basin and its relationships to the Chile Triple Junction. Journal of Geological society of London 151, 251–268.

Flynn, J.J., Novacek, M.J., Dodson, H.E., Frassinetti, D., McKenna, M.C., Norell, M.A., Sears, K.E., Swisher, C.C. III, Wyss, A.R., 2002. A new fossil mammal assemblage from the southern Chilean Andes: Implications for geology, geochronology and tectonics. Journal of Southern American Earth Sciences 15, 285–302.

Garret, S.W., Storey, B.C., 1987. Lithospheric extension on the Antarctica Peninsula during Cenozoic subduction. In Extension in the basin and Range Province and East Pacific Margin, Geological Society, London, Special Publications, v. 28, pp. 419–431.

Ghiglione, M.C., Ramos, V.A., 2005. Progression of deformation and sedimentation in the southernmost Andes. Tectonophysics 405, 25–46.

Gorring, M., Kay, S., 2001. Mantle processes and sources of Neogene slab window magmas from Southern Patagonia, Argentina. Journal of Petrology 42, 1067–1094.

Gorring, M., Kay, S., Zeitler, P., Ramos, V., Rubiolo, D., Fernandez, M., Panza, J., 1997. Neogene Patagonian plateau lavas: continental magmas associated with ridge collision at the Chile Triple Junction. Tectonics 16, 1–17.

Gorring, M., Singer, B., Gowers, J., Kay, S., 2003. Plio-Pleistocene basalts from the Meseta del Lago Buenos Aires, Argentina: evidence for asthenosphere-lithosphere interactions during slab-window magmatism. Chemical Geology 193, 213–235.

Gripp, A.E., Gordon, R.G., 1990. Current plate velocities relative to the hotspots incorporating the NUVEL-1 global plate motion model. Geophysical Research Letters 17, 1109–1112.

Guerstein, G.R., Guler, M.V., and Casadio, S., 2004. Polynostratigraphy and paleoenvironments across the Oligocene-Miocene boundary within the Centinela Formation, southwestern Argentina. Geological Society, London, Special Publications, 230, 325–343.

Guivel, C., Lagabrielle, Y., Bourgois, J., Maury, R.C., Fourcade, S., Martin, H., Arnaud, N., 1999. New geochemical constraints for the origin of ridge-subduction-related plutonic and volcanic suites from the Chile Triple Junction (Taitao Peninsula and site 862, LEG ODP141 on the Taitao Ridge). Tectonophysics 311, 83–111.

Guivel, C., Morata, D., Pelleter, E., Espinoza, F., Maury, R.C., Lagabrielle, Y., Polvé, M., Bellon, H., Cotten, J., Benoit, M., Suarez, M., de la Cruz, R., 2006. Miocene to Late Quaternary Patagonian basalts (46°-47°S): Geochronometric and geochemical evidence for slab tearing due to active spreading ride subduction. Journal of Volcanology and Geothermal Research 149, 346–370.

Gutiérrez, F., Gioncada, A., Gonzalez Ferran, O., Lahsen, A., Mazzuoli, R., 2005. The Hudson volcano and surrounding monogenetic centres (Chilean Patagonia): An example of volcanism associated with ridge-trench collision environment. Journal of Volcanology and Geothermal Research 145, pp. 207–233.

Haeussler, P.J., Bradley, D., Goldfarb, R., Snee, L., Taylor, C., 1995. Link between ridge subduction and gold mineralization in southern Alaska. Geology 23 (11), 995–998.

Haeussler, P.J., Dwight, C., Bradley, D., Goldfarb, R.J., 2003. Brittle deformation along the Gulf of Alaska margin in response to Paleocene-Eocene triple junction migration. In Sisson, V.B., Roeske, S.M., and Palvis, T.L., eds., Geology of a transpressional orogen developed during ridge-trench interaction along the North pacific margin: Boulder, Colorado, Geological Society of America Special paper 371, pp. 119–141.

Hamza, V.M., Munoz, M., 1996. Heat flow map of South America. Geothermics 25(6), 599–646.

Heintz, M., Debayle, E., Vauchez, A., 2005. Upper mantle structure of the South America continent and neighboring oceans from surface wave tomography. Tectonophysics 406, 115–139.

Herron, E.M., Cande, S.C., Hall, B.R., 1981. An active spreading center collides with a subduction zone; a geophysical survey of the Chile margin triple junction. Memoir-Geological Society of America 154, 683–701.

Hervé, F., 1994. The Southern Andes between 39°S and 44°S latitude: the geological signature of a transpressive tectonic regime related to a magmatic arc. In: Reutter, K-J., Scheuber, E., Wigger, P.J., eds., Tectonics of the Southern central Andes. Springer, Berlin, pp. 243–248.

Hibbard, J.P., Karig, D.E., 1990. Structural and magmatic responses to spreading ridge subduction: an example from southwest Japan. Tectonics 9 (2), 207–230.

Hole, M.J., Larter, R.D., 1993. Trench-proximal volcanism following ridge crest-trench collision along the Antarctic Peninsula. Tectonics 12, 897–901.

Johnson, R.W., Jaques, A.L., Langmuir, C.H., Perfit, M.R., Dunkley, P.N., Chappell, B.W., Taylor, S.R., 1987. Ridge subduction and forearc volcanism: petrology and geochemistry of rocks dredged from the western Solomon arc and Woodlark basin. 155–226.

Kraemer, P.E., 2003.Orogenic shortening and the origin of the Patagonian orocline (56° S.lat). Journal of South American Earth Sciences 15, 731–748.

Kusky, T.M., Young, C.P., 1999. Emplacement of the ressurection Peninsula ophiolite in the southernalaska forearc during a ridge-trench encounter. Journal of Geophysical Research, vol. 104, NoB12, p. 29.025–29.054.

Kusky, T.M., Bradley, D., Donley, D.T., Rowley, D., Haeussler, P., 2003. Controls on intrusion of near-trench magmas of the Sanak-Baranof Belt, Alaska, during Paleogene ridge subduction, and consequences for forearc evolution. In Sisson, V.B., Roeske, S.M., and Palvis, T.L., eds., Geology of a transpressional orogen developed during ridge-trench interaction along the North pacific margin: Boulder, Colorado, Geological Society of America Special paper 371, p. 269–292.

Lagabrielle, Y., LeMoigne, J., Maury, R.C., Cotton, J., Bourgois, J., 1994. Volcanic record of the subduction of an active spreading ridge, Taitao peninsula (southern Chile). Geology 22, 515–518.

Lagabrielle, Y., Guivel, C., Maury, R.C., Bourgois, Y., Fourcade, S., Martin, H., 2000. Magmatic-tectonic effects of high thermal regime at the site of active spreading ridge subduction: the Chile Triple Junction model. Tectonophysics 326 (3–4), 215–228.

Lagabrielle, Y., Suarez, M., Rossello, E.A., Hérail, G., Martinod, J., Régnier, M., de la Cruz, R., 2004. Neogene to Quaternary tectonic evolution of the Patagonian Andes at the latitude of the Chile Triple Junction. Tectonophysics 385, 211–241.

Lagabrielle, Y., Suarez, M., Malavieille, J., Morata, D., Espinoza, F., Maury, R.C., Scalabrino, B., Barbero, L., de la Cruz, R., Rossello, E., Bellon, H., 2007. Pliocene extensional tectonics in Eastern Central Patagonian Cordillera: geochronological constraints and new field evidence. Terra Nova in press.

Lawver, L.A., Gahagan, L.M., 2003. Evolution of Cenozoic seaways in the circum-Antarctic region. Palaeogeography, Palaeoclimatology, Palaeoecology 198, 11–37.

Linares, E., Gonzalez, R., 1990. Catalogo de edades radimetricas de la Republica argentina: 1957–1987, Publicaciones Especiales de la Asociacion Geologica Argentina, serie B, vol. 19, 628 pp.

Livermore, R., Nankivell, A., Eagles, G., Morris, P., 2005. Paleogene opening of Drake Passage. Earth and Planetary Science Letters, 236, 459–470.

Livermore, R., Hillenbrand, C.D., Meredith, M., Eagles, G., 2007. Drake Passage and Cenozoic climate: An open and shut case? Geochemistry, Geophysics, Geosystems 8, 1. Q01005, doi:10,1029/2005GC001224.

Maeda, J., and Kagami, H., 1996 Interaction of a spreading ridge and an accretionary prism: implications for MORB magmatism in the Hidaka magmatic zone, Hokkaido, Japan. Geology 24 (1), 31–34.

Marenssi, S.A., Casadio, S., Santillana, S.N., 2003. Estratigrafia y sedimentologia de las unidades del Cretacico superior-Paleogeno aflorantes en la margen sureste del lago Viedma, provincia de Santa Cruz, Argentina. Revista de la Asociacion Geologica Argentina 58 (3), 403–416.

McCarron, J.J., Larter, D.D., 1998. Late cretaceous to Early Tertiary Subduction history of the Antarctic Peninsula. Journal of Geological Society, London 155, 255–268.

Meglioli, A., 1992. Glacial geology and geochronology of southernmost patagonia and Tierra del Fuego, Argentina and Chile. Ph.D. dissertation. Leigh University, Bethlehem, PA, USA.

Mercer, J.H., 1976. Glacial history of southernmost South America. Quat. Res. 6, 125–166.

Melnick, D., Folguera, A., Roseneau, M., Echtler, H., Potent, S., 2002. Tectonics from the northern segment of the Liquine-Ofqui fault system (37°S-39°S), Patagonian Andes, paper presented at Fifth International Symposium of Andean Geodynamics, IRD, Toulouse, France.

Mercer, J.H., Sutter, J.F., 1982. Late Miocene-Earliest Pliocene Glaciation in Southern Argentina: implications for global ice-sheet history. Paleogeography, Paleoclimatology, Paleoecology 38, 185–206.

Morata, D., Barbero, L., Suarez, M., de la Cruz, R., 2002. Early Pliocene magmatism and high exhumation rates in the Patagonian Cordillera (46°40'S): K-Ar, and fission track data. ISAG.

Murdie, R., Styles, P., Prior, D.J., Daniel, A.J., 2000. A new gravity map of southern Chile and its preliminary interpretation. Revista Geologica de Chile 27 (1), 49–63.

Nelson, E.P., Forsythe, R.F., 1989. Ridge collision at convergent margins: implications for Archean and post Archean crustal growth. Tectonophysics 161, 307–315.

Nelson, E., Forsythe, R., Diemer, J., Allen, M., Urbina, O., 1994. Taitao ophiolite: a ridge collision ophiolite in the forearc of the southern Chile (46°S). Revista Geologica de Chile 20(2), 137–165.

Osazawa, A., 1992. Double ridge subduction recorded in the Shimanto accretionary complex, Japan, and plate reconstruction. Geology 20, 939–942.

Pankhurst, R.J., Riley, T.R., Fanning, C.M., Kelley, S.P., 2000. Episodic silicic = volcanism in Patagonian and Antarctic Peninsula: chronology of magmatism associated with break-up of Gondwana. Journal of Petrology 41, 605–625.

Parras, A., Casadio, S., Feldmann, R., Griffin, M., Schweitzer, C.E., 2004. Age and paleogeography of the marine transgression at the Paleogene-Neogene boundary in Patagonia, southern Argentina. Denver Annual Meeting, Geological Society of America Abstracts with programs 16(5), 364.

Pardo-Casas, F., Molnar, P., 1987. Relative motion of the Nazca (Farallon) and South America plates since Late Cretaceous time. Tectonics 6(3), 233–248.

Pelleter, E., 2003. Diversité géochimique et géochronologique des basaltes de la région du lac Général Carrera-Buenos Aires: nouvelles contraintes sur la subduction de la dorsale du Chili. Unpublished DEA Thesis; Université de Bretagne Occidentale, Brest, France.

Petford, N., Cheadle, M., Barreiro, B., 1996. Age and origin of southern flood basalts, Chile Chico region (46°45'S). Third International Symposium on Andean Geology (ISAG), St. Malo, France, 629–632 (extended abstract).

Petford, N., Turner, P., 1996. Reconnaissance 40Ar/39Ar age and paleomagnetic study of igneous rocks around Coyhaique,

S. Chile. Third International Symposium on Andean Geology (ISAG), St. Malo, France, 625–627.

Ramos, V.A., 1989. Andean foothills structures in the northern Magallanes Basin, Argentina. American Association of Petroleum Geologist Bulletin 73, 887–903.

Ramos, V.A., 2005. Ridge collision and topography: foreland deformation in the Patagonian Andes. Tectonophysics 399, 73–86.

Ramos, V., Kay, S.M., 1992. Southern Patagonian plateau basalts and deformation: backarc testimony of ridge collision. Tectonophysics 205, 261–282.

Ray, F.M., 1996. Stratigraphical and structural evolution of Tertiary backarc basins in Southern Chile. University of Liverpool, England, 208 pp; Unpublished Thesis.

Roseneau, M., Melnick, D., Echtler, H., 2006. Kinematic constraints on intra-arc shear and strain partitioning in the southern Andes between 38°S and 42°S latitude. Tectonics vol. 25, TC4013, doi:10.1029/2005TC001943.

Sakagushi, A., 1996. High paleogeothermal gradient with ridge subduction beneath the cretaceous Shimanto accretionary prism, southwest Japan. Geology 24(9), 795–798.

Scalabrino, B., Lagabrielle, Y., Malavieille, J., Dominguez, S., Morata, D., Espinoza, F., Suarez, M., Rossello, E., 2007. Impact of spreading ridge subduction on the morphotectonic evolution of the Patagonian Cordillera since 15 Ma (latitude of the Chile Triple Junction, CTJ). Poster, GEOSUR 2007 International Geological Congress on the Southern hemisphere, Santiago, Chile.

Shapiro, N.M., Ritzwoller, M.H., 2004. Inferring surface heat flux distributions guided by a global seismic model: particular application to Antarctica. Earth and Planetary Earth Letters 223, 213–224.

Singer, B.S., Ton-That, T., Vincze, T., Rabassa, J., Roig, C., Brunstad, K., 1997. Timescale of late Cenozoic climate change in the southern hemisphere from 40Ar/39Ar dating of patagonia lavas. Terra Abs., Eur. Union Geosci. 9(9), 65–66.

Sisson, V.B., and Palvis, T.L., 1993. Geologic consequences of plate reorganization: an example from the Eocene southern Alaska forearc. Geology 21, 913–916.

Sisson, V.B., Poole, A.R., Harris, N.R., Cooper Burner, H., Palvis, T.L., Copeland, P., Donelick, R.A., and McLelland, W.C., 2003. Geochemical and geochronologic constraints for genesis of a tonalite-trondhjemite suite and associated mafic intrusive rocks in the eastern Chugach Mountains, Alaska: a record of ridge-transform subduction. In Sisson, V.B., Roeske, S.M., and Palvis, T.L., eds., Geology of a transpressional orogen developed during ridge-trench interaction along the North pacific margin: Boulder, Colorado, Geological Society of America Special paper 371, p. 293–326.

Stern, C.R., 2004. Active Andean volcanism: its geological and tectonic setting. Revista Geologica de Chile, vol. 31, no. 2, pp. 161–206.

Stern, C.R., Kilian, R., 1996. Role of the subducted slab, mantle wedge and continental crust in the generation of adakites from the Andean Austral Volcanic Zone. Contributions to Mineralogy and Petrology 123, 263–281.

Stern, C.R., Zartman, F.A., Futa, K., Zartman, R.E., Peng, Z., Kyser, T.K., 1990. Trace-element and Sr, Nd, Pb, and O isotopic composition of Pliocene and Quaternary alkali basalts of the Patagonian Plateau lavas of the Southernmost South America. Contributions to Mineralogy and Petrology 104, 294–308.

Stern, C.R., Kilian, R., 1996. Role of the subducted slab, mantle wedge and continental crust in the generation of adakites from the Andean Austral Volcanic Zone. Contributions to Mineralogy and Petrology 123, 263–281.

Suarez, M., de la Cruz, R., 2001. Jurassic to Miocene K-Ar dates from eastern central Patagonian Cordillera plutons, Chile (45°-48°S). Geological Magazine 138(1), 53–66.

Suarez, M., de la Cruz, R., 2000. Tectonics in the eastern central Patagonian Cordillera (45°30'-47°30'S). Journal of the Geological Society (London) 157, 995–1001.

Suarez, M., de la Cruz, R., Troncoso, A., 2000a. Tropical/subtropical Upper Paleocene-Lower Eocene fluviatil deposits in eastern Patagonia, Chile (46°45'S). Journal of South American Earth Sciences 13, 527–536.

Suarez, M., de la Cruz, R., Bell, C.M., 2000b. Timing and origin of deformation along the Patagonian fold and thrust belt. Geological Magazine 137, 345–353.

Tebbens, S.F., Cande, S.C., 1997. Southeast Pacific tectonic evolution from the early Oligocene to Present. Journal of Geophysical Research 102, 12061–12084.

Tebbens, S.F., Cande, S.C., Kpovacs, L., Parra, J.C., Labrecque, J.L., Vergara, H., 1997. The Chile Ridge: a tectonic framework. Journal of Geophysical Research 102, 12035–12059.

Thomson, S.N., Hervé, F., Brix, M., and Stockhert, B., 2001. The Mesozoic-Cenozoic denudation history of the southern Chilean Andes and its correlation to different subduction processes. Tectonics 20, 693–711.

Thorkelson, D.J., 1990. Tectonic and magmatic aspects of slab window, Geological Association of Canada Program and Abstract, 15, A105.

Welkner, D., 1999. Geologia del area del Cerro de San Lorenzo: cordillera patagonica oriental, XI region de Aysen, Chile (47°25'-47°50'S). Memoria Titulo Departemento de Geologia. Universitad de Chile. Unpublished.

Zumsteg, C.L., Himmelberg, G.R., Karl, S.M., Hauessler, P.J., 2003. Metamorphism within the Chugach accretionary complex on southern Baranof Island, southeastern Alaska. In Sisson, V.B., Roeske, S.M., and Palvis, T.L., eds., Geology of a transpressional orogen developed during ridge-trench interaction along the North pacific margin: Boulder, Colorado, Geological Society of America Special paper 371, p. 253–269.

Configuration of the Colombian Caribbean Margin: Constraints from 2D Seismic Reflection data and Potential Fields Interpretation

Adriana Maria Mantilla-Pimiento, Gerhard Jentzsch, Jonas Kley and Carlos Alfonso-Pava

Abstract The active Colombian Caribbean margin has evolved since late Cretaceous time, resulting in an intricate deformation history involving oblique subduction – accretion, extension, and tectonic inversion during the Cenozoic. The combined interpretation of 2D seismic reflection, gravity and magnetic data provides new insights into the configuration of the Colombian Caribbean margin and the crust-types present along it. The margin displays the morphological and tectonic characteristics of a typical accretion-dominated subduction complex. The 3D gravity modelling suggests that the Caribbean Plate is subducting beneath NW Colombia at a low angle of about 5° in E to SE direction. The major tectonic domains forming the margin include, from west to east: trench, active accretionary prism, outer high and forearc basins. The trench axis coincides with the toe of the active accretionary prism. The active prism corresponds to the external part of the Sinú-Colombia Accretionary Wedge. The outer high domain includes the major structural complex formed by the easternmost part of the Sinú-Colombia Accretionary Wedge and the San Jacinto Fold Belt. It represents the fossil part of the accretionary prism, which today acts as a dynamic backstop to the active accretionary prism. The outer high comprises several small sedimentary basins containing post-kinematic Plio-Pleistocene deposits, which fossilize the complex outer high structure. The landward boundary of the outer high is marked by the well-developed positive flower structure of the Romeral North or San Jacinto Fault System, which represents a structural break between the smaller basins deformed by mud diapirism to the west and the main and deeper forearc San Jorge Basin to the east. The basement-type present along outer high-forearc transition (San Jacinto Fold Belt-San Jorge Basin) is still discussed controversially. The results of 3D gravity and 3D magnetic modelling in this study support the presence of an oceanic "basement complex" (mixture of basalts and sediments) underlain by a continental tectonic wedge (CTW) which belongs to the overriding South American Plate. The emplacement of oceanic affinity rocks over continental basement is interpreted as offscraping and backthrusting of Caribbean material onto the continental margin during the initiation of the oblique subduction of Caribbean crust beneath NW South America. The continental crust of the South American arc framework provided the static backstop, against which the material of the later dynamic backstop

A.M. Mantilla-Pimiento
Fugro Gravity & Magnetics Servuces (Fugro Middle East) P.O.
Box 2863 Dubai, United Arab Emirates, AMantilla@fugro.
com, adriana_maria_mantilla@yahoo.com

S. Lallemand and F. Funiciello (eds.), *Subduction Zone Geodynamics*,
DOI 10.1007/978-3-540-87974-9, © Springer-Verlag Berlin Heidelberg 2009

was accreted. The existence of a continental tectonic wedge (CTW) beneath the oceanic "basement complex" in the San Jacinto Fold Belt (west of the Romeral Zone) indicates that the Romeral North Fault System does not form the tectonic boundary between oceanic crust to the west and continental crust to the east. The Romeral Fault and its associated structures form a dextrally transpressive fault system developed within continental crust (the static backstop) as a result of the oblique convergence between Caribbean and South American plates. From the crustal model proposed here, two tectonic limits are identified as ocean-continental crust boundaries: an upper one, which represents a major east-directed backthrust that detached oceanic basalts and sediments from the downgoing slab, emplacing them over the continental tectonic wedge (CTW), and a lower one, which corresponds to the east-dipping boundary between the downgoing Caribbean and overriding South American plates.

Keywords Romeral Fault System • Structural Configuration of the Colombian Caribbean Margin • 3D Gravity and Magnetic Modelling • Tectonic Evolution of the San Jacinto and Sinu Fold Belts.

1 Introduction

Since the 1960s the Andean margin of the South America and the circum-Caribbean region have been areas of particular interest to geoscientists (Pennington, 1981). Plate tectonic reconstructions (e.g. Pindell et al., 1998) have determined that the convergence of the Caribbean Plate and the NW South America Plate has been responsible for the diachronous accretion of allocthonous terranes and development of fold-thrust-belts along the Colombian and Venezuelan Caribbean margins.

The Colombian Caribbean margin extends from the Urabá Gulf (Panama border) in the SW to Guajira Peninsula (Venezuela border) in the NE. The area of interest of the present work represents the Caribbean coast of Colombia, situated in the north of the culmination of the Colombian Andes and southwest of the Santa Marta Massif (Figs. 1, 2 and 3). This region

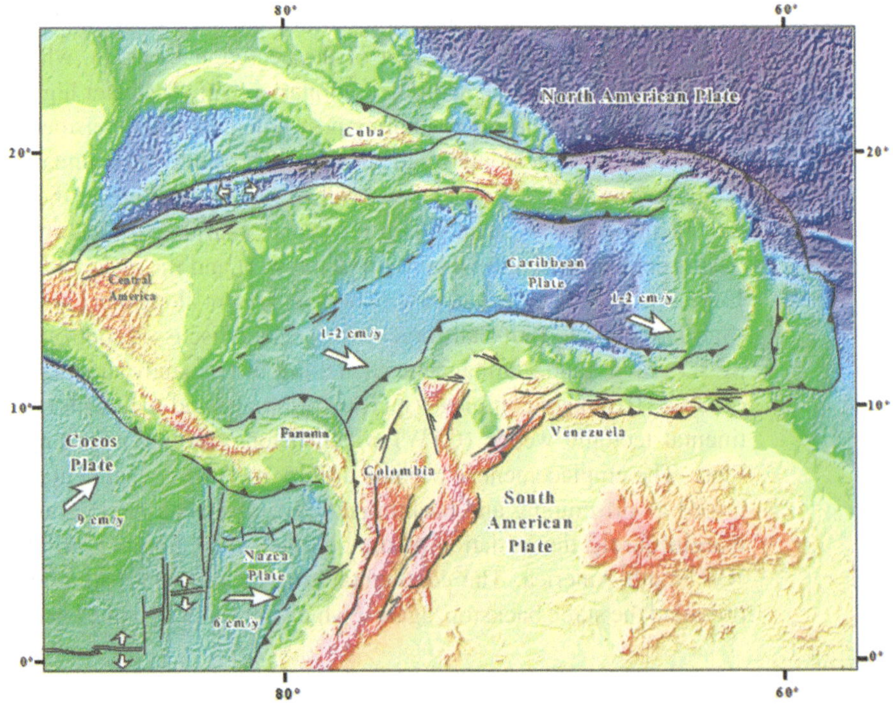

Fig. 1 Present day tectonic map of the northern South America and Caribbean region, indicating the plate boundaries and main fault systems. Map sources: Pindell and Barrett (1990); Taboada et al. (2000); topography from http://topex.ucsd.edu

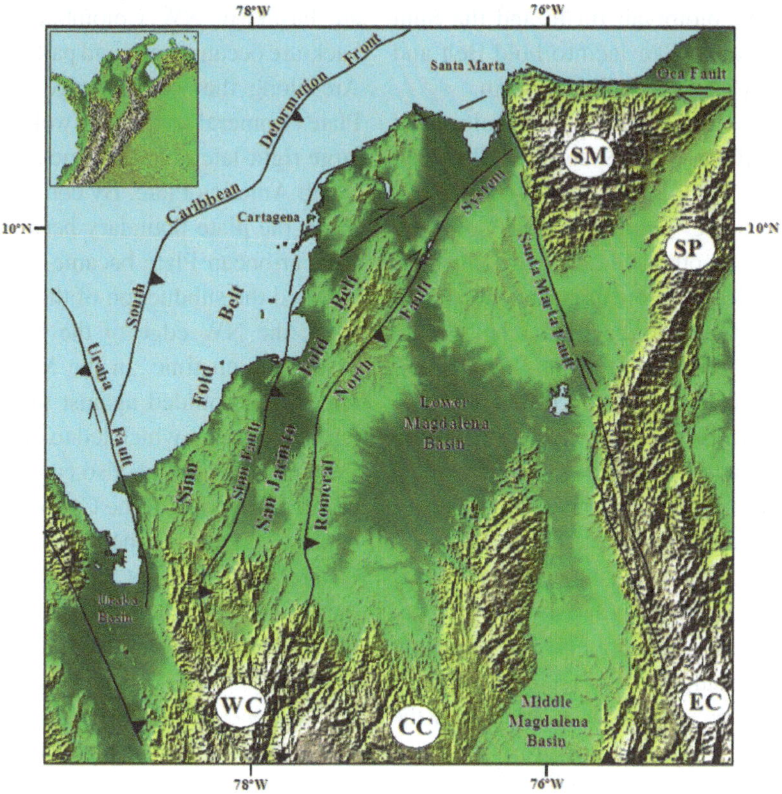

Fig. 2 Tectonic map of NW Colombia (simplified from Cediel and Cáceres, 2000). *WC* Western Cordillera; *CC* Central Cordillera; *EC* Eastern Cordillera; *SP* Sierra de Perijá; *SM* Santa Marta Massif

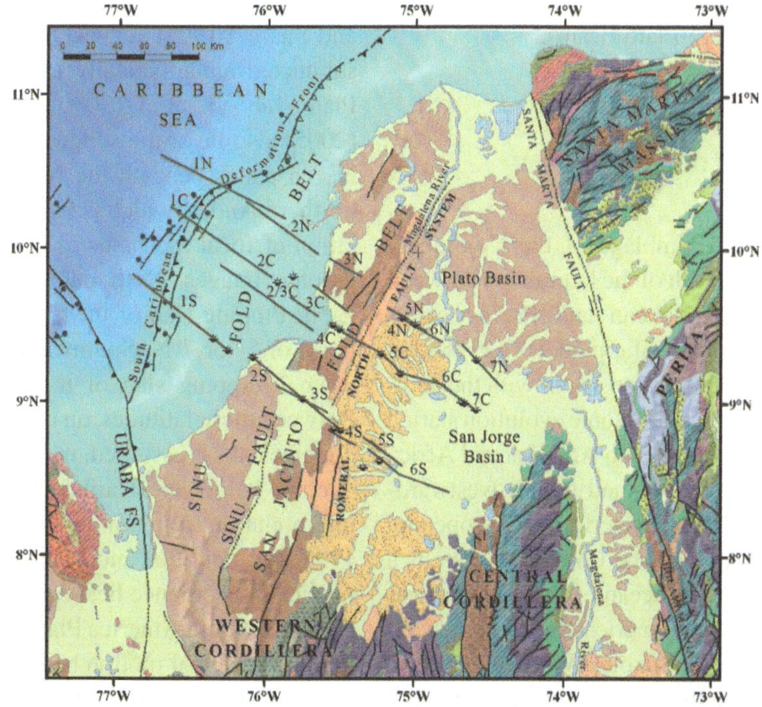

Fig. 3 Simplified Geologic Map of NW Colombia (French and Schenk, 2004). *Grey lines and black symbols* indicate the seismic lines and wells used for the interpretation

includes the offshore Colombian Basin and the Sinú Fold Belt, and the onshore San Jacinto Fold Belt and the Lower Magdalena (San Jorge-Plato) Basin.

The investigations carried out by the Colombian Geological Survey (INGEOMINAS) and the Colombian Geographical Survey (IGAC) as well as the joint venture projects between the Colombian Oil Company (ECOPETROL S.A) and other oil and gas companies in this area have provided the more important integrated geological and geophysical studies. Nonetheless, the crustal composition of NW Colombia and the responsible tectonic mechanisms of the current tectonic assemblage continue to be debated. The identification of the deep crustal elements as well as the effects of the deformation has important implications for understanding the tectonic evolution of the area. The main objective of this work is to characterize the crustal structure of the Colombian Caribbean margin using a combination of 2D seismic reflection, gravity and magnetic data. The new interpretation of the tectonic domains along the margin, here presented, is constrained by the 2D seismic dataset. The crustal structure is resolved satisfactorily by 3D gravity modelling. Additional 3D magnetic modelling is used to confirm the basement type (obtained from the 3D gravity modelling) involved along the Romeral Fault System, which has been historically defined as the limit between ocean crust to the west (San Jacinto-Sinú Fold Belts) and continental crust to the east (San Jorge-Plato Basin).

2 Caribbean Tectonic Overview

The complex tectonic assemblage of the Colombian Caribbean region is the result of the interaction between the South American, Caribbean and Pacific plates (Fig. 1). According to the Pacific models (e.g. Pindell and Kennan, 2001), which are consistent with the geology of the northern Andes, the tectonic evolution started with the separation of North-South America and Africa in late Jurassic time and continued with a westward-directed movement of the two Americas by the opening of the Atlantic since Aptian. The Caribbean Plate is a part of the Pacific Farallón oceanic Plate, which was displaced between the two Americas by these general plate movements during Mesozoic. With the acceleration of the Atlantic spreading, resulting in a relative NE movement of the Caribbean Plate, closure of the back-arc basins in NW Colombia led to the accretion of back-arc oceanic crust and part of the Greater Caribbean Arc along the western edge of the South American Plate (Romeral terrane) as well as to the formation of a large right-lateral shear zone along the NW part of the South America Plate. By continued eastward displacement, the plate boundary between NW Colombia and the Caribbean Plate became more compressive which initiated the subduction of the Caribbean oceanic crust along the NW edge of the South American Plate in Maastrichtian time. In the Miocene, the Panamanian Arc Ridge collided against the western border of the northern Andes, which led to the main Andean phase in Colombia, but which also caused the dramatic slowing of the subduction of the Caribbean Plate beneath South America. During the Miocene, formation of the Galapagos hot spot resulted in the break-up of the Farallón Plate into the Cocos and Nazca plates. The Nazca Plate is currently subducted along the western part of the South America Plate.

The geodetic observations indicate that convergence along the Colombian Caribbean margin is still occurring (Fig. 1). The calculated rates of convergence are 1.7 cm per year (Kellog and Bonini, 1982), 1 cm per year (Freymüller et al., 1993), 1.3 ± 0.3 cm per year (Van der Hilst and Mann, 1994) and 2.0 cm per year Cortés and Angelier (2005). Seismicity studies (Pennington, 1981; Kellog and Bonini, 1982; Malavé and Suárez, 1995) suggest that the Caribbean Plate is subducted amagmatically under the South American Plate. Taboada et al. (2000), and Cortes and Angelier (2005) also favor the subduction of the Caribbean and Nazca plates beneath the NW and W wedges of the northern Andes, which is based on integrated interpretation of focal mechanisms of earthquakes and tomographic images. Both publications assume a gently E-NE dipping slab of the Caribbean Plate at latitudes north of 5° or 7°N. Southward of 7°N the moderately steep E-dipping slab of the Nazca Plate is present. Between these latitudes, an overlapping of the two subducting slabs is assumed, where the Nazca Plate is subducting below the Caribbean Plate. In contrast, the interpretation from tomography images (Van der Hilst, 1990; Van der Hilst and Mann, 1994) indicates no significant convergence between the Caribbean Plate and the entire South America Plate. These authors conclude that the principal motion between these plates is right-lateral strike-slip, where the Caribbean Plate is underthrusted beneath the Maracaibo Block due to its northward directed movement. Nevertheless, it must be noted that the discussion is restricted to the interaction between the Caribbean Plate and the Maracaibo Block only. The identified subducted "Bucaramanga slab" is interpreted to be a part of the oceanic Nazca Plate, but which can be seen only in the tomographic images for latitudes from 7 to 10°N. The type of the continental margin at the NW Colombian margin (Sinú-San Jacinto accretion zone) is not discussed by these authors.

3 Tectonic Setting of NW Colombia

The Colombian Caribbean margin is part of the northern Colombian Andes region. A first overview of the evolution of this region based on surface geology and well information was published by Duque-Caro (1972, 1973, 1975, 1976 and 1979). This author identified three structural provinces, which include from west to east (Figs. 2 and 3): (1) the Sinú Fold Belt, (2) the San Jacinto Fold Belt and (3) the Lower Magdalena Basin. Two major structures separate these provinces: (1) the Romeral Fault System, which is considered as a major Paleo-suture between the Lower Magdalena Valley and the San Jacinto Fold Belt, and (2) the Sinú Fault between the San Jacinto Fold Belt and the Sinú Fold Belt.

The San Jacinto and Sinú Fold Belts have been interpreted as two prisms accreted to the NW Colombian margin (Duque-Caro, 1984). The San Jacinto Fold Belt, which is a narrow and complex structural zone developed adjacent to the platform, includes a complex of basalts and ocean-floor sediments of upper Cretaceous - Paleocene (?) age. Paleomagnetic (Brock and Duque-Caro in Cediel et al., 2003) and geochemical data (Kerr et al., 1996) suggest an origin southwest of its current position and an association with the tholeiitic MORB-type basalts of the Pacific Dagua-Piñón terrane (south part of the Western Cordillera). The Paleocene to Eocene turbiditic sequences of continental affinity and the middle Eocene erosional unconformity indicates a pre-Oligocene accretion of the San Jacinto terrane to the South American Plate along the Romeral North Fault System. The Sinú Fold Belt corresponds to the east part of the Sinú-Colombia Accretionary Wedge, which lies between the present day offshore Colombia trench and the San Jacinto Fold Belt. Accretion of the Sinú terrane on the San Jacinto terrane took place since Miocene (Cediel et al., 2003).

The Sinú Fault marks the boundary between the San Jacinto and Sinú Fold Belts. Duque-Caro (1979) defined this limit by its surface expression, controlling the Sinú River, and by the change of the structural style between these two fold belts. Additionally, the San Jacinto Fold Belt has no mud diapirism or mud volcanoes, which are common to the west in the Sinú Fold Belt. Active mud volcanoes are present in both areas Sinú River to the south and Cartagena-Barranquilla to the north. This phenomenon is well described and discussed by Case (1974), Duque-Caro (1979), Vernette (1986), Töto and Kellog (1992), Vernette et al. (1992) and others. The age assigned to this mud is late Oligocene to early Miocene based on foraminifera determinations from the heterogeneous mixture extruded at the surface (Duque-Caro, 1979).

The Romeral Fault System was initially recognized by Grosse (1926) in the southwest of the Antioquia State. Barrero et al. (1969) assigned this name to a west-vergent fault system, which separates the Western and Central Cordilleras. They assumed also that this fault is the contact between oceanic crust to the west and continental crust to the east. This fault system extends from southwestern Ecuador, where it has been mapped as the Peltec-Giron-Portovelo faults (CODIGEM, 1993a in Cediel et al., 2003), to the northern Colombian Andes. Duque-Caro (1975) extended this fault system 140 km further to the north into the Colombian Caribbean area to separate two distinct geologic environments: a continental one to the east and an oceanic one to the west (Figs. 1, 2 and 3). The Romeral Fault system was thus not interpreted as just a fault zone, but as a major structural contact between oceanic and continental crusts at the complete western part of Colombia. The Romeral Fault System in the Colombian Caribbean region has been later interpreted as the early-middle Eocene tectonic boundary or suture between continental crust to the east and the accreted San Jacinto terrain to the west (ESRI-ILEX, 1995; Laverde, 2000). The obduction model presented by Flinch et al. (2000), Flinch (2003) and Flinch (2003a, b) shows the Romeral Fault as a structure with a poliphase transpressional and transtensional evolution that offsets the west-dipping obduction suture. Cediel et al. (2003) use the name San Jacinto Fault to refer to the "Romeral North Fault", interpreting it and its associated structures as a dextral-oblique accretion record of the San Jacinto and later Sinú terranes to the continental margin.

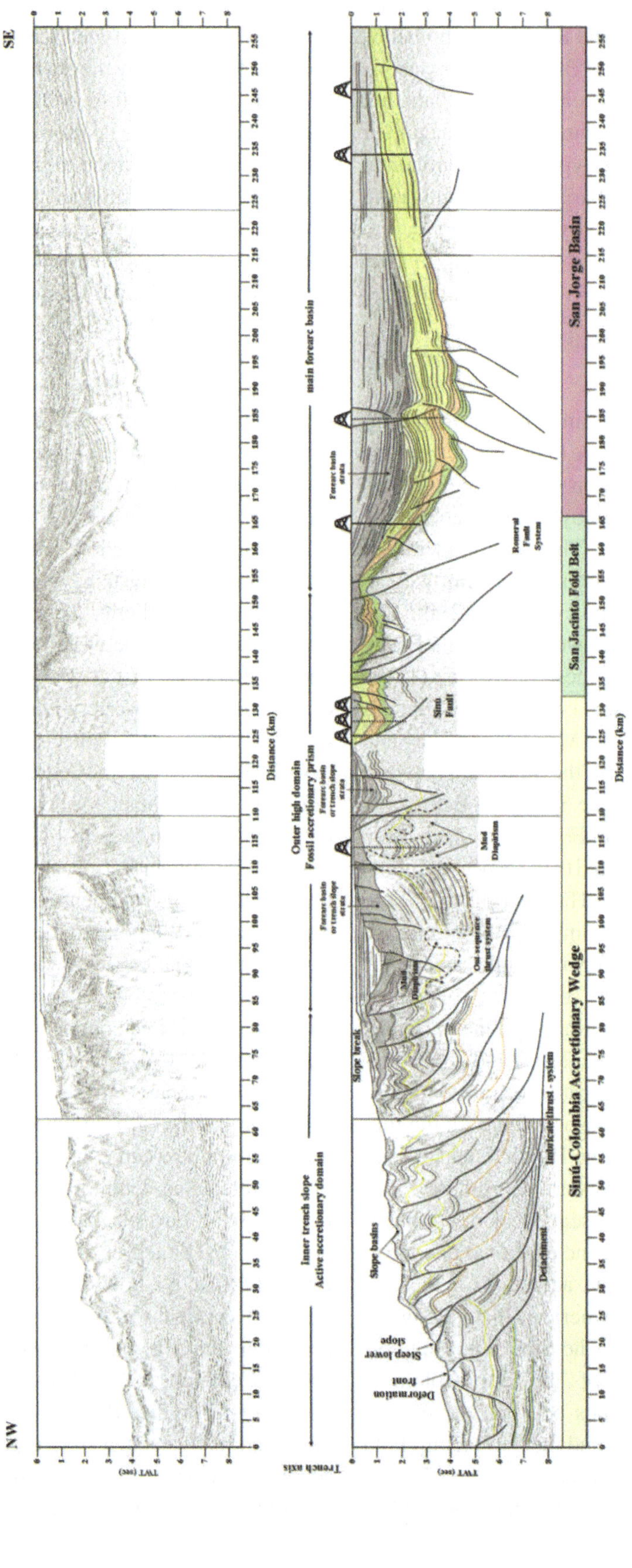

Fig. 4 Composite seismic profile displays the typical morphological elements and tectonic domains of the subduction complex along the Colombian Caribbean Margin. See Fig. 3 for location

4 2D Structural Configuration

The interpretation of 850 km of 2D seismic reflection data throughout the study area, integrating the surface geology and well-log data allows the identification of the morphological elements and tectonic domains of a typical subduction complex. Figure 3 indicates the location of the 3 composite 2D seismic profiles on the geologic map. Each of them is about 250 km long and covers both offshore and onshore areas of the NW Colombian Caribbean margin. Four major tectonic domains from west to east can be defined (Fig. 4): trench, active accretionary prism, outer high and forearc basins.

4.1 Colombian Caribbean Basin-trench Domain

The trench domain is particularly well defined in the westernmost part of the northern seismic profile (Fig. 5). The trench axis coincides with the toe of the active accretionary prism. The Colombian Caribbean Basin extends from the structural front of the accretionary prism. The basin infill is characterized by sub-parallel, subhorizontal, continuous, landward-divergent reflections. Seismic-stratigraphic analysis of the multi-channel seismic-reflection profiles (Bowland, 1993) and DSDP sampled sediments (in Bowland, 1993) suggest that a very thick sedimentary sequence, locally over 6 km thick, was deposited at the top of the oceanic basement. This sequence includes late Cretaceous-Eocene biogenic pelagic sediments, Eocene - late Miocene clastic turbidites and late Miocene–Holocene clastic sediments from turbidite-fan systems (Bowland, 1993).

4.2 The Active Accretionary Domain (Frontal Prism)

A clear image of the front prism is shown in Figs. 5 and 6. This domain is limited seaward by the trench and landward by the outer high domain and represents the younger part of the Sinú-Colombia Accretionary Wedge. The area is characterized by a zone of active deformation and strong shortening, resulting in a complex seaward-vergent imbricate thrust system of short wavelength folds and complicated patterns of syn-depositional deformation. This imbricate thrust system involves the complete sedimentary sequence. Its basal detachment could be localized near to the top of the oceanic basement. Because this detachment is deeper than the seismic record length (more than 8 seconds), its definition is based on structural modelling and geometrical constrains.

A very steep slope characterizes the sea floor topography at the frontal structures of the active accretionary prism. Several small basins can be recognized here. Pelagic sediments cover the thin-skinned structures, and pounded Pleistocene sediments are trapped in these small basins.

The seismic information shows a change in the structural style of the active front along strike, which varies from an imbricate thrust system (Fig. 5) to a triangle zone (Fig. 6). These geometric structural changes could be the result of some stratigraphic or some deeper-seated mechanical control.

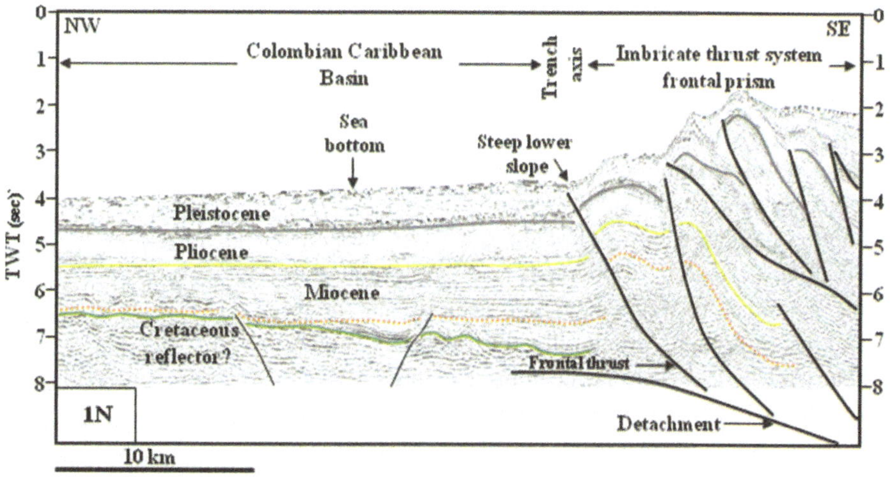

Fig. 5 Seismic image of the trench domain in the western part of the seismic line 1N (see location in Fig. 3). The trench infill is characterized by sub-parallel, sub-horizontal, continuous, landward-divergent reflections that onlaps the interpreted basement indicated by green reflector

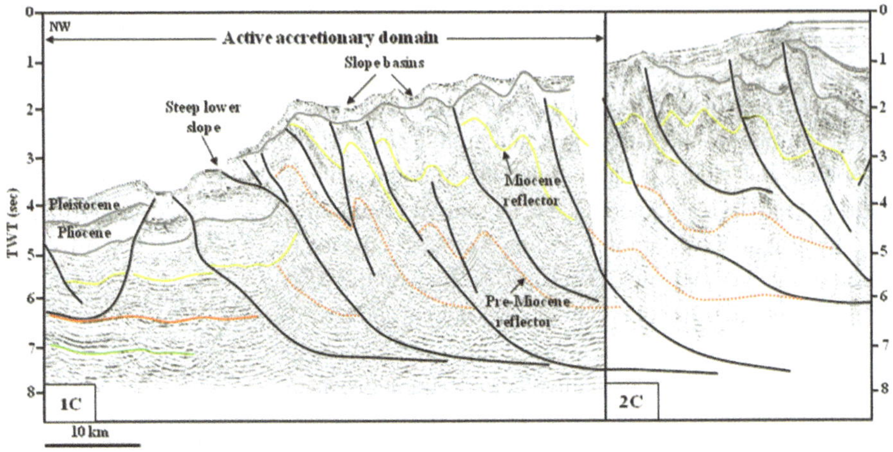

Fig. 6 Seismic expression of the active accretionary domain in seismic line 1C (see location in Fig. 3). This domain represents the younger part of the Sinú–Colombia Accretionary Wedge and is limited seaward by the trench and landward by the outer high domain

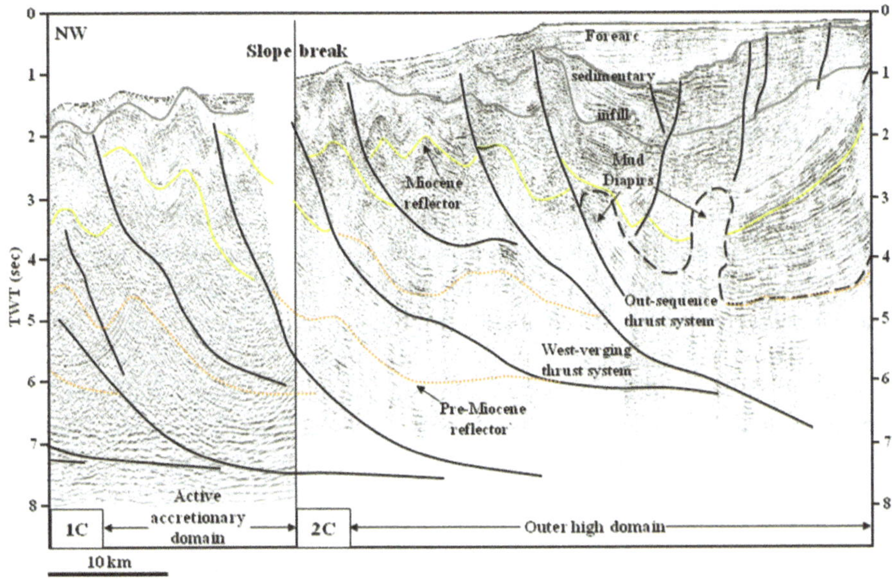

Fig. 7 Seismic expression of the boundary between the active accretionary prism and outer high domain in seismic line 2C (see location in Fig. 3). A west-verging imbricate thrust system and a backward breaking out-sequence thrust system characterize this zone

4.3 Outer High (Older Accretionary Domain)

The transition between the active accretionary prism and the forearc domain is marked by the outer high domain (Byrne et al., 1993). A seismically well-imaged deformation zone is located 80 km landward from the trench in the landward direction (Fig. 5). This zone coincides with an elevated zone and the slope break at the seafloor topography and is interpreted as the boundary between the active accretionary prism and the outer high. Structurally, this boundary is characterized by a west-verging imbricate thrust system and by a backward breaking out-of-sequence thrust system. Some thrusts forming these systems cut through to the top of the forearc basin deposits (Fig. 7). Younger deformation is evidenced by faulting and folding present in the forearc sedimentary fill. Towards the east, the zone of this boundary is characterized by normal faulting, growth folding and mud diapirism (Fig. 8). The geometry of the pre-kinematic sedimentary units (beneath the yellow reflector) is relatively well defined, but the high deformation and mud diapirism masks the lower units which characterize the internal structure of this domain. Shortening is concentrated in anticlines, where the underlying ductile mud can flow towards the anticline cores leading to the formation of compressional diapirism. These mud diapirs do not necessarily extrude up to the surface, but can be stopped and sealed by further sedimentation (forearc basin sedimentary fill). Flanked to

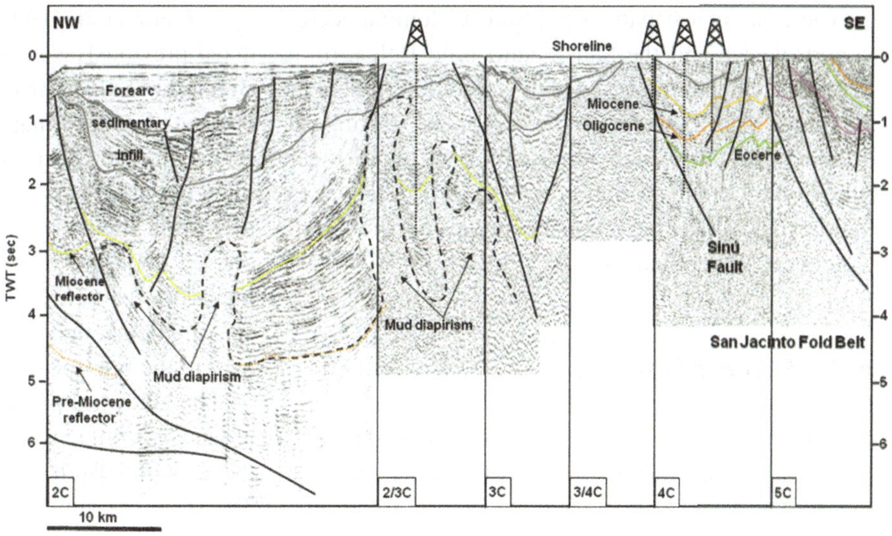

Fig. 8 Outer high domain displayed in the central composite seismic profile (see location in Fig. 3). Normal faulting, growth folding and mud diapirism characterizes this domain

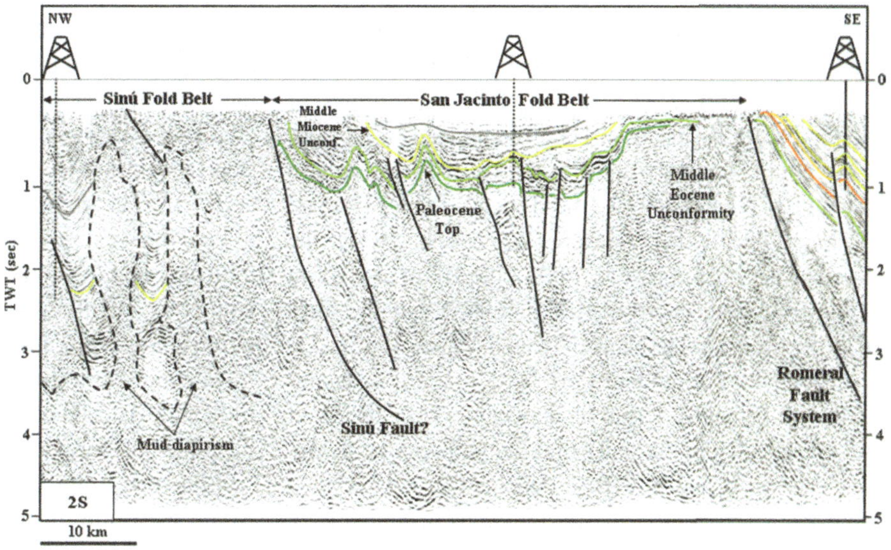

Fig. 9 Seismic line 2S (see location in Fig. 3) displays the best seismic image of the San Jacinto Fold Belt. Normal faults affecting the pre-middle Miocene sequence, steep thrust faults and some inversion structures characterize this zone

the compressional diapirism structures, syn-tectonic sedimentation can be recognized. The layers of syn-tectonic strata (middle Miocene – early Pliocene) deposited during uplift of the anticlines and diapirism display fan-like patterns with progressive onlaps onto the limbs of anticlines, leading to syn-tectonic unconformities with overlying sediments. Normal faulting is present 15 km landward from the slope break. These faults show a typical fan-like pattern of deposition, indicating a contemporaneous fault-control subsidence and syn-tectonics deposition.

Landward, where the limit between the Sinú-Colombia Accretionary Wedge and San Jacinto Fold Belt has been defined, the seismic character changes from well defined reflection geometry of the San Jacinto Fold Belt to chaotic seismic patterns of the Sinú Fault (Figs. 8 and 9). This loss in resolution indicates the region has undergone a high degree of deformation in this zone. Duque-Caro (1979) interpreted the Sinú Fault as the feature that marks a change in the structural style between the Sinú Fold belt in the west and the San Jacinto Fold Belt to the east, as well as the east-limit of the mud diapirism, which is only present in the Sinú Area.

Because the San Jacinto Fold Belt is poorly defined by seismic images, it was only possible to identify the middle Eocene, middle Miocene and intra-Pliocene unconformities but not the basement top (Fig. 9). Normal faults affecting the pre-middle Miocene sequence are common, while the compressional deformation is represented by steep thrust faults and some inversion structures.

The easternmost part of the Sinú-Colombia accretionary wedge and the San Jacinto Fold Belt form a major structural complex, which is here interpreted as the outer high domain at the Colombia Caribbean margin. The outer high represents part of an older accretionary prism, which now acts as a backstop to the frontal active accretionary prism of the Sinú-Colombia Accretionary Wedge (Fig. 4).

4.4 Forearc Domain

Oceanographic investigations and experimental modelling have revealed that the process of tectonic accretion in active margins is an efficient mechanism to develop forearc basins (Larroque et al., 1995). These basins are bounded seaward by the structural high of the accretionary system and landward by a relative rigid backstop against which their sediments are deformed (Dickinson, 1973, 1974; Karig & Sharman, 1975). Examples of these forearc basins have been recognized in western Colombia (Mountney and Westbrook, 1997), northern Colombia (Hardy, 1991), the Tobago Trough forearc basin off the Lesser Antilles (Westbrook, 1975), the Sunda Margin (Kopp et al., 2001), the Hikurangi Basin – New Zealand (Larroque et al, 1995) and the Tosa Basin adjacent to the Nankai Trough (Nasu, 1982).

Along the Colombian Caribbean margin, a seismically well-imaged series of depocenters, in the shelf sector, fossilizing the outer high domain can be recognized (Figs. 4 and 8). These depocenters have a wide spread surface extension, over hundreds of square kilometers, with an over 1 km thick sedimentary cover. They are elongated, oriented parallel to the regional structural trend, and they are particularly developed behind the Neogene accretionary prism and characterized as a series of forearc sub-basins. These sub-basins are filled by post-kinematic Plio-Pleistocene deposits and seismically characterized by parallel to sub-parallel and lightly continuous reflectors onlapping the intra-Pliocene unconformity (Figs. 4 and 8). Growth folding and deformation caused by mud diapirism are evident. The deformation phase that generated the development of the active accretionary prism in the front of the Caribbean convergent margin was contemporaneous with the deposition of the upper Miocene – lower Pliocene (?) sequence, and had little effects on the post-kinematic forearc deposits, resulting in the preservation of almost flat-lying sequences of Pliocene – Pleistocene strata.

Vernette et al. (1992) and Ruiz et al. (2000) recognized previously some of these depocenters (e.g. San Bernardo Sub-basin), indicating the influence of oblique-slip deformation along its trend within the Sinú Fold Belt. Within the forearc basins, recent tectonic movements are documented by folded sedimentary strata and active faults, some of which crop out at the sea floor and seem to be active, because the sediments of the sea floor are affected. A well defined active extensional fault trend characterizes the sector near to the present shoreline (Figs. 4 and 8).

The landward part of the forearc domain is characterized by a bulge which coincides with the zone where the Romeral Fault System has been traced (Fig. 10). The structural uplift generated by the development of the Romeral Fault System (RFS) is the most prominent characteristic at the transition between the highly deformed forearc domain to the west and the deep San Jorge-Plato Basin to the east (Figs. 4, 10 and 11). Upon the evidence on hand the Romeral Fault System is characterized by a seismically well defined, regional scale, asymmetric, positive flower structure (Figs. 4 and 10). The thinning of the early Miocene strata toward the western border of the San Jorge-Plato Basin indicates rotation and uplift along the fault zone, which could be contemporaneous with the Miocene sedimentation (Figs. 9, 10 and 11).

Seismic stratigraphic investigations (ICP, 2000) indicate a late Paleogene age for the oldest sediment layers in the forearc basins, corresponding to Eocene – Oligocene strata, which are thickening progressively to the west due to the Romeral structural uplift. The Pliocene to recent deposits reveal that the development of the Plato – San Jorge Basin is characterized by continuous sedimentation and related subsidence. As in the adjacent basins, more than 4 km of sediments have been accumulated. The San Jorge-Plato Basin infill consists of horizontally layered strata, roughly parallel and mainly undisturbed (Fig. 11). Exploratory drilling has penetrated alternating sequences of sand, silty mudstone and limestones as well as metamorphic and igneous rock of continental basement. Structurally, the basin is characterized by initial extensional structures and later by thrusting and inversion, generated by a major transpressional tectonic regime. Interpreting the Plio-Pleistocene sequence of the San Jorge-Plato Basin as forearc sedimentary fill, the Romeral Fault System represents a major structural break separating the smaller basins

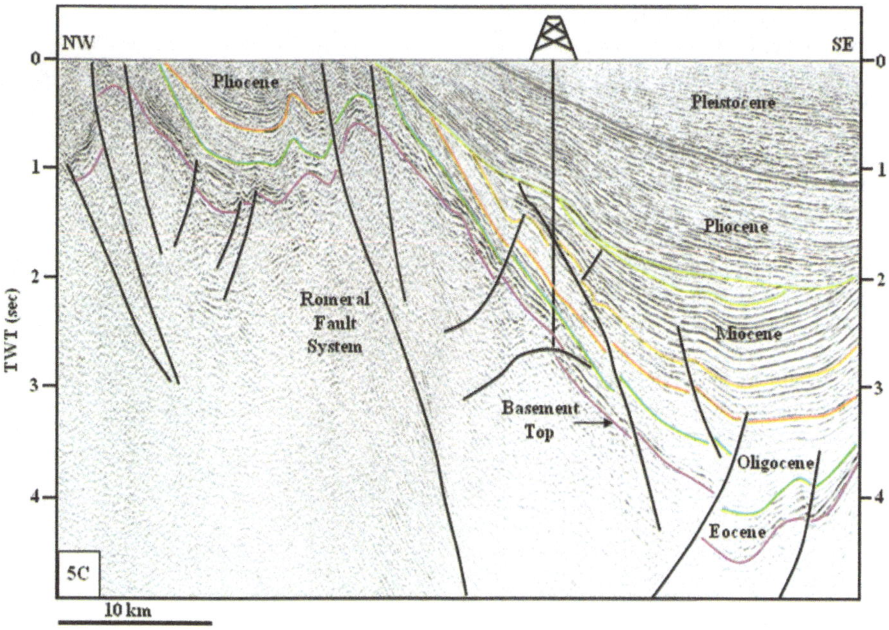

Fig. 10 Line 5C (see location in Fig. 3) displays the seismic expression of the Romeral Fault System. A regional scale, asymmetric and positive flower structure is well defined

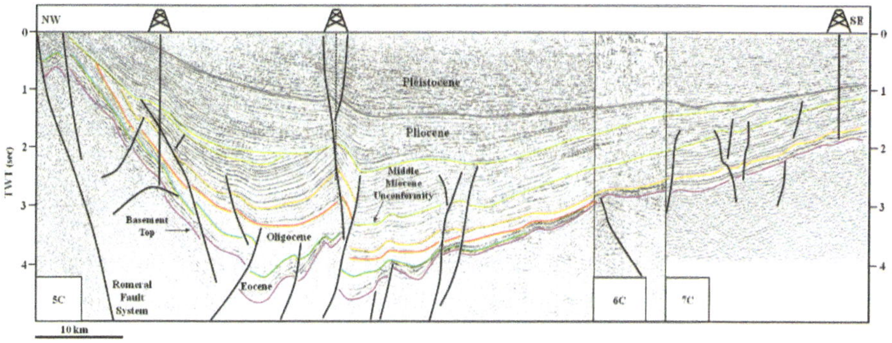

Fig. 11 Seismic lines 5C, 6C and 7C (for location refer to Fig. 3) cover fully the San Jorge Basin. Structurally, the basin is characterized by initial extensional structures and later by thrusting and inversion

from the main Plato-San Jorge Basin in the forearc domain (Fig. 4).

According to weak reflections present beneath the basin infill, which show low frequencies and high amplitudes, the continental basement continues westward across the Romeral Zone and probably below the entire forearc basin.

5 Potential Field Interpretation

5.1 *Bouguer Anomaly map*

The gravity data set for NW Colombia was compiled from available land surveys from Colombia, gravity marine surveys acquired on the Colombia platform and adjacent areas, and complemented with satellite-derived gravity data for the offshore area (Cerón, 2002). The data set was reduced to complete Bouguer anomaly using a density of 2.67×10^3 kg m^{-3} onshore and 2.2×10^3 kg m^{-3} offshore (GETECH, 2001). Figure 12 shows the Bouguer anomaly map in conjunction with the major structural features of northwestern Colombia.

Strongly positive Bouguer anomalies, which decrease toward the continental margin, characterize the Colombian Caribbean Basin. The region east of the South Caribbean deformation front is characterized by local high and low gravity anomalies which follow the trend of the Sinú and San Jacinto Fold Belts. The Sinú Fault (Duque-Caro, 1979), which separates the two fold belts, has no significant gravity expression, which indicates that rocks on both sides of the fault have

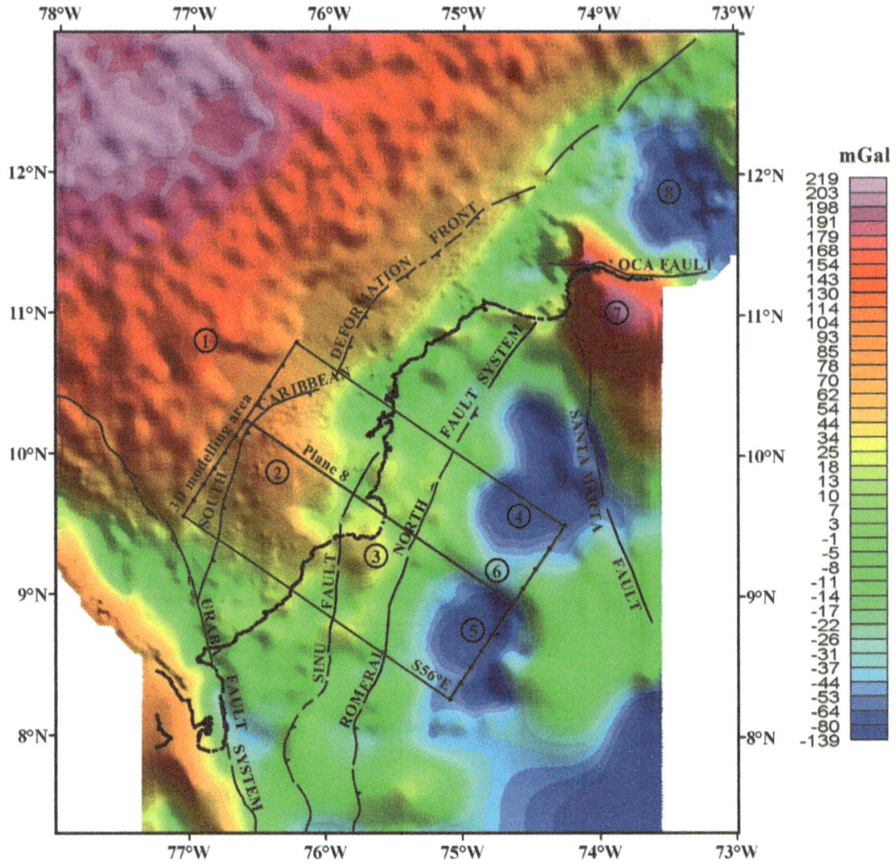

Fig. 12 Bouguer anomaly map of the NW Colombia and major structural features. (1) Colombian Caribbean Basin, (2) Sinú–Colombia Accretionary Wedge, (3) San Jacinto Fold Belt, (4) Plato Sub-basin, (5) San Jorge Sub-basin, (6) Cicuco High, (7) Santa Marta Massif, (8) Ranchería Offshore and Chuchupa-Ballenai Basins. The *black rectangle* indicates the 3D gravity modelling area. The plane 8 corresponds to the seismic composite profile in Fig. 4

similar average densities. The Romeral Zone corresponds to a NE-SW striking positive anomaly. The Plato and San Jorge sub-basins are characterized by gravity lows down to -90 mGal. These basins are separated by a local gravity high (–30 to 10 mGal), which correlates with the Cicuco High. A steep gradient is found along the Uramita (Urabá) Fault, which was interpreted by Duque-Caro (1990) as the structural boundary between the Urabá Basin and the Sinú Accretionary Wedge. The strongly positive anomalies of up to 190 mGals at the Santa Marta Massif, which rises to 5.776 m, indicate that the area is out of isostatic balance. An uncompensated excess mass must exist under the massif at relatively shallow depth (Case and McDonald, 1973). The massif is limited to the north by the Oca Fault which is correlated with a steep gradient that marks the density contrast between the crystalline rocks of the Santa Marta Massif and sedimentary deposits of the Ranchería offshore and Chuchupa-Ballena basins. Another very steep gravity gradient characterizes the western limit of the massif. This gradient coincides with the mapped trace of the Santa Marta – Bucaramanga Fault Zone, which represents the boundary between the massif and the San Jorge-Plato Basin, where the anomaly values decrease substantially.

5.2 3D Density Model

A large scale 3D density model was constructed to characterize the crustal structure and density distribution of the Colombian Caribbean margin. The 3D gravity modelling was carried out with the interactive modelling program IGMAS (Schmidt and Götze, 1995). The 3D modelling region covers an area of 258.7 × 163 sq. km and strikes with S56°E approximately parallel to the direction of the tectonic transport (Fig. 12). Eleven planes build the 3D model (Fig. 13). The planes 4, 8 and 10 correspond to the south, central and north 2D structural cross-sections (Fig. 3), for which seismic and well information are available. A very good fit between the observed and calculated

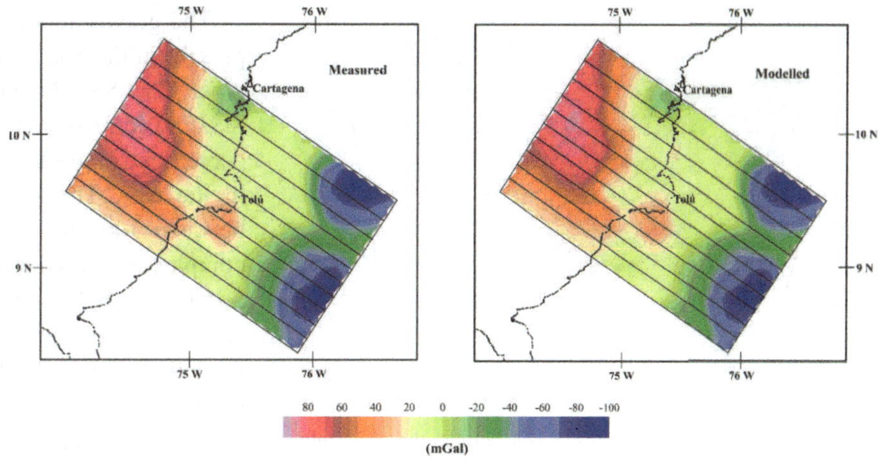

Fig. 13 Observed and calculated gravity fields in the 3D modelling area (contour interval 5 mGal)

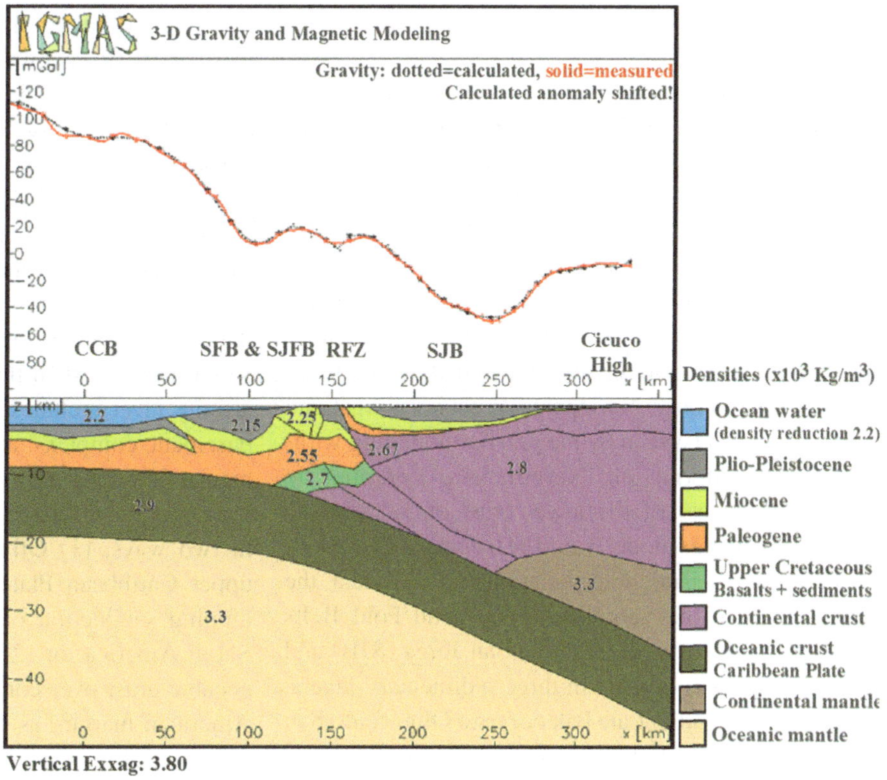

Fig. 14 Plane 8 of the 3D density model (see Fig. 12 for location). *CCB* Colombian Caribbean basin; *SFB* Sinú fold belt; *SJFB* San Jacinto Fold Belt; *RFZ* Romeral Fault Zone; *SJB* San Jorge Basin

curves was obtained for each plane along the different tectonic domains. Figure 13 shows a comparison between the calculated and the observed gravity field. Figure 14 shows the plane 8 of the model. The model illustrates very well the current subduction complex, where the Caribbean Plate dips at a low angle of about 5° in ESE direction (Fig. 15).

Due to use of the complete Bouguer anomaly in both offshore and onshore areas, the density of the water is replaced by the reduction density at the ocean side (Nettleton, 1971), which is 2.2×10^3 kg m^{-3}. The Caribbean Plate was assumed to be normal oceanic crust and was modelled with a density of 2.9×10^3 kg m^{-3}. The 3D density model shows that more than 6 km of sediments overlie the oceanic basement of the Caribbean Plate, which is consistent with interpretations of Edgar et al. (1971), Lu and McMillen (1982), and Bowland (1993) of refraction and multi-channel reflection seismic data from the Colombian Caribbean Basin (CCB). A density of 3.3×10^3 kg m^{-3} was

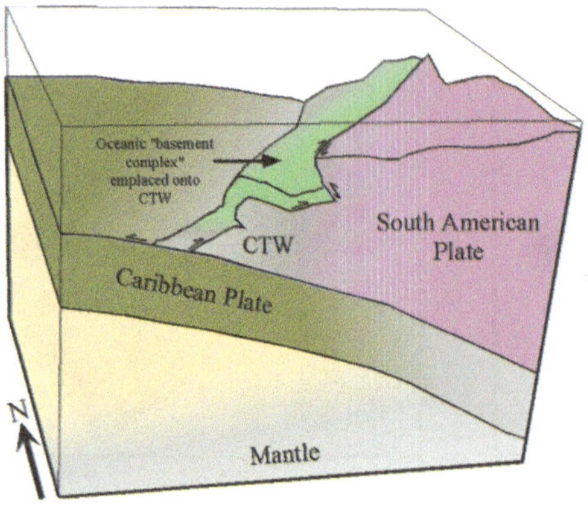

Fig. 15 Distribution of crust-types along the Colombian Caribbean margin (3D view of the density model). CTW: continental tectonic wedge

selected for both oceanic and continental lithospheric mantle. The depth of the crust – upper mantle boundary of ≈25 km was initially taken from the publications of Case et al. (1984, 1990). However, the exact location of the Moho-boundary could not be determined by this modelling. Any changes of the lower continental crust by granulite-eclogite transition, ultramafic underplating or -intrusions were not considered because these changes lead to a rise of the density from 3.0 to 3.3 × 10^3 kg m only. No separation between crust and lithosphere can be seen by the gravity modelling without further information. Nevertheless, assuming an average thickness of the lower crust of 5–8 km, the Moho-limit in this model would be located at 23–26 km. According to seismic interpretation and well data, the upper crust in the San Jacinto and Sinú Fold Belts (SFB & SJFB) as well as in the San Jorge (SJB) and Plato Basins was divided in three sedimentary layers, which correspond to Late Paleocene to Oligocene (San Jacinto and Sinú Fold Belts), Eocene to Oligocene (San Jorge and Plato Basins), Miocene and Plio – Pleistocene sequences. The applied densities for these polygons were initially taken from density logs. The computed gravity anomaly was then compared with the observed data and the densities and boundaries of the layers were adjusted, until a satisfactory fit was achieved. Thus, in the resultant model a density gradient for the sedimentary cover can be observed, which increases from 2.13 to 2.55 × 10^3 kg m^{-3}. Average densities of 2.8 × 10^3 kg m^{-3} for middle crust and 2.67 × 10^3 kg m^{-3} for upper crust in the San Jorge (SJB) and Plato Basins were used.

The gravity modelling confirms that the basement present in the San Jorge-Plato Basin is of continental type. The San Jacinto and Sinú Fold Belts (SFB &SJFB) correspond to a local positive anomaly, which decreases towards the San Jorge (SJB) and Plato Basins in the east. The variations of the gravity response between 100 and 200 km (Fig. 14) suggest a more complex deformation of the basement in the Sinú and San Jacinto Fold Belts (SFB & SJFB). Just as it can be seen in the Bouguer anomaly map (Fig. 12), also in the modelled planes the Sinú Fault cannot be resolved, because no density differences occur between the two areas. Therefore, the location of this lineament is based on surface geology and seismic data only. Several different interpretations of the composition of the basement were tested for the San Jacinto Fold Belt (SJFB), west of the Romeral Fault Zone (RFZ). Finally two crustal bodies were included: an upper one, which represents lithologies of oceanic affinity with an average density of 2.70 × 10^3 kg m^{-3} and the other lower one (CTW) with a density 2.8 × 10^3 kg m^{-3}, which is modelled as a block that contains middle continental crust. The upper layer presumably consists of a mixture of deep ocean floor sediments and oceanic crust, and therefore has an average density of 2.70 × 10^3 kg m^{-3}, which is lower than that of pure basalt. This oceanic affinity layer includes the upper Cretaceous "oceanic basement complex" and the deep ocean floor sediments. The emplacement of mixtured oceanic rocks over continental basement (CTW) can be interpreted in two ways: (1) Offscraping of materials from the upper Caribbean Plate during the initiation of east-dipping subduction of Caribbean crust beneath NW South America, or (2) obduction of proto-Caribbean oceanic crust over continental basement of the South American margin, as proposed by Flinch (2003) and Flinch et al. (2003a, b). The density model does not allow the differentiation of these two cases. Flinch (2003) and Flinch et al. (2003a, b) assumed obduction of the proto-Caribbean oceanic crust during the upper Cretaceous (Campanian – Maastrichtian), followed by subduction and accretion. These authors identified ophiolites thrust onto the basement of the South American Plate, which are interpreted as obducted fragments of the Great Caribbean Arc of Burke (1988) or even parts of the over-thickened Caribbean Plateau crust. In their interpretation, the Romeral Fault System offsets the westward dipping obduction suture. However, the 3D gravity modelling results obtained here do not support the presence of layers with oceanic-type density east of the Romeral Fault Zone, in the San Jorge-Plato Basin (Fig. 14). Apart from the discussion of the emplacement-mechanism of these basalts (either obduction or early subduction related), the absence of rocks with oceanic affinity east of the Romeral Fault Zone (San Jorge-Plato Basin) could be explained by their erosion caused by the hanging wall uplift of the subduction system during the earlier phase of development of the Romeral Fault System. The minimum and maximum depth of the top of the upper oceanic layer ($\rho = 2.70 \times 10^3$ kg m^{-3}) is 5.5 and 13 km, respectively. This unit has a minimum thickness of about 2.6 km in the San Jacinto Fold Belt and it extends westernmost into the Sinú Fold Belt, with an approximate maximum thickness of 4 km. Because these lower boundaries cannot be identified from the seismic data, their approximation is based solely on the results of the gravity modelling. The lower layer consists of middle continental crust and is interpreted as a continental tectonic wedge (CTW), which is part of the overriding South American Plate. This is partially in accordance with the interpretation of Ruiz et al. (2000): according to them, continental crust is present west of the Romeral Fault System in the northern part of the San Jacinto Fold Belt, whereas the role of the Romeral Fault System as an oceanic-continental crust boundary is restricted to the south of the Canoas Fault Zone.

However, the results of the 3D gravity modelling from this study show that continental crust underlies the oceanic "basement complex" of the San Jacinto Fold Belt along the whole modelled region, which includes the area south of the Canoas Fault Zone. The lack of any density contrast between middle continental crust of the CTW and the continental crust of the South American Plate, which occurs almost at the same depths, indicates that both together form the continental margin.

According to the modelling results, the Romeral Fault Zone (RFZ) coincides with an east-dipping limit, which separates the upper continental crust (2.67×10^3 kg m^{-3}) to the east from the upper oceanic crust (2.70×10^3 kg m^{-3}) to the west at least to a depth of 10 km. The presence of blocks that contain middle continental crust beneath a layer of oceanic type rocks in the San Jacinto Fold Belt (SJFB) indicates that the Romeral Fault System does not represent a paleo-suture, along which oceanic crust has been accreted. Apparently, the Romeral Fault System had cut the block of continental crust, on which the San Jacinto Fold Belt (SJFB) was formed. Therefore, the Romeral Fault System is not an oceanic-continental crust boundary as has been interpreted elsewhere, but represents a transcurrent fault system, which is developed on the continental side. This implies a new definition of the ocean-continental crust boundary. Two tectonic limits are identified: an upper one, which represents the west-ward dipping contact between the continental tectonic wedge (CTW) and oceanic basalts decoupled from the downgoing slab, and a lower one, which represents the east-ward dipping boundary between the subducting Caribbean Plate and overriding South American Plate.

5.3 Reduced-to-Pole map (RTP)

Magnetic surveys both from aerial and marine platforms were compiled into a single grid using a leveling technique to bring different data patches to a common equivalent elevation of 1,000 m barometric at average sea level (Cerón, 2002). The data coverage in the Colombian Caribbean Basin (offshore region) is not sufficient to analyze the character of the anomalies in this area. Therefore, the description presented below is restricted to the onshore portion and the most eastern offshore part of the study area. LCT-Fugro software was used for the reduction to pole process (RTP) to obtain the map shown in Fig. 16. The parameters of the Earths regional magnetic field are determined from a 13th order polynomial approximation of magnetic satellite readings and calculated for the time of the survey. The value for the Earth's magnetic intensity of 28,400 nT and the magnetic inclination and declination for the survey area of 39.5°N and 3.5°W respectively were used as magnetic parameters for the merged onshore and offshore area. Anomalies of medium to high amplitude and high-relief, trending NW–SE and E–W are present across the San Jorge and Plato sub-basins (east of the Romeral Fault System). The Cicuco High, which separates the San Jorge and Plato sub-basins, follows NW-SE linear, strong and positive anomalies. The anomalies present east of the Romeral Zone must be generated by the basement (intra-basement source) that underlies the San Jorge and Plato sub-basins, which has been drilled by several wells. The basement consists of Silurian to Paleozoic metamorphic rocks (greenschists, amphibolites, metabasites, granulitic migmatites and

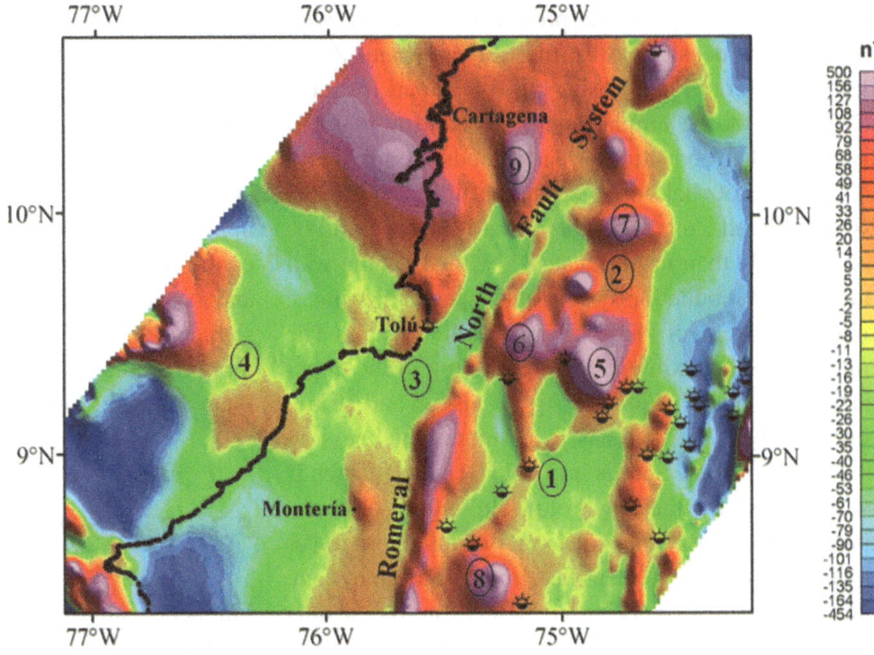

Fig. 16 Magnetic Reduced-to-pole anomaly map of NW Colombia (1) San Jorge Sub-basin, (2) Plato Sub-basin, (3) San Jacinto Fold Belt, (4) Sinú Fold Belt, (5) Cicuco High, (6) Ayhombe High (7) Apure High, (8) Cábano-Tablón High, and (9) positive anomaly southeast to Cartagena city. The *symbols* indicate the location of some wells, which have drilled the basement

gneisses), which were intruded by Paleozoic mafic to ultramafic rocks and Cretaceous stocks of tonalites and granodiorites (INGEOMINAS, 1997). Ultramafic to mafic intrusions occur frequently in the western part of the Cordillera Central. It can be assumed that the high amplitude and high relief anomalies (in Fig. 16 designated with 5 and 7) are caused by these local lithologic variations within the basement, as these rocks contain higher amounts of magnetic minerals. In the RTP-map (Fig. 16), the location of the Romeral Fault System can be much better identified when compared with the Bouguer anomaly map (Fig. 12). The Romeral Zone is characterized by long, NE-SW striking anomalies with steep gradients at their borders. The strong positive anomalies indicate that this fault system has displaced strongly magnetized rocks near to the surface. This assumption is confirmed by the occurrence of positive magnetic anomalies at the Cicuco, Ayhombe and Apure highs, where wells have drilled shallow basement. Reyes-Santos et al. (2000) interpreted the region, east to the Romeral Fault System, as a group of transtensional and rotating basins, which are underlain by basements blocks. These blocks are bordered by normal faults which had controlled the thickness of the overlaying sedimentary units. The San Jacinto Fold Belt, west of the Romeral Fault Zone, is generally characterized by large positive to negative anomalies of medium magnetic relief with smooth gradients. This indicates higher sediment load, where the magnetized rocks of the basement are very deep. Small local positive anomalies present to the east of Monteria and Tolú cities could be related to highs of oceanic basement, which is exposed in the southern part of the San Jacinto Fold Belt. The positive anomaly up to 190 nT, located southeast of Cartagena city, is interpreted to be related with the continental basement present in the San Jorge and Plato basins. This would indicate that toward north, the Romeral Fault System is propagating to the west and its frontal structures are involving also here continental basement.

5.4 3 D Magnetic Model

The IGMAS software was also used to obtain a 3D magnetic model. Figure 17 indicates the location of the modelled areas in the RTP-map. The confirmation of the basement-type involved by the Romeral Fault System (obtained from the 3D gravity modelling) constitutes the main objective of the magnetic modelling.

Because remanent intensities as well as the direction of magnetization are unknown for the rocks in the study area, only apparent susceptibilities can be estimated (sum of induced and remanent rock magnetization). Constraints from the 3D density model and seismic interpretation, as well as the basement-depth information from some wells and magnetic data (Mantilla-Pimiento, 2007) are used. For the sedimentary cover, apparent magnetic susceptibility is set to 0. This value is supported by the general observation from the TMT-map (Mantilla-Pimiento, 2007), as magnetic intensity is inversely correlated with the sediment load on the basement surface. Therefore, contribution to the magnetic profiles from sediments is negligible. For the continental basement, only the depth-values previously calculated by Peter's method (Mantilla-Pimiento, 2007) are set to be constant during modelling with the software IGMAS. The remaining geometry of the basement-surface and the magnetic susceptibilities are modified, until a satisfying fit between calculated and observed anomalies is achieved. Apparent susceptibility of the layer "oceanic basement complex (basalts + sediments)" of San Jacinto Fold Belt is adjusted without modifying the basement geometry, which is controlled by depth estimations (see Table 2 in Mantilla-Pimiento, 2007) and 3D density model. The susceptibility value of this layer is lower than the average value of pure basalt. This is in accordance with the assumption that this layer represents a mixture between deep-ocean sediments and offscraped basalts from the Caribbean Plate. The position of the Caribbean Plate is not changed from that one of the 3D density model. An apparent susceptibility of average basalt is assumed. Nevertheless, the impact on the calculated anomaly is very small due to its depth. The oceanic-continental crust boundaries are taken from the 3D density model and only small adjustments were made.

Two areas were selected for the magnetic modelling (north and south model, Fig. 17) to confirm the basement type involved by the Romeral Fault System and to explain the origin of larger differences observed in the residuals from the 3D density model. Magnetic profiles for the two areas are similar as three main anomalies can be identified, which corresponds from west to east to the layers "basalts + sediments", Romeral Fault Zone and to some anomalies within the basement, which will be explained in the following (Figs. 18 and 19).

Superimposed on the main anomalies are several smaller ones, especially for the layer "basalts+sediments", which indicate a strong structural deformation and inhomogeneous composition for the north as well as south model area. Magnetic modelling indicates also the presence of continental basement (CTW) below this layer, which is in accordance with the results from the 3D density model.

The Romeral Fault Zone appears as strong positive magnetic anomaly in both model areas, which is caused by the uplift of the basement as explained before. In difference to the south model, where the Romeral Fault System can be modelled as a narrow fault zone, the presence of several magnetic suprabasement peaks indicates the complex configuration of the Romeral Fault System in the North model (Fig. 18). Final modelling resulted in a more than 20 km wide fault zone, which extends into the San Jacinto Fold Belt. It can be concluded that the Romeral Fault Zone is trending almost N-S within the complete study area and that its width is increasing from south to north. Reyes-Santos et al. (2000) suggest that toward the south the compressive component along this fault zone is stronger, whereas the lateral movement has a larger contribution towards the north.

The origin of the anomalies within the basement is different for the north and south model. In case of the south model (Fig. 19), the anomaly can be explained as suprabasement source only. This is also supported by the identification of a basement configuration from seismic profiles. In case of the north model (Fig. 18), the anomaly is caused by an intrabasement source, because depth variation of the basement top cannot explain the strong magnetic anomaly. Lower Palaeozoic mafic to ultramafic intrusives occur frequently in the Silurian to Ordovician metabasalts of the Central Cordillera. These rocks build up the western part of the basement of the San Jorge Basin and Cicuco High. It can be concluded that this anomaly, but also most of the others in the eastern part of the San-Jorge-Plato Basin, which have strong magnetic expression but limited spatial extension, are caused by these intrusions.

6 Integration of Results and Discussion

Based on the integrated analysis of different geophysical methods, the structural architecture of the Colombian Caribbean margin has been examined. Three important aspects of the results in this study are discussed in the following.

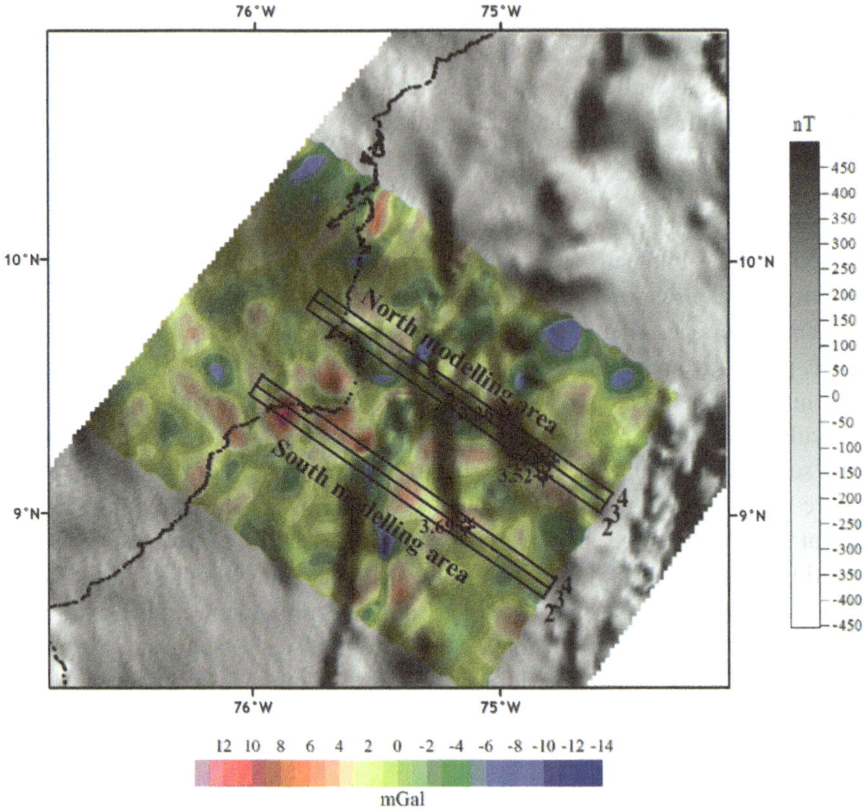

Fig. 17 Combination of the RTP-map (Fig 16) and residual map of resultant 3D density model. The *black lines* depict the planes of the magnetic modelling areas; the red symbols indicate the location of wells used for the modelling with their respectively basement-top depth in km

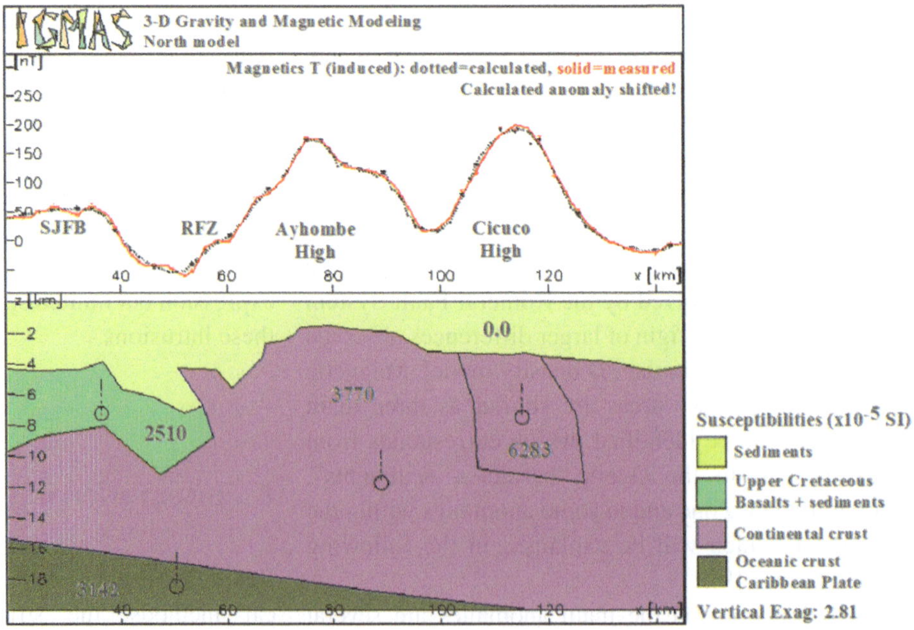

Fig. 18 Plane 3 of the 3D magnetic north model (see Fig. 17 for location). *SJFB* San Jacinto fold belt; *RFZ* Romeral fault zone

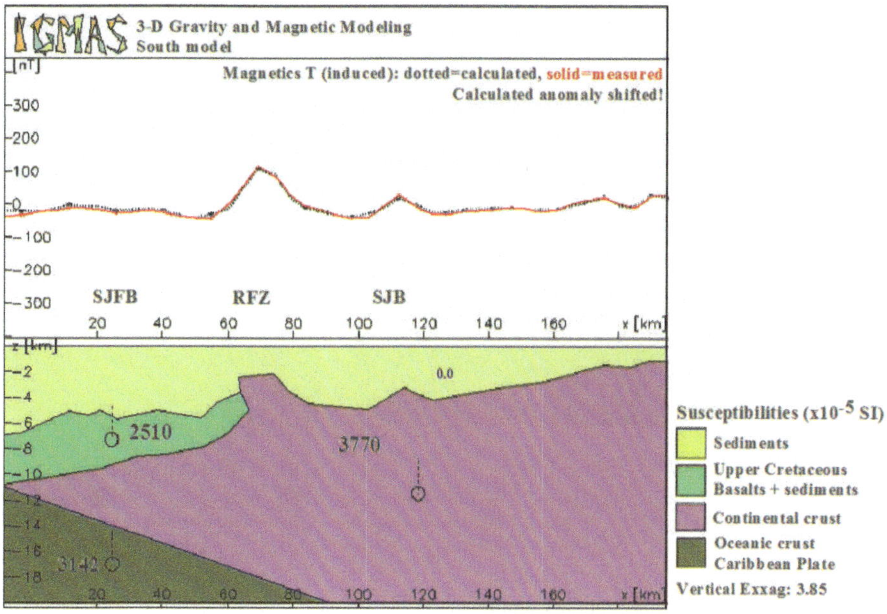

Fig. 19 Plane 3 of the 3D magnetic south model (see Fig. 17 for location). *SJFB* San Jacinto fold belt; *RFZ* Romeral fault zone; *SJB* San Jorge basin

6.1 Crustal Structure

An accretion-dominated subduction model is proposed for the Colombian Caribbean margin. The major tectonic features of the margin are shown in Figure 20. The gravity modelling defines the Caribbean Plate as an 11 km thick slab, which is subducting beneath NW Colombia in ESE direction. The dip angle of the plate is 2.5° between the deformation front and the outer high, reaching more than 4° beneath the San Jorge-Plato Basin.

The active accretionary prism is defined by Byrne et al. (1993), as the material accreted relative recently to a backstop of stronger rocks. In this model, the active prism corresponds to the external part of the Sinú-Colombia Accretionary Wedge. The development of seaward-vergent thrusting and accretionary ridges in the frontal part of the margin indicates that the accretion is ongoing.

The outer high domain, which includes the major structural complex formed by the easternmost part of the Sinú-Colombia Accretionary Wedge and the San Jacinto Fold Belt (SJFB), results from the gradual increase in both strength and bulk density as the material is compacted with increasing depth and distance from the deformation front (Byrne et al., 1993). Thus, this domain represents the fossil part of the accretionary prism, which acts as a dynamic backstop to the frontal active accretionary prism. This backstop is named dynamic according to the definition of Kopp and Kukowski (2003) because its trenchward termination will progress seaward as new material is frontally accreted and lithified. Seaward of the outer high, normal faulting, growth folding and mud diapirism can be well identified. Depth of the seismic data and its resolution are decreasing landward of this structural complex. Consequently, the detailed configuration of the pre-Oligocene units cannot be resolved.

The Sinú Fault has been interpreted as the boundary between the early-middle Eocene San Jacinto and middle Miocene to recent Sinú-Colombia accretionary prisms. However, from the examination of the density model (Fig. 14) as well as the Bouguer anomaly Map (Fig. 12) it can be observed that the fault zone does not correspond to a sharp lateral density contrast, as would be expected at the limit between two distinct units accreted in different events. In this study, the outer high is considered as the result of a continuous accretion from upper Cretaceous to late Miocene time, when the plate convergence changed from oblique to orthogonal, leading to the development of the accretionary prism active today. Therefore, it can be concluded that this fault has no regional tectonic significance.

The limit between the active accretionary prism and the outer high domain corresponds to a structural backstop, whose definition is commonly based on strong acoustic reflectors and zones of higher seismic velocity.

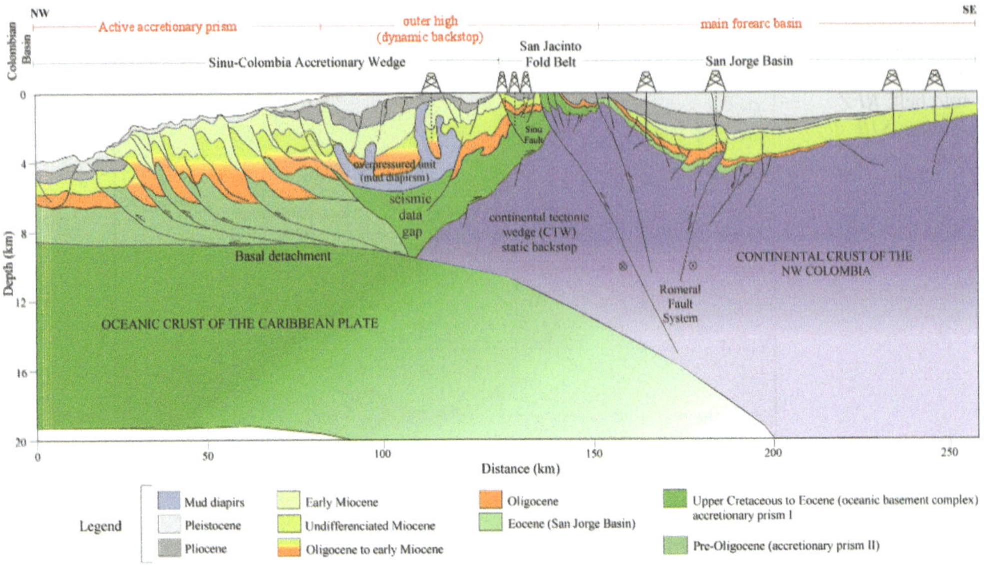

Fig. 20 Major tectonic features of the Colombian Caribbean margin. The accretionary system between the forearc region and deformation front is divided into two zones (active accretionary prism and outer high) with different morphological and structural characteristics. The outer high initially evolved by accretion against the static backstop (continental crust of the arc framework). Later, the outer high acts as dynamic backstop for the younger accretionary prism that remains active to the present day

These geometries are well documented by Silver et al. (1985), Brandon et al. (1984), Byme and Hibbard (1987), Byrne et al. (1993), Davis (1996) and others. In the proposed model, this boundary is interpreted be located at the structurally elevated zone, which coincides with a topographic high and the slope break at the seafloor. This zone, which is characterized by a west-verging imbricate thrust system and by a backward breaking out-of-sequence thrust system, suggest a landward-dipping geometry for the dynamic backstop. This geometry corresponds to a backstop type II of Byrne et al. (1993), which implies that much of the deformation involved dragging of weaker material beneath the backstop. The density model (Fig. 14) shows that the lower end of the limit between the active accretionary prism and the outer high domain coincides with the seaward termination of the oceanic "basement complex" present in the San Jacinto Fold Belt. The downward limit of the dynamic backstop (outer high) is marked by the subducting Caribbean Plate, while its landward termination is provided by the continental basement of the overriding South American Plate.

To the present day, the composition of the basement at the outer high domain is discussed controversially. Basalts of upper Cretaceous age are exposed together with ocean-floor sediments in fault slices and have also been drilled by some wells in the onshore area of the San Jacinto Fold Belt. Although the contact of this "basement complex" with the overlying sedimentary strata is impossible to identify from the existent seismic data, the 3D gravity (Fig. 14) and 3D magnetic (Figs. 18 and 19) modelling supports the presence of an oceanic affinity unit, which includes a mixture of the basalt and sediments (density: 2.70×10^3 kg m^{-3}, susceptibility: 2.51×10^{-2} SI units). The oceanic "basement complex" of the outer high is underlain by a continental tectonic wedge (CTW), which is modelled with a density of 2.80×10^3 kg m^{-3} and a susceptibility of 3.77×10^{-2} (SI units), and which is interpreted to be part of the South American margin.

According to the proposed model, the continental basement of the arc framework provides the static backstop, against which the material of the dynamic backstop was previously accreted. While a static backstop supports most of the plate boundary stress, a region of low stress is developed above it (Byrne et al. 1993). This explains the moderate deformation undergone by the overlying Plio-Pleistocene forearc basin infill, which fossilizes most structures of the outer high domain.

The forearc domain exhibits several sub-basins filled by Plio-Pleistocene deposits. Seaward, these post-kinematic deposits fossilize the complexly deformed outer high. The landward part of the forearc domain comprises the bulge generated by the Romeral Fault System.

The geometry of the static backstop in the study area is impossible to determine from the available seismic data due to its limited depth. However, a seaward-dipping geometry of the static backstop can be interpreted from the density model (Fig. 14). This geometry indicates a backstop type I of Byrne et al. (1993), whose toe is located on the top of the Caribbean Plate, approximately 130 km from the trench. In this static backstop type I, much of the weaker rock lies above the backstop, where it forms a large outer arc high and a forearc basin. Byrne et al. (1993) suggest that the understuffing of wedge sediments beneath the high causes a slight landward tilt of the outer high, which was initially developed as a relatively symmetric pop-up structure, over the toe of the static backstop.

6.2 Emplacement of the Oceanic "Basement Complex"

The emplacement mechanism of mafic and ultramafic rocks along western Colombia is still a subject of discussion. Besides the units here termed collectively oceanic "basement complex" (basalts + deep ocean sediments), isolated bodies of ultramafic rocks are also exposed in the southern part of the San Jacinto Fold Belt. These rocks, which contain the main iron–nickel ore deposits in Colombia, have been interpreted (e.g. Mejía and Durango, 1981–1982) to be part of the ophiolite sequences located along the Romeral mélange zone to the west of the Central Cordillera. The rocks located in the Romeral mélange were accreted during the Aptian – Albian collision of the Romeral terrane (Cediel et al., 2003; Amaime terrane in Pindell publications) with the Colombian Central Cordillera. However, as mentioned by Cediel et al. (2003), the San Jacinto Fold Belt (west of the Romeral North Fault Zone) does not exhibit evidences of a tectonic mélange similar to that of the Romeral mélange zone.

Although peridotites and ultramafic to mafic sequences are found in the San Jacinto Fold Belt (e.g. Ortiz, 2004), most of the members of an ophiolite suite (Coleman, 1977) are missing and contacts between its members are not well understood due to strong tectonic deformation. The oceanic "basement" of the San Jacinto Fold Belt is only defined by thrust sheets or blocks of oceanic sediments, basalts and mafic to ultramafic rocks, which are intercalated within the accretionary prism. The lack of a complete ophiolite trinity (e.g. ultramafic rocks - layered gabbros and pillow basalts - pelagic sediments) supports the assumption that these complexes cannot be interpreted as evidence for wholesale obduction of oceanic crust, but represent offscraped material from an oceanic crust during subduction. Offscraping is defined as the removal of topographic highs from the upper part of a downgoing oceanic slab. Ultramafic rocks within accretionary prisms may be offscraped remnants of peridotite-cored uplifts at ridge-transform intersections (Coleman, 2000).

Palinspastic reconstructions (Tectonic Analysis Inc., 1995; Pindell and Tabbutt, 1995) suggest that after accretion of the Amaime (Romeral) terrane the convergence of the South American and Caribbean plates in the northern Andes was accommodated by the initiation of the east-dipping subduction of the Caribbean crust beneath the Amaime terrane and rest of Colombia. This subduction led to the westward emplacement of offscraped upper Caribbean crust onto the Amaime terrane to the south (Dagua terrane) and on top of the continental crust (CTW) along the NW Colombian margin. Thus, the presence of oceanic affinity rocks over continental basement (CTW: current static backstop) is here interpreted as material detached from the downgoing Caribbean slab by backthrusting, which is developed in the internal part of the prism as result of the oblique convergence.

6.3 Romeral Fault System and Ocean-Continent Crust Boundary

The existence of a continental tectonic wedge (CTW) beneath the oceanic "basement complex" in the San Jacinto Fold Belt indicates that the Romeral Fault System does not represent a paleo-suture or tectonic boundary between oceanic crust to the west and continental crust to the east. The main reason why this interpretation had been proposed and for many years accepted is the presence of upper Cretaceous basalts in the San Jacinto Fold Belt (west of the Romeral Zone). However, the results of 3D gravity and magnetic modelling suggest that the Romeral Fault System originated within the block of continental crust on which the San Jacinto Fold Belt was formed. The absence of oceanic "basement complex" units (present in the San Jacinto Fold Belt) east of the Romeral north zone (western flank of the San Jorge-Plato Basin) is explained by their erosion due to tectonic uplift along the Romeral Fault Zone before the sedimentation of the middle (?)-late Eocene units. A clear characteristic, which can be observed in Figure 20, is the configuration of the Romeral Fault System as a positive flower structure, along which blocks of continental crust have been displaced following the major fault rise. Thus, the Romeral North Fault Zone (San Jacinto of Cediel et al., 2003) and its associated structures are here interpreted as a transcurrent fault system developed on the continental side of the forearc domain (static backstop) as a result of the oblique collision between the Caribbean and South American plates (Fig. 21). The evolution of transcurrent fault systems in zones of oblique convergence is amply discussed by Fitch (1972). Callasou et al. (1993) presented analogue experiments that show strike-slip faulting developing flower structures near the static backstop.

It should be noted that the Colombian continental margin has a long and complex history. From south to north Colombia, the Romeral-Peltetec-San Jacinto Fault System has been considered as the continental margin along which Jurassic (?) and Cretaceous oceanic rocks were accreted. However, Cediel et al. (2003) highlight the differences in style and associated characteristics observed in the Romeral-Peltetec System (Romeral South) and the San Jacinto Fault System (Romeral North). The migration of the Caribbean Plate along the northern margin of the South American craton during the Cretaceous-Cenozoic is documented by the transpressive strike-slip movement along the Romeral-Peltetec and San Jacinto faults accompanied with the east-west growth of the Oca–El Pilar transform system (Pindell, 1993; Cediel et al., 1994; Maresch et al., 2000). Since Miocene time, collision of the Panama Arc (Cañas Gordas and Baudó terranes) with NW Colombia has truncated the southern San Jacinto System and the faults of the Romeral-Peltetec system (Cediel et al., 2003).

The new interpretation of the Romeral Fault System as discussed above implies a new definition of the ocean-continental crust boundary. Two tectonic limits are identified from the proposed crustal model: an upper one, which represents the west-ward dipping contact between the continental tectonic wedge (CTW) and oceanic basalts detached from the downgoing slab, and a lower one, which represents the eastward dipping boundary between the subducting Caribbean Plate and the overriding South American Plate. The exact architecture of the upper ocean-continental crust boundary is difficult to define from this model. New, higher resolution seismic data and a detailed modelling along this zone, applying modern interpretation techniques, which allow the loading of the seismic data as a backdrop during the gravity and magnetic modelling, are necessary to distinguish the basement features and structures confined to the sedimentary cover. The lower ocean-continental crust boundary can be identified in the resultant model, where a density contrast between the subducted Caribbean Plate and the crust of the overriding South American Plate

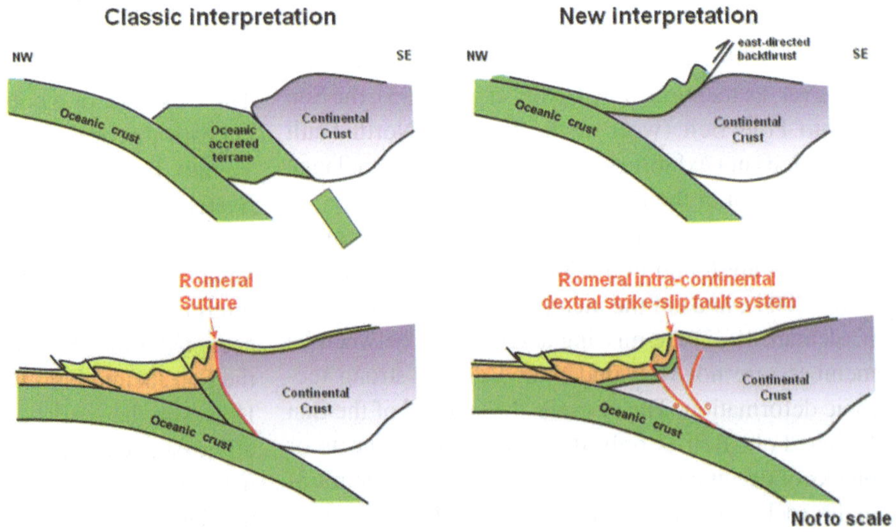

Fig. 21 Interpretation of the Romeral North (San Jacinto) fault system as an intracontinental transpressive fault system. *Left side* shows the classic, right side the new interpretation of this study. Presence of oceanic affinity rocks in the San Jacinto Fold Belt is interpreted as material swept from the downgoing Caribbean slab by backthrusting

(static backstop) occurs. In the modelled region, this zone is approximately 180 km long and starts at a depth of 14 until the crust-mantle transition at a depth of 24 km.

7 Conclusions

The present-day crustal structure of the Colombian Caribbean margin is a result of the E-NE migration of the Caribbean Plate and its subsequent subduction beneath Colombia, which has generated strike-slip fault systems, lateral displacements, rotation and inversion during the Cenozoic.

The 3D gravity modelling defines a shallow dipping Caribbean Plate that subducts at ~5° beneath NW Colombia.

The continued Caribbean-South American plate convergence since upper Cretaceous has formed two accretionary prisms: the fossil accretionary prism (outer high) which includes the easternmost part of the Sinú-Colombia Accretionary Wedge and the San Jacinto Fold Belt, is assumed to be of upper Cretaceous to middle Miocene age, and the active frontal accretionary prism of late Miocene to Recent age, which corresponds to the external part of the Sinú-Colombia Accretionary Wedge.

A seaward-vergent imbricate thrust system, which is developed in the frontal prism by the active deformation and strong shortening, is well defined in the seismic data.

The older (fossil) accretionary prism is initially formed by upper Cretaceous basalts and ocean-floor sediments detached from the downgoing Caribbean Plate and subsequently emplaced on top of the continental basement of the South American Plate through backthrusting. Blocks of this oceanic "basement complex" are exposed and has been drilled in the San Jacinto Fold Belt. Occurrence of ultramafic rocks in the south of the San Jacinto Fold Belt is interpreted to be offscraped remnants of peridotite-cored uplifts, but not obducted oceanic crust.

The older prism becomes an outer high as it grows and uplifts. Seaward, the outer high is characterized by normal faulting, growth folding and mud diapirism. Landward, insufficient seismic coverage and its decreasing resolution do not allow determining the configuration of the pre-Oligocene units.

The continental crust of the overriding South American Plate acted as a static-rigid backstop for the accreted older prism. 3D gravity and magnetic modelling suggest a seaward-dipping geometry for the static backstop.

The complexly deformed outer high, which is approximately 80 km wide and more than 10 km deep, behaves as the dynamic backstop for the younger accretionary prism. A landward-dipping limit between the active and fossil (dynamic backstop) accretionary prisms is inferred from the seismic data.

The Sinú Fault does not represent a limit between two distinct units accreted in different events because no sharp lateral density contrast from the 3D density model can be identified.

The San Jorge-Plato Forearc Basin is formed between the older accretionary complex and arc framework. Some small depocenters developed behind the modern accretionary prism in the shelf sector are also interpreted as part of the forearc domain. The transition between the outer high and the main forearc basin is marked by the kilometer-wide Romeral Fault Zone, which represents a structural break between the smaller basins that fossilizes the outer high to the west and the main San Jorge Forearc Basin to the east. The evolution of the forearc basins is controlled by the tectonic uplift, sediment supply and basement subsidence by loading. Tectonic subsidence mechanisms, which were not discussed in this study, could be related to crustal extension of the static backstop or localized tectonic erosion.

3D gravity and magnetic modelling indicate that the Romeral Fault System is involving continental basement-type along the transition from San Jorge-Plato Basin to San Jacinto Fold Belt, which allows to conclude that Romeral does not represent a boundary between ocean and continental crust. Romeral and its associated structures form a dextral transpressive fault system, which is developed within continental crust (the static backstop) as a result of the oblique convergence between Caribbean and South American plates.

From the proposed crustal model, two ocean-continental crust boundaries can be identified: an upper one, which represents the highly deformed east-vergent backthrust that emplaced the oceanic material onto the continental basement of the arc framework, and a lower one, which represents the eastward dipping boundary between the downgoing Caribbean Plate and the overriding South American Plate.

Acknowledgements This paper was written while A. Mantilla-Pimiento was a PhD student at the University of Jena. She is very grateful to the German Academic Exchange Service (DAAD) for the financial support. The authors wish to thank the Colombian Oil Company (ECOPETROL S. A.) for the permission to publish this interpretation and for the use of its facilities during the compilation of the data and seismic interpretation. We wish to express our appreciation to Pedro Restrepo and Diego Carvajal as well as to the remaining members of the staff of the Exploration Vice-Presidency. Special thanks to John Cerón for providing the gravity and magnetic grid-files and for his invaluable contribution during the gravity modelling. Sabine Schmidt and Yvonne Hofmann are kindly thanked for their support with IGMAS software. We thankfully appreciate the contribution derived from discussion with Fabio Cediel, Carlos Cáceres, Hermann Duque-Caro, Dario Barrero, Jaime Galvis, Germán Y. Ojeda, Myriam Caro and Eduardo Rossello. The idea to develop this project was born in INGEOMINAS. Adolfo Alarcón and Georgina Guzmán are acknowledged for their institutional support. Special thanks also to Eduardo López, Wilson Quintero and Carlos Rey for their technical suggestions. Thanks to Serge Lallemand, Alfredo Taboada, Robin Leinster and Richard Brumm for reviewing this paper.

References

Barrero, D., Alvarez, J., and Kassem, T. 1969. Actividad ígnea y tectónica en la Cordillera Central, Boletín Geológico de INGEOMINAS 17 (1–3): 145–173.

Bowland, C. L., 1993. Depositional history of the western Colombian, Caribbean Sea, revealed by seismic stratigraphy, GSA Bulletin 105: 1321–1345.

Brandon, M. T., Johnson, S. Y., Stewart, R. J., and McClain, K. J. (1984). Cross section B-B′-B″, western Washington: In Kulm et al., eds., Ocean margin drilling program, regional atlas Series 1, Marine Science International Woods Hole, MA: 19–20

Burke, K., 1988. Tectonic evolution of the Caribbean: Annual Revue of Earth and Planetary Sciences 16: 201–230.

Byme, T., and Hibbard, J., 1987. Landward vergence in accretionary prisms: The role of the backstop and thermal history, Geology 15: 1163–1167

Byrne, D. E., Wang, W., and Davis, D., 1993. Mechanical role of backstops in the growth of forearcs, Tectonics 12 (1): 123–144.

Callasou, S., Larroque, C., and Malavieille, J., 1993. Transfer zones of deformation in thrust wedges: an experimental study, Tectonophysics, 221: 325–344.

Case, J. E., 1974. Major Basins along the Continental Margin of Northern South America: In Burk, C. A, and C. L. Drake, eds., The Geology of Continental Margins, Springer, New York: 733–742.

Case, J. E., and McDonald, W. D., 1973. Regional gravity anomalies and crustal structure in Northern Colombia, GSA Bulletin 84: 2905–2916.

Case, J. E., Holcome, T. L., and Martin, R. G., 1984. Map of geologic provinces in the Caribbean region, In: Bonini W. E, Hargraves R. B., and Hagam R, eds., The Caribbean-South America plate boundary and regional tectonics, 1–30.

Case, J. E., Shagam, R., and Giegengack, R. F., 1990. Geology of the northern Andes; an overview, In: Dengo G., and Case, J. E., eds., The Caribbean region: GSA, The geology of North America. Vol H: 177–200.

Cediel, F., Etayo, F., and Cáceres, C., 1994. Facies distribution and tectonic setting through the Phanerozoic of Colombia: INGEOMINAS, ed., Geotec Ltd., Bogota (17 time-slices/maps in scale 1:2000.000).

Cediel, F., Shaw, R., and Cáceres, C., 2003. Tectonic Assembly of the Northern Andean Block., in Bartolini, C, Buffer, R. T., and Blickwede, J, eds., The Circum-Gulf of Mexico and the Caribbean: Hydrocarbon habitats, basin formation, and plate tectonics: AAPG Memoir 79: 815–848.

Cediel, F., and Cáceres, C., 2000. Geological Map of Colombia, 3rd Edition: Geotec Ltd., Bogota, digital format with legend and tectonostratigraphic chart.

Cerón, J., 2002. Evaluación regional de la Cuenca Caribe – Offshore Colombia, Technical Report of ECOPETROL: 33 p.

Coleman, R.G., 1977. Ophilolites: ancient oceanic lithosphere? (Mineral and rocks). Springer, Berlin:229.

Coleman, R.G., 2000: Prospecting for ophiolites along the California continental margin. In: Dilek, Y. D., Moores, E. M., Elthon, D., and Nicolas, A., eds., Ophiolites and oceanic crust: New insights from field studies and the ocean drilling program. GSA, Special Papers 349: 351–364.

Cortes, M., and Angelier, J., 2005. Current states of stress in the northern Andes as indicated by focal mechanisms of earthquakes, Tectonophysics 403: 29–58.

Davis, D. M., 1996. Accretionary mechanics with properties that vary in space and time, In: Bedout G. E., et al. eds., Subduction: Top to bottom, Geophysics Monogr. Series 96: 39–48.

Dickinson, W. R., 1973. Widths of modern arc-trench gaps proportional to past duration of igneous activity in associated magmatic arcs, Journal Geophysical Research, 78: 3376–3389.

Dickinson, W. R., 1974. Plate tectonics and sedimentation. In: Dickinson W. R., eds., Tectonics and sedimentation, Soc. Econ. Paleont. Miner. Spec. Pub., 22: 1–27.

Duque-Caro, H., 1972. Ciclos tectónicos y sedimentarios en el norte de Colombia y sus relaciones con la paleo-ecología: Boletín Geológico de INGEOMINAS 19 (3): 1–23.

Duque-Caro, H., 1973. The geology of the Montería area: Colombian society of petroleum geologists and geophysicists 14th Annual Field Conference, Guidebook, Bogotá, Colombia: 397–431.

Duque-Caro, H., 1975. Los foraminíferos planctónicos y el terciario de Colombia, Revue Española Micropaleontol 7 (3): 403–427.

Duque-Caro, H., 1976. Características estratigráficas y sedimentarias del Terciario marino de Colombia: Memorias del 2nd Congreso Latinoamericano de Geología, Caracas, 1973: 945–964.

Duque-Caro, H., 1979. Major structural elements and evolution of northwestern Colombia, In: Watkins, J. S., Montadert, L., and Dickerson, P. W., eds., Geological and geophysical investigations of continental margins, AAPG Memoir 29: 329–351.

Duque-Caro, H., 1984. Structural Style, diapirism and accretionary episodes of the Sinú-San Jacinto terrane, southwestern Caribbean borderland, GSA Memoir 162: 303–316.

Duque-Caro, H., 1990. The Chocó Block in the northwestern corner of South American: Structural tectonostratigraphic and paleographic implications, Journal of South American Earth Sciences 3: 71–84.

Edgar, T. N., Ewing J., and Hennion J., 1971. Seismic refraction and reflection in Caribbean Sea, AAPG Bulletin 55 (6): 833–870.

ESRI (Earth Sciences Resources Institute at University of South Carolina) - ILEX Ltd., 1995. Evaluación Geológica Regional de la Cuenca de Sinú – San Jacinto, Technical Report 3851 of ECOPETROL S.A.

French, C., and Schenk, C., 2004. Map showing geology, oil and gas fields, and geologic provinces of the Caribbean region. Open-File Report 97-470-K USGS.

Fitch, T., 1972. Plate convergence, transcurrent faults, and internal deformation adjacent to Southeast Asia and the Western Pacific. Journal of Geophysical Research 77 (23): 4432–4460.

Flinch, J. F., 2003. Structural evolution of the Sinú-Lower Magdalena Area (Northern Colombia), In: Bartolini, C., Buffler, R. T., and Blickwede, J. eds., The circum-Gulf of Mexico and the Caribbean: Hydrocarbon habitats, basin formation, and plate tectonics: AAPG Memoir 79: 776–796.

Flinch, J. F., Grand, M. V., and Casero, P. 2000. Accretion and obduction along the Sinú-Lower Magdalena area (Northern Colombia), Memorias VII Simposio Bolivariano, Exploración Petrolera en las Cuencas Subandinas, Sociedad Venezolana de Geólogos, Caracas, Venezuela, September, 2000: 218–229.

Flinch, J. F., Amaral J., Doulcet, A., Mouly, B., Osorio, C., and Pince, J. M., 2003a. Onshore-offshore structure of the Northern Colombia accretionary complex, Memory AAPG International Conference Barcelona, Spain, September, 2003: 1–5.

Flinch, J., Amaral, J., Doulcet, A., Mouly, B., Osorio, C., and Pince, J. M., 2003b. Structure Of The Offshore Sinú Accretionary Wedge, Northern Colombia, Memorias VIII Simposio Bolivariano - Exploración Petrolera en las Cuencas Subandinas, Asociación Colombiana de Geólogos y Geofísicos del Petróleo, Cartagena, Colombia, September, 2003: 76–83.

Freymüller, J., Kellog, J., and Vega, V., 1993. Plate motions in the North Andean Region, Journal of Geophysical Research 98 (B12): 21,853–21,863.

GETECH (Geophysical Exploration Technology Ltd. – School of Earth Sciences at University of Leeds), 2001. Colombia Satellite Gravity and Bathymetry (Atlantic Region), Technical Report of ECOPETROL S.A: 30 p.

Grosse, E., 1926. Estudio Geológico del Terciario Carbonífero de Antioquia en la parte occidental de la Cordillera Central de Colombia, entre el río Arma y Sacaojal, ejecutado en los años de 1920–1923: Dictrich Reimer, Berlín: 361.

Hardy, N. C., 1991. A marine geophysycal study of the Pacific margin of Colombia and south east Panama. PhD thesis, University of Birmingham.

ICP (Instituto Colombiano del Petroleo – ECOPETROL S.A.), 2000. Evaluación Regional Integrada Cuenca Valle Inferior del Magdalena, Technical Report of ECOPETROL S.A (CDROM 2).

INGEOMINAS, 1997. Atlas Geológico Digital de Colombia, scale 1:500.000, plates 1, 3, 4 and 6.

Karig, D. F., and Sharman, G. F., 1975. Subduction and accretion at trenches, Bulletin Geological Society of America, 86: 377–389.

Kellog, J., and Bonini, W. E., 1982. Subduction of the Caribbean Plate and basement uplifts in the overriding South America Plate, Tectonics 1: 251–276.

Kerr, A. C., Taney, G. F., Marriner, F., Nivia, A., Klaver, G. T., and Saunders, A. D., 1996. The geochemistry and tectonic setting of late Cretaceous Caribbean and Colombian volcanism: Journal of South America Earth Sciences 9: 111–120.

Kopp, H., Flueh, E. R., Klaeschen, D., Bialas, J., and Reichert, C., 2001. Crustal Structure of the central Sunda margin at the onset of oblique subduction, Geophysical Journal International, 147: 449–474.

Kopp, H., and Kukowski, N. 2003. Backstop geometry and accretionary mechanics of the Sunda margin, Tectonics 22: 11-1 – 11-15.

Larroque, C., Calassou, S., Malavielle, J., and Chaniert, F., 1995. Experimental modelling of the forearc basin development during accretionary wedge growth, Basin Research 7: 255–268.

Laverde, F., 2000. The Caribbean Basin of Colombia, a composite Cenozoic Accretionary Wedge with Under-Explored Hydrocarbon Potential, Memorias del VII Simposio Bolivariano de Exploración Petrolera en Cuencas Subandinas, Caracas, Venezuela, September, 2000: 394–410.

Lu, R. S., and McMillen, K. J., 1982. Multichannel seismic survey of the Colombia basin and adjacent margins, In Watkins, J. S., and Drake, C. L., eds, Studies in continental margin geology: AAPG Memoir 34: 395–410.

Malavé, G., and Suárez, G., 1995. Intermediate-depth seismicity in northern Colombia and western Venezuela and its relationship to Caribbean Plate subduction, Tectonics 14: 617–628.

Mantilla-Pimiento, A. M., 2007. Crustal structure of the SW Colombian Caribbean margin: Geological interpretation of geophysical data. Doctoral Dissertation, Institute of Geosciences, Friedrich-Schiller Universität Jena, Germany. URL: http://www.db-thueringen.de/servlets/DocumentServlet?id=10710

Maresch, W. V., Stoeckhert, B., Baumann, A., Kaiser, C., Kluge, R., Krueckhans-Lueder, G., Brix, M. R., and Thomson, M., 2000. Crustal history and plate tectonic development in the southern Caribbean: Zeitschrift für Angewandte Geologie, 1: 283–289.

Mejía V. M., and Durango, J. R., 1981–1982. Geología de las lateritas niquelíferas de Cerromatoso, Boletín de Geología de la Universidad Industrial de Santander 15 (29): 117–123.

Mountney, N. P., and Westbrook, G., 1997. Quantitative analysis of Miocene to recent forearc basin evolution along the Colombian convergent margin, Basin Research 9: 177–196.

Nasu, N., 1982. Multi-channel seismic reflection data across the Nakai Through. IPOD-Japan Basic Data Series, 4, Ocean Research Insitute, University of Tokio.

Nettleton, L. L., 1971. Elementary gravity and magnetics for geologists and seismologists. In: Wuenschel, P. C., eds., Monograph Series, Number 1, Society of Exploration Geophysicists: 121.

Ortiz, F., 2004. Guías para la localización de metales preciosos en ofiolitas colombianas, Informe de Avance Proyecto CYTED XIII.1. Ofiolitas: Características mineralógicas y petrográficas del Yacimiento de Níquel de Cerro Matoso. Dyna, Medellín, Colombia, Año 71 (142): 11 – 23.

Pennington, W. D., 1981. Subduction of the Eastern Panama Basin and Seismotectonics of Northwestern South America. Journal of Geophysical Research 86 (B11): 10753–10770.

Pindell, J. L., 1993. Regional synopsis of Gulf of Mexico and Caribbean evolution: Gulf Coast Section, Society for Sedimentary Geology (SEPM), 13th Annual Research Conference Proceedings: 251–274.

Pindell, J. L., and Barrett, S. F., 1990. Geological evolution of the Caribbean region; a plate-tectonic perspective, In: Dengo, G., and Case, J. E., eds., The Caribbean region, The Geology of North America, GSA H: 405–432.

Pindell, J. L., and Tabbutt, K. D., 1995. Mesozoic-Cenozoic Andean paleogeography and regional controls on hydrocarbon systems, In: Tankard, A. J., Suárez S. R., and Weisink, H. J., eds., Petroleum basins of South America: AAPG Memoir 62: 101–128.

Pindell, J.., Higgs, R., and Dewey, J. F., 1998. Cenozoic palinspastic reconstruction, paleogeographic evolution, and hydrocarbon setting of the northern margin of South America, In: Pindell, J. L., and Drake, C. L., eds., Palaeogeographic evolution and Non-glacial Eustasy, northern South America: Special Publication, SEPM (Society for Sedimentary Geology): 45–86.

Pindell, J., and Kennan, L., 2001. Kinematic evolution of the Gulf of Mexico and Caribbean, in petroleum system of deepwater basins: Global and Gulf of Mexico experience, Proceedings, Gulf Coast Section, SEPM, 21st Annual Research Conference, December 2-5: Houston, Texas, Society for Sedimentary Geology (SEPM):193–220.

Reyes-Santos, J. P., Mantilla, M., and González, J. S., 2000. Regiones Tectono-Sedimentarias del Valle Inferior del Magdalena, Colombia, Memorias VII Simposio Bolivariano, Exploración Petrolera en las Cuencas Subandinas, Sociedad Venezolana de Geólogos, Caracas, Venezuela, September 2000: 310–333.

Ruiz, C., Davis, N., Bentham, P., Price, A., and Carvajal, D., 2000. Structure and tectonic evolution of the South Caribbean Basin, southern offshore Colombia: A progressive accretionary system, Memorias VII Simposio Bolivariano, Exploración Petrolera en las Cuencas Subandinas, Sociedad Venezolana de Geólogos, Caracas, September 2000: 334–355.

Schmidt, S., and Götze, H. J., 1995. IGMAS 3D gravity and magnetic modelling program, Publication on the internet at: http://userpage.fu-berlin.de/ sschmidt/Sabine_IGMAS.html

Silver, E.A., Ellis, M. J., Breen, N. A., Shipley, T. H., 1985: Comments on the growth of accretionary wedges, Geology 13: 6–9

Taboada, A., Rivera, L. A., Fuenzalida, A., Cisternas, A., Philip, H., Bijwaard, H., Olaya, J., and Rivera, C., 2000. Geodynamics of the northern Andes: Subductions and intracontinental deformation (Colombia), Tectonics 19 (5): 787–813.

Tectonic Analysis Inc., 1995. The paleographic evolution of Colombia: implications for hydrocarbons, Technical Report of ECOPETROL S.A.

Töto, E. A., and Kellog, J., 1992. Structure of the Sinú – San Jacinto fold belt – an active accretionary prism in Northern Colombia. Journal of South America Earth Sciences 5 (2): 211–222.

Van der Hilst, R., 1990. Tomography with P, PP and pP delaytime data and three-dimensional mantle structure below the Caribbean region: Doctoral Dissertation, Institute of Earth Sciences, University of Utrecht, Netherlands, No. 67: 257.

Van der Hilst, R., and Mann, P. 1994. Tectonic implications of tomography images of subducted lithosphere beneath northwestern South America: Geology 22: 451–454.

Vernette, G., 1986. Le diapirisme argileux et ses consequences sur les caracteres morphologiques et sedimentaires de la marge colombienne des caraibes. Bull. Inst. Géol. Bash d'Aquiíah,Bordeaux 40: 35–51.

Vernette, G., Mauffret, A., Bobier, C., Briceño, L., and Gayet, J., 1992. Mud diapirism, fansedimentation and strike-slipe faulting, Caribbean Colombian margin, Tectonophysics 202: 335–349.

Westbrook, G. K., 1975. The structure of the crust and upper mantle in the region of Barbados and the Lesser Antilles. Geophysical Journal of the Royal Astronomical Society 36: 57–90.

Index

The manufacturer's authorised representative in the EU is Springer Nature Customer Service Centre GmbH, Europaplatz 3, 69115 Heidelberg, Germany. If you have any concerns regarding our products, please contact ProductSafety@springernature.com

Printed and bound by CPI Group (UK) Ltd, Croydon, CR0 4YY
30/09/2026
02225516-0003